AF597482

Industrial Flow Measurement

Industrial Flow Measurement

David W. Spitzer

Resources for Measurement and Control Series

Instrument Society of America

Printed in the United States of America.

INSTRUMENT SOCIETY OF AMERICA
67 Alexander Drive
P.O. Box 12277
Research Triangle Park
North Carolina 27709

Library of Congress Cataloging-in-Publication Data

Spitzer, David W.
Industrial flow measurement / David W. Spitzer. -- 2nd ed.
p. cm. -- (Resources for measurement and control series)
Includes bibliographical references (p.).
ISBN 1-55617-243-5
1. Flow meters. I. Title. II. Series.
TJ935.S58 1990 90-4368
681'.2--dc20 CIP

3rd Printing November 1995

This work is dedicated to my children,
Deborah Goldie and Michael James,
in the hope that it will inspire them
and others to continue their education
and their pursuit of knowledge; and
to my wife Ida for understanding
why I was so early to rise and
late to bed for so many months.

Contents

Preface
Why Measure Flow?

Virtually all technical books contain introductions that expound upon the importance and applicability of the contents. In reality, these sections are rarely read, because the general information adds little or nothing to the overall technical content of the book.

Often, that which is *not* presented can be a book's most valuable lesson. Consider giving a person clear, verbal directions on how to get from one location to another, understanding that the route may not be the shortest. Teach that same person how to read a map, and he or she can determine the best route to anywhere in the world.

The content of this book is a "map" comprised of numerous pieces of information that supply the answers to questions that provide an understanding of the subject. Knowing when to read this "map" entails intimate knowledge of the process and the laws that govern it.

Most individuals succumb to pressure and are content to find the quick solution to a perceived problem. This approach usually yields less than optimum results and may even camouflage real problems.

To effectively apply information, one must *think*. Even though most individuals are capable of logical thought, the process of defining the real problem before finding a solution occurs far too infrequently. The real problem and a good solution thereto must be determined before one can decide whether the content of this book is truly applicable and useful.

About the Author

DAVID WILLIAM SPITZER is the Lead Utility and Instrumentation Engineer at Nepera, Inc., Harriman, NY, where he is responsible for technical support and direction in the electrical, instrumentation, and utility areas on a plant-wide basis. He has proposed many energy-saving projects which are in planning, design, construction, and operational stages.

Mr. Spitzer serves on the American Society of Mechanical Engineers' Committee on Measurement of Fluid Flow in Closed Conduits. He is also a member of the Instrument Society of America and the Association of Energy Engineers.

A frequent instructor at the ISA Training Center in Raleigh, NC, and throughout the United States and internationally in the Short Course Program, Mr. Spitzer is the author of a revised edition of *Variable Speed Drives, Principles and Applications for Energy Cost Savings*. He is also Volume Editor of *Flow*, part of ISA's Practical Guides for Instrumentation and Control series.

About the Book

Modern flowmeters handle many more applications than could have been imagined centuries ago. Today's flow measurements encompass operating conditions that range from capillary blood flow, flows over spillways, flow of gases, plasmas, pseudo-plastics, solids, and corrosives, to name but a few. This text reviews the important concepts of flow measurement and provides explanations, practical considerations, illustrations, and examples of existing flowmeter technologies. It presents a rational procedure for flowmeter selection that is based on factual information. This text is directed to technical personnel involved with flow measurement from technicians through engineers and managers, to scientists. Sales engineers and instructors at community colleges and technical schools may also benefit.

Acknowledgments

My involvement in the development of this text and the "Industrial Flow Measurement" Short Course offered by ISA is a result of some degree of luck, coincidence, and lots of long hours and hard work. David Bartran of Spectrum Engineering, Inc., thoroughly and frankly reviewed the rough draft and offered much in the way of incisive, constructive criticism and suggestions. I would like to thank Mobay Chemical Corporation and Nepera, Inc., for making possible the use of their word processing systems before and after working hours.

At times mention is made of specific products and manufacturers, and references to particular instrument types are included. This was done in an effort to explain measurement techniques and the principles involved. The choices made were based on personal experience and knowledge and should in no way be construed as suggesting endorsement of any instrument type or manufacturer by either the author or ISA.

Industrial Flow Measurement

1

Introduction

Introduction Flow measurement technology has evolved rapidly in recent decades. Some technologies have survived, while others have fallen by the wayside or have never been commercially developed. Physical phenomena discovered centuries ago have been the starting point for many viable flowmeter designs. In recent years, technical developments in other fields, namely in optics, acoustics, and electromagnetism, have resulted not only in improved sensor designs but also in new flowmeter concepts.

This technology "explosion" has enabled modern flowmeters to handle many more applications than could have been imagined centuries ago. Today's flow measurements encompass operating conditions that range from capillary blood flow to flows over spillways, flows of gases, plasmas, pseudo-plastics, solids, and corrosives, to name but a few.

Effective flowmeter selection requires a thorough understanding of flowmeter technology in addition to a practical knowledge of the process and the fluid being measured. The difficulty in bringing these two facets of flow measurement to bear on a practical application is challenging even to experienced engineers, technicians, and sales personnel.

Objectives The primary objective of this text is to review the important concepts of flow measurement and to provide explanations, practical considerations, illustrations, and examples of existing flowmeter technologies. The ultimate goal is to present a rational procedure for flowmeter selection based on factual information. The title, *Industrial Flow Measurement*, was chosen to emphasize the goal of presenting the knowledge with which practical and precise industrial measurements can be made. A purely mathematical treatment of flowmeters is avoided in favor of heuristic explanations of the principles and installation considerations involved and how they apply to a given flowmeter.

Prerequisites and Audience This text is intended as an introduction to flowmeter technology and has application for sales personnel, technicians, engineers, instructors, and those who are endeavoring to broaden their knowledge of industrial flow measurement. It can be used by persons without engineering or scientific training; however, it may be necessary to accept some principles on faith rather than through rigorous mathematical development. Knowledge of algebra is necessary, but calculus is avoided to keep within the text's practical framework.

Learning Objectives Understanding of basic flowmeter concepts is essential to technical evaluations of flowmeter options and the selection of equipment based on technical merits. Decisions made without this technical basis are likely to result in misapplication of flowmeter devices and in unacceptable flowmeter errors. The costs associated with misapplication can range from additional engineering to replacement of the instruments.

In presenting the basics of flowmeter technology and application, this text is not intended as a substitute for experience or for specific flow measurement handbooks. It is hoped that upon completion of this study, the reader will be in a position to face flowmeter problems with a broadened perspective and to be more knowledgeable of alternatives and constraints of the specific applications.

2

Fluid Flow Fundamentals

Introduction Fundamental to an investigation of the operation and attributes of the various flowmeter technologies is a working knowledge of the physical properties used to describe liquids and gases, as well as a basic understanding of some of the physical phenomena associated with flow in pipes. These physical properties need only be studied in a practical sense in order to understand the operation and limitations of various flowmeter technologies.

Units commonly used to describe physical properties of fluids are generally a combination of the English system, the SI system, and other unique systems often common only to particular industries. Vendor technical data on flow ranges, size, and the like, are typically expressed using the English system unless the manufacturer distributes the same literature in international markets, in which case SI information is also available. If the flow range is sufficiently small, it is often expressed in SI units, although the remainder of the data will probably be in the English system. A hybrid but commonly used system of units is used throughout this text so that a clear picture of the subject matter can be maintained in the discussions that ensue.

Temperature For the purpose of describing flow measurement, it is sufficient to state that temperature is a measure of relative hotness or coldness. In the SI system, temperature is expressed in degrees Celsius (°C) with 0°C and 100°C corresponding to the freezing and boiling points of water, respectively. At times, the absolute temperature, that is, the temperature referenced to lowest theoretical temperature, is required. Absolute temperature is measured in kelvins (K) and can be calculated by adding 273.15 to the temperature in degrees Celsius. The English equivalents are degrees Fahrenheit (°F), where 32°F and 212°F represent the freezing and boiling points of water, respectively, and degrees Rankine (°R) for expressing absolute temperature.

The following equaltions may be useful in converting units of temperature.

$$°C = \frac{5\,(°F - 32)}{9}$$

$$K = °C + 273.15$$

$$°R = °F + 460$$

EXAMPLE 2-1

Problem: Convert 320°F to kelvins.

Solution: Convert to degrees Celsius and then to kelvins as follows:

$$°C = 5\,(320 - 32)\,/\,9 = 160°C$$

$$K = 160 + 273$$

$$= 433\ K$$

EXAMPLE 2-2

Problem: Convert 233 K to degrees Fahrenheit.

Solution: Convert to degrees Celsius and then to degrees Fahrenheit as follows:

$$°C = 233 - 273 = -40°C$$

$$°F = (9 \times -40\,/\,5) + 32$$

$$= -40°F$$

Pressure Pressure is defined as the ratio of a force divided by the area over which it is exerted.

$$P = \frac{F}{A}$$

The commonly used English units to express pressure are pounds per square inch (psi). If pressure is referenced to atmospheric pressure, it is termed gage pressure. If it is referenced to a perfect vacuum, it is termed absolute pressure. To convert from gage to absolute units, atmospheric pressure is simply added to the gage pressure.

The following conversions may be useful to convert units of pressure.

1 "standard" atmosphere (atm) = 14.696 psi = 1013.25 mbar

1 inch of mercury (in. Hg) = 0.491154 psi

1 inch of water (in. WC) = 0.03609 psi

1 kilogram per square centimeter (kg/cm^2) = 14.2233 psi

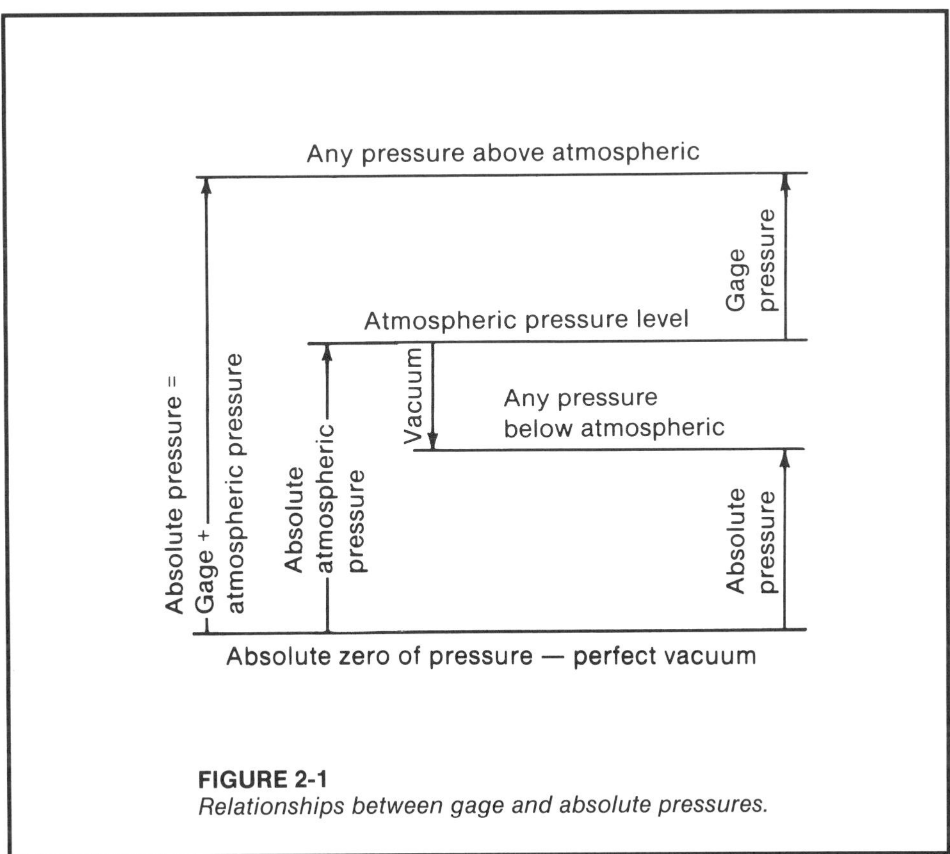

FIGURE 2-1
Relationships between gage and absolute pressures.

1 bar = 14.5038 psi

1 kilopascal (kPa) = 0.145038 psi

EXAMPLE 2-3

Problem: Determine the pressure exerted under a 2-inch cube weighing 5 pounds placed on a table.

Solution:

$$P = \frac{F}{A} = \frac{5 \text{ lb}}{4 \text{ in.}^2} = 1.25 \text{ psi}$$

If a 0.1-inch diameter metal rod were pushed into the table with a force of 5 pounds, the pressure exerted under the rod would be considerably higher than the above, as follows:

$$P = \frac{F}{A} = \frac{5 \text{ lb}}{\pi \times 0.01 \text{ in.}^2/4} = 636 \text{ psi}$$

EXAMPLE 2-4

Problem: Convert 3 kg/cm^2 to psia.

Solution: Convert the pressure to psig and then to psia as follows:

$$3\ kg/cm^2 \times (14.2233\ psig/kg/cm^2) = 42.67\ psig$$

$$42.67 + 14.696 = 57.366\ psia$$

EXAMPLE 2-5

Problem: Convert 100 feet of water column (WC) to psig.

Solution:

$$100\ feet\ WC \times (12\ inches/foot) = 1200\ inches\ WC$$

$$1200\ inches\ WC \times (0.03609\ psi/inch\ WC) = 43.308\ psig$$

The result of dividing 100 feet of water column by 43.308 psig is a useful conversion factor of 2.31 feet of water column, which is equivalent to 1 psi. As a rule of thumb, 2 feet of water per psi can be used for rough calculations.

EXAMPLE 2-6

Problem: Calculate the dynamic head of a fan with 7-inch WC vacuum and 1-inch WC pressure on the inlet and outlet of the fan, respectively.

Solution:

$$(+1\ in.\ WC) - (-7\ in.\ WC) = +8\ in.\ WC$$

Density The density of a fluid at given operating conditions is its mass per unit volume.

$$\rho = m / V$$

In the English system, density is expressed in pound mass per cubic foot (lb/ft^3), while common SI units are grams per cubic centimeter (g/cm^3).

The following conversion may be useful to convert units of density at 60°F.

$$1\ lb/ft^3 = 0.0160262\ g/cm^3$$

EXAMPLE 2-7

Problem: What is the density of a liquid in g/cm^3, 100 pounds of which at 60°F occupies 1.53 cubic feet of a 2.04 cubic foot container?

Solution:

$$\rho = m / V = 100\ lb/1.53\ ft^3 = 65.359\ lb/ft^3$$

$$= 65.359\ lb/ft^3 \times 0.0160262\ g/cm^3 / 1\ lb/ft^3$$

$$= 1.0475\ g/cm^3$$

EXAMPLE 2-8

Problem: A 3.2 cubic foot air cylinder at 68°F is measured to be 28.2 pounds completely empty and 32.4 pounds after filling. Determine the density of the air before and after filling.

Solution: When the cylinder is empty and open to atmosphere, the density of air is 0.07528 lb/ft^3. The mass of air in the cylinder before filling is

$$3.2\ \text{ft}^3 \times 0.07528\ \text{lb/ft}^3 = 0.24\ \text{lb}$$

and the amount added during filling is

$$32.4 - 28.2 = 4.2\ \text{lb}$$

The total mass in the cylinder after filling is the sum of the mass of the air in the cylinder before filling and the air added to the cylinder, such that

$$\rho = m / V = (0.24\ \text{lb} + 4.2\ \text{lb})/3.2\ \text{ft}^3$$
$$= 1.38\ \text{lb/ft}^3$$

Expansion of Liquids and Solids The density of a fluid will vary with both operating pressure and operating temperature. Since most liquids are nearly incompressible, the effects of pressure are often negligible and can be readily ignored. The effects of temperature on density are small compared to gases and except when the operating temperature is significantly different from the temperature at which density measurements are available or when a high degree of accuracy is desired. Volumetric expansion, which affects the density of the liquid, can be expressed as

$$V = V_0(1 + \beta[\Delta t])$$

where β is the cubical coefficient of expansion of the liquid that is consistent with the temperature units used.

EXAMPLE 2-9

Problem: What will be the change in the density of a liquid due to a 10°C temperature rise if the liquid has a cubical expansion factor of 0.9×10^{-3} per degree Celsius?

Solution:

$$V = V_0(1 + [0.9 \times 10^{-3}/°\text{C}] \times [10°\text{C}])$$
$$= 1.009\ V_0$$

As the mass is the same before and after the temperature rise, the change in density is inversely proportional to the change in volume and can be expressed as

$$\rho/\rho_0 = V_0/V$$
$$= (1.009)^{-1}$$
$$= 0.991$$

Therefore, the net decrease in density is 0.9 percent.

Expansion in solids is described by the same equation as for liquids using the following relation:

$$\beta = 3 \times \alpha$$

where α is the coefficient of linear expansion of the solid.

Boyle's Law The density of a gas will vary significantly with absolute pressure, and variations of more than a few percent typically cannot be ignored. Increasing the pressure of a gas at constant temperature causes the gas to be compressed. This decreases the volume the gas occupies, thereby increasing the density of the gas, as the same mass occupies a smaller volume. Boyle's Law states that for any ideal gas or mixture of ideal gases at constant temperature, the volume is inversely proportional to the absolute pressure.,

$$V = \frac{\text{constant}}{P}$$

Boyle's Law can be stated in the following form, which is more useful in comparing the volumes of an ideal gas at constant temperature and at different pressures:

$$\frac{V}{V_0} = \frac{P_0}{P}$$

EXAMPLE 2-10

Problem: How is the volume of an ideal gas at constant temperature and a pressure of 28 psig affected by a 5-psig increase in pressure?

Solution:

$$\frac{V}{V_0} = \frac{P_0}{P} = \frac{(28 + 14.7)}{(28+5 + 14.7)}$$
$$= 0.895$$

Therefore, there is a 10.5 percent decrease in volume.

Charles' Law The density of a gas will vary significantly with absolute temperature, and variations of more than a few percent typically cannot be ignored. Increasing the temperature of a gas at constant pressure causes the gas molecules to increase their activity and motion in relation to each other. This increased activity requires a larger volume in which to move, thereby decreasing the density of the gas, as the same mass now occupies a larger volume. Charles' Law states that for any ideal gas or mixture of ideal gases at constant pressure, the volume is proportional to the absolute temperature.

$$V = \text{constant} \times T$$

Charles' Law can be stated in the following form, which is more useful in comparing the volumes of an ideal gas at constant pressure and at different temperatures:

$$\frac{V}{V_0} = \frac{T}{T_0}$$

EXAMPLE 2-11

Problem: How is the volume of a gas at constant pressure and a temperature of 15°C affected by 10°C fall in temperature?

Solution:

$$\frac{V}{V_0} = \frac{T}{T_0} = \frac{(273 + 15 - 10)}{(273 + 15)}$$

$$= 0.965$$

Therefore, there is a decrease of 3.5 percent in volume.

Ideal Gas Law Charles' and Boyle's Laws can be combined to yield the Ideal Gas Law where the constants of proportionality are the number of moles of gas and a gas constant as follows:

$$P \times V = n \times R \times T$$

where R is the universal gas constant in consistent units and n is the number of moles, which can be expressed as:

$$n = m / M_w$$

where m is the mass of the gas and M_w is its molecular weight. An "ideal" gas is defined as a gas that follows the Ideal Gas Law.

The Ideal Gas Law can also be expressed in the following form, which is more useful for discussion and calculations:

$$\frac{V}{V_0} = \frac{P_0 \times T}{P \times T_0}$$

When variations in pressure and temperature are small, the temperature and pressure act almost independently of each other and estimates of reasonable accuracy can be obtained by adding the percentage temperature and pressure deviations from a given set of conditions.

EXAMPLE 2-12

Problem: What is the net change in volume due to a 10.5 percent decrease in volume caused by increased pressure and a 3.5 percent decrease in volume caused by a decrease in temperature? (See Examples 2-10 and 2-11.)

Solution:

$$\frac{V}{V_0} = 0.895 \times 0.965$$

$$= 0.864$$

Therefore, the net change in volume is a decrease of 13.6 percent. This could have been estimated by adding the individual percentage deviations, that is, minus 3.5 percent and minus 10.5 percent, yielding a 14 percent estimated decrease in volume, which is close to the calculated change in volume.

Non-Ideal Gas Law Many gases do not act as ideal gases at certain conditions such as at high pressures, low temperatures, and under saturated conditions. These gases are termed "non-ideal" gases and their behavior may be accounted for by modifying the Ideal Gas Law as follows:

$$P \times V = n \times Z \times R \times T$$

The Z factor (or compressibility factor) is defined as the volume of a real gas divided by the volume occupied by the same mass of an ideal gas at the same pressure and temperature. This equation reduces to the Ideal Gas Law if the Z factor is equal to unity. It should be noted that some common gases are non-ideal even at standard conditions. Therefore, the Z factor should be considered whenever the gas or vapor density is calculated.

Rewriting the above equation in a more convenient form shows the effects of the Z factor, which can be numerically different under different operating conditions.

$$\frac{V}{V_0} = \frac{P_0 \times T \times Z}{P \times T_0 \times Z_0}$$

The Z factor is a function of the reduced pressure and reduced temperature at the respective operating conditions and can be read from generalized compressibility charts with a reasonable degree of accuracy, where:

$$P_R = \frac{P}{P_{critical}}$$

$$T_R = \frac{T}{T_{critical}}$$

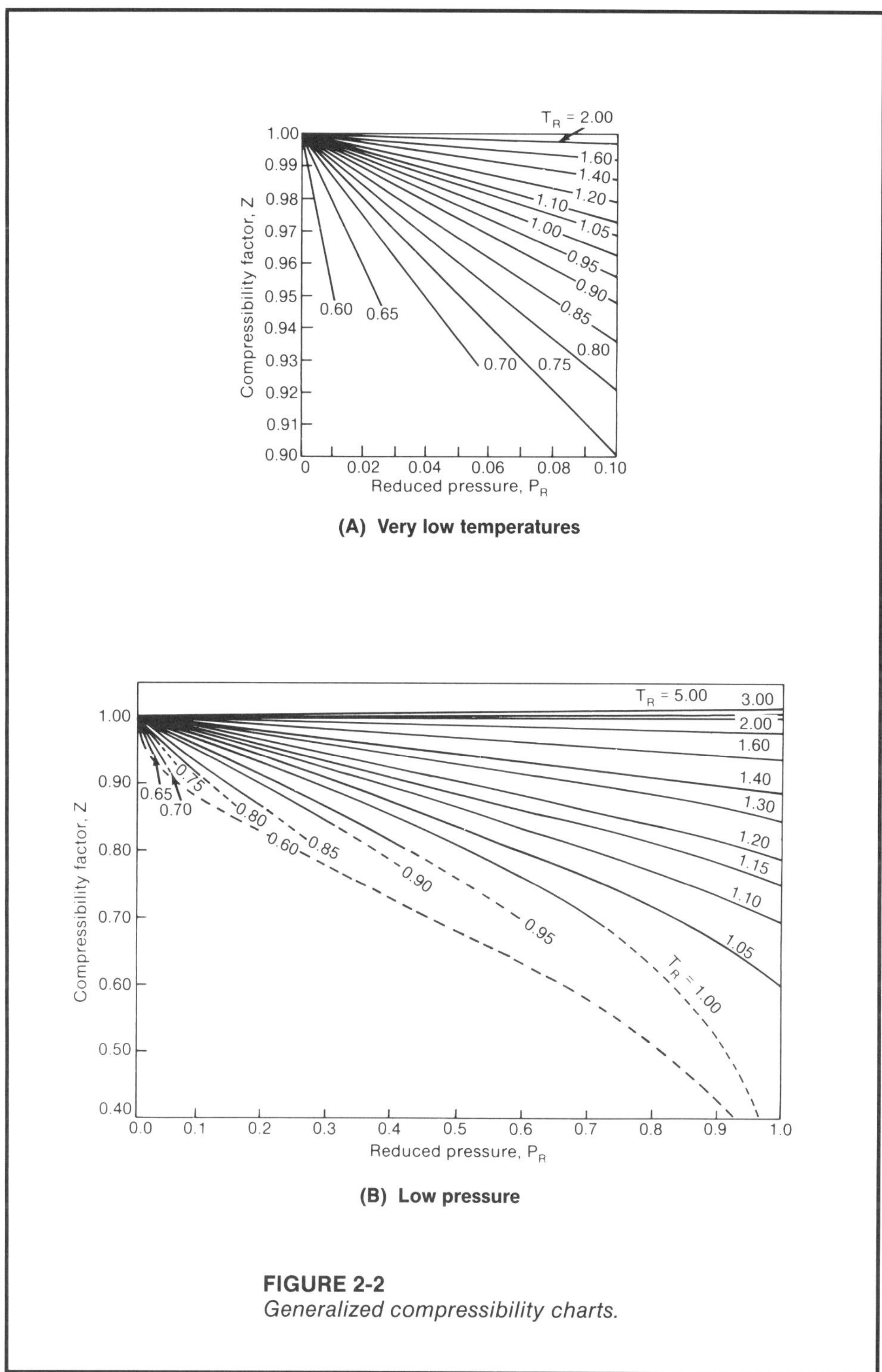

(A) Very low temperatures

(B) Low pressure

FIGURE 2-2
Generalized compressibility charts.

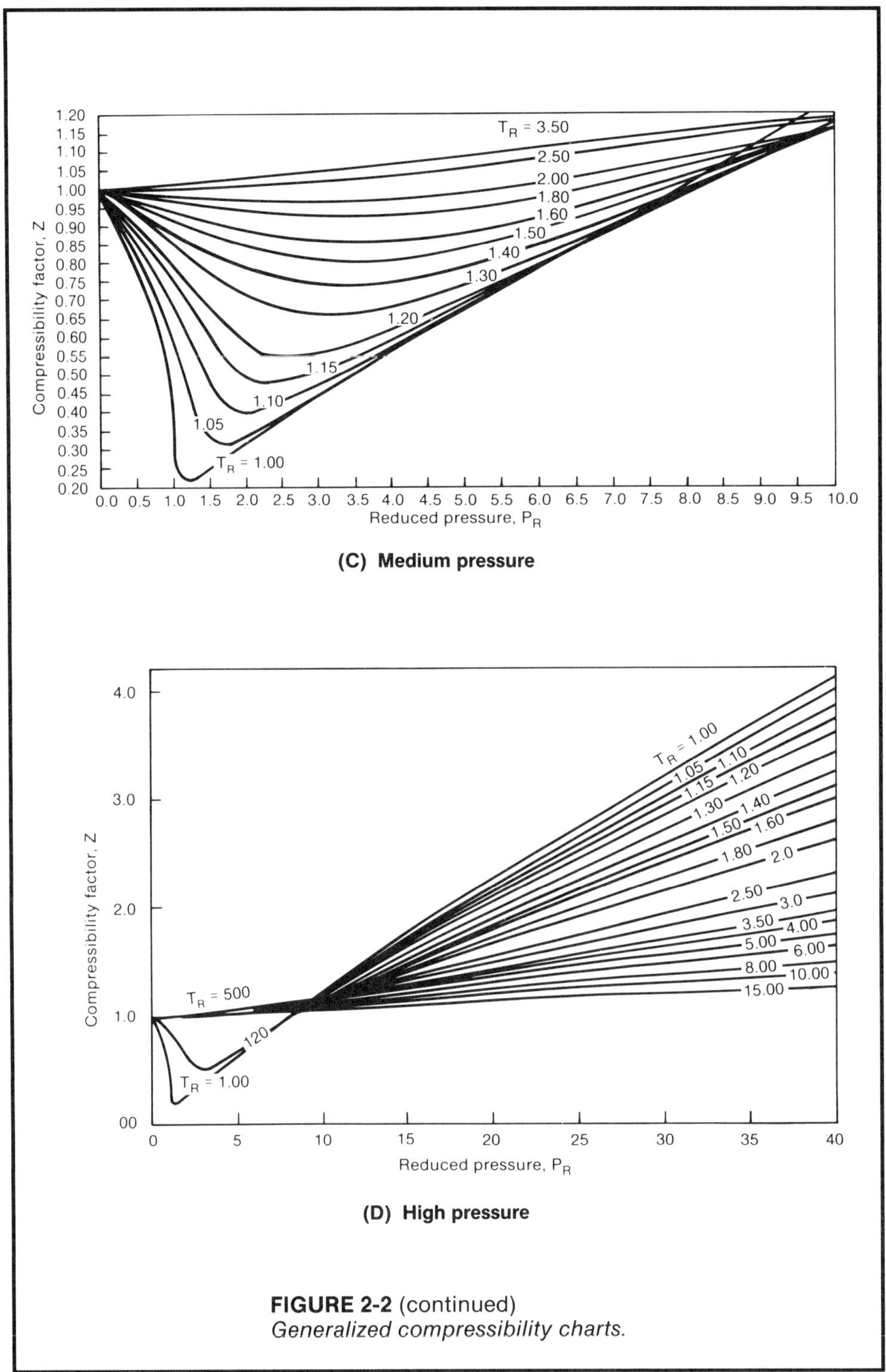

(C) Medium pressure

(D) High pressure

FIGURE 2-2 (continued)
Generalized compressibility charts.

EXAMPLE 2-13

Problem: Use the generalized compressibility charts to determine the compressibility factor of a gas at 28 psig (42.7 psia) and 15°C (288 K) if the critical pressure and temperature are 1136 psia and 751°R, respectively.

Solution: Calculate the reduced pressure and temperature:

$$P_R = \frac{42.7}{1136}$$

$$= 0.038$$

$$T_R = \frac{288\ \text{K}}{751\,°\text{R}\left(\dfrac{5\ \text{K}}{9\,°\text{R}}\right)}$$

$$= 0.69$$

Reading from the proper graph,

$$Z = 0.95 \text{ approximately}$$

Therefore, there is a 5 percent variance from the Ideal Gas Law.

EXAMPLE 2-14

Problem: Given that the gas in the previous example is nominally at 28 psig and 15°C, calculate the change in volume caused by a 5-psi increase in pressure and a 10°C decrease in temperature.

Solution: From the previous example, the Z factor at the nominal operating conditions is 0.95. The reduced pressure and reduced temperature at the actual operating conditions are 0.042 and 0.666, respectively, so that the Z factor at the operating conditions is approximately 0.93. Then,

$$V = \frac{Z\,T\,P_0}{Z_0\,T_0\,P}\ V_0$$

$$= \frac{(0.93)\,(273°\text{C} + 5°\text{C})\,(28\ \text{psi} + 14.7\ \text{psi})}{(0.95)\,(273°\text{C} + 15°\text{C})\,(33\ \text{psi} + 14.7\ \text{psi})}\ V_0$$

$$= 0.846\ V_0$$

The actual volume of the gas at operating conditions must be corrected by a factor of 1/0.846 or approximately 1.182 to correspond to the correct volume of gas at nominal operating conditions. Therefore, the gas under the new operating conditions has decreased in volume by approximately 18 percent.

Specific Gravity The operating specific gravity (SG) of a liquid or a gas is the ratio of its operating density to that of water or air at standard conditions, respectively, expressed in the same units.

$$SG = \frac{\rho_{\text{liquid}}}{\rho_{\text{water at standard conditions}}}$$

$$SG = \frac{\rho_{\text{gas}}}{\rho_{\text{air at standard conditions}}}$$

It should be noted that the specific gravity is a pure, dimensionless number, as the units cancel in the above equation, and that different industries often use different standard conditions that would result in different numerical values of specific gravity. Therefore, care must be taken to define standard conditions and conditions under which the density of the fluid was measured so as not to introduce error, especially if the density is to be calculated from the specific gravity. For this reason, the density is more commonly used to describe gases. Liquid specific gravity is usually referenced to 60°F (15.6°C) and 14.696 psia (101.325 kPa); however, other standard temperatures may be used and would result in slightly different numerical values of specific gravity.

The following conversions may be useful to convert specific gravities at different temperatures:

Density of water at 60°F = 62.33630 lb/ft^3 (0.9990121 g/cm^3)

Density of water at 68°F = 62.31572 lb/ft^3 (0.9982019 g/cm^3)

Density of air at 60°F (15.6°C) and 14.696 psia = 0.0764 lb/ft^3 (1.2236 kg/m^3)

Density of air at 68°F (20°C) and 14.696 psia = 0.07528 lb/ft^3 (1.2057 kg/m^3)

EXAMPLE 2-15

Problem: Calculate the density in g/cm^3 and specific gravity referenced to 60°F where 10 cm^3 of a liquid that is measured to have a mass of 10.95 grams at 60°F and 10.94 grams at 68°F.

Solution:

$$\rho_{60°F} = 10.95 \text{ g} / 10 \text{ cm}^3 = 1.095 \text{ g/cm}^3$$

$$SG = \rho_{60°F} / \rho_{\text{water at } 60°F} = \frac{1.095 \text{ g/cm}^3}{0.9990121 \text{ g/cm}^3}$$

$$= 1.096$$

Similarly,

$$SG = \frac{\rho_{68°F}}{\rho_{\text{water @ } 60°F}} = \frac{1.094 \text{ g/cm}^2}{0.9990121 \text{ g/cm}^3} = 1.095$$

From this example, it can be seen that changes in temperature have effects on the specific gravity and hence the density of a liquid.

EXAMPLE 2-16

Problem: Determine the specific gravity of air at 68°F with a density of 1.38 lb/ft^3.

Solution:

$$SG = (1.38\ lb/ft^3) / (0.07528\ lb/ft^3) = 18.33$$

A hydrometer may be used to measure liquid specific gravity in laboratory samples. The three commonly used hydrometer scales are the API scale for oils and two Baume scales.

The following conversions may be useful to convert units of specific gravity:

$$SG = 141.5 / (131.5 + °API)$$

$$SG = 140 / (130 + °Baume) \text{ (liquids lighter than water)}$$

$$SG = 145 / (145 + °Baume) \text{ (liquids heavier than water)}$$

Flow

Flow can be defined as the actual volume of fluid that passes a given point in a pipe per unit time. This can be expressed as:

$$Q = A \times v$$

and can be used to calculate the actual flow of liquid where A is the cross-sectional area of the pipe and v is the average fluid velocity. From the actual flow, the mass flow of a gas can be calculated by using

$$W = Q \times \rho$$

when the density is known at the actual operating conditions and where W is the mass flow. When the density is not known, it may be estimated from the density at known conditions using handbook data, laboratory analyses, the gas laws, or liquid expansion factors to correct the available data to the new conditions. Gas laws illustrate the significant changes in density that can occur at different operating conditions; however, it should be clearly understood that the density of a liquid can vary significantly over a sufficiently large temperature range.

In an attempt to standardize expressions of gas flow, the gas flow at operating conditions is often referred to standard temperature and pressure conditions. Gas flow expressed in standard units is the amount of gas at standard conditions that would be required to effect the same mass flow. The rationale for this approach is to relate the volumetric flow to mass flow at a given operating condition. While the mass of a given volume of gas at 10 psig is very different from that of the same volume of gas at 1000 psig due to increased density at elevated pressures, volumes expressed in standard conditions infer the mass flow, independent of the operating conditions.

Standard conditions are assumed to be 14.696 psia (101.325 kPa) and 59°F (15°C). Recognized standard conditions can and do vary from industry to industry, so caution must be taken to carefully define which conditions are considered standard and consistently maintained. It is good practice to define these matters at the outset to avoid misunderstandings and the need for revised calculations. Commonly encountered pressures and temperatures are combinations of 14.696 psia, 14.7 psia, 14.4 psia, 4 oz pressure, 1 bar, 59°F (15°C), 60°F, 68°F, 70°F, and 0°C.

Another frequently encountered problem is the nonideal behavior of vapors and gases. Errors in the calculation of the density of a gas at operating conditions using the Ideal Gas Laws can result in inaccuracies when the gas flow is converted to standard units. This error can be minimized by using accepted experimental correlations or tables for the fluid (such as steam tables or other experimental data) at conditions that correspond as closely as possible to the operating conditions of the gas.

Commonly used units for liquid and gas flow in the English system are gallons per minute (gpm) and cubic feet per minute (cfm). The SI system commonly uses cubic meters per hour (m^3/hr) for liquid flows, while small flows are usually expressed in cubic centimeters per minute (cc/min) in both systems for convenience. These units are often prefixed with the words standard, normal, or actual to denote to which conditions the flow is referred.

The following conversions may be useful to convert units of flow:

1 cubic meter per hour (m^3/hr) = 4.403 gpm

1 cubic foot per minute (ft^3/min) = 7.48052 gpm

1 gallon of water = 8.337 pounds of water

EXAMPLE 2-17

Problem: Derive an equation for the flow through a pipe in gallons per minute (gpm).

Solution:

$$Q = A \times v = \tfrac{1}{4}\,\pi \times D^2 \times v$$

As the diameter of the pipe (D) and the velocity are commonly expressed in inches and feet per second, respectively, the following conversions are used:

1 minute = 60 seconds

1 cubic foot = 7.48052 gallons

1 foot = 12 inches

$$Q_{\text{gpm}} = \tfrac{1}{4}\,\pi \times D^2 \times (\text{ft}/12\text{ in.})^2 \times (7.48052\text{ gal/ft}^3) \times v \times (60\text{ sec/min})$$

$$= 2.448\,D^2 v$$

EXAMPLE 2-18

Problem: Derive an equation for flow through a pipe in actual cubic feet per minute (acfm).

Solution:

$$Q = A \times v = \tfrac{1}{4}\,\pi \times D^2 \times v$$

As the diameter of the pipe and the velocity are commonly expressed in inches and feet per second, respectively, the following conversions are used:

$$1 \text{ minute} = 60 \text{ seconds}$$

$$1 \text{ foot} = 12 \text{ inches}$$

$$Q_{\text{acfm}} = \tfrac{1}{4}\,\pi \times D^2 \times (\text{ft}/12 \text{ in.})^2 \times v \times (60 \text{ sec/min})$$

$$= 0.3272\, D^2 \times v$$

Inside Pipe Diameter In the English system, the units commonly used for measuring line size are inches. The nominal size of pipes 14 inches and larger represents the nominal outside diameter of the pipe, while smaller nominal pipe sizes are rough approximations of the inside diameter of commonly used pipe. The pipe wall thickness, which is determined by the pipe schedule, can vary substantially for a given pipe size, while the outside diameter remains constant. Therefore, the inside pipe diameter can vary significantly and must be obtained from tables, such as presented in Crane's *Flow of Fluids through Valves, Fittings, and Pipe*, Technical Paper No. 410. In general, as the schedule number increases, the wall thickness increases and the inside diameter decreases, as shown in Table 2-1.

Kinematic Viscosity Simply stated, viscosity is a measure of how freely a fluid flows. It can be thought of as the internal friction of the fluid or the ability of the fluid to flow over itself, which can be highly temperature dependent.

Various experimental means have been developed to measure viscosity from which evolved a number of units of measurement. The kinematic or measured viscosity (ν) is expressed in centistokes (cSt) in the SI system of units. Figure 2-3 may be useful in illustrating the effects of temperature.

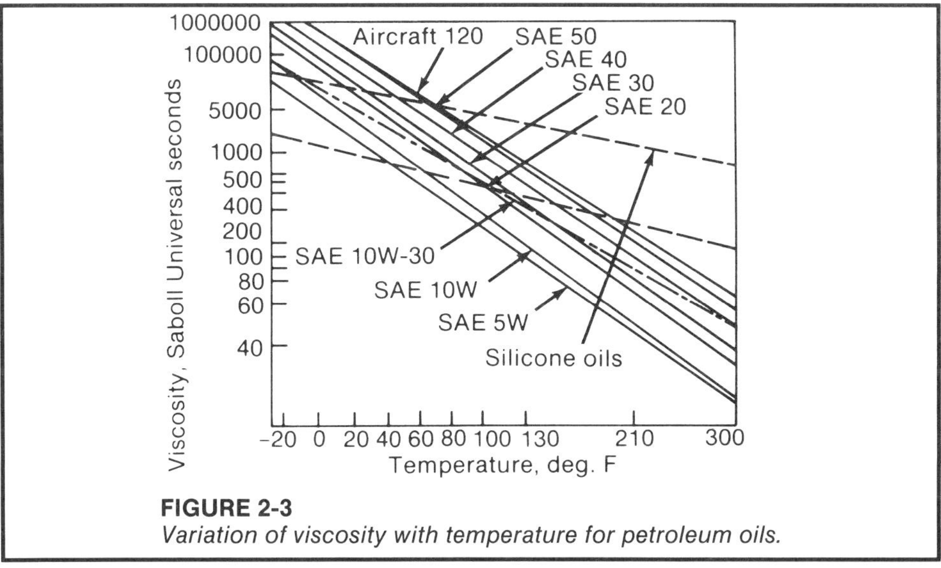

FIGURE 2-3
Variation of viscosity with temperature for petroleum oils.

Dynamic (Absolute) Viscosity The dynamic or absolute viscosity is commonly expressed in centipoise (cP) and in the SI system of measurement is calculated using

$$\mu_{cP} = \nu_{cSt} \times \rho$$

As the density of water is unity and the specific gravity is equal to the density of the liquid, the above equation reduces to

$$\mu_{cP} = \nu_{cSt} \times SG$$

Dynamic viscosity in centipoise is the most convenient unit of expressing fluid viscosity for the purpose of discussing flow measurement, and viscosities expressed in other viscosity measurement systems should be converted to centipoise.

Liquid viscosity can be highly temperature dependent and order of magnitude changes can occur over relatively small temperature ranges. An example of this is honey, which is significantly more viscous (thicker) when stored in the refrigerator than at ambient temperature. To obtain a "feel" for the relative magnitudes of viscosity, water, honey, and peanut butter have viscosities on the order of 1 cP, 300 cP and 10,000 cP, respectively.

A generalized viscosity curve for liquids in Figure 2-4 illustrates this property and can be used to estimate the viscosity of a liquid at a given temperature when no experimental data are available and the viscosity of the liquid is known at one temperature.

EXAMPLE 2-19

Problem: Estimate the viscosity of a liquid at 0°C using the generalized viscosity curve for liquids if the viscosity at 50°C is 25 cP.

Solution: The generalized viscosity curve intersects 25 cP at the third division of the horizontal axis and therefore corresponds to 50°C. As each division is 50°C, the viscosity read at the second division is approximately 400 cP, which illustrates the relatively large viscosity variation that can occur as a result of relatively small temperature changes.

EXAMPLE 2-20

Problem: Estimate the viscosity of a liquid at 50°C using the generalized viscosity curve for liquids if the viscosity at 85°C is 0.6 cP.

Solution: Using the same procedure as in the preceding example, the viscosity at 50°C is approximately 1.5 cP.

Graphs of the viscosities of various gases are shown in Figure 2-5 and 2-6.

Viscosity variations due to pressure effects are small and are typically neglected. It should be noted that gas viscosities typically vary over less than a 3:1 range from 0 to 1000°F, such that temperature effects are smaller than observed with liquid flows. Also note that, contrary to expectations, gas viscosity increases with temperature.

TABLE 2-1
PIPE DATA (Carbon and Alloy Steel — Stainless Steel)

Nominal pipe size, in.	Outside diam, in.	Identification: Steel: Iron pipe size	Identification: Steel: Sched. no.	Identification: Stainless steel sched. no.	Wall thickness (t), in.	Inside diam, (d), in.	Area of metal $in.^2$	Transverse internal area (a) $in.^2$	Transverse internal area (A) ft^2	Moment of inertia (I), in.	Weight pipe, lbs/ft	Weight water, lbs/ft of pipe	External surface, ft^2 per foot of pipe	Section modulus, $\left(2 \frac{1}{O.D.}\right)$
1/8	0.405	—	—	10S	0.049	0.307	0.0548	0.0740	0.00051	0.00088	0.19	0.032	0.106	0.00437
		STD	40	40S	0.068	0.269	0.0720	0.0568	0.00040	0.00106	0.24	0.025	0.106	0.00523
		XS	80	80S	0.095	0.215	0.0925	0.0364	0.00025	0.00122	0.31	0.016	0.106	0.00602
1/4	0.540	—	—	10S	0.065	0.410	0.0970	0.1320	0.00091	0.00279	0.33	0.057	0.141	0.01032
		STD	40	40S	0.088	0.364	0.1250	0.1041	0.00072	0.00331	0.42	0.045	0.141	0.01227
		XS	80	80S	0.119	0.302	0.1574	0.0716	0.00050	0.00377	0.54	0.031	0.141	0.01395
3/8	0.675	—	—	10S	0.065	0.545	0.1246	0.2333	0.00162	0.00586	0.42	0.101	0.178	0.01736
		STD	40	40S	0.091	0.493	0.1670	0.1910	0.00133	0.00729	0.57	0.083	0.178	0.02160
		XS	80	80S	0.126	0.423	0.2173	0.1405	0.00098	0.00862	0.74	0.061	0.178	0.02554
1/2	0.840	—	—	5S	0.065	0.710	0.1583	0.3959	0.00275	0.01197	0.54	0.172	0.220	0.02849
		—	—	10S	0.083	0.674	0.1974	0.3568	0.00248	0.01431	0.67	0.155	0.220	0.03407
		STD	40	40S	0.109	0.622	0.2503	0.3040	0.00211	0.01709	0.85	0.132	0.220	0.04069
		XS	80	80S	0.147	0.546	0.3200	0.2340	0.00163	0.02008	1.09	0.102	0.220	0.04780
		—	160	—	0.187	0.466	0.3836	0.1706	0.00118	0.02212	1.31	0.074	0.220	0.05267
		XXS	—	—	0.294	0.252	0.5043	0.050	0.00035	0.02424	1.71	0.022	0.220	0.05772
3/4	1.050	—	—	5S	0.065	0.920	0.2011	0.6648	0.00462	0.02450	0.69	0.288	0.275	0.04667
		—	—	10S	0.083	0.884	0.2521	0.6138	0.00426	0.02969	0.86	0.266	0.275	0.05655
		STD	40	40S	0.113	0.824	0.3326	0.5330	0.00371	0.03704	1.13	0.231	0.275	0.07055
		XS	80	80S	0.154	0.742	0.4335	0.4330	0.00300	0.04479	1.47	0.188	0.275	0.08531
		—	160	—	0.219	0.612	0.5698	0.2961	0.00206	0.05269	1.94	0.128	0.275	0.10036
		XXS	—	—	0.308	0.434	0.7180	0.148	0.00103	0.05792	2.44	0.064	0.275	0.11032
1	1.315	—	—	5S	0.065	1.185	0.2553	1.1029	0.00766	0.04999	0.87	0.478	0.344	0.07603
		—	—	10S	0.109	1.097	0.4130	0.9452	0.00656	0.07569	1.40	0.409	0.344	0.11512
		STD	40	40S	0.133	1.049	0.4939	0.8640	0.00600	0.08734	1.68	0.375	0.344	0.1328
		XS	80	80S	0.179	0.957	0.6388	0.7190	0.00499	0.1056	2.17	0.312	0.344	0.1606
		—	160	—	0.250	0.815	0.8365	0.5217	0.00362	0.1251	2.84	0.230	0.344	0.1903
		XXS	—	—	0.358	0.599	1.0760	0.282	0.00196	0.1405	3.66	0.122	0.344	0.2136

(continued)

TABLE 2-1
PIPE DATA (Carbon and Alloy Steel — Stainless Steel) (continued)

		—	—	5S	0.065	1.530	0.3257	1.839	0.01277	0.1038	1.11	0.797	0.435	0.1250
		—	—	10S	0.109	1.442	0.4717	1.633	0.01134	0.1605	1.81	0.708	0.435	0.1934
		STD	40	40S	0.140	1.380	0.6685	1.495	0.01040	0.1947	2.27	0.649	0.435	0.2346
1-1/4	1.660	XS	80	80S	0.191	1.278	0.8815	1.283	0.00891	0.2418	3.00	0.555	0.435	0.2913
		—	160	—	0.250	1.160	1.1070	1.057	0.00734	0.2839	3.76	0.458	0.435	0.3421
		XXS	—	—	0.382	0.896	1.534	0.630	0.00438	0.3411	5.21	0.273	0.435	0.4110
		—	—	5S	0.065	1.770	0.3747	2.461	0.01709	0.1579	1.28	1.066	0.497	0.1662
		—	—	10S	0.109	1.682	0.6133	2.222	0.01543	0.2468	2.09	0.963	0.497	0.2598
		STD	40	40S	0.145	1.610	0.7995	2.036	0.01414	0.3099	2.72	0.882	0.497	0.3262
1-1/2	1.900	XS	80	80S	0.200	1.500	1.068	1.767	0.01225	0.3912	3.63	0.765	0.497	0.4118
		—	160	—	0.281	1.338	1.429	1.406	0.00976	0.4824	4.86	0.608	0.497	0.5078
		XXS	—	—	0.400	1.100	1.885	0.950	0.00660	0.5678	6.41	0.42	0.497	0.5977
		—	—	5S	0.065	2.245	0.4717	3.958	0.02749	0.3149	1.61	1.72	0.622	0.2652
		—	—	10S	0.109	2.157	0.7760	3.654	0.02538	0.4992	2.64	1.58	0.622	0.4204
		STD	40	40S	0.154	2.067	1.075	3.355	0.02330	0.6657	3.65	1.45	0.622	0.5606
2	2.375	XS	80	80S	0.218	1.939	1.477	2.953	0.02050	0.8679	5.02	1.28	0.622	0.7309
		—	160	—	0.344	1.687	2.190	2.241	0.01556	1.162	7.46	0.97	0.622	0.979
		XXS	—	—	0.436	1.503	2.656	1.774	0.01232	1.311	9.03	0.77	0.622	1.104
		—	—	5S	0.083	2.709	0.7280	5.764	0.04002	0.7100	2.48	2.50	0.753	0.4939
		—	—	10S	0.120	2.635	1.039	5.453	0.03787	0.9873	3.53	2.36	0.753	0.6868
		STD	40	40S	0.203	2.469	1.704	4.788	0.03322	1.530	5.79	2.07	0.753	1.064
2-1/2	2.875	XS	80	80S	0.276	2.323	2.254	4.238	0.02942	1.924	7.66	1.87	0.753	1.339
		—	160	—	0.375	2.125	2.945	3.546	0.02463	2.353	10.01	1.54	0.753	1.638
		XXS	—	—	0.552	1.771	4.028	2.464	0.01710	2.871	13.69	1.07	0.753	1.997
		—	—	5S	0.083	3.334	0.8910	8.730	0.06063	1.301	3.03	3.78	0.916	0.7435
		—	—	10S	0.120	3.260	1.274	8.347	0.05796	1.822	4.33	3.62	0.916	1.041
		STD	40	40S	0.216	3.068	2.228	7.393	0.05130	3.017	7.58	3.20	0.916	1.724
3	3.500	XS	80	80S	0.300	2.900	3.016	6.606	0.04587	3.894	10.25	2.86	0.916	2.225
		—	160	—	0.438	2.624	4.205	5.408	0.03755	5.032	14.32	2.35	0.916	2.876
		XXS	—	—	0.600	2.300	5.466	4.155	0.02885	5.993	18.58	1.80	0.916	3.424
		—	—	5S	0.083	3.834	1.021	11.545	0.08017	1.960	3.48	5.00	1.047	0.9799
		—	—	10S	0.120	3.760	1.463	11.104	0.07711	2.755	4.97	4.81	1.047	1.378
3-1/2	4.000	STD	40	40S	0.226	3.548	2.680	9.886	0.06870	4.788	9.11	4.29	1.047	2.394
		XS	80	80S	0.318	3.364	3.678	8.888	0.06170	6.280	12.50	3.84	1.047	3.140

Identification, wall thickness and weights are extracted from ANSI B36.10 and B36.19. The notations STD, XS, and XXS indicate Standard, Extra Strong, and Double Extra Strong pipe, respectively.

Transverse internal area values listed in "square feet" also represent volume in cubic feet per foot of pipe length.

(From *Flow of Fluids through Valves, Fittings, and Pipe*, Technical Paper No. 410, The Crane Company, NY. Used with permission.)

(continued)

TABLE 2-1
PIPE DATA (Carbon and Alloy Steel — Stainless Steel) (continued)

Nominal pipe size, in.	Outside diam, in.	Identification: Steel: Iron pipe size	Identification: Steel: Sched. no.	Stainless steel sched. no.	Wall thickness (t), in.	Inside diam, (d), in.	Area of metal in.^{2}	Transverse internal area (a) in.^{2}	Transverse internal area (A) ft^2	Moment of inertia (I), in.	Weight pipe, lbs/ft	Weight water, lbs/ft of pipe	External surface, ft^2 per foot of pipe	Section modulus, $\left(2\frac{1}{O.D.}\right)$
4	4.500	—	—	5S	0.083	4.334	1.152	14.75	0.10245	2.810	3.92	6.39	1.178	1.249
		—	—	10S	0.120	4.260	1.651	14.25	0.09898	3.963	5.61	6.18	1.178	1.761
		STD	40	40S	0.237	4.026	3.174	12.73	0.08840	7.233	10.79	5.50	1.178	3.214
		XS	80	80S	0.337	3.826	4.407	11.50	0.07986	9.610	14.98	4.98	1.178	4.271
		—	120	—	0.438	3.624	5.595	10.31	0.0716	11.65	19.00	4.47	1.178	5.178
		—	160	—	0.531	3.438	6.621	9.28	0.0645	13.27	22.51	4.02	1.178	5.898
		XXS	—	—	0.674	3.152	8.101	7.80	0.0542	15.28	27.54	3.38	1.178	6.791
5	5.563	—	—	5S	0.109	5.345	1.868	22.44	0.1558	6.947	6.36	9.72	1.456	2.498
		—	—	10S	0.134	5.295	2.285	22.02	0.1529	8.425	7.77	9.54	1.456	3.029
		STD	40	40S	0.258	5.047	4.300	20.01	0.1390	15.16	14.62	8.67	1.456	5.451
		XS	80	80S	0.375	4.813	6.112	18.19	0.1263	20.67	20.78	7.88	1.456	7.431
		—	120	—	0.500	4.563	7.953	16.35	0.1136	25.73	27.04	7.09	1.456	9.250
		—	160	—	0.625	4.313	9.696	14.61	0.1015	30.03	32.96	6.33	1.456	10.796
		XXS	—	—	0.750	4.063	11.340	12.97	0.0901	33.63	38.55	5.61	1.456	12.090
6	6.625	—	—	5S	0.109	6.407	2.231	32.24	0.2239	11.85	7.60	13.97	1.734	3.576
		—	—	10S	0.134	6.357	2.733	31.74	0.2204	14.40	9.29	13.75	1.734	4.346
		STD	40	40S	0.280	6.065	5.581	28.89	0.2006	28.14	18.97	12.51	1.734	8.496
		XS	80	80S	0.432	5.761	8.405	26.07	0.1810	40.49	28.57	11.29	1.734	12.22
		—	120	—	0.562	5.501	10.70	23.77	0.1650	49.61	36.39	10.30	1.734	14.98
		—	160	—	0.719	5.187	13.32	21.15	0.1469	58.97	45.35	9.16	1.734	17.81
		XXS	—	—	0.864	4.897	15.64	18.84	0.1308	66.33	53.16	8.16	1.734	20.02
8	8.625	—	—	5S	0.109	8.407	2.916	55.51	0.3855	26.44	9.93	24.06	2.258	6.131
		—	—	10S	0.148	8.329	3.941	54.48	0.3784	35.41	13.40	23.61	2.258	8.212
		—	20	—	0.250	8.125	6.57	51.85	0.3601	57.72	22.36	22.47	2.258	13.39
		—	30	—	0.277	8.071	7.26	51.16	0.3553	63.35	24.70	22.17	2.258	14.69
		STD	40	40S	0.322	7.981	8.40	50.03	0.3474	72.49	28.55	21.70	2.258	16.81
		—	60	—	0.406	7.813	10.48	47.94	0.3329	88.73	35.64	20.77	2.258	20.58
		XS	80	80S	0.500	7.625	12.76	45.66	0.3171	105.7	43.39	19.78	2.258	24.51
		—	100	—	0.594	7.437	14.96	43.46	0.3018	121.3	50.95	18.83	2.258	28.14
		—	120	—	0.719	7.187	17.84	40.59	0.2819	140.5	60.71	17.59	2.258	32.58
		—	140	—	0.812	7.001	19.93	38.50	0.2673	153.7	67.76	16.68	2.258	35.65
		XXS	—	—	0.875	6.875	21.30	37.12	0.2578	162.0	72.42	16.10	2.258	37.56
		—	160	—	0.906	6.813	21.97	36.46	0.2532	165.9	74.69	15.80	2.258	38.48

(continued)

TABLE 2-1
PIPE DATA (Carbon and Alloy Steel — Stainless Steel) (continued)

		—	—	5S	0.134	10.482	4.36	86.29	0.5992	63.0	15.19	37.39	2.814	11.71
		—	—	10S	0.165	10.420	5.49	85.28	0.5922	76.9	18.65	36.95	2.814	14.30
		—	20	—	0.250	10.250	8.24	82.52	0.5731	113.7	28.04	35.76	2.814	21.15
		—	30	—	0.307	10.136	10.07	80.69	0.5603	137.4	34.24	34.96	2.814	25.57
		STD	40	40S	0.365	10.020	11.90	78.86	0.5475	160.7	40.48	34.20	2.814	29.90
10	10.750	XS	60	80S	0.500	9.750	16.10	74.66	0.5185	212.0	54.74	32.35	2.814	39.43
		—	80	—	0.594	9.562	18.92	71.84	0.4989	244.8	64.43	31.13	2.814	45.54
		—	100	—	0.719	9.312	22.63	68.13	0.4732	286.1	77.03	29.53	2.814	53.22
		—	120	—	0.844	9.062	26.24	64.53	0.4481	324.2	89.29	27.96	2.814	60.32
		XXS	140	—	1.000	8.750	30.63	60.13	0.4176	367.8	104.13	26.06	2.814	68.43
		—	160	—	1.125	8.500	34.02	56.75	0.3941	399.3	115.64	24.59	2.814	74.29
		—	—	5S	0.156	12.438	6.17	121.50	0.8438	122.4	20.98	52.65	3.338	19.2
		—	—	10S	0.180	12.390	7.11	120.57	0.8373	140.4	24.17	52.25	3.338	22.0
		—	20	—	0.250	12.250	9.82	117.86	0.8185	191.8	33.38	51.07	3.338	30.2
		—	30	—	0.330	12.090	12.87	114.80	0.7972	248.4	43.77	49.74	3.338	39.0
		STD	—	40S	0.375	12.000	14.58	113.10	0.7854	279.3	49.56	49.00	3.338	43.8
		—	40	—	0.406	11.938	15.77	111.93	0.7773	300.3	53.52	48.50	3.338	47.1
		XS	—	80S	0.500	11.750	19.24	108.43	0.7528	361.5	65.42	46.92	3.338	56.7
12	12.75	—	60	—	0.562	11.626	21.52	106.16	0.7372	400.4	73.15	46.00	3.338	62.8
		—	80	—	0.688	11.374	26.03	101.64	0.7058	475.1	88.63	44.04	3.338	74.6
		—	100	—	0.844	11.062	31.53	96.14	0.6677	561.6	107.32	41.66	3.338	88.1
		XXS	120	—	1.000	10.750	36.91	90.76	0.6303	641.6	125.49	39.33	3.338	100.7
		—	140	—	1.125	10.500	41.08	86.59	0.6013	700.5	139.67	37.52	3.338	109.9
		—	160	—	1.312	10.126	47.14	80.53	0.5592	781.1	160.27	34.89	3.338	122.6
		—	—	5S	0.156	13.688	6.78	147.15	1.0219	162.6	23.07	63.77	3.665	23.2
		—	—	10S	0.188	13.624	8.16	145.78	1.0124	194.6	27.73	63.17	3.665	27.8
		—	10	—	0.250	13.500	10.80	143.14	0.9940	255.3	36.71	62.03	3.665	36.6
		—	20	—	0.312	13.376	13.42	140.52	0.9758	314.4	45.61	60.89	3.665	45.0
		STD	30	—	0.375	13.250	16.05	137.86	0.9575	372.8	54.57	59.75	3.665	53.2
		—	40	—	0.438	13.124	18.66	135.28	0.9394	429.1	63.44	58.64	3.665	61.3
14	14.00	XS	—	—	0.500	13.000	21.21	132.73	0.9217	483.8	72.09	57.46	3.665	69.1
		—	60	—	0.594	12.812	24.98	128.96	0.8956	562.3	85.05	55.86	3.665	80.3
		—	80	—	0.750	12.500	31.22	122.72	0.8522	678.3	106.13	53.18	3.665	98.2
		—	100	—	0.938	12.124	38.45	115.49	0.8020	824.4	130.85	50.04	3.665	117.8
		—	120	—	1.094	11.812	44.32	109.62	0.7612	929.6	150.79	47.45	3.665	132.8
		—	140	—	1.250	11.500	50.07	103.87	0.7213	1027.0	170.28	45.01	3.665	146.8
		—	160	—	1.406	11.188	55.63	98.31	0.6827	1117.0	189.11	42.60	3.665	159.6

Identification, wall thickness and weights are extracted from ANSI B36.10 and B36.19. The notations STD, XS, and XXS indicate Standard, Extra Strong, and Double Extra Strong pipe, respectively.

Transverse internal area values listed in "square feet" also represent volume in cubic feet per foot of pipe length.

(From *Flow of Fluids through Valves, Fittings, and Pipe,* Technical Paper No. 410, The Crane Company, NY. Used with permission.)

(continued)

TABLE 2-1
PIPE DATA (Carbon and Alloy Steel — Stainless Steel) (continued)

Nominal pipe size, in.	Outside diam, in.	Identification: Steel: Iron pipe size	Identification: Steel: Sched. no.	Identification: Stainless steel sched. no.	Wall thickness (t), in.	Inside diam, (d), in.	Area of metal in.^{2}	Transverse internal area (a) in.^{2}	Transverse internal area (A) ft^2	Moment of inertia (I), in.	Weight pipe, lbs/ft	Weight water, lbs/ft of pipe	External surface, ft^2 per foot of pipe	Section modulus, $\left(2\frac{1}{O.D.}\right)$
		—	—	5S	0.165	15.670	8.21	192.85	1.3393	257.3	27.90	83.57	4.189	32.2
		—	—	10S	0.188	15.624	9.34	191.72	1.3314	291.9	31.75	83.08	4.189	36.5
		—	10	—	0.250	15.500	12.37	188.69	1.3103	383.7	42.05	81.74	4.189	48.0
		—	20	—	0.312	15.376	15.38	185.09	1.2895	473.2	52.27	80.50	4.189	59.2
		STD	30	—	0.375	15.250	18.41	182.65	1.2684	562.1	62.58	79.12	4.189	70.3
16	16.00	XS	40	—	0.500	15.000	24.35	176.72	1.2272	731.9	82.77	76.58	4.189	91.5
		—	60	—	0.656	14.688	31.62	169.44	1.1766	932.4	107.50	73.42	4.189	116.6
		—	80	—	0.844	14.312	40.14	160.92	1.1175	1155.8	136.61	69.73	4.189	144.5
		—	100	—	1.031	13.938	48.48	152.58	1.0596	1364.5	164.82	66.72	4.189	170.5
		—	120	—	1.219	13.562	56.56	144.50	1.0035	1555.8	192.43	62.62	4.189	194.5
		—	140	—	1.438	13.124	65.78	135.28	0.9394	1760.3	223.64	58.64	4.189	220.0
		—	160	—	1.594	12.812	72.10	128.96	0.8956	1893.5	245.25	55.83	4.189	236.7
		—	—	5S	0.165	17.670	9.25	245.22	1.7029	367.6	31.43	106.26	4.712	40.8
		—	—	10S	0.188	17.624	10.52	243.95	1.6941	417.3	35.76	105.71	4.712	46.4
		—	10	—	0.250	17.500	13.94	240.54	1.6703	549.1	47.39	104.21	4.712	61.1
		—	20	—	0.312	17.376	17.34	237.13	1.6467	678.2	58.94	102.77	4.712	75.5
		STD	—	—	0.375	17.250	20.76	233.71	1.6230	806.7	70.59	101.18	4.712	89.6
		—	30	—	0.438	17.124	24.17	230.30	1.5990	930.3	82.15	99.84	4.712	103.4
18	18.00	XS	—	—	0.500	17.000	27.49	226.98	1.5763	1053.2	93.45	98.27	4.712	117.0
		—	40	—	0.562	16.876	30.79	223.68	1.5533	1171.5	104.67	96.93	4.712	130.1
		—	60	—	0.750	16.500	40.64	213.83	1.4849	1514.7	138.17	92.57	4.712	168.3
		—	80	—	0.938	16.124	50.23	204.24	1.4183	1833.0	170.92	88.50	4.712	203.8
		—	100	—	1.156	15.688	61.17	193.30	1.3423	2180.0	207.96	83.76	4.712	242.3
		—	120	—	1.375	15.250	71.81	182.66	1.2684	2498.1	244.14	79.07	4.712	277.6
		—	140	—	1.562	14.876	80.66	173.80	1.2070	2749.0	274.22	75.32	4.712	305.5
		—	160	—	1.781	14.438	90.75	163.72	1.1369	3020.0	308.50	70.88	4.712	335.6

(continued)

TABLE 2-1
PIPE DATA (Carbon and Alloy Steel — Stainless Steel) (continued)

20	20.00	—	—	5S	0.188	19.624	11.70	302.46	2.1004	574.2	39.78	131.06	5.236	57.4
		—	—	10S	0.218	19.564	13.55	300.61	2.0876	662.8	46.06	130.27	5.236	66.3
		—	10	—	0.250	19.500	15.51	298.65	2.0740	765.4	52.73	129.42	5.236	75.6
		STD	20	—	0.375	19.250	23.12	290.04	2.0142	1113.0	78.60	125.67	5.236	111.3
		XS	30	—	0.500	19.000	30.63	283.53	1.9690	1457.0	104.13	122.87	5.236	145.7
		—	40	—	0.594	18.812	36.15	278.00	1.9305	1703.0	123.11	120.46	5.236	170.4
		—	60	—	0.812	18.376	48.95	265.21	1.8417	2257.0	166.40	114.92	5.236	225.7
		—	80	—	1.031	17.938	61.44	252.72	1.7550	2772.0	208.87	109.51	5.236	277.1
		—	100	—	1.281	17.438	75.33	238.83	1.6585	3315.2	256.10	103.39	5.236	331.5
		—	120	—	1.500	17.000	87.18	226.98	1.5762	3754.0	296.37	98.35	5.236	375.5
		—	140	—	1.750	16.500	100.33	213.82	1.4849	4216.0	341.09	92.66	5.236	421.7
		—	160	—	1.969	16.062	111.49	202.67	1.4074	4585.5	379.17	87.74	5.236	458.5
22	22.00	—	—	5S	0.188	21.624	12.88	367.25	2.5503	766.2	43.80	159.14	5.760	69.7
		—	—	10S	0.218	21.564	14.92	365.21	2.5362	884.8	50.71	158.26	5.760	80.4
		—	10	—	0.250	21.500	17.08	363.05	2.5212	1010.3	58.07	157.32	5.760	91.8
		STD	20	—	0.375	21.250	25.48	354.66	2.4629	1489.7	86.61	153.68	5.760	135.4
		XS	30	—	0.500	21.000	33.77	346.36	2.4053	1952.5	114.81	150.09	5.760	117.5
		—	60	—	0.875	20.250	58.07	322.06	2.2365	3244.9	197.41	139.56	5.760	295.0
		—	80	—	1.125	19.75	73.78	306.35	2.1275	4030.4	250.81	132.76	5.760	366.4
		—	100	—	1.375	19.25	89.09	291.04	2.0211	4758.5	302.88	126.12	5.760	432.6
		—	120	—	1.625	18.75	104.02	276.12	1.9175	5432.0	353.61	119.65	5.760	493.8
		—	140	—	1.875	18.25	118.55	261.59	1.8166	6053.7	403.00	113.36	5.760	550.3
		—	160	—	2.125	17.75	132.68	247.45	1.7184	6626.4	451.06	107.23	5.760	602.4
24	24.00	—	—	5S	0.218	23.564	16.29	436.10	3.0285	1151.6	55.37	188.98	6.283	96.0
		—	10	10S	0.250	23.500	18.65	433.74	3.0121	1315.4	63.41	187.95	6.283	109.6
		STD	20	—	0.375	23.250	27.83	424.56	2.9483	1942.0	94.62	183.95	6.283	161.9
		XS	—	—	0.500	23.000	36.91	415.48	2.8853	2549.5	125.49	179.87	6.283	212.5
		—	30	—	0.562	22.876	41.39	411.00	2.8542	2843.0	140.68	178.09	6.283	237.0
		—	40	—	0.688	22.624	50.31	402.07	2.7921	3421.3	171.29	174.23	6.283	285.1
		—	60	—	0.969	22.062	70.04	382.35	2.6552	4652.8	238.35	165.52	6.283	387.7
		—	80	—	1.219	21.562	87.17	365.22	2.5362	5672.0	296.58	158.26	6.283	472.8
		—	100	—	1.531	20.938	108.07	344.32	2.3911	6849.9	367.39	149.06	6.283	570.8
		—	120	—	1.812	20.376	126.31	326.08	2.2645	7825.0	429.39	141.17	6.283	652.1
		—	140	—	2.062	19.876	142.11	310.28	2.1547	8625.0	483.12	134.45	6.283	718.9
		—	160	—	2.344	19.312	159.41	292.98	2.0346	9455.9	542.13	126.84	6.283	787.9

Identification, wall thickness and weights extracted from ANSI B36.10 and B36.19. The notations STD, XS, and XXS indicate Standard, Extra Strong, and Double Extra Strong pipe, respectively.

Transverse internal area values listed in "square feet" also represent volume in cubic feet per foot of pipe length.

(From *Flow of Fluids through Valves, Fittings, and Pipe,* Technical Paper No. 410, The Crane Company, NY. Used with permission.)

(continued)

TABLE 2-1
PIPE DATA (Carbon and Alloy Steel — Stainless Steel) (continued)

Nominal pipe size, in.	Outside diam, in.	Identification: Steel: Iron pipe size	Identification: Steel: Sched. no.	Identification: Stainless steel sched. no.	Wall thickness (t), in.	Inside diam, (d), in.	Area of metal in.^{2}	Transverse internal area (a) in.^{2}	Transverse internal area (A) ft^2	Moment of inertia (I), in.	Weight pipe, lbs/ft	Weight water, lbs/ft of pipe	External surface, ft^2 per foot of pipe	Section modulus, $\left(2\frac{1}{O.D.}\right)$
		—	10	—	0.312	25.376	25.18	505.75	3.5122	2077.2	85.60	219.16	6.806	159.8
26	26.00	STD	—	—	0.375	25.250	30.19	500.74	3.4774	2478.4	102.63	216.99	6.806	190.6
		XS	20	—	0.500	25.000	40.06	490.87	3.4088	3257.0	136.17	212.71	6.806	250.5
		—	10	—	0.312	27.376	27.14	588.61	4.0876	2601.0	92.26	255.07	7.330	185.8
		STD	—	—	0.375	27.250	32.54	583.21	4.0501	3105.1	110.64	252.73	7.330	221.8
28	28.00	XS	20	—	0.500	27.000	43.20	572.56	3.9761	4084.8	146.85	248.11	7.330	291.8
		—	30	—	0.625	26.750	53.75	562.00	3.9028	5037.7	182.73	243.53	7.330	359.8
		—	—	5S	0.250	29.500	23.37	683.49	4.7465	2585.2	79.43	296.18	7.854	172.3
		—	10	10S	0.312	29.376	29.10	677.76	4.7067	3206.3	98.93	293.70	7.854	213.8
30	30.00	STD	—	—	0.375	29.250	34.90	671.96	4.6644	3829.4	118.65	291.18	7.854	255.3
		XS	20	—	0.500	29.000	46.34	606.52	4.5869	5042.2	157.53	286.22	7.854	336.1
		—	30	—	0.625	28.750	57.63	649.18	4.5082	6224.0	196.08	281.31	7.854	414.9
		—	10	—	0.312	31.376	31.06	773.19	5.3694	3898.9	105.59	335.05	8.378	243.7
		STD	—	—	0.375	31.250	37.26	766.99	5.3263	4658.5	126.66	332.36	8.378	291.2
32	32.00	XS	20	—	0.500	31.000	49.48	754.77	5.2414	6138.6	168.21	327.06	8.378	383.7
		—	30	—	0.625	30.750	61.60	742.64	5.1572	7583.4	209.43	321.81	8.378	474.0
		—	40	—	0.688	30.624	67.68	736.57	5.1151	8298.3	230.08	319.18	8.378	518.6
		—	10	—	0.344	33.312	36.37	871.55	6.0524	5150.5	123.65	377.67	8.901	303.0
		STD	—	—	0.375	33.250	39.61	868.31	6.0299	5599.3	134.67	376.27	8.901	329.4
34	34.00	XS	20	—	0.500	33.000	52.62	855.30	5.9396	7383.5	178.89	370.63	8.901	434.3
		—	30	—	0.625	32.750	65.53	842.39	5.8499	9127.6	222.78	365.03	8.901	536.9
		—	40	—	0.688	32.624	72.00	835.92	5.8050	9991.6	244.77	362.23	8.901	587.7
		—	10	—	0.312	35.376	34.98	982.90	6.8257	5569.5	118.92	425.92	9.425	309.4
		STD	—	—	0.375	35.250	41.97	975.91	6.7771	6658.9	142.68	422.89	9.425	369.9
36	36.00	XS	20	—	0.500	35.000	55.76	962.11	6.6813	8786.2	189.57	416.91	9.425	488.1
		—	30	—	0.625	34.750	69.46	948.42	6.5862	10868.4	236.13	417.22	9.425	603.8
		—	40	—	0.750	34.500	83.06	934.82	6.4918	12906.1	282.35	405.09	9.425	717.0

Identification, wall thickness and weights are extracted from ANSI B36.10 and B36.19. The notations STD, XS, and XXS indicate Standard, Extra Strong, and Double Extra Strong pipe, respectively.

Transverse internal area values listed in "square feet" also represent volume in cubic feet per foot of pipe length.

(From *Flow of Fluids through Valves, Fittings, and Pipe,* Technical Paper No. 410, The Crane Company, NY. Used with permission.)

TABLE 2-2
Viscosity Equivalents

Equivalents of Kinematic and Saybolt Universal Viscosity			*Equivalents of Kinematic and Saybolt Furol Viscosity*		
Kinematic viscosity, centistokes	*Equivalent Saybolt universal viscosity, sec*		*Kinematic viscosity, centistokes*	*Equivalent Saybolt furol viscosity, sec*	
ν	*At 100° F basic values*	*At 210° F*	ν	*At 122° F*	*At 210° F*
1.83	32.01	32.23	48	25.3	
2.0	32.62	32.85	50	26.1	25.2
4.0	39.14	39.41	60	30.6	29.8
6.0	45.56	45.88	70	35.1	34.4
8.0	52.09	52.45	80	39.6	39.0
10.0	58.91	59.32	90	44.1	43.7
15.0	77.39	77.93	100	48.6	48.3
20.0	97.77	98.45	125	60.1	60.1
25.0	119.3	120.1	150	71.7	71.8
30.0	141.3	142.3	175	83.8	83.7
35.0	163.7	164.9	200	95.0	95.6
40.0	186.3	187.6	225	106.7	107.5
45.0	209.1	210.5	250	118.4	119.4
50.0	232.1	233.8	275	130.1	131.4
55.0	255.2	257.0	300	141.8	143.5
60.0	278.3	280.2	325	153.6	155.5
65.0	301.4	303.5	350	165.3	167.6
70.0	324.4	326.7	375	177.0	179.7
75.0	347.6	350.0	400	188.8	191.8
80.0	370.8	373.4	425	200.6	204.0
85.0	393.9	396.7	450	212.4	216.1
90.0	417.1	420.0	475	224.1	228.3
95.0	440.3	443.4	500	235.9	240.5
100.0	463.5	466.7	525	247.7	252.8
120.0	556.2	560.1	550	259.5	265.0
140.0	648.9	653.4	575	271.3	277.2
160.0	741.6		600	283.1	289.5
180.0	834.2		625	294.9	301.8
200.0	926.9		650	306.7	314.1
220.0	1019.6		675	318.4	326.4
240.0	1112.3		700	330.2	338.7
260.0	1205.0		725	342.0	351.0
280.0	1297.7		750	353.8	363.4
300.0	1390.4	Saybolt	775	365.5	375.7
320.0	1483.1	seconds	800	377.4	388.1
340.0	1575.8	equal	825	389.2	400.5
360.0	1668.5	centistokes	850	400.9	412.9
380.0	1761.2	times 4.6673	875	412.7	425.3

(continued)

TABLE 2-2
Viscosity Equivalents
(continued)

Equivalents of Kinematic and Saybolt Universal Viscosity

Kinematic viscosity, centistokes ν	*Equivalent Saybold universal viscosity, sec* *At 100° F basic values*	*At 210° F*
400.0	1853.9	
420.0	1946.6	
440.0	2039.3	
460.0	2132.0	
480.0	2224.7	
500.0	2317.4	
Over 500	Saybolt seconds equal centistokes times 4.6347	

Note: To obtain the Saybolt universal viscosity equivalent to a kinematic viscosity determined at *t*, multiply the equivalent Saybolt Universal viscosity at 100° F by 1 + (*t* – 100) 0.000 064.

For example, 10 ν at 210° F are equivalent to 58.91 multiplied by 1.0070 or 59.32 sec Saybolt Universal at 210° F.

Equivalents of Kinematic and Saybolt Furol Viscosity

Kinematic viscosity, centistokes ν	*Equivalent Saybolt furol viscosity, sec* *At 122° F*	*At 210° F*
900	424.5	437.7
925	436.3	450.1
950	448.1	462.5
975	459.9	474.9
1000	471.7	487.4
1025	483.5	499.8
1050	495.2	512.3
1075	507.0	524.8
1100	518.8	537.2
1125	530.6	549.7
1150	542.4	562.2
1175	554.2	574.7
1200	566.0	587.2
1225	577.8	599.7
1250	589.5	612.2
1275	601.3	624.8
1300	613.1	637.3
Over 1300	*	†

*Over 1300 centistokes at 122° F: Saybolt fluid sec = centistokes × 0.4717.

†Over 1300 centistokes at 210° F: Log (Saybolt furol sec – 287) = 1.0276 [log (centistokes)] – 0.3975.

(From *Flow of Liquids through Valves, Fittings, and Pipe,* Technical Paper No. 410, The Crane Company, NY. Used with permission.)

EXAMPLE 2-21

Problem: Determine the viscosity of sulfur dioxide gas (SO_2) at 50°F and 200°F.

Solution: From Figure 2-5, the viscosities at 50°F and 200°F are 0.012 cP and 0.016 cP, respectively. Note that this relatively large temperature difference results in a 25 percent change in gas viscosity.

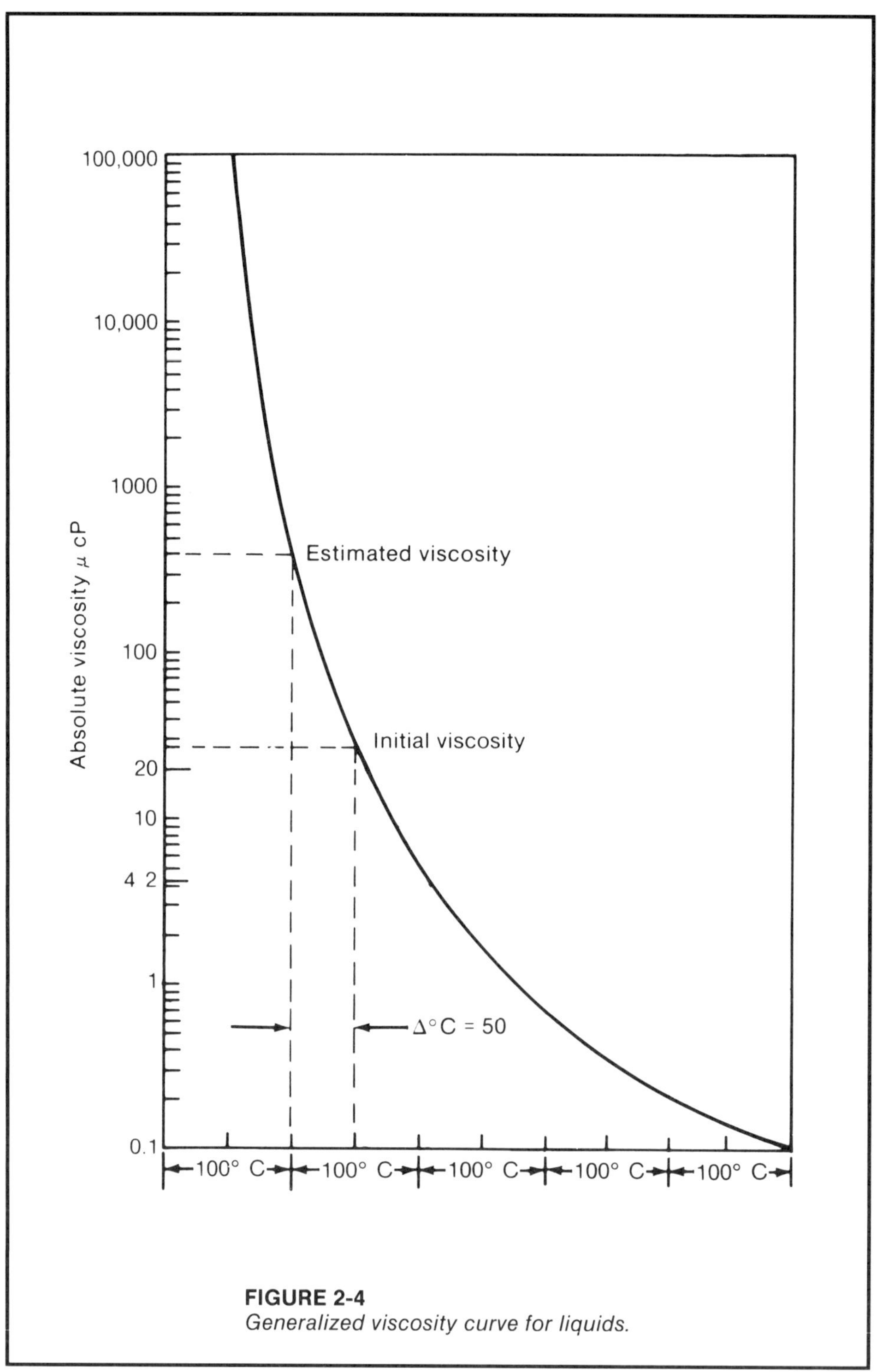

FIGURE 2-4
Generalized viscosity curve for liquids.

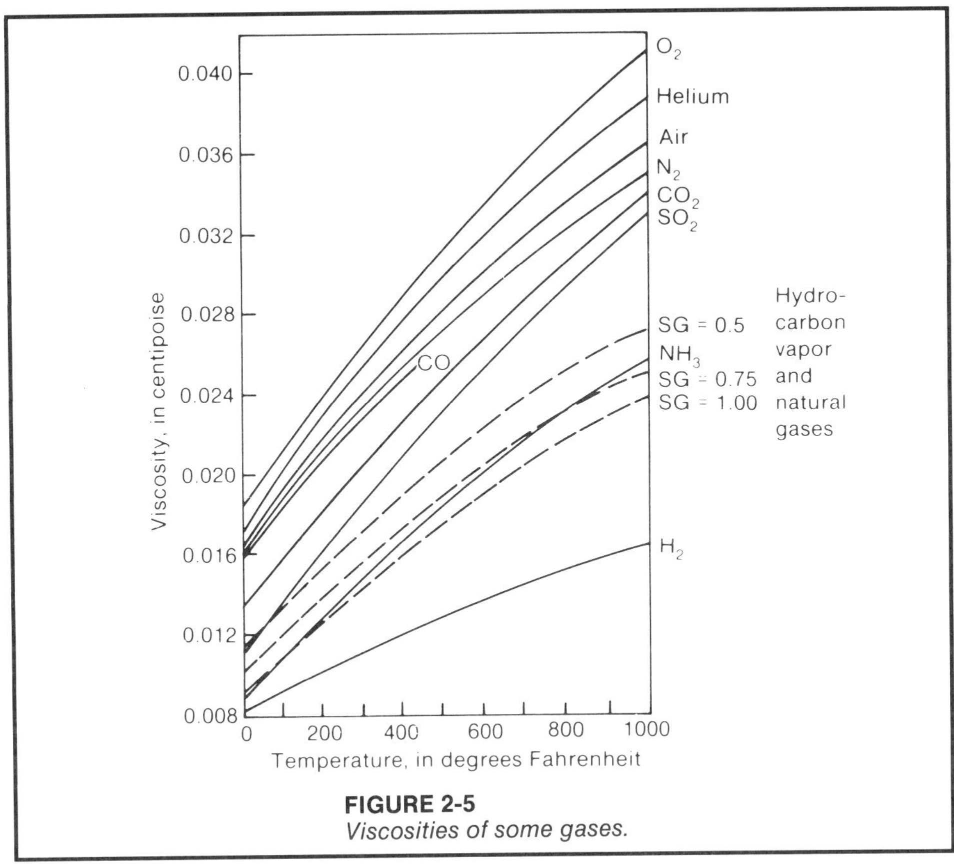

FIGURE 2-5
Viscosities of some gases.

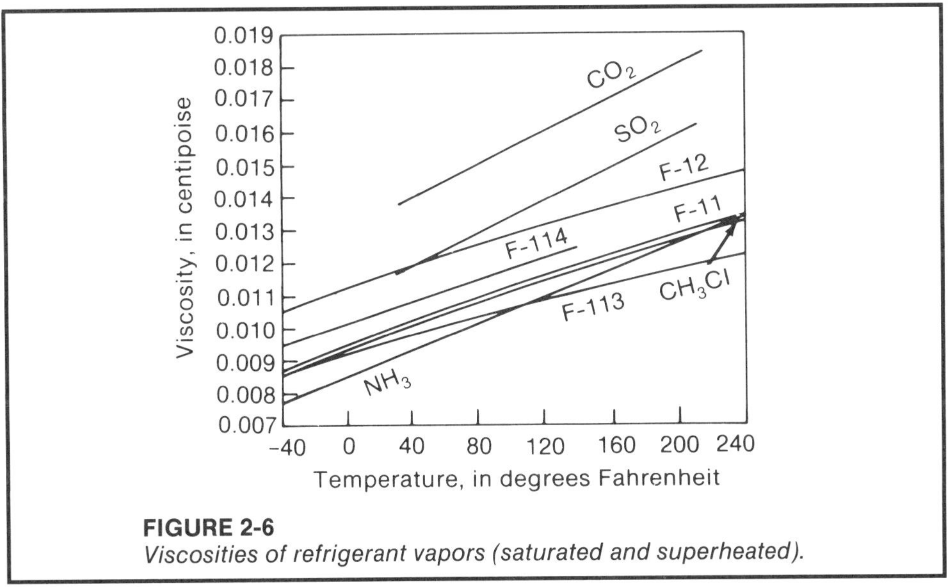

FIGURE 2-6
Viscosities of refrigerant vapors (saturated and superheated).

Velocity Profile and Reynolds Number The shape of the velocity profile of flow through a pipe is dependent upon the inertial or momentum forces of the fluid in the pipe, which tend to make the fluid travel through the pipe, and the viscous forces of the fluid, which tend to slow the fluid as it passes over the pipe wall and itself. Bends in the pipe, valves, line restrictions, roughness of the interior pipe surface, and the like can also affect the shape of the flow profile and how fast it changes in recovering from flow disturbances.

In laminar flow, the relatively large viscous forces have the effect of causing the fluid to slow as it passes near the walls of the pipe. Theoretically, the velocity profile in the laminar region is essentially parabolic; more flow travels through the center of the pipe due to a slowdown near the pipe walls.

This does not happen to the same extent in turbulent flow, where the inertial forces are large in relation to the viscous forces and are thereby less affected by pipe wall effects. Therefore, turbulent flow exhibits a more uniform velocity profile than laminar flow due to the relatively small viscous forces present. The fluid layer next to the wall remains laminar even for the case of fully developed turbulent flow.

The velocity profile in the transition between laminar and turbulent flow regimes can be unstable and difficult to predict, as the flow may exhibit properties of both the laminar and the turbulent regions or oscillate between them.

Experiment has shown that there is a combination of four factors that determines whether the flow of a fluid through a pipe is in the laminar, transition, or turbulent regime. This combination is called pipe Reynolds number and is a good gage of flow regime only when the flow profile is allowed to set up properly. Reynolds number is defined as

$$R_D = \frac{\text{inertia forces}}{\text{viscous forces}} = \frac{\rho \times v \times D}{\mu}$$

Reynolds number is a pure, dimensionless number and therefore its value will be the same in any consistent set of units. The following equations are used to more conveniently calculate Reynolds number for liquid and gas flow through a pipe.

$$R_D = \frac{3160 \times Q_{gpm} \times SG}{\mu_{cP} \times D} \quad \text{(Liquid)}$$

$$R_D = \frac{379 \times Q_{acfm} \times \rho}{\mu_{cP} \times D} \quad \text{(Gas)}$$

where ρ is in pounds per cubic foot and D is in inches.

Flow is in the laminar region when Reynolds number is less than 2000, while Reynolds numbers greater than 4000 are generally accepted as being in the turbulent region. Transition flow occurs in the range of 2000 to 4000. As Reynolds number only reflects liquid effects and disregards such factors as pipe roughness, pipe obstructions, and pipe bends, the boundaries of the laminar, transition, and turbulent regions are merely estimates that are suggested for practical applications as opposed to controlled laboratory experiments.

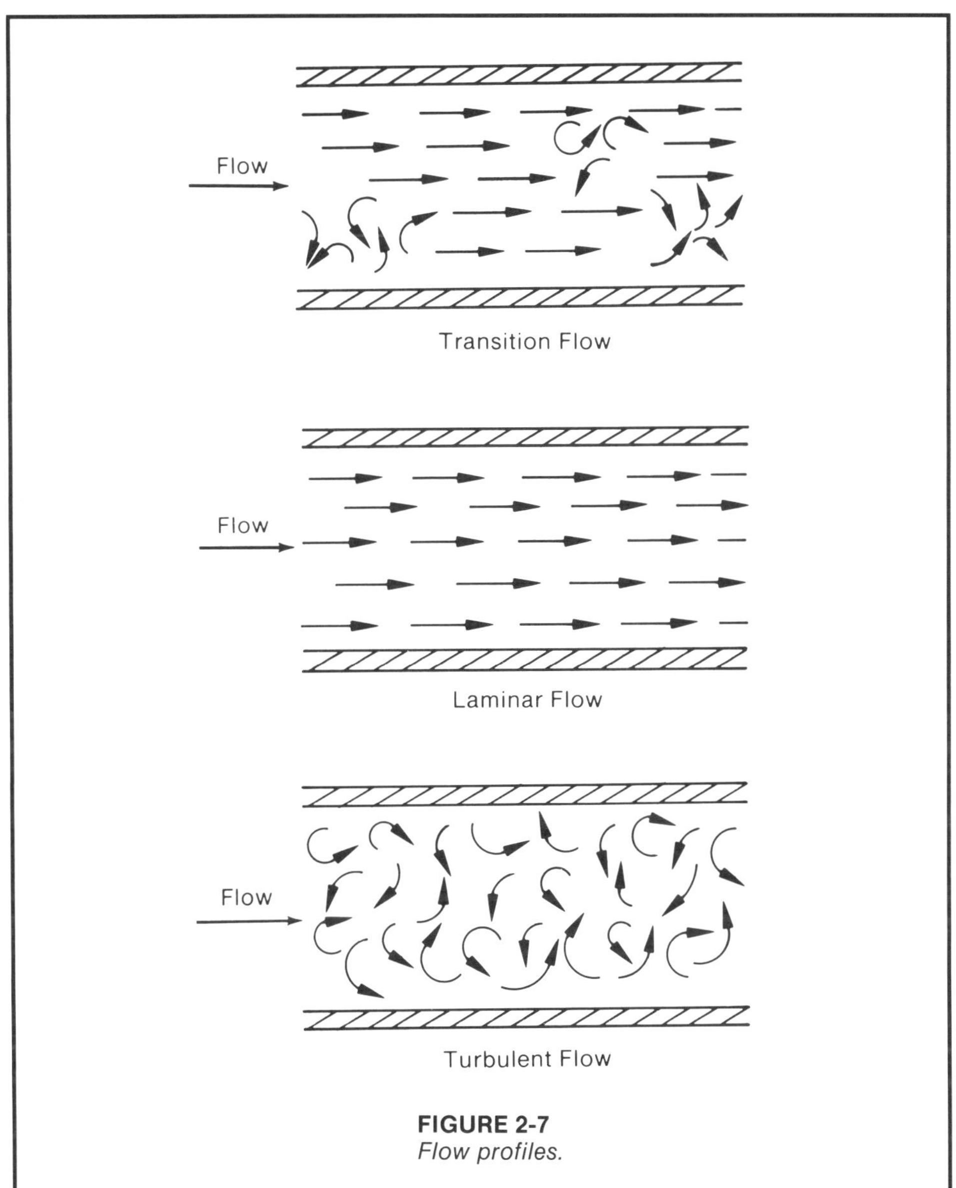

FIGURE 2-7
Flow profiles.

In the equation for Reynolds number for liquid flow, the flow can vary nominally over a known 10:1 range, the specific gravity is virtually constant between 0.8 and 1.2, the pipe diameter is constant, and viscosity can vary from less than one to thousands of centipoise. Therefore, the factor that can affect Reynolds number the most is the viscosity and is typically the least defined of the three applicable fluid properties. While a viscosity can be well defined for a particular liquid under certain operating conditions, relatively small changes in

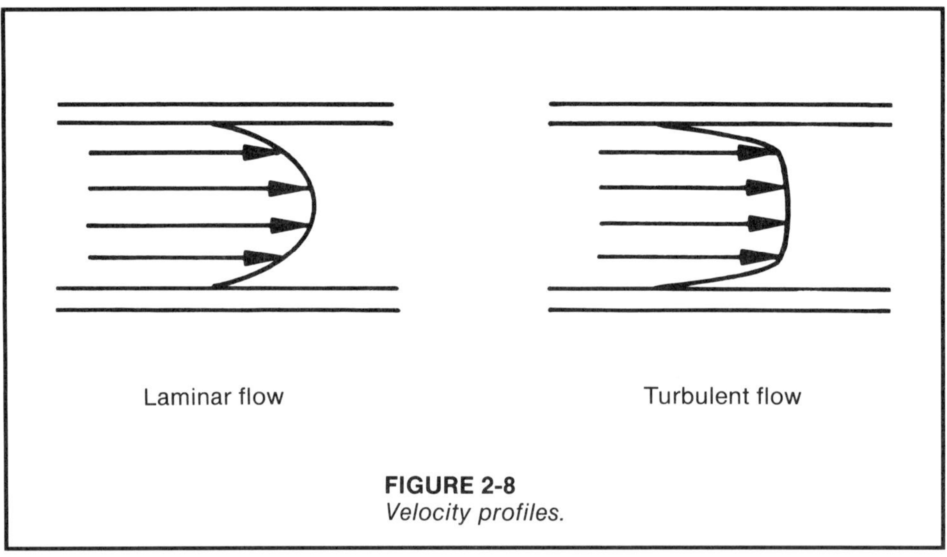

FIGURE 2-8
Velocity profiles.

temperature can cause order of magnitude changes in viscosity, which cause order of magnitude changes in Reynolds number, and can affect whether the flow is in the laminar, transition, or turbulent flow regime.

EXAMPLE 2-22

Problem: Determine whether 30 gpm of a liquid flowing in a 2-in. schedule 40 pipe having a specific gravity of 1.05 and a viscosity of 4 cP is in the laminar, transition, or turbulent flow regime.

Solution: Using 2.067 in. as the inside diameter of the pipe, Reynolds number is calculated:

$$R_D = \frac{3160 \times 30 \text{ gpm} \times 1.05}{4 \text{ cP} \times 2.067 \text{ in.}}$$

$$R_D = 12039$$

Reynolds number is greater than 4000 and the flow is in the turbulent region.

If the viscosity can vary from 4 to 30 cP due to different operating temperatures, Reynolds number could vary from 1605 to 12039 due to viscosity changes. Therefore, if the viscosity varies, the flow could be laminar, transition, or turbulent.

In the equation for Reynolds number for gas flow, the flow can vary nominally over a known 10:1 range, the density typically varies over a range of less than 2:1, the pipe diameter is constant, and viscosity is small and virtually constant in most applications. The factors that affect Reynolds number most are the flow and density, which are usually well defined. Therefore, Reynolds number and hence the flow regime are usually well defined in gas applications. Due to the low viscosity of gases, gas flow in properly sized pipe is typically in the turbulent regime.

EXAMPLE 2-23

Problem: Determine whether 10 acfm of a gas flowing in a 2-in. schedule 40 pipe having a density of 0.336 lb/ft^3 and a viscosity of 0.019 cP is in the laminar, transition, or turbulent flow regime.

Solution: Using 2.067 in. as the inside diameter of pipe, Reynolds number is calculated:

$$R_D = \frac{379 \times 10 \text{ acfm} \times 0.356 \text{ lb/ft}^3}{0.019 \text{ cP} \times 2.067 \text{ in.}}$$

$$R_D = 32425$$

Reynolds number is greater than 4000 and the flow is in the turbulent region. The flow would enter the transition regime at flows below

$$\frac{4000 \times 10 \text{ acfm}}{32425} = 1.23 \text{ acfm}$$

Therefore, flow will be in the turbulent regime over better than an 8:1 flow range.

Newtonian and Non-Newtonian Liquids

The dynamic (absolute) viscosity is the proportionality factor between the applied shear stress and the rate of deformation of the liquid. If the relation between the applied shear stress and the rate of deformation is linear, the liquid is said to be Newtonian. For the purpose of this discussion, deformation and flow will be used interchangeably. Therefore, the viscosity is the slope of a stress versus flow curve. For a Newtonian fluid, the stress versus flow curve is a straight line beginning at the origin.

If the relation is nonlinear, the liquid is said to be non-Newtonian. Ideal plastics will not start to flow until a threshold stress level is reached, beyond which the viscosity is a constant, while a thixotropic liquid is highly viscous at the initiation of flow and becomes more Newtonian at higher flows. Inverted plastics have low viscosity at low flow and become increasingly more viscous as flow increases.

Friction Losses

Friction losses are pressure drops that occur due to the friction caused by movement of the fluid through the pipe and restrictions in the pipe such as valves, elbows, fittings, and the like. Friction losses are dependent upon pipe size, pipe roughness, flow, viscosity, and Reynolds number, as well as the type, size, and quantity of elbows, fittings, valves, and the like that comprise the piping system.

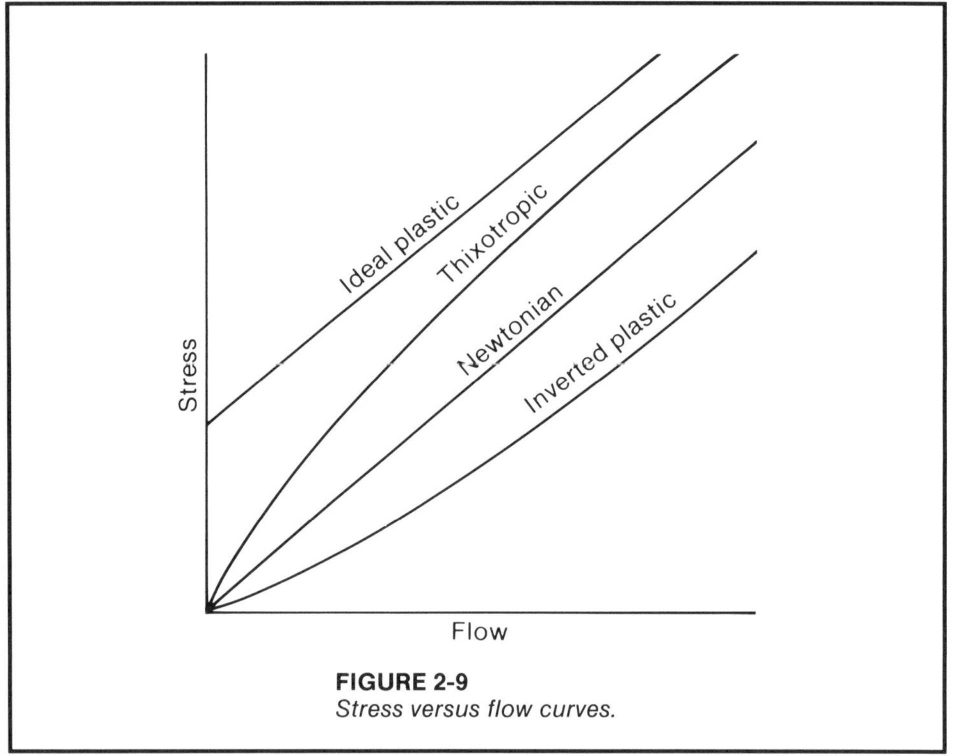

FIGURE 2-9
Stress versus flow curves.

Friction losses are normally calculated in order to size the pipe so as to ensure delivery of the liquid with the available motive pressure. For flowmeter calculation purposes, friction losses are proportional to the square of the flow and are usually calculated at maximum flow and assumed constant. Pressure drops for water service in various pipe sizes are tabulated in Table 2-3 from which pressure drops can be estimated for low viscosity service.

Friction losses can be significant in high viscosity liquid service, since pressure drop is a linear function of viscosity and flow. When it is suspected that the piping system will not produce enough pressure upstream of the flowmeter to ensure proper operation of the flowmeter, documents such as Crane's *Flow of Fluids through Valves, Fittings, and Pipe*, Technical Paper No. 410 should be consulted. This paper presents the theory and empirical data necessary to calculate friction and other piping losses to ensure that there is sufficient pressure for proper flowmeter operation.

Friction losses in gas service can be significant under certain conditions, but they are typically neglected for the purposes of flowmeter discussions. Table 2-4 is included to illustrate the losses that can occur.

EXAMPLE 2-24

Problem: Calculate friction losses to the inlet of the flowmeter for the water system shown in Figure 2-10.

Solution: As the water is flowing in the turbulent flow regime, the equivalent length method of calculating friction losses may be applied. If a pipe elbow is estimated to be equivalent to 15 feet of straight pipe, the equivalent lengths of straight pipe in each size are calculated to be:

$$L_{2\text{ in.}} = 30 + 50 + 40 + (2 \times 15) = 150 \text{ feet}$$

$$L_{3\text{ in.}} = 10 \text{ feet}$$

The pressure drops per 100 feet of straight pipe can be read from the table for each pipe size and used to calculate the overall friction losses.

$$\Delta P_{2\text{ in.}} = 150 \text{ ft} \times (7.59 \text{ psi}/100 \text{ ft}) = 11.385 \text{ psi}$$

$$\Delta P_{3\text{ in.}} = 10 \text{ ft} \times (1.05 \text{ psi}/100 \text{ ft}) = 0.105 \text{ psi}$$

$$\text{Total } \Delta P \qquad 11.490 \text{ psi}$$

EXAMPLE 2-25

Problem: Determine the friction losses associated with 175 feet of straight 2-in. schedule 40 pipe in which 500 scfm of air flows at 150 psi and 90°F.

Solution: The friction losses of 500 scfm of air in 100 feet of straight pipe at 60° F can be read from the table as 1.55 psi. Pressure and temperature differences are taken into account as follows:

$$\Delta P = \frac{1.55 \text{ psi}}{100 \text{ feet}} \times \frac{100 \text{ psi} + 14.7 \text{ psi}}{150 \text{ psi} + 14.7 \text{ psi}} \times \frac{460°\text{F} + 90°\text{F}}{460°\text{F} + 60°\text{F}} \times 175 \text{ feet}$$

$$= 1.96 \text{ psi}$$

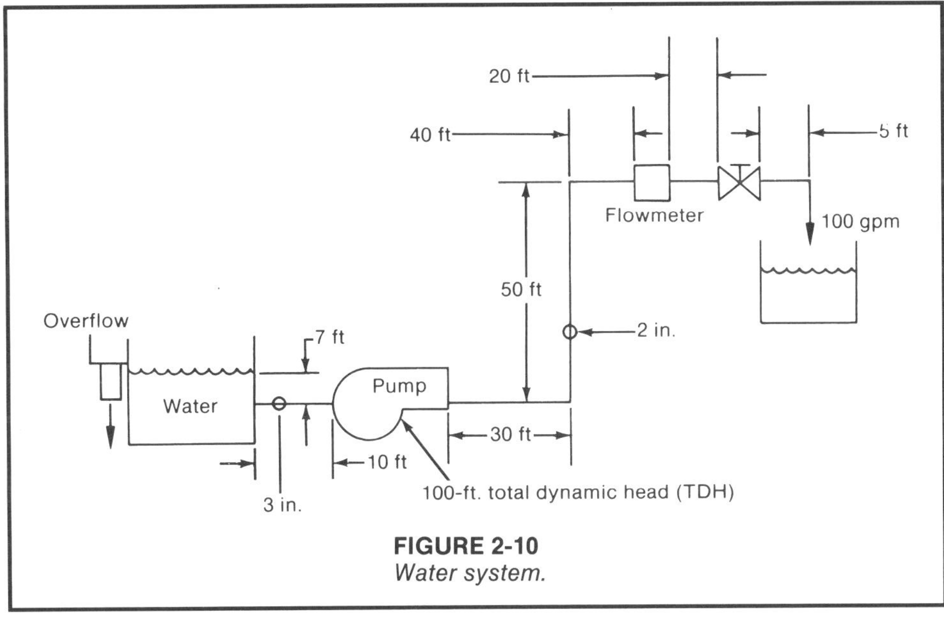

FIGURE 2-10
Water system.

TABLE 2-3
Flow of Water through Schedule 40 Steel Pipe

Discharge		Pressure drop per 100 feet and velocity in schedule 40 pipe for water at 60° F															
Gallons per min	Cubic ft per sec	Velocity, ft per sec	Pressure drop, lbs per in²	Velocity, ft per sec	Pressure drop, lbs per in²	Velocity, ft per sec	Pressure drop, lbs per in²	Velocity, ft per sec	Pressure drop, lbs per in²	Velocity, ft per sec	Pressure drop, lbs per in²	Velocity, ft per sec	Pressure drop, lbs per in²	Velocity, ft per sec	Pressure drop, lbs per in²	Velocity, ft per sec	Pressure drop, lbs per in²
		1/8 in.		1/4 in.		3/8 in.		1/2 in.									
0.2	0.000446	*1.13*	1.86	*0.616*	0.359												
0.3	0.000668	*1.69*	4.22	*0.924*	0.903	*0.504*	0.159	*0.317*	0.061	3/4 in.							
0.4	0.000891	*2.26*	6.98	*1.23*	1.61	*0.672*	0.345	*0.422*	0.086								
0.5	0.00111	*2.82*	10.5	*1.54*	2.39	*0.840*	0.539	*0.528*	0.167	*0.301*	0.033						
0.6	0.00134	*3.39*	14.7	*1.85*	3.29	*1.01*	0.751	*0.633*	0.240	*0.361*	0.041						
0.8	0.00178	*4.52*	25.0	*2.46*	5.44	*1.34*	1.25	*0.844*	0.408	*0.481*	0.102	1 in.		1-1/4 in.			
1	0.00223	*5.65*	37.2	*3.08*	8.28	*1.68*	1.85	*1.06*	0.600	*0.602*	0.155	*0.371*	0.048			1-1/2 in.	
2	0.00446	*11.29*	134.4	*6.16*	30.1	*3.36*	6.58	*2.11*	2.10	*1.20*	0.526	*0.743*	0.164	*0.429*	0.044		
3	0.00668			*9.25*	64.1	*5.04*	13.9	*3.17*	4.33	*1.81*	1.09	*1.114*	0.336	*0.644*	0.090	*0.473*	0.043
4	0.00891			*12.33*	111.2	*6.72*	23.9	*4.22*	7.42	*2.41*	1.83	*1.49*	0.565	*0.858*	0.150	*0.630*	0.071
5	0.01114	2 in.				*8.40*	36.7	*5.28*	11.2	*3.01*	2.75	*1.86*	0.835	*1.073*	0.223	*0.788*	0.104
6	0.01337	*0.574*	0.044	2-1/2 in.		*10.08*	51.9	*6.33*	15.8	*3.61*	3.84	*2.23*	1.17	*1.29*	0.309	*0.946*	0.145
8	0.01782	*0.765*	0.073			*13.44*	91.1	*8.45*	27.7	*4.81*	6.60	*2.97*	1.99	*1.72*	0.518	*1.26*	0.241
10	0.02228	*0.956*	0.108	*0.670*	0.046			*10.56*	42.4	*6.02*	9.99	*3.71*	2.99	*2.15*	0.774	*1.58*	0.361
15	0.03342	*1.43*	0.224	*1.01*	0.094	3 in.				*9.03*	21.6	*5.57*	6.36	*3.22*	1.63	*2.37*	0.755
20	0.04456	*1.91*	0.375	*0.34*	0.158	*0.868*	0.056	3-1/2 in.		*12.03*	37.8	*7.43*	10.9	*4.29*	2.78	*3.16*	1.28
25	0.05570	*2.39*	0.561	*1.68*	0.234	*1.09*	0.083	*0.812*	0.041	4 in.		*9.28*	16.7	*5.37*	4.22	*3.94*	1.93
30	0.06684	*2.87*	0.786	*2.01*	0.327	*1.30*	0.114	*0.974*	0.056			*11.14*	23.8	*6.44*	5.92	*4.73*	2.72
35	0.07798	*3.35*	1.05	*2.35*	0.436	*1.52*	0.151	*1.14*	0.704	*0.882*	0.041	*12.99*	32.2	*7.51*	7.90	*5.52*	3.64
40	0.08912	*3.83*	1.35	*2.68*	0.556	*1.74*	0.192	*1.30*	0.095	*1.01*	0.052	*14.85*	41.5	*8.59*	10.24	*6.30*	4.65
45	0.1003	*4.30*	1.67	*3.02*	0.668	*1.95*	0.239	*1.46*	0.117	*1.13*	0.064			*9.67*	12.80	*7.09*	5.85
50	0.1114	*4.78*	2.03	*3.35*	0.839	*2.17*	0.288	*1.62*	0.142	*1.26*	0.076	5 in.		*10.74*	15.66	*7.88*	7.15
60	0.1337	*5.74*	2.87	*4.02*	1.18	*2.60*	0.406	*1.95*	0.204	*1.51*	0.107			*12.89*	22.2	*9.47*	10.21
70	0.1560	*6.70*	3.84	*4.69*	1.59	*3.04*	0.540	*2.27*	0.261	*1.76*	0.143	*1.12*	0.047			*11.05*	13.71
80	0.1782	*7.65*	4.97	*5.36*	2.03	*3.47*	0.687	*2.60*	0.334	*2.02*	0.180	*1.28*	0.060			*12.62*	17.59
90	0.2005	*8.60*	6.20	*6.03*	2.53	*3.91*	0.861	*2.92*	0.416	*2.27*	0.224	*1.44*	0.074	6 in.		*14.20*	22.0
100	0.2228	*9.56*	7.59	*6.70*	3.09	*4.34*	1.05	*3.25*	0.509	*2.52*	0.272	*1.60*	0.090	*1.11*	0.036	*15.78*	26.9
125	0.2785	*11.97*	11.76	*8.38*	4.71	*5.43*	1.61	*4.06*	0.769	*3.15*	0.415	*2.01*	0.135	*1.39*	0.055	*19.72*	41.4
150	0.3342	*14.36*	16.70	*10.05*	6.69	*6.51*	2.24	*4.87*	1.08	*3.78*	0.580	*2.41*	0.190	*1.67*	0.077		
175	0.3899	*16.75*	22.3	*11.73*	8.97	*7.60*	3.00	*5.68*	1.44	*4.41*	0.774	*2.81*	0.253	*1.94*	0.102		
200	0.4456	*19.14*	28.8	*13.42*	11.68	*8.68*	3.87	*6.49*	1.85	*5.04*	0.985	*3.21*	0.323	*2.22*	0.130	8 in.	
225	0.5013	—	—	*15.09*	14.63	*9.77*	4.83	*7.30*	2.32	*5.67*	1.23	*3.61*	0.401	*2.50*	0.162	*1.44*	0.043
250	0.557	—	—	—	—	*10.85*	5.93	*8.12*	2.84	*6.30*	1.46	*4.01*	0.495	*2.78*	0.195	*1.60*	0.051
275	0.6127	—	—	—	—	*11.94*	7.14	*8.93*	3.40	*6.93*	1.79	*4.41*	0.583	*3.05*	0.234	*1.76*	0.061
300	0.6684	—	—	—	—	*13.00*	8.36	*9.74*	4.02	*7.56*	2.11	*4.81*	0.683	*3.33*	0.275	*1.92*	0.072
325	0.7241	—	—	—	—	*14.12*	9.89	*10.53*	4.09	*8.19*	2.47	*5.21*	0.797	*3.61*	0.320	*2.08*	0.083
350	0.7798			—	—	—	—	*11.36*	5.41	*8.82*	2.84	*5.62*	0.919	*3.89*	0.367	*2.24*	0.095
375	0.8355			—	—	—	—	*12.17*	6.18	*9.45*	3.25	*6.02*	1.05	*4.16*	0.416	*2.40*	0.108
400	0.8912			—	—	—	—	*12.98*	7.03	*10.08*	3.68	*6.42*	1.19	*4.44*	0.471	*2.56*	0.121
425	0.9469			—	—	—	—	*13.80*	7.89	*10.71*	4.12	*6.82*	1.33	*4.72*	0.529	*2.73*	0.136
450	1.003	10 in.		—	—	—	—	*14.61*	8.80	*11.34*	4.60	*7.22*	1.48	*5.00*	0.590	*2.89*	0.151

TABLE 2-3
Flow of Water through Schedule 40 Steel Pipe
(continued)

475	1.059	*1.93*	0.054			—	—	—	—	*11.97*	5.12	*7.62*	1.64	*5.27*	0.653	*3.04*	0.166
500	1.114	*2.03*	0.059			—	—	—	—	*12.60*	5.65	*8.02*	1.81	*5.55*	0.720	*3.21*	0.182
550	1.225	*2.24*	0.071			—	—	—	—	*13.85*	6.79	*8.82*	2.17	*6.11*	0.861	*3.53*	0.219
600	1.337	*2.44*	0.083			—	—	—	—	*15.12*	8.04	*9.63*	2.55	*6.66*	1.02	*3.85*	0.258
650	1.448	*2.64*	0.097			—	—	—	—	—	—	*10.43*	2.98	*7.22*	1.18	*4.17*	0.301
				12 in.													
700	1.560	*2.85*	0.112	*2.01*	0.047			—	—	—	—	*11.23*	3.43	*7.78*	1.35	*4.49*	0.343
750	1.671	*3.05*	0.127	*2.15*	0.054			—	—	—	—	*12.03*	3.92	*8.33*	1.55	*4.81*	0.392
800	1.782	*3.25*	0.143	*2.29*	0.061			—	—	—	—	*12.83*	4.43	*8.88*	1.75	*5.13*	0.443
						14 in.											
850	1.894	*3.46*	0.160	*2.44*	0.068	*2.02*	0.042	—	—	—	—	*13.64*	5.00	*9.44*	1.96	*5.45*	0.497
900	2.005	*3.66*	0.179	*2.58*	0.075	*2.13*	0.047	—	—	—	—	*14.44*	5.58	*9.99*	2.18	*5.77*	0.554
950	2.117	*3.86*	0.198	*2.72*	0.083	*2.25*	0.052			—	—	*15.24*	6.21	*10.55*	2.42	*6.09*	0.613
1 000	2.228	*4.07*	0.218	*2.87*	0.091	*2.37*	0.057			—	—	*16.04*	6.84	*11.10*	2.68	*6.41*	0.675
1 100	2.451	*4.48*	0.260	*3.15*	0.110	*2.61*	0.068			—	—	*17.65*	8.23	*12.22*	3.22	*7.05*	0.807
								16 in.									
1 200	2.674	*4.88*	0.306	*3.44*	0.128	*2.85*	0.080	*2.18*	0.042	—	—	—	—	*13.33*	3.81	*7.70*	0.948
1 300	2.896	*5.29*	0.355	*3.73*	0.150	*3.08*	0.093	*2.36*	0.048	—	—	—	—	*14.43*	4.45	*8.33*	1.11
1 400	3.119	*5.70*	0.409	*4.01*	0.171	*3.32*	0.107	*2.54*	0.056					*15.55*	5.13	*8.98*	1.28
1 500	3.342	*6.10*	0.466	*4.30*	0.195	*3.56*	0.122	*2.72*	0.063					*16.66*	5.85	*9.62*	1.46
1 600	3.565	*6.51*	0.527	*4.59*	0.219	*3.79*	0.138	*2.90*	0.071					*17.77*	6.61	*10.26*	1.65
										18 in.							
1 800	4.010	*7.32*	0.663	*5.16*	0.276	*4.27*	0.172	*3.27*	0.088	*2.58*	0.050			*19.99*	8.37	*11.54*	2.08
2 000	4.456	*8.14*	0.808	*5.73*	0.339	*4.74*	0.209	*3.63*	0.107	*2.87*	0.060			*22.21*	10.3	*12.82*	2.55
2 500	5.570	*10.17*	1.24	*7.17*	0.515	*5.93*	0.321	*4.54*	0.163	*3.59*	0.091					*16.03*	3.94
												20 in.					
3 000	6.684	*12.20*	1.76	*8.60*	0.731	*7.11*	0.451	*5.45*	0.232	*4.30*	0.129	*3.46*	0.075			*19.24*	5.59
3 500	7.798	*14.24*	2.38	*10.03*	0.982	*8.30*	0.607	*6.35*	0.312	*5.02*	0.173	*4.04*	0.101			*22.44*	7.56
														24 in.			
4 000	8.912	*16.27*	3.08	*11.47*	1.27	*9.48*	0.787	*7.26*	0.401	*5.74*	0.222	*4.62*	0.129	*3.19*	0.052	*25.65*	9.80
4 500	10.03	*18.31*	3.87	*12.90*	1.60	*10.67*	0.990	*8.17*	0.503	*6.46*	0.280	*5.20*	0.162	*3.59*	0.065	*25.87*	12.2
5 000	11.14	*20.35*	4.71	*14.33*	1.95	*11.85*	1.21	*9.08*	0.617	*7.17*	0.340	*5.77*	0.199	*3.99*	0.079	—	—
6 000	11.37	*24.41*	6.74	*17.20*	2.77	*14.23*	1.71	*10.89*	0.877	*8.61*	0.483	*6.93*	0.280	*4.79*	0.111	—	—
7 000	15.60	*28.49*	9.11	*20.07*	3.74	*16.60*	2.31	*12.71*	1.18	*10.04*	0.652	*8.08*	0.376	*5.59*	0.150	—	—
8 000	17.82	—	—	*22.93*	4.84	*18.96*	2.99	*14.52*	1.51	*11.47*	0.839	*9.23*	0.488	*6.38*	0.192	—	—
9 000	20.05	—	—	*25.79*	6.09	*21.34*	3.76	*16.34*	1.90	*12.91*	1.05	*10.39*	0.608	*7.18*	0.242	—	—
10 000	22.28	—	—	*28.66*	7.46	*23.71*	4.61	*18.15*	2.34	*14.34*	1.28	*11.54*	0.739	*7.98*	0.294	—	—
12 000	26.74	—	—	*34.40*	10.7	*28.45*	6.59	*21.79*	3.33	*17.21*	1.83	*13.85*	1.06	*9.58*	0.416	—	—
14 000	31.19	—	—	—	—	*33.19*	8.89	*25.42*	4.49	*20.08*	2.45	*16.16*	1.43	*11.17*	0.562	—	—
16 000	35.65	—	—	—	—	—	—	*29.05*	5.83	*22.95*	3.18	*18.47*	1.85	*12.77*	0.723	—	—
18 000	40.10	—	—	—	—	—	—	*32.68*	7.31	*25.82*	4.03	*20.77*	2.32	*14.36*	0.907	—	—
20 000	44.56	—	—	—	—	—	—	*36.31*	9.03	*28.69*	4.93	*23.08*	2.86	*15.96*	1.12	—	—

For pipe lengths other than 100 feet, the pressure drop is proportional to the length. Thus, for 50 feet of pipe, the pressure drop is approximately one-half the value given in this table . . . for 300 feet, three times the given value, etc.

Velocity is a function of the cross-sectional flow area; thus, it is constant for a given flow rate and is independent of pipe length.

(From *Flow of Liquids through Valves, Fittings, and Pipe,* Technical Paper No. 410, The Crane Company, NY. Used with permission.)

TABLE 2-4
Flow of Air through Schedule 40 Steel Pipe

Free air q'_m ft³/min at 60° F and 14.7 psia	*Compressed air, ft³/min at 60° F and 100 psig*	*Pressure drop of air in lbs per in.² per 100 feet of Schedule 40 pipe for air at 100 lbs per in.² gage pressure and 60° F temperature*								
		1/8 in.	1/4 in.	3/8 in.	1/2 in.					
1	0.128	0.361	0.083	0.018						
2	0.256	1.31	0.285	0.064	0.020					
3	0.384	3.06	0.605	0.133	0.042	3/4 in.				
4	0.513	4.83	1.04	0.226	0.071					
5	0.641	7.45	1.58	0.343	0.106	0.027				
							1 in.			
6	0.769	10.6	2.23	0.408	0.148	0.037				
8	1.025	18.6	3.89	0.848	0.255	0.062	0.019			
10	1.282	28.7	5.96	1.26	0.356	0.094	0.029	1-1/4 in.	1-1/2 in.	
15	1.922	—	13.0	2.73	0.834	0.201	0.062			
20	2.563	—	22.8	4.76	1.43	0.345	0.102	0.026		
25	3.204	—	35.6	7.34	2.21	0.526	0.156	0.039	0.019	
30	3.845	—	—	10.5	3.15	0.748	0.219	0.055	0.026	
35	4.486	—	—	14.2	4.24	1.00	0.293	0.073	0.035	
40	5.126	—	—	18.4	5.49	1.30	0.379	0.095	0.044	
45	5.767	—	—	23.1	6.90	1.62	0.474	0.116	0.055	2 in.
50	6.408			28.5	8.49	1.99	0.578	0.149	0.067	0.019
60	7.690	2-1/2 in.		40.7	12.2	2.85	0.819	0.200	0.094	0.027
70	8.971			—	16.5	3.83	1.10	0.270	0.126	0.036
80	10.25	0.019		—	21.4	4.96	1.43	0.350	0.162	0.046
90	11.53	0.023		—	27.0	6.25	1.80	0.437	0.203	0.058
100	12.82	0.029	3 in.		33.2	7.69	2.21	0.534	0.247	0.070
125	16.02	0.044			—	11.9	3.39	0.825	0.380	0.107
150	19.22	0.062	0.021		—	17.0	4.87	1.17	0.537	0.151
175	22.43	0.083	0.028		—	23.1	6.60	1.58	0.727	0.205
200	25.63	0.107	0.036	3-1/2 in.	—	30.0	8.54	2.05	0.937	0.264
225	28.84	0.134	0.045	0.022		37.9	10.8	2.59	1.19	0.331
250	32.04	0.164	0.055	0.027		—	13.3	3.18	1.45	0.404
275	35.24	0.191	0.066	0.032		—	16.0	3.83	1.75	0.484
300	38.45	0.232	0.078	0.037		—	19.0	4.56	2.07	0.573
325	41.65	0.270	0.090	0.043	4 in.	—	22.3	5.32	2.42	0.673
350	44.87	0.313	0.104	0.050		—	25.8	6.17	2.80	0.776
375	48.06	0.356	0.119	0.057	0.030	—	29.6	7.05	3.20	0.887
400	51.26	0.402	0.134	0.064	0.034	—	33.6	8.02	3.64	1.00
425	54.47	0.452	0.151	0.072	0.038	—	37.9	9.01	4.09	1.13
450	57.67	0.507	0.168	0.081	0.042	—	—	10.2	4.59	1.26
475	60.88	0.562	0.187	0.089	0.047		—	11.3	5.09	1.40
500	64.08	0.623	0.206	0.099	0.052		—	12.5	5.61	1.55
550	70.49	0.749	0.248	0.118	0.062		—	15.1	6.79	1.87
600	76.90	0.887	0.293	0.139	0.073		—	18.0	8.04	2.21
650	83.30	1.04	0.342	0.163	0.086	5 in.	—	21.1	9.43	2.60

(continued)

TABLE 2-4
Flow of Air through Schedule 40 Steel Pipe
(continued)

Free air q'_m ft³/min at 60° F and 14.7 psia	*Compressed air, ft³/min at 60° F and 100 psig*	*Pressure drop of air in lbs per in.² per 100 feet of Schedule 40 pipe for air at 100 lbs per in.² gage pressure and 60° F temperature*								
700	89.71	1.19	0.395	0.188	0.099	0.032		24.3	10.9	3.00
750	96.12	1.36	0.451	0.214	0.113	0.036		27.9	12.6	3.44
800	102.5	1.55	0.513	0.244	0.127	0.041		31.8	14.2	3.90
850	108.9	1.74	0.576	0.274	0.144	0.046		35.9	16.0	4.40
900	115.3	1.95	0.642	0.305	0.160	0.051	6 in.	40.2	18.0	4.91
950	121.8	2.18	0.715	0.340	0.178	0.057	0.023	—	20.0	5.47
1 000	128.2	2.40	0.788	0.375	0.197	0.063	0.025	—	22.1	6.06
1 100	141.0	2.89	0.948	0.451	0.236	0.075	0.030	—	26.7	7.29
1 200	153.8	3.44	1.13	0.533	0.279	0.089	0.035	—	31.8	8.63
1 300	166.6	4.01	1.32	0.626	0.327	0.103	0.041	—	37.3	10.1
1 400	179.4	4.65	1.52	0.718	0.377	0.119	0.047			11.8
1 500	192.2	5.31	1.74	0.824	0.431	0.136	0.054			13.5
1 600	205.1	6.04	1.97	0.932	0.490	0.154	0.061	8 in.		15.3
1 800	230.7	7.65	2.50	1.18	0.616	0.193	0.075			19.3
2 000	256.3	9.44	3.06	1.45	0.757	0.237	0.094	0.023	10 in.	23.9
2 500	320.4	14.7	4.76	2.25	1.17	0.366	0.143	0.035		37.3
3 000	384.5	21.1	6.82	3.20	1.67	0.524	0.204	0.051	0.016	
3 500	448.6	28.8	9.23	4.33	2.26	0.709	0.276	0.068	0.022	
4 000	512.6	37.6	12.1	5.66	2.94	0.919	0.358	0.088	0.028	
4 500	576.7	47.6	15.3	7.16	3.69	1.16	0.450	0.111	0.035	12 in.
5 000	640.8	—	18.8	8.85	4.56	1.42	0.552	0.136	0.043	0.018
6 000	769.0	—	27.1	12.7	6.57	2.03	0.794	0.195	0.061	0.025
7 000	897.1	—	36.9	17.2	8.94	2.76	1.07	0.262	0.082	0.034
8 000	1025	—	—	22.5	11.7	3.59	1.39	0.339	0.107	0.044
9 000	1153	—	—	28.5	14.9	4.54	1.76	0.427	0.134	0.055
10 000	1282	—	—	35.2	18.4	5.60	2.16	0.526	0.164	0.067
11 000	1410	—	—	—	22.2	6.78	2.62	0.633	0.197	0.081
12 000	1538	—	—	—	26.4	8.07	3.09	0.753	0.234	0.096
13 000	1666	—	—	—	31.0	9.47	3.63	0.884	0.273	0.112
14 000	1794	—	—	—	36.0	11.0	4.21	1.02	0.316	0.129
15 000	1922	—	—	—	—	12.6	4.84	1.17	0.364	0.148
16 000	2051	—	—	—	—	14.3	5.50	1.13	0.411	0.167
18 000	2037	—	—	—	—	18.2	6.96	1.68	0.520	0.213
20 000	2563	—	—	—	—	22.4	8.60	2.01	0.642	0.260
22 000	2820	—	—	—	—	27.1	10.4	2.50	0.771	0.314
24 000	3076	—	—	—	—	32.3	12.4	2.97	0.918	0.371
26 000	3332	—	—	—	—	37.9	14.5	3.49	1.12	0.435
28 000	3588	—	—	—	—	—	16.9	4.04	1.25	0.505
30 000	3845	—	—	—	—	—	19.3	4.64	1.42	0.520

(From *Flow of Liquids through Valves, Fittings, and Pipe,* Technical Paper No. 410, The Crane Company, NY. Used with permission.)

(continued)

TABLE 2-4
Flow of Air through Schedule 40 Steel Pipe
(continued)

For lengths of pipe other than 100 feet, the pressure drop is proportional to the length. Thus, for 50 feet of pipe, the pressure drop is approximately one-half the value given in the table (for 300 feet, three times the given value, etc.).

The pressure drop is also inversely proportional to the absolute pressure and directly proportional to the absolute temperature.

Therefore, to determine the pressure drop for inlet or average pressures other than 100 psi and at temperature other than 60° F, multiply the values given in the table by the ratio:

$$\left(\frac{100 + 14.7}{P + 14.7}\right)\left(\frac{460 + t}{520}\right)$$

where P is the inlet or average gage pressure in pounds per square inch and t is the temperature in degrees Fahrenheit under consideration.

The cubic feet per minute of compressed air at any pressure is inversely proportional to the absolute pressure and diectly proportional to the absolute temperature.

To determine the cubic feet per minute of compressed air at any temperature and pressure other than standard conditions, multiply the value of cubic feet per minute of free air by the ratio:

$$\left(\frac{14.7}{14.7 + P}\right)\left(\frac{460 + t}{520}\right)$$

Calculations for Pipe Other than Schedule 40

To determine the velocity of water or the pressure drop of water or air through pipe other than Schedule 40, use the following formulas:

$$\nu = \nu_{40}\left(\frac{d_{40}}{d_a}\right)^2$$

$$\Delta P_a = \Delta P_{40}\left(\frac{d_{40}}{d_a}\right)^3$$

Subscript *a* refers to the schedule of pipe through which velocity or pressure drop is desired. Subscript 40 refers to the velocity or pressure drop through Schedule 40 pipe.

Miscellaneous Hydraulic Phenomena

Upstream Pressure Most flowmeter applications require the calculation of the upstream or inlet pressure of the flowmeter. This is done by summing applicable pressures related to the piping configuration. Pressure differences occur due to differences in elevation, pressure-generating devices (such as pumps), and friction losses.

EXAMPLE 2-26

Problem: Calculate the upstream pressure of the flowmeter in the piping system illustrated in Figure 2-10.

Solution: The pressure at the inlet of the flowmeter is calculated in feet of water column as follows:

+ 7 ft	static pressure at pump inlet
+ 100 ft	gain through pump
- 50 ft	loss due to elevation
-26.54 ft	pipe friction losses (from preceding example, 11.49 psi × 2.31 ft/psi)
30.46 ft	pressure at flowmeter inlet

The pressure at the inlet to the flowmeter is

$$30.46 \text{ ft WC} \times (12 \text{ in./ft WC}) \times (0.03609 \text{ psi/in. WC}) = 13.19 \text{ psi}$$

Vapor Pressure The vapor pressure of a liquid at a given temperature can be defined as the pressure at which a liquid and its vapor can exist in equilibrium. An example is that the vapor pressure of water at 100°C is one atmosphere, in that at one atmosphere both water and water vapor (steam) can exist in equilibrium. When the vapor pressure of a liquid equals the pressure exerted on the liquid, the liquid is at its boiling point. Vapor pressure is usually an exponential function of temperature.

EXAMPLE 2-27

Problem: Will water at 100°C boil at 13.2 psia?

Solution: The vapor pressure of water at 100°C is one atmosphere or 14.7 psia. If the water is maintained at a pressure above 14.7 psia, the water will remain in the liquid state. In this example, the designated pressure is 1.5 psig vacuum, which is less than the vapor pressure, and the water will boil.

Cavitation When a flowing liquid is depressurized, the local pressure at a restriction of the pipe (such as a flowmeter, valve, or fitting) can be less than the downstream pressure of the fluid. If this local pressure drops below the vapor pressure of the liquid, it is possible for the fluid to develop a two-phase flow consisting of vapor cavities and liquid. Cavitation occurs when the pressure downstream of the restriction is greater than the vapor pressure and these vapor cavities collapse, generating intense local shock waves that result in cavitation noise and material damage. Flashing occurs if the downstream pressure is low enough to result in vaporization of the liquid. Cavitation in flowmeters is usually remedied by increasing the upstream pressure or downstream pressure sufficiently to eliminate cavitation, or by cooling the liquid in order to lower vapor pressure sufficiently to eliminate the cavitation.

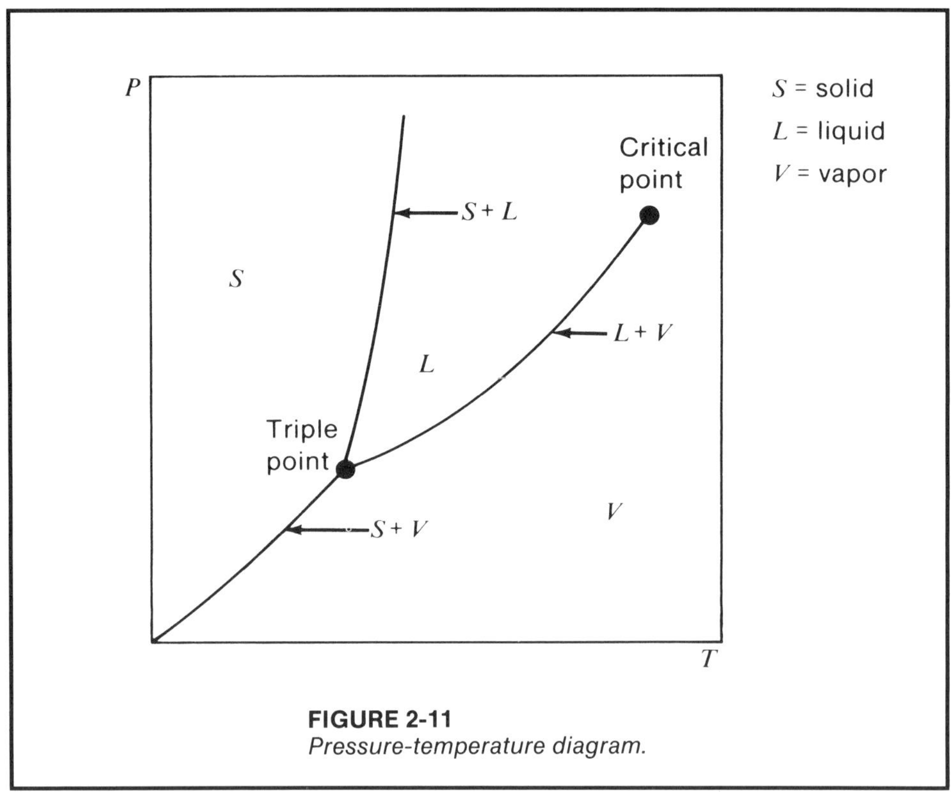

FIGURE 2-11
Pressure-temperature diagram.

Saturated Vapor A saturated vapor is a vapor that is in equilibrium with its liquid, but is totally a vapor. Saturated vapors at various pressures are shown conceptually on the pressure-temperature diagram in Figure 2-11 by the line between the triple point, where all three states can exist simultaneously, and the critical point, beyond which the liquid and vapor states are indiscernible.

Superheated Vapor A superheated vapor is a saturated vapor that has a higher temperature than a saturated vapor at the same pressure. This is usually expressed as degrees superheat, which represents the number of degrees the vapor is above the saturation temperature at its operating pressure. The higher the degree of superheat, the closer the vapor behaves as an ideal gas.

EXAMPLE 2-28

Problem: Calculate the degrees superheat of steam at 125°C and 1 atmosphere.

Solution: At 1 atmosphere, steam is saturated at 100°C, which is its boiling point at that pressure. The steam is superheated 125 – 100, or 25°C.

EXERCISES

2.1 Convert 100°F, 298 K and 500°R to degrees centigrade.

2.2 Convert 3 psi vacuum, 100 mm Hg vacuum, 17 psig, 42 feet of water column, and 15 bar to psia.

2.3 Calculate Reynolds number and the pressure upstream of the flowmeter illustrated in the figure below. The liquid has a viscosity of 1 cP and a specific gravity of 1.05. Assume that the heat exchanger is equivalent to 200 feet of straight 6-inch pipe.

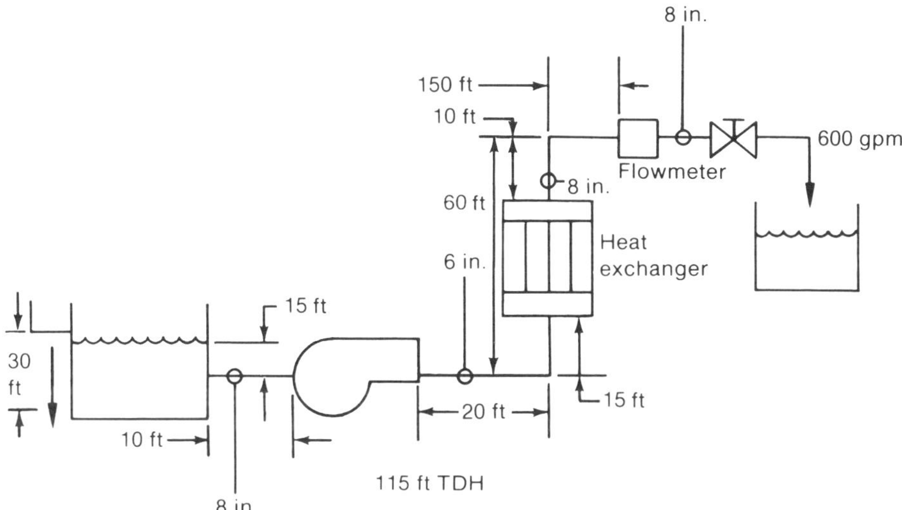

2.4 Calculate the pressure upstream of the flowmeter illustrated in the figure below if the viscosity of the liquid is 1 cP and the density is 1.09 g/cm^3. Disregard pipe friction losses.
Is the liquid flowing? Why or why not?

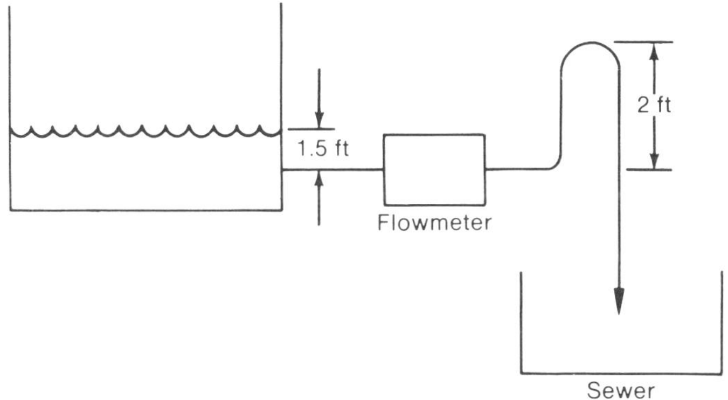

2.5. Calculate the specific gravity of the liquid and the pressure upstream of the flowmeter illustrated in the figure below if the viscosity of the liquid is 1 cP and the density is 76.75 lb/ft^3.
Is the liquid flowing? If not, what can be done to make the fluid flow?

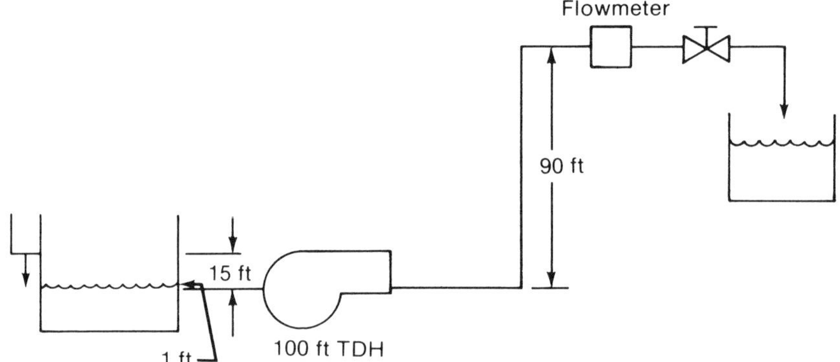

2.6. Determine the minimum flow necessary for flow to be in the turbulent flow regime in a 2-inch pipe filled with a liquid with a viscosity of 5 cSt.

2.7. Determine the average velocity in a 2-inch schedule 40 pipe in which 80 gpm is flowing.

2.8. Calculate Reynolds number and flow in standard units of a gas flowing at 28 psig and 90°F in a 4-inch schedule 40 pipe at 50 feet per second. Density at standard conditions is 0.1869 lb/ft^3 and compressibility is 0.987 and 0.955 at standard and operating conditions, respectively. Assume the viscosity to be 0.017 cP.

2.9. Calculate the flow in standard units of an ideal gas flowing at 50 psi and 30°C in a 4-inch schedule 80 pipe at 60 feet per second.

3 Performance Measures

Introduction Practical flowmeter systems require careful consideration not only of the flowmeter technology itself but also of its application to the process at hand. Process requirements dictate the amount of effort required for a successful application. A few of the criteria that must be considered are:

- Installation complexity and cost
- Maintenance
- Accuracy
- Linearity
- Repeatability
- Dependence on fluid properties
- Operating costs
- Hydraulic characteristics of the flowmeter and of the fluid
- Reliability
- Safety

As the physical realities involved with all of these criteria must be dealt with in flowmeter design, installation, and operation, the development of a perfect flowmeter is a virtual impossibility. Flowmeter performance is one of the key criteria in flowmeter selection that must be examined in detail.

Performance Statements Measures of flowmeter element performance represent the difference between how an ideal flowmeter would perform and how the real flowmeter actually performs. The most common measures of performance are percentage of rate, percentage of full scale, and percentage of meter capacity, although some flowmeter specifications are stated in terms of accuracy at a particular point. Often, flowmeter specifications do not state to which of these measures a percentage refers. Due to the significant difference in performance between these methods of expression, the manufacturer should be asked to clarify the specification.

There is a significant difference between a specification as a percentage of rate and one of a percentage of full scale. A specification expressed as a percentage of rate is defined such that the error is equal to the percentage times the actual flow and is, hence, a relative error.

$$\text{measurement error} = \%\ \text{rate} \times \text{actual measurement}$$

A specification expressed as a percentage of full scale (FS) means that the error associated with that measurement is equal to that percentage times the full scale flow and is, hence, an absolute error.

$$\text{measurement error} = \%\ \text{FS} \times \text{full scale flow}$$

It can be seen that the absolute measurement error associated with a percentage of rate specification will decrease as flow decreases, while that of a full scale specification will be constant for all applicable flows. As a result, as flow is decreased, the percentage error of rate of a percentage of full scale specification increases, as illustrated in the following graph.

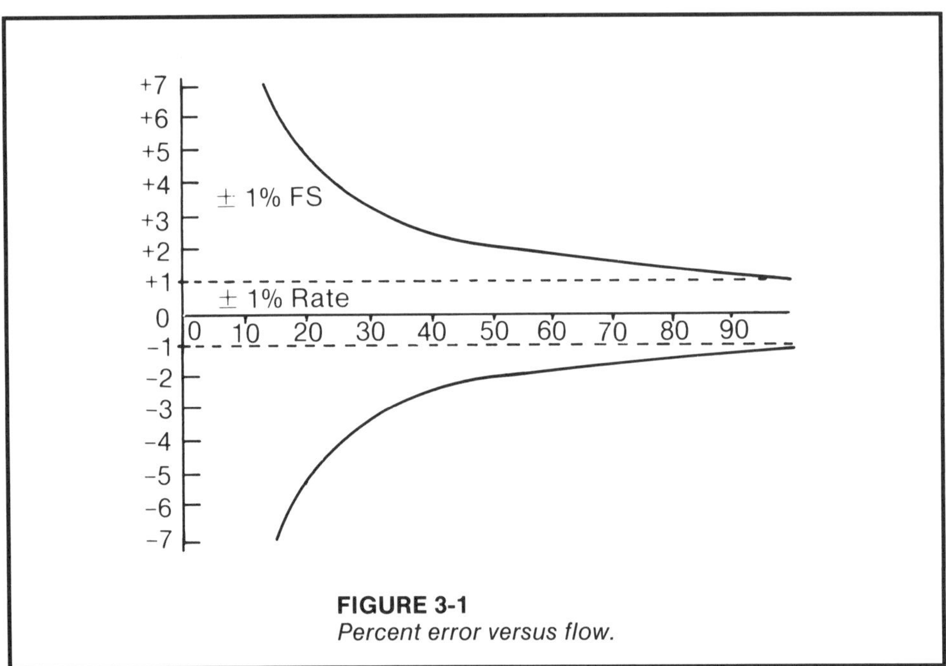

FIGURE 3-1
Percent error versus flow.

EXAMPLE 3-1

Problem: Determine which specification is preferable, 0.5% FS or 1% rate.

Solution: At first glance, the full scale specification would appear to be tighter than the rate specification; however, at low scale readings the full scale specification actually produces the larger measurement error. The point at which they are equal is where the absolute errors are equal, which is given by:

$$\% \text{ rate} \times \text{actual flow} = \% \text{ FS} \times \text{full scale flow}$$

Solving for the actual flow and substituting known quantities:

$$\text{actual flow} = (0.5\% / 1\%) \times \text{full scale flow}$$

the errors are equal at 50 percent of full scale. The full scale specification is superior from 50 to 100 percent of scale, while the rate specification is superior over the remainder of the range. It should be noted that when flow is above 50 percent of scale, the maximum difference between these specifications is 0.5 percent of rate (at 100 percent of scale), while at, for example, 10 percent of scale the difference is 9.5 percent of rate. Therefore, if flow will always be between 50 percent and 100 percent of scale, the full scale specification is superior; but if the flow will be varied throughout the flow range, the rate specification is superior.

The meter capacity error is the product of the percentage of meter capacity specification and the maximum flow that the flowmeter can handle. This can be thought of as a variation of the percent of full scale specification where full scale is fixed at meter capacity:

$$\text{measurement error} = \% \text{ meter capacity} \times \text{maximum flowmeter flow}$$

As the maximum flow that the flowmeter can handle is generally larger than the full scale flow, the measurement error is larger than an identical percentage of full scale specification.

EXAMPLE 3-2

Problem: Compare the errors associated with 0.5 percent FS and 0.25 percent meter capacities if full scale flow is 25 percent of flowmeter capacity.

Solution: As full scale is 25 percent of flowmeter capacity, the measurement error as a percentage of full scale can be expressed as:

$$\text{measurement error} = 0.25\% \text{ meter capacity} \times (\text{FS}/0.25 \text{ flowmeter capacity}) = 1\% \text{ FS}$$

Therefore, the meter capacity specification is equivalent to twice the full scale specification, even though numerically the meter capacity specification appears superior.

Some manufacturers specify flowmeters as a function of performance at one point under defined operating conditions such that the flowmeter achieves a stated accuracy, which can be significantly better than the performance of the flowmeter over its operating range. As there is considerable difficulty in reproducing exact design conditions in an industrial environment, performance stated as a function of one point can be misleading as to the expected performance in real industrial applications.

From the above discussion, it should be noted that there are significant differences between the ways in which errors are expressed and what their true meaning is. There should be no hesitation in seeking clarification of any performance specification that is not clearly defined. Consideration of flowmeter specifications should be performed with all specifications on a common basis so that performance can be properly and fairly compared. Usually the most convenient basis to use is the percentage of rate statement, as it readily indicates the error in the measurement as a function of the measured variable instead of a number or value dependent upon the flowmeter.

Performance statements should not be considered absolute in nature, as each flowmeter may not be individually tested. The performance statement is a measure of how the flowmeter will perform, often subject to some degree of certainty, typically 95 percent.

Repeatability and Hysteresis Flowmeter repeatability is the ability of a flowmeter to reproduce a measurement each time a set of conditions is repeated. It is not to be implied that the indicated flow is correct, but rather that the indication is the same each time.

The characteristics of a nonrepeatable flowmeter are shown in Figure 3-2. Note that the same flow produces different outputs each time a measurement is taken. If experimentation is performed to determine the flow desired to satisfy process constraints, the flow cannot be reproduced well.

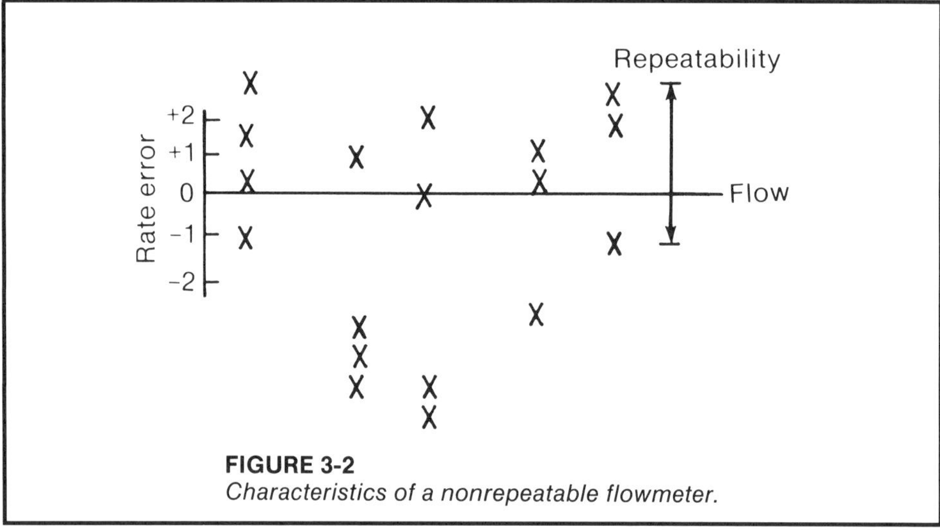

FIGURE 3-2
Characteristics of a nonrepeatable flowmeter.

Flowmeters can exhibit hysteretic error, where identical flows are measured differently when the flow is traversed upscale and downscale. A dead band can occur when the flowmeter is insensitive to a small change in the flow, as shown in Figure 3-3.

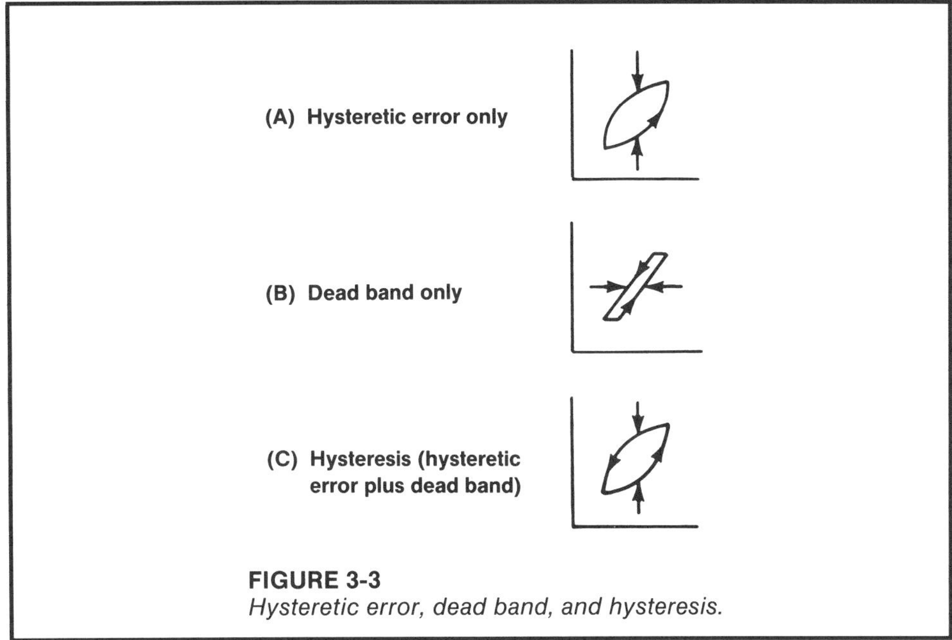

FIGURE 3-3
Hysteretic error, dead band, and hysteresis.

Figure 3-4 illustrates the characteristics of a repeatable flowmeter. Each time the same flow is put through the flowmeter, the readings closely approximate each other. This means that if a known percentage of flow produces a desired effect on the process, it can be set each time within a small tolerance of error even though the actual value of the flow may not be known.

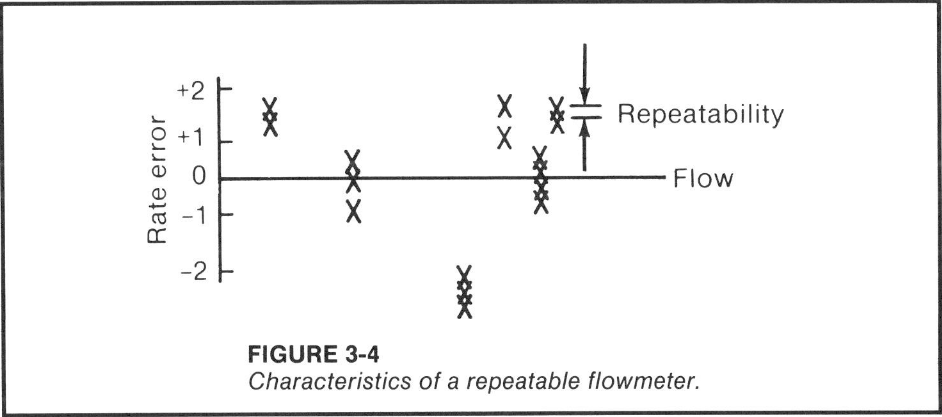

FIGURE 3-4
Characteristics of a repeatable flowmeter.

A flowmeter with ideal repeatability has measurements that exactly coincide each time the measurement is repeated. The above graphs have been normalized to show the repeatability as a rate error deviation. Care must be taken to determine what percentage is specified, as there are significant differences in performance associated with different methods of expression.

EXAMPLE 3-3

Problem: Two flowmeters with ranges of 0 to 100 gpm are used to measure water flow at no flow conditions. Determine which is more repeatable given the following data:

Measurement	Device 1	Device 2
1	0.8 gpm	0.5 gpm
2	0.9	-0.3
3	0.7	-0.1
4	0.8	-1.1
5	0.7	0.8

Solution: Graphically representing the data in a target representation with 0 gpm in the middle, it is seen that Device 1 is more repeatable than Device 2. Although Device 1 does not measure exactly 0 gpm, the data points are clustered in a small area, indicating that the device is repeatable (see Figure 3-5).

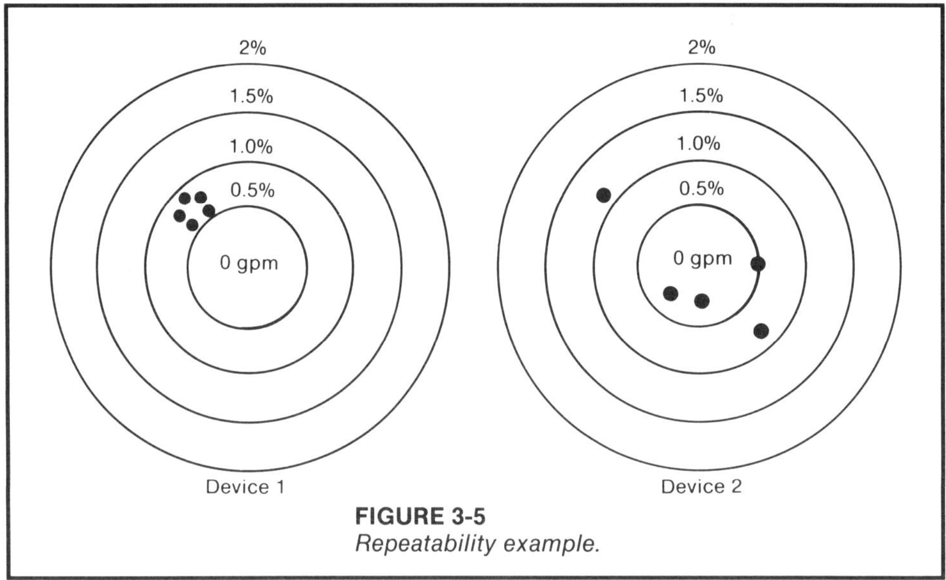

FIGURE 3-5
Repeatability example.

Linearity Linearity is the ability of the relationship between flow and flowmeter output, often called the characteristic curve or signature of the flowmeter, to approximate a linear relationship, as illustrated in Figure 3-6. Nonlinearities are often difficult to detect on graphs that are presented with large scales, such as shown in the figure. While there are several representations of linearity, the term linearity is assumed to represent the independent linearity of the flowmeter where a straight line is positioned to minimize the maximum deviation of the actual characteristic.

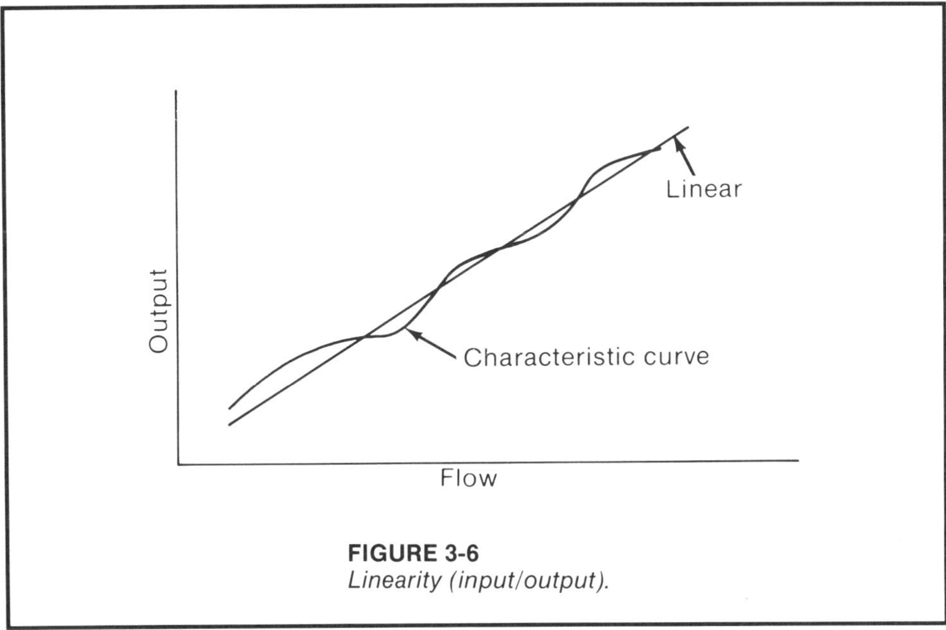

FIGURE 3-6
Linearity (input/output).

The graph in Figure 3-7 has been normalized to show the linearity as a rate error deviation from an ideal curve, which would be a straight horizontal line with zero error. Care must be taken to determine how the linearity is specified in order to determine the true linearity of a flowmeter due to the significant differences between methods of expression.

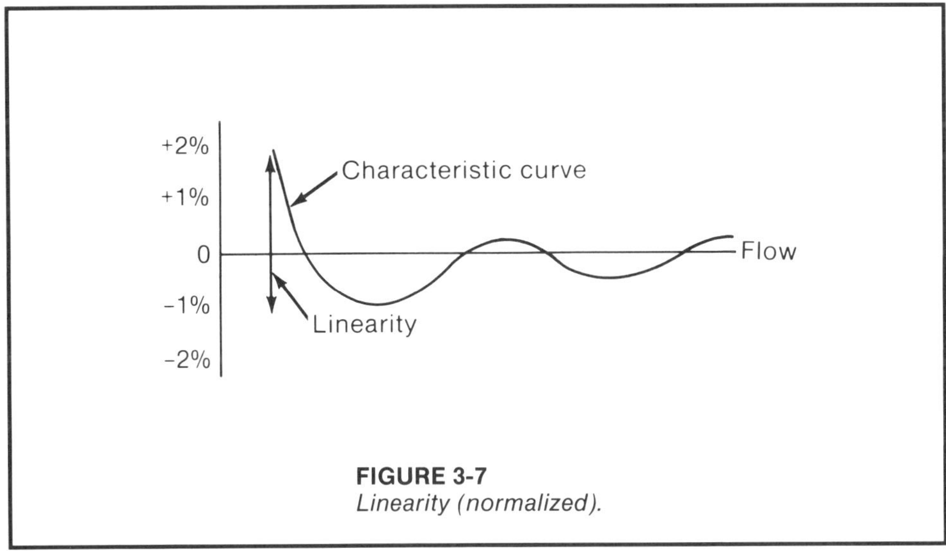

FIGURE 3-7
Linearity (normalized).

EXAMPLE 3-4

Problem: Two flowmeters with ranges of 0 to 100 gpm are used to measure water flow. Determine which is more linear given the following data:

Measurement	Device 1	Device 2
0	0.5 gpm	0.0 gpm
20	20.5	19.5
40	40.5	40.2
60	60.6	60.6
80	80.4	79.4
100	100.5	99.5

Solution: Graphically representing the data with a large scale does not illustrate the nonlinearities that are present as clearly as does the normalized graph. It is seen that Device 1 is more linear than Device 2, although Device 1 does have an offset and does not measure exactly (see Figure 3-8).

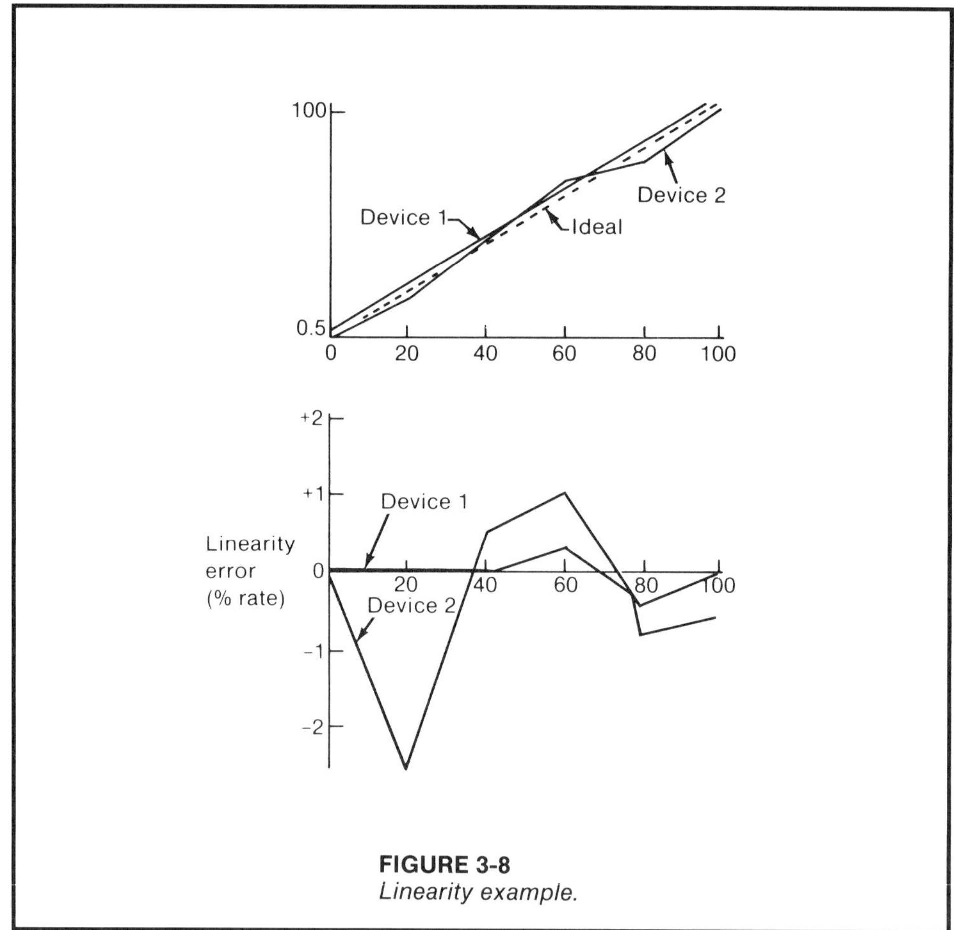

FIGURE 3-8
Linearity example.

Accuracy The accuracy of a flowmeter is its ability to produce an output that corresponds to its characteristic curve.

The flow data points of a flowmeter with poor accuracy are graphed in Figure 3-9. The measurements do not fall on or near the characteristic curve of the flowmeter as they do in Figure 3-10, the graph of a more accurate flowmeter.

A flowmeter with ideal accuracy would have all of its flow data points exactly coinciding with the characteristic curve. Because of the differences in methods of expression, care must be taken to determine how the accuracy is specified in order to determine the flowmeter's accuracy.

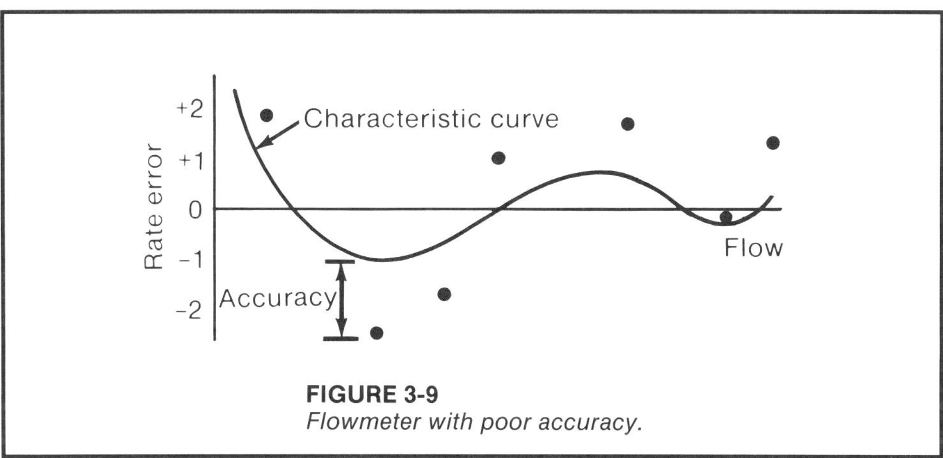

FIGURE 3-9
Flowmeter with poor accuracy.

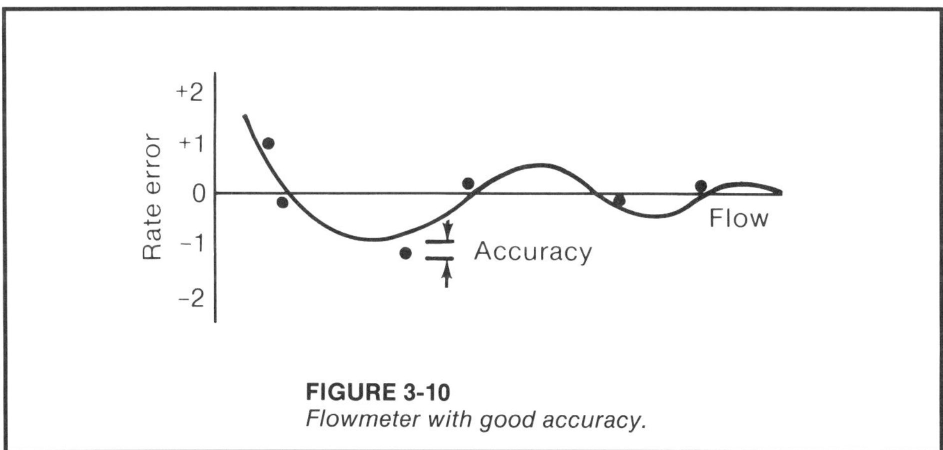

FIGURE 3-10
Flowmeter with good accuracy.

EXAMPLE 3-5

Problem: Two flowmeters with ranges of 0 to 100 gpm are used to measure water flow at no flow conditions. Determine which is more accurate given the following data:

Measurement	Device 1	Device 2
1	0.2 gpm	1.0 gpm
2	0.3	-0.9
3	-0.3	0.7
4	0.0	0.9
5	-0.1	0.9

Solution: Graphically representing the data in a target representation with 0 gpm in the middle, it is seen that Device 1 is more accurate than Device 2. Although Device 1 does not measure exactly 0 gpm, the data points are clustered in a small area near 0 gpm, indicating that the device is accurate (see Figure 3-11).

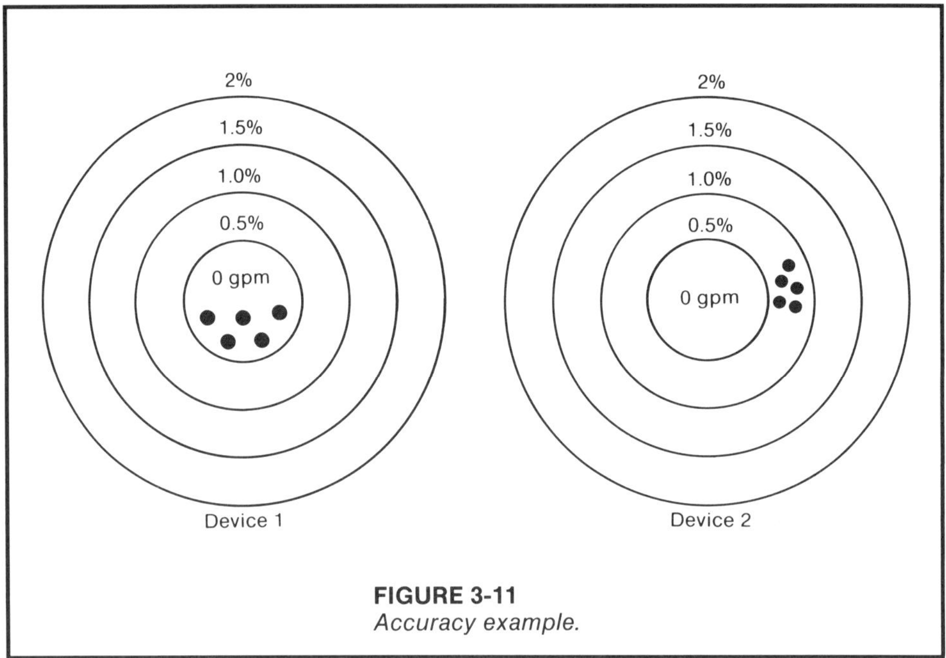

FIGURE 3-11
Accuracy example.

Flowmeter accuracy is usually stated for conditions where the fluid is Newtonian with a steady, homogeneous (single-phase), fully developed, nonswirling velocity profile symmetric about the center of the pipe at a reference temperature, usually 25°C. As it is difficult if not impossible to duplicate these conditions in an industrial environment, stated flowmeter accuracy may not be achieved. Therefore, an important factor in achieving more accurate flow measurement is the proper application of correction factors to account for the actual flowing conditions. In many flowmeter applications, the most significant sources of error are the correctness of fluid properties and variances in the installation, which cause the measurement to deviate from the measurement that would occur at reference conditions.

Composite Accuracy A composite accuracy statement for a flowmeter is a measure of the combined effects of repeatability, linearity, and accuracy. Unfortunately, it is often termed the "accuracy of the flowmeter," thereby causing confusion.

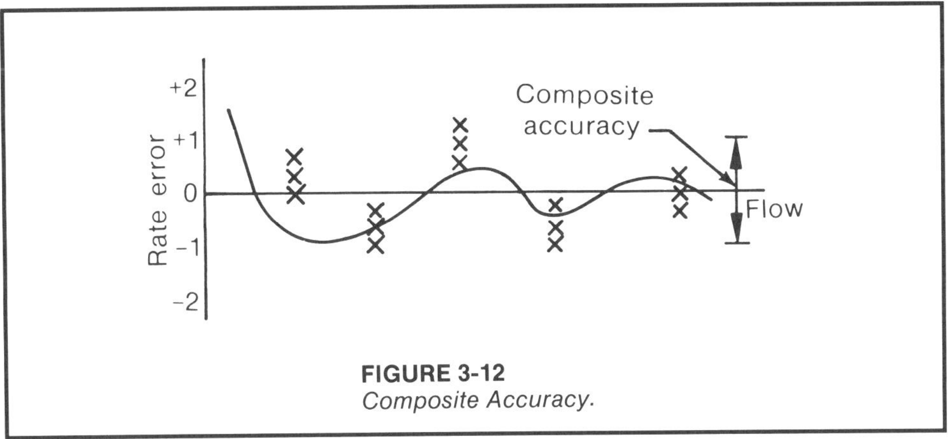

FIGURE 3-12
Composite Accuracy.

A flowmeter that is highly nonlinear can have excellent repeatability and accuracy but a poor composite accuracy due to the nonlinearity of the flowmeter. Manufacturers of such a flowmeter may choose to publish repeatability and accuracy specifications, or only the accuracy specification, and conveniently forget about the linearity specification. If the meaning of what is presented is not understood, it may be falsely assumed that the accuracy specification presented is the composite accuracy. If a specification is unclear, the manufacturer should be questioned. Depending on application, a nonlinear flowmeter may be satisfactory when repeatability is important.

The relationship between repeatability and accuracy is best explained by Figure 3-13.

Turndown The turndown is defined herein as the ratio of the maximum flow that the flowmeter will measure within the stated accuracy, usually the full scale flow, to the minimum flow that can be measured within the stated accuracy.

Each component of the system, such as the flowmeter element and the flowmeter secondary, has an associated turndown that may limit the turndown of the flowmeter system. Process conditions can severely limit the turndown of the flowmeter primary.

EXAMPLE 3-6

Problem: Calculate the turndown of a flowmeter that can measure from 20 to 100 percent of its scale within a given accuracy.

Solution: The maximum and minimum accurate flows are 100 percent and 20 percent, respectively. The turndown is then 100/20, or 5:1.

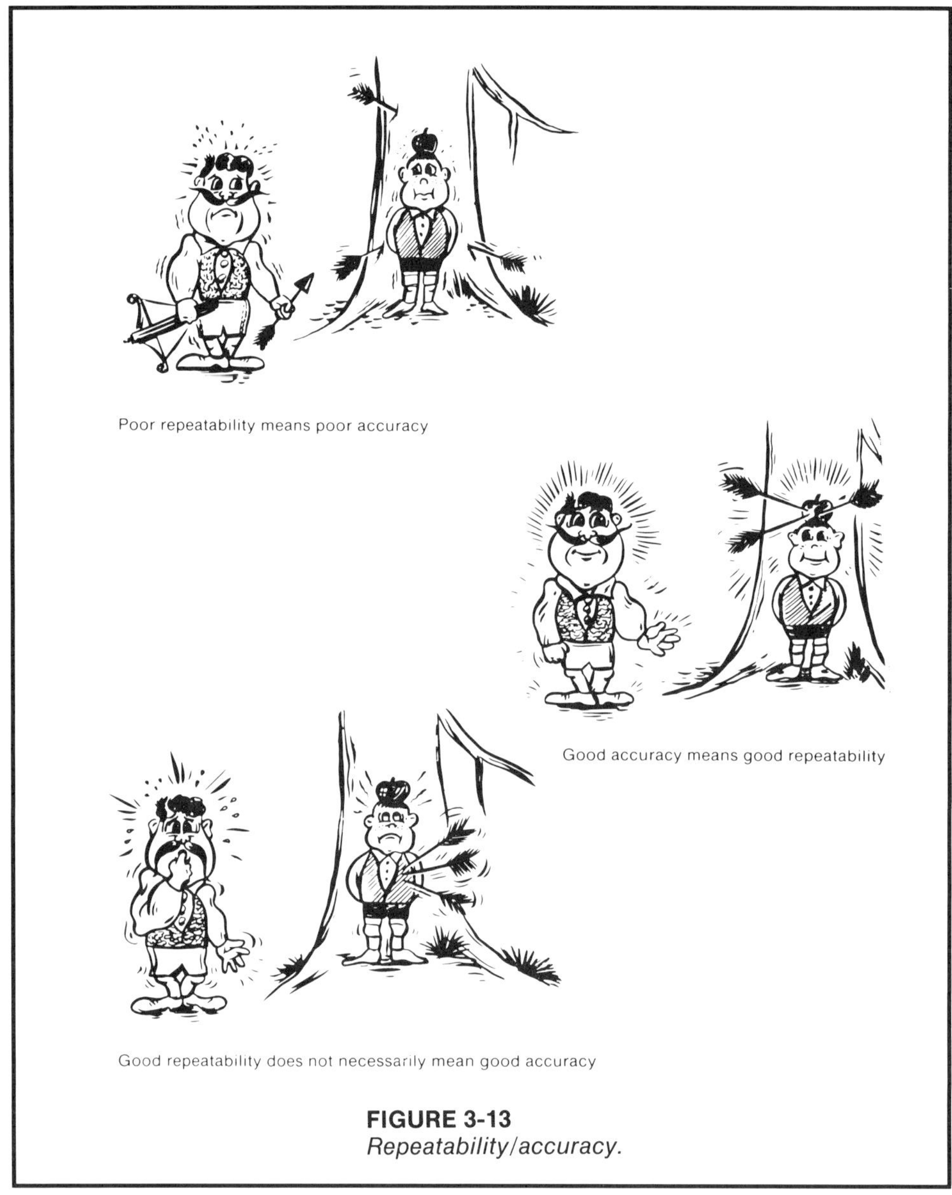

FIGURE 3-13
Repeatability/accuracy.

Rangeability Rangeability is defined herein as a measure of how much the range can be adjusted without major modification. It is the ratio of the maximum full scale range to the minimum full scale range of the flowmeter. This measure is an indication of how versatile the flowmeter is in relation to range changes that are often necessary during startup and plant expansion. Other factors, such as fluid properties and flowmeter installation, can be the limiting factors in the determination of rangeability of a flowmeter installation.

EXAMPLE 3-7

Problem: Calculate the rangeability of a flowmeter in which full scale can be adjusted from 33 to 100 percent of meter capacity.

Solution: The rangeability is 100/33, or 3:1.

EXERCISES

3.1 Do ideal flowmeters exist? Why or why not?

3.2 Which is preferable, a flowmeter that performs with an accuracy of 2 percent rate, 2 percent full scale, or 2 percent meter capacity? Why?

3.3 What does a stated accuracy of 1 percent mean?

3.4 Would a repeatable flowmeter be applicable when an exact amount of fluid is required to be added to a chemical reaction? Why or why not?

3.5 Which is preferable, a flowmeter with repeatability of 1 percent rate, an accuracy of 1 percent rate, or a composite accuracy of 1 percent rate? Why?

3.6 What is the turndown of a flowmeter that measures accurately from 30 to 100 percent of scale?

3.7 What is the rangeability of a flowmeter whose full scale can be adjusted from 50 to 175 percent of its present full scale?

3.8 Given the following applications:

1. Maintain cooling water
2. Chemical reactor feed
3. Fill storage tank
4. Boiler feedwater

which of the following flowmeters would be applicable, all else being equal?

	Performance	Cost
a	±1% rate	$1500
b	±Better of ±0.5 F.S. or ±1% rate	2500
c	±0.25% meter capacity	1700
d	±5% F.S.	900

4

Linearization and Compensation

Introduction Compensation for parameters that vary, such as the operating conditions (which are dynamic) and those that are fixed (such as flowmeter characteristic curves), can often be compensated for by the use of electronic devices. Although the most common form of compensation performed is pressure and/or temperature compensation to effect more accurate gas flow measurements, available also are flow computers that linearize flowmeter curves and compensate for operating conditions with the net result being a significant increase in accuracy.

Once linearized, the flow that has passed through the flowmeter over a period of time can be totalized. This can be achieved by integrating the flow; however, advances in digital technology are improving the methods and resultant accuracy with which flow signals are totalized.

Linear and Nonlinear Flowmeters A linear flowmeter is one whose output varies directly with flow. This means that a given percentage output corresponds to the same percentage of flow, as illustrated in Figure 4-1. Therefore, the output of the flowmeter changes by the same percentage as the flow through the flowmeter.

If the output does not correspond to flow in the above manner, the flowmeter is termed nonlinear. The most common nonlinear flowmeter is one that approximates a squared output, as shown in Figure 4-2.

With this relationship between flow and the flowmeter output, doubling the flow will result in four times the original output. As a result, at low flows small output changes correspond to large changes in flow, while at higher flows large output changes correspond to small changes in flow.

While a more complex representation of the input/output relationship may be appropriate for increased accuracy, squared output flowmeters are typically

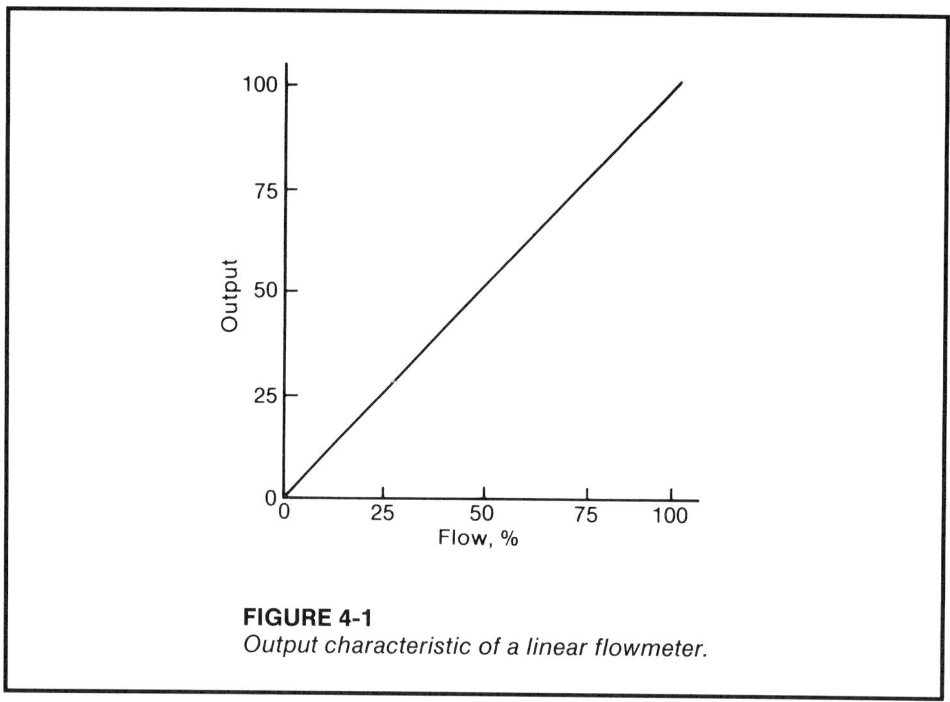

FIGURE 4-1
Output characteristic of a linear flowmeter.

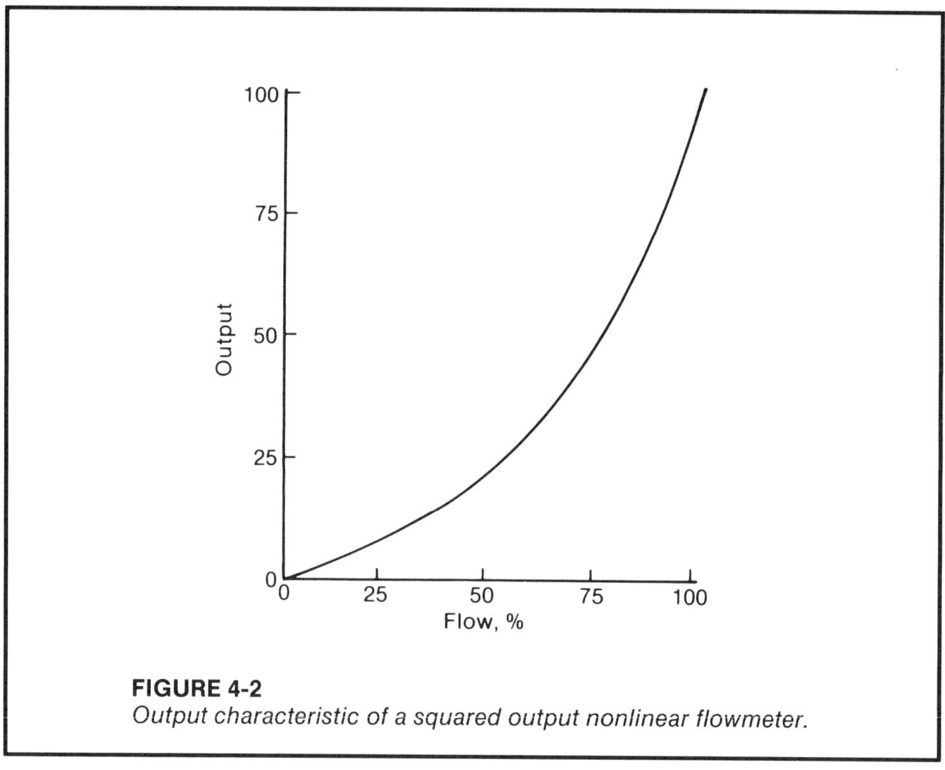

FIGURE 4-2
Output characteristic of a squared output nonlinear flowmeter.

linearized with a square root extractor, which has the following convenient input/output relationship:

$$\text{output}_{\%} = (\text{input}_{\%})^{1/2}$$

Square root extractors have large gains at the lower end of the scale, making any shifts in the flowmeter calibration, square root extractor calibration, or pipe perturbations that cause a small flowmeter output to have a relatively large effect on the output of the square root extractor and, hence, the linearized flow signal. It is not uncommon for a squared output flowmeter indication to be bouncy under no flow conditions. This can be removed by eliminating flow noise and vibration or by calibrating the zero of the square root extractor slightly below true zero, which usually eliminates the bouncy zero at the expense of a relatively small flow error at higher flows.

EXAMPLE 4-1

Problem: Calculate the percentage of full scale flow of a linear and a squared output flowmeter when the output of the primary flowmeter element is 0, 1, 10, 25, 50, 75, and 100 percent.

Solution: The flow through a linear flowmeter is identical to the output of the flowmeter primary, while the flow through a squared output flowmeter is proportional to the square root of the output of the flowmeter primary as follows:

Flowmeter Output, %	Linear Flowmeter Flow, %	Squared Output Flow, %
0	0	0.0
1	1	10.0
10	10	31.6
25	25	50.0
50	50	70.7
75	75	86.6
100	100	100.0

While flowmeters with nonlinearities other than those stated above can be found and similarly linearized by performing the inverse function of the nonlinearity, flowmeters with linear and squared outputs represent the characteristics of the great majority of flowmeter technologies.

Gas Flow Pressure and Temperature Compensation In gas service, most flowmeters measure actual volume or infer the actual volume while assuming that the gas is flowing at nominal operating conditions. Significant changes in actual volume can occur when operating conditions vary from the nominal operating conditions and will result in significant uncertainties in the flow measurement. One way to avoid this problem is to measure or calculate the density of the flowing fluid and use the following relationship to calculate the mass flow:

$$W = \rho \times Q$$

As the measurement of density of a flowing fluid is relatively expensive to effect and to maintain as well as being subject to many sources of error, density is usually inferred from the measured pressure and/or temperature of the fluid.

Applying the Non-Ideal Gas Law, the relationship between the volumetric flow at nominal operating conditions and actual operating conditions can be expressed as:

$$V_{nom} = \frac{Z_{nom} \times P \times T_{nom}}{Z \times P_{nom} \times T} \times V$$

where the nominal operating conditions are known and the equation takes the form

$$V_{nom} = \text{constant} \times \frac{P}{(Z \times T)} \times V$$

where P, Z, and T represent the pressure, compressibility, and temperature at actual flowing conditions. The $P/(Z \times T)$ term in the above equation compensates for the variation in density between actual and nominal operating conditions to calculate the volume required at nominal operating conditions to effect the same flow at actual operating conditions. This means that, for example, if $P/(Z \times T)$ is 1.10, the gas is 1.10 times as dense as the gas at nominal operating conditions, and 10 percent more gas actually flows through a linear flowmeter than is measured, assuming nominal operating conditions.

At nominal operating conditions, the $P/(Z \times T)$ factor is used to correct the actual volume before flowmeter nonlinearities are compensated for; hence these factors are treated in the same manner as density in flowmeter equations. As a result, when flow varies nonlinearly with gas density, it varies in the same nonlinear relationship with the $P/(Z \times T)$ factor.

Pressure and temperature compensation in its most commonly applied form uses the assumption that the compressibility factor is constant at operating conditions near the nominal operating conditions and neglects compressibility effects. Implementation of pressure and temperature compensation is achieved by applying hardware that divides the absolute pressure by the absolute temperature and multiplies the result by the measured flow signal before linearizing the flowmeter output.

When pressure compensation only is desired, the temperature is assumed to be equal to its nominal value and becomes lumped into the constant. Compensation for pressure is implemented by multiplying the absolute pressure by the measured flow and a constant before linearizing the flowmeter output.

When temperature compensation only is desired, the pressure is assumed to be equal to its nominal value and becomes lumped into the constant. Compensation for temperature is implemented by multiplying the measured flow by a constant and dividing the result by the absolute temperature before linearizing the flowmeter output.

The effects of variations in temperature and pressure are presented for linear flowmeters in Table 4-1. Squared flowmeters have approximately one-half of the deviation shown.

These effects can also be represented graphically for linear and squared flowmeters, as illustrated in Figure 4-3.

TABLE 4-1
Effects of Variations in Temperature and Pressure for Linear Flowmeters

Gas measurement errors if meter is not temperature compensated

Flowing temp.	*Error*	*Flowing temp.*	*Error*
0	-13%	60	0
10	-11%	70	+ 2%
20	- 8%	80	+ 4%
30	- 6%	90	+ 5%
40	- 4%	100	+ 7%
50	- 2%	110	+ 9%
60	0	120	+10%

If temperature is constant, apply a factor to the meter reading in lieu of using a temperature compensated meter to eliminate the above indicated errors.

Gas measurement errors if meter is not pressure compensated

Nominal metering pressure	*Tolerance around nominal metering pressure*					
	±.25 psig	*±.5 psig*	*±1 psig*	*±2 psig*	*±5 psig*	*±10 psig*
0.25 psig	±1.7%	N/A	N/A	N/A	N/A	N/A
2 psig	±1.5%	±3.0%	±6.1%	±12.2%	N/A	N/A
5 psig	±1.3%	±2.6%	±5.2%	±10.3%	±25.8%	N/A
10 psig	±1.0%	±2.0%	±4.1%	± 8.2%	±20.5%	±41.0%
20 psig	±0.7%	±1.5%	±2.9%	± 5.8%	±14.5%	±29.1%
50 psig	±0.4%	±0.8%	±1.6%	± 3.1%	± 7.8%	±15.5%
75 psig	±0.3%	±0.6%	±1.1%	± 2.2%	± 5.6%	±11.2%
100 psig	±0.2%	±0.4%	±0.9%	± 1.7%	± 4.4%	± 8.7%
125 psig	±0.2%	±0.4%	±0.7%	± 1.4%	± 3.6%	± 7.2%

If pressure is relatively constant (within acceptable limit per the above table), apply a factor to the meter reading in lieu of using a pressure compensated meter to eliminate the above indicated errors.

Data presented for linear flowmeters. Errors are approximately 1/2 of the above for squared output flowmeters.

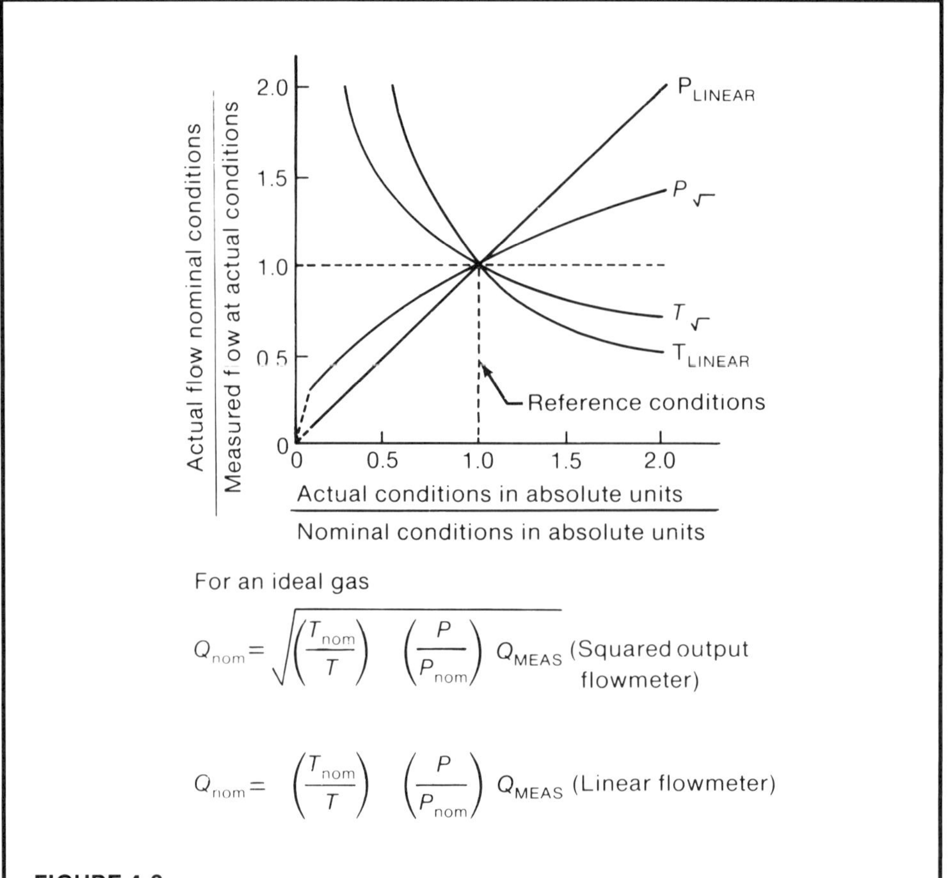

FIGURE 4-3
Effects of variations in temperature and pressure for linear and squared flowmeters.

EXAMPLE 4-2

Problem: Calculate the effective change in measurement of an ideal gas at 50 psig and 15°C, measured with a linear and a squared output flowmeter, when the pressure falls to 40 psig and the temperature rises to 25°C.

Solution: Examination of the graph in Figure 4-4, which shows greater detail in the area of interest, yields a correction factor of approximately 0.82, where

$$P / P_{nom} = (40 \text{ psi} + 14.7 \text{ psi}) / (50 \text{ psi} + 14.7 \text{ psi}) = 0.845$$

The correction factor can be calculated as follows:

$$\frac{V_{nom}}{V} = \frac{P \times T_{nom}}{T \times P_{nom}} = \frac{(40 \text{ psi} + 14.7 \text{ psi})}{(273°\text{C} + 25°\text{C})} \times \frac{(273°\text{C} + 15°\text{C})}{(50 \text{ psi} + 14.7 \text{ psi})} = 0.817$$

The flowmeter output is multiplied by 0.817 to correct the flow at nominal conditions for variation in operating conditions. From Figure 4-4, a squared output flowmeter would have a factor of approximately 0.90 under the same conditions, where the square root of the result of the above example is calculated as follows:

$$V_{nom} / V = (0.817)^{1/2} = 0.904$$

The errors involved are 18.3 percent and 9.6 percent for linear and squared outputs, respectively, which illustrates that the squared output error is approximately one-half of the linear error.

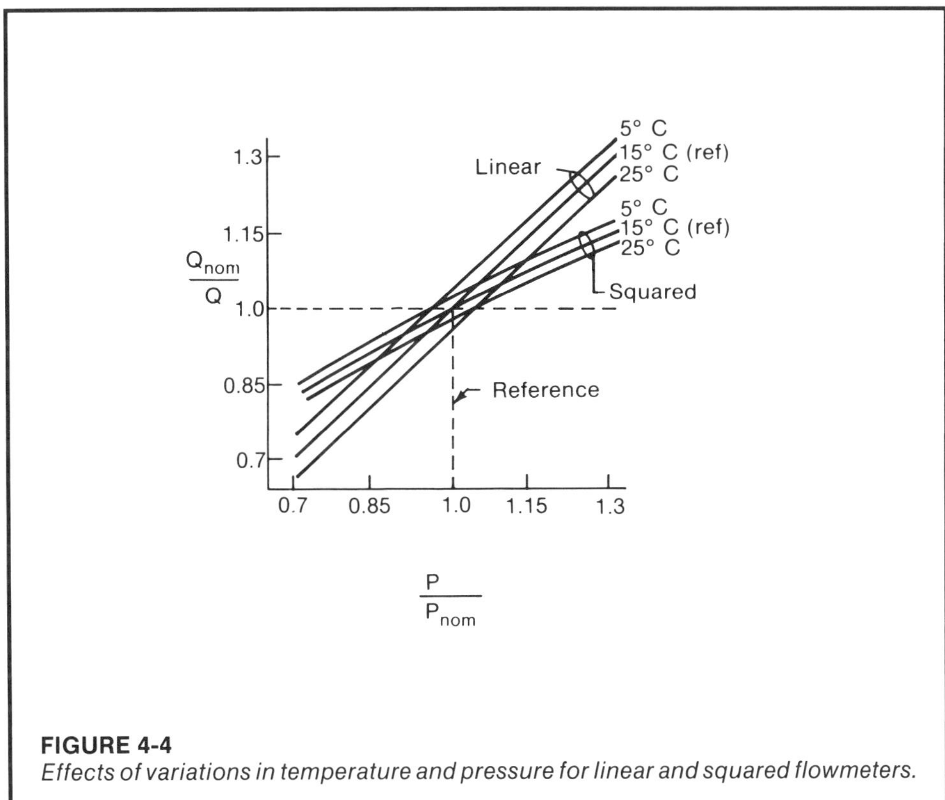

FIGURE 4-4
Effects of variations in temperature and pressure for linear and squared flowmeters.

Liquid Temperature Compensation Requirements for accuracy that necessitate compensation for density variations caused by variations in liquid temperature are few in number; however, it should be recognized that liquid density can and does change, however slightly, as illustrated in Figure 4-5.

If the accuracy of the flow measurement is critical in a given application, a mass flowmeter, a temperature compensating flowmeter, or implementation of temperature compensation should be considered.

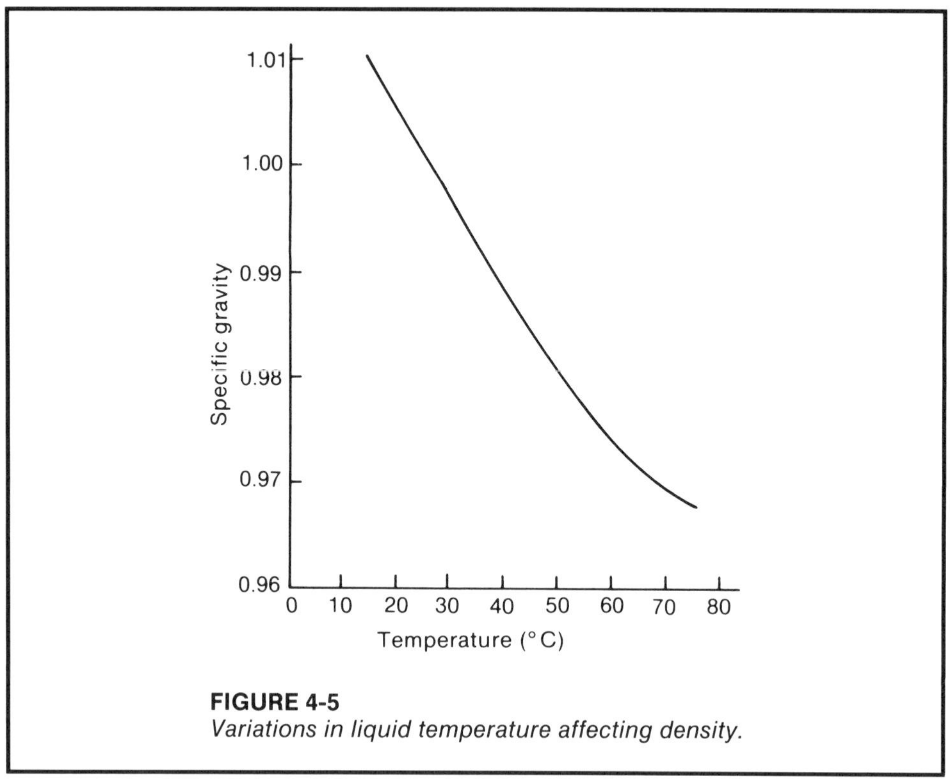

FIGURE 4-5
Variations in liquid temperature affecting density.

EXAMPLE 4-3

Problem: Based upon the above graph, estimate the density change caused by a change in temperature from 20 to 40°C.

Solution: From the above graph, the density is 1.0032 at 20°C and 0.9882 at 40°C. The ratio of these densities is 1.0152, which indicates that there is a 1.5 percent change in density due to a 20°C temperature change.

Pressure and Temperature Tap Location Pressure and temperature measurement taps must be properly located for each type of flowmeter to minimize error in the overall measurement.

Pressure tap location is critical since local pressures in the flowmeter can vary significantly. Ideally, the pressure tap location would be at the point that minimizes flow error, usually upstream of the flowmeter, unless special corrections are being applied. The pressure should correspond to the undisturbed flow upstream or downstream of the flowmeter. Care should be taken to avoid tap location where pressure fluctuations are known to occur. A few flowmeters have pressure taps located in their bodies that minimize error for that particular flowmeter.

Temperature measurement taps are not as critical as pressure measurement taps since the temperature varies only slightly between the inlet and the outlet of the flowmeter. Temperature taps are typically located a prescribed distance (e.g., 8 to 10 pipe diameters) downstream of the flowmeter so as not to cause turbulence at the flowmeter inlet, as would be the case if it were located upstream of the flowmeter.

While pressure and/or temperature compensation may not be implemented in a given application, it makes good sense to investigate if any pressure or temperature measurements are to be made upstream or downstream of the flowmeter and, if so, to properly locate the taps such that compensation can be implemented in the future without relocating any instruments. It may be appropriate to install a test well and a pressure gage connection to verify actual, though representative, flow conditions.

Flow Computers

Flowmeters have accuracy, repeatability, and linearity specifications associated with their operation. In a repeatable flowmeter, the accuracy specification can be an order of magnitude poorer than the repeatability specification, while the characteristic curve of the meter might be highly nonlinear. The flowmeter, however, can be linearized using a flow computer if enough information is known about the flowmeter characteristics and operating conditions.

When a repeatable flowmeter is flow tested, a relatively large number of flow data points can be obtained that are repeatable under the same operating conditions, such as flow, pressure, temperature, and Reynolds number. This data, which can be thought of as the characteristic curve or signature of the flowmeter, can be programmed into a flow computer that uses process inputs to determine exactly where the flowmeter is operating on its characteristic curve. Assuming that the operating conditions or parameters can be accurately determined from process inputs, the flow can be determined dependent upon the repeatability specification of the flowmeter, virtually eliminating the effects of accuracy and linearity specifications.

Similarly, complex mathematical or experimentally developed relationships between measurable process variables and parameters that affect flowmeter output, such as fluid density, supercompressibility factors, flowmeter expansion coefficients, and flowmeter correction coefficients, can be programmed into flow computers to correct flowmeter output and improve accuracy even further, especially in gas applications.

EXERCISES

4.1 Calculate the output of a square root extractor if the input is 6.25 percent, 56.25 percent, and 81.0 percent of full scale.

4.2 Calculate the effective change in measurement of an ideal gas at 100 psig and 15°C, measured with a linear and a squared output flowmeter, when the pressure rises to 120 psig and the temperature falls to 5°C.

4.3 Can the specific gravity of liquids be affected by more than 1 percent due to temperature changes?

4.4 Can pressure taps for pressure compensation of the flow measurement be located at any arbitrary point upstream of the flowmeter? Why or why not?

4.5 Can flow computers be used to linearize flowmeters when process conditions are not well defined? Why or why not?

5

Totalization

Introduction Flowmeters are commonly used to totalize flows, most often for charging batches, for internal custody transfer, and for billing purposes. In the industrial environment, totalized values of raw materials, utilities, and finished products are essential for determining process yields and conversion efficiencies. In custody transfer applications, flow totalization provides the only basis for the cost of the total fluid transfer. Flow totalization systems can be made more accurate by applying flowmeter designs that are more applicable to flow totalization due to the nature of their operation.

Analog and Digital Flowmeters Primary flowmeter elements may be classified according to their having either an analog or a digital oscillatory output. Each type requires significantly different technologies for manipulating the raw signal and totalizing the flows.

Analog flowmeter elements have a continuous output. The signal generated by the flowmeter may be electrical, such as voltage, or mechanical, such as a differential pressure or force, and may be a nonlinear function of flow. The output of a squared output flowmeter from 0 to 100 percent flow is illustrated in Figure 5-1.

Digital flowmeter elements, on the other hand, produce outputs that are oscillatory or pulsed in nature, such as mechanical vibrations or electrical oscillations or pulses, as illustrated in Figure 5-2.

The frequency of the oscillations is a function of flow and the *K* factor of the flowmeter at operating conditions. The *K* factor is the number of pulses generated by the flowmeter per unit volume of fluid that pass through the flowmeter. The meter factor is the inverse of the *K* factor and is often used to describe the output of the flowmeter.

Digital flowmeters typically have a fixed minimum, accurately measurable flow that is dependent on meter size, fluid viscosity, and operating conditions. As a result, turndown will vary depending on the full scale flow range selected.

100
75
50
25
0
Output, %
Output is
proportional
to flow2
0 25 50 75 100
Flow, %

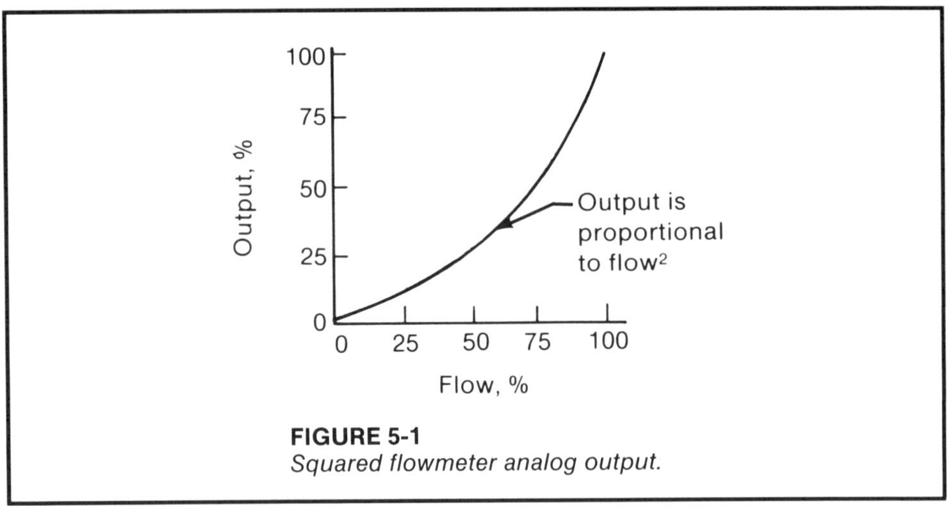

FIGURE 5-1
Squared flowmeter analog output.

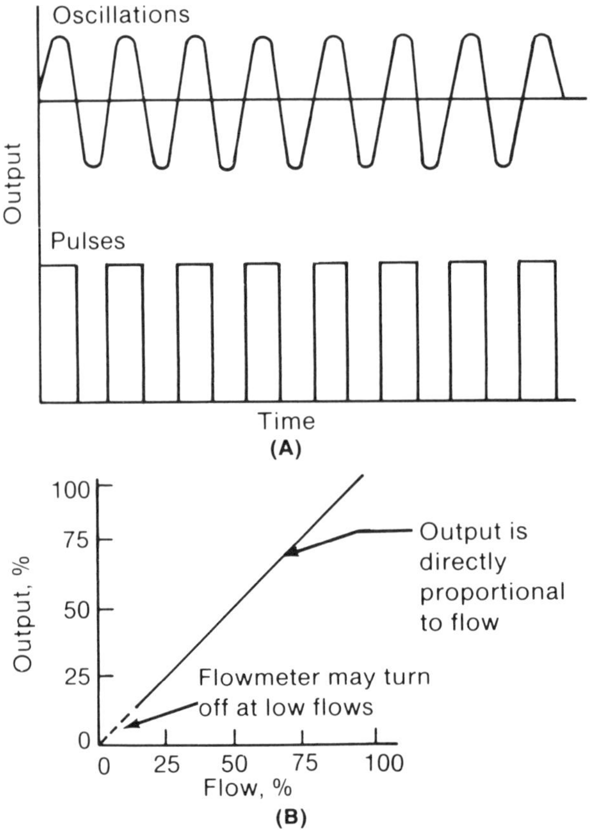

Oscillations
Output
Pulses
Time
(A)
100
75
50
25
0
Output, %
Output is
directly
proportional
to flow
Flowmeter may turn
off at low flows
0 25 50 75 100
Flow, %
(B)

FIGURE 5-2
Digital flowmeter outputs.

Accuracy statements are most often (but not always) expressed as a percent of rate, since the output at zero flow is usually well defined where there are no oscillations or pulses. Digital flowmeter elements can be further subdivided into those that operate at all flows, which can have percentage of rate or full scale accuracy, and those that cease operating at low flows, typically due to hydraulic considerations, which typically have accuracies expressed as a percentage of rate.

Analog flowmeter elements typically (although not always) have a turndown that is virtually independent of the full scale flow range and is usually limited by hydraulic conditions at low flows. Accuracy statements are typically expressed as a percentage of rate, as the output at zero flow is usually well defined. The transmitter associated with these flowmeters has a zero stability specification that is specified in terms of a full scale error, which may significantly contribute to inaccuracy at low flows.

Implementation Different techniques are used to implement flow totalization of analog and digital flowmeters. Analog flowmeters inherently generate an analog signal that can be directly totalized using an integrator that typically has an accuracy of ±0.5 percent of rate over a 10:1 range, as illustrated in Figure 5-3.

Digital flowmeters on the other hand produce pulses, each of which represents a known amount of flow. Counting and scaling the pulses to engineering units provides a totalization of the flow with a ±1 pulse error, which is negligible over the long term (see Figure 5-4). The net result is a system that is primarily dependent upon the accuracy of the primary flowmeter element and virtually independent of the signal processing devices. If flow indication is also required, then an analog output converter can be used to generate the signal for indication.

Often, a combination of the above systems is implemented as illustrated in Figure 5-5, where a digital flowmeter is purchased with an analog output, which is in turn digitized again to perform the totalization.

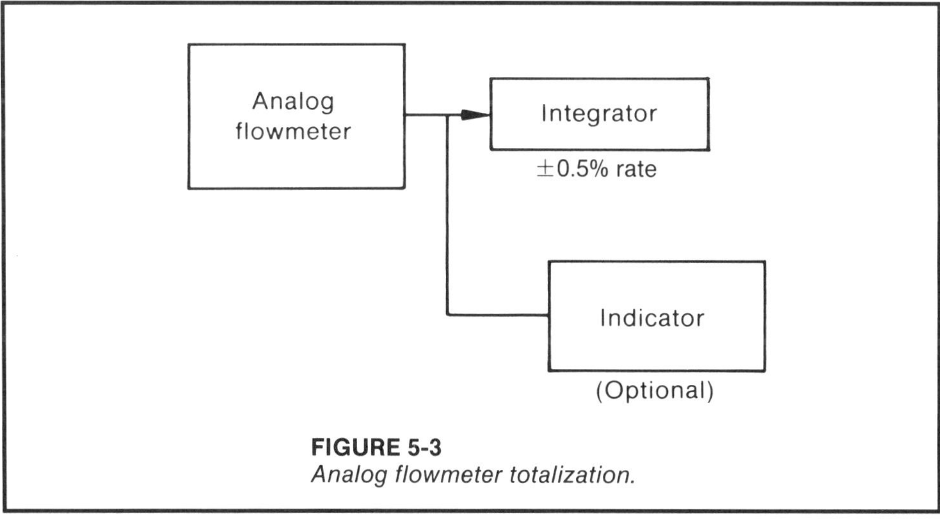

FIGURE 5-3
Analog flowmeter totalization.

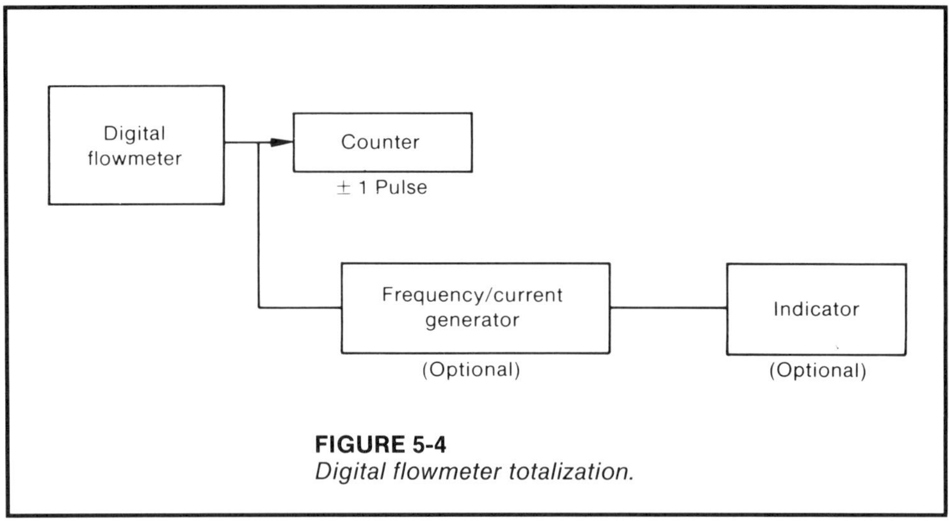

FIGURE 5-4
Digital flowmeter totalization.

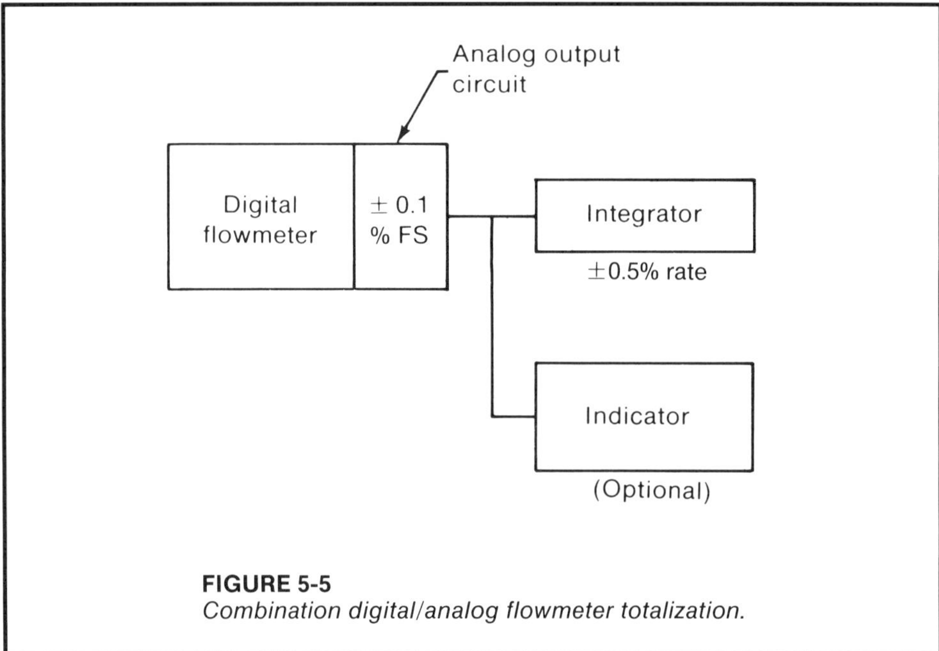

FIGURE 5-5
Combination digital/analog flowmeter totalization.

From the above, primary flowmeter specifications being equal, the most accurate method of totalizing flows is to apply a digital flowmeter. When applicable, digital flowmeters are usually more desirable when digital control systems are used, due to reduced conversion errors versus analog inputs. The digital input of a digital control system typically totalizes the pulses directly so as to effect an accurate total. An analog input must be converted to digital form, which produces some error.

EXERCISES

5.1 Why is it possible for a digital flowmeter to have smaller operating range and turndown than those of an analog flowmeter?

5.2 What does each pulse produced by a digital flowmeter signify?

5.3 Is the flow at which a digital flowmeter ceases to operate a function of full scale flow, meter capacity, and/or operating conditions? Explain.

6

Flowmeter Calibration

Introduction Flowmeter calibration is a conceptually simple matter of performing adjustments to the flowmeter such that it measures flow within predetermined accuracy constraints. Ideally, this is performed under operating conditions. However, it would be prohibitively expensive to construct flow facilities for each application. Depending upon design, flowmeter calibration is more practically achieved by utilizing a flow laboratory or a flow calibration facility, by verifying physical dimensions according to established empirical correlations, or by simulation of flow by electronic means.

Calibration Techniques Flow calibration is generally done to certify meter factor accuracy by measurement of the flowmeter output under flow conditions that are hydraulically similar to the actual installation, i.e., Reynolds number equivalence. This does not necessarily guarantee overall system accuracy, but rather that the primary metering element has a known degree of uncertainty. Adjustments that compensate for differences in production flowmeters can be made in the flowmeter element in some designs. In many designs, however, adjustments are made electronically in the transmitter once the performance of the flowmeter element is known.

While liquid measurement requires that measurements other than flow be taken to effect accurate measurement, such as temperature, pressure and viscosity, the compressible nature of gas makes accurate control of these secondary parameters more critical. The net result is that accurate gas flow measurement is more difficult to achieve than liquid flow measurement, and liquid flow measurement accuracy is generally superior to gas flow measurement accuracy.

Liquid and gas flow calibration facilities can accurately measure flows that operate at Reynolds numbers of up to approximately 10^6 and 7×10^6, respectively. Most liquid applications in small and medium pipe sizes at reasonable velocities operate at Reynolds numbers of less than 10^6 while typical gas applications operate at Reynolds numbers in excess of 10^6 (and often well in excess of 7×10^6, due to the relatively low viscosity of gas). The net result is considerable performance uncertainty at high Reynolds numbers where flowmeter performance is a function of Reynolds number, which can perhaps be defined theoretically or by extrapolation but cannot be accurately verified by experimental means. Under low Reynolds number conditions, the liquid is not necessarily Newtonian. These conditions are similarly difficult to simulate accurately.

Flow Laboratory A flow laboratory is a facility constructed for the purpose of measuring flow through a pipe with extreme accuracy. As a result of practical constraints, most flow laboratories utilize water or air as the flowing medium for liquid or gas applications, respectively, due to the large amount of precise experimental data available. For service other than water or air, performance at other operating conditions is adjusted by correction factors based upon the fluid properties of the actual fluid relative to the fluid on which the flowmeter was calibrated. This method introduces some flow measurement uncertainty for substances other than those used to test the flowmeter; however, when the properties of the fluid to be measured are well defined, uncertainties can be kept to a minimum.

For precise applications, flowmeter calibration is performed in a flow laboratory in which the laboratory equipment is maintained at a composite accuracy of better than approximately 10 times the accuracy of the flowmeter being calibrated. These applications are typically associated with laboratory standards and custody transfer applications, which represent only a few percent of all flowmeters. Most flow laboratories are configured and maintained per industry standards and are traceable to the National Institute of Standards and Technology (NIST). Flow laboratories are usually operated and maintained by flowmeter manufacturers, but independent flow laboratory facilities exist and perform flowmeter testing and calibration. Independent flow laboratories are often more extensive and versatile than those maintained by manufacturers for their own use, as the applications encountered are usually not limited as to application or manufacturer. Accuracies in the order of 0.1 percent of rate or better can be achieved in liquid (water) flow laboratories, while accuracies of only 0.5 percent of rate can be achieved in gas flow laboratories.

Unless there are a considerable number of high accuracy gas applications so that a gas flow laboratory is economically feasible, or unless gas flow measurement represents the primary product line and is required for product development, most manufacturers with a need for a flow laboratory maintain a liquid flow laboratory, and precise gas calibration or testing are performed as necessary in independent flow laboratories with adequate facilities.

Determination of the performance of a flowmeter to be used for gas service can be determined from calibration using liquid flow data, with some resultant error.

Flow Calibration Facility A flow calibration facility consists of a master flowmeter system, the performance of which is traceable to calibration in a certified flow laboratory. Differences in the properties of the fluid used to perform the calibration and the fluid at reference conditions are compensated for by making secondary measurements, such as pressure, temperature, viscosity, and the like, and performing calculations to determine the flow through the master flowmeter. Most modern flow calibration facilities are automated and utilize a computer to sense the measured variables, calculate flow, and document the performance of the flowmeter being calibrated.

Flow calibration facilities are usually maintained by flowmeter manufacturers for production flowmeter calibration purposes and are usually traceable to the National Institute of Standards and Techology (NIST), since the performance of the master flowmeter is determined in a flow laboratory. Sufficient calibration accuracy is achieved in a flow calibration facility for the great majority of industrial flowmeter applications. Flowmeter calibration in a flow calibration facility is typically referred to as a hydraulic calibration. Such "wet" calibrations are typically performed with water or air, but special calibrations are available from some manufacturers to simulate special operating conditions, such as low Reynolds number in the case of an oil calibration.

Depending upon the nature of the flowmeter, the flowmeter primary or both the flowmeter primary and the transmitter may be calibrated. While it is conceptually more appealing for both to be calibrated together as a system to achieve stated performance, in practice it is more desirable for the individual components of the system to be calibrated separately to higher tolerances to achieve the same performance. Replacement of a component, perhaps due to failure, would require a system calibration to again achieve stated accuracy if the flowmeter had been originally calibrated as a system, while calibration of the new component is all that is necessary if each component had been calibrated separately.

Dry Calibration

Dry calibration is a calibration check performed without subjecting the flowmeter to a flowing medium. Lack of a wet calibration effectively causes the flowmeter installation to be inferential in nature, regardless of flowmeter design, increasing the uncertainties associated with the flow measurement. This is due to the nature of an inferential device, in that an uncalibrated flowmeter is inserted into the flowstream and is assumed to perform in a certain manner, inferring the flow measurement.

A dry calibration assumes that the meter element is accurately described by empirical correlations developed from hydraulically similar meters in several flow laboratories. A wet calibration reduces the uncertainty of correlation between the actual meter dimensions and the relationship between the flow and the flowmeter output. The dry calibration is effectively an electronic (or pneumatic) calibration of the transmitter and not of the primary flow element itself.

Physical Dimensions Many flowmeters rely upon tight manufacturing tolerances to achieve sufficiently small deviations in production flowmeters such that all production flowmeters perform nearly equally and hence do not require wet calibration. Verification of flowmeter dimensions is virtually the only method of verifying calibration of inferential flowmeters short of having the flowmeter element hydraulically tested.

Electronic Techniques Electronic calibration techniques are used to calibrate flowmeter transmitters, independent of the primary flowmeter element. The assumption made here is that the primary element is already calibrated, either hydraulically or inferentially, and the flowmeter electronics need only be adjusted, within certain tolerances, to give an accurate output. Two adjustments, zero and span, are usually used to calibrate the flowmeter electronics, although others such as for scaling factors or an analog output may be necessary.

Zero Adjustment

The zero adjustment of the transmitter is used to adjust the output of the flowmeter to a signal that is equivalent to zero flow. This adjustment is performed by simulating an input to the transmitter that corresponds to a zero flow condition at the operating pressure and adjusting the zero potentiometer. Zero adjustment is critical to the proper operation of the flowmeter since inaccuracies in setting zero, as well as drift of the zero adjustment, will cause a shift in the flowmeter output for all flows. This results in a constant error over the entire flow range and can be expressed as an error that is a function of the full scale flow.

EXAMPLE 6-1

Problem: Determine the error attributable to zero adjustment and drift when the zero of a linear flowmeter can be adjusted and maintained to within ±0.5 percent of full scale.

Solution: Due to the nature of the zero adjustment, any error in setting the zero adjustment will cause a shift that will be constant at all flows. Therefore, the zero adjustement error is ±0.5 percent of full scale, the effects of which are illustrated in Figure 6-1.

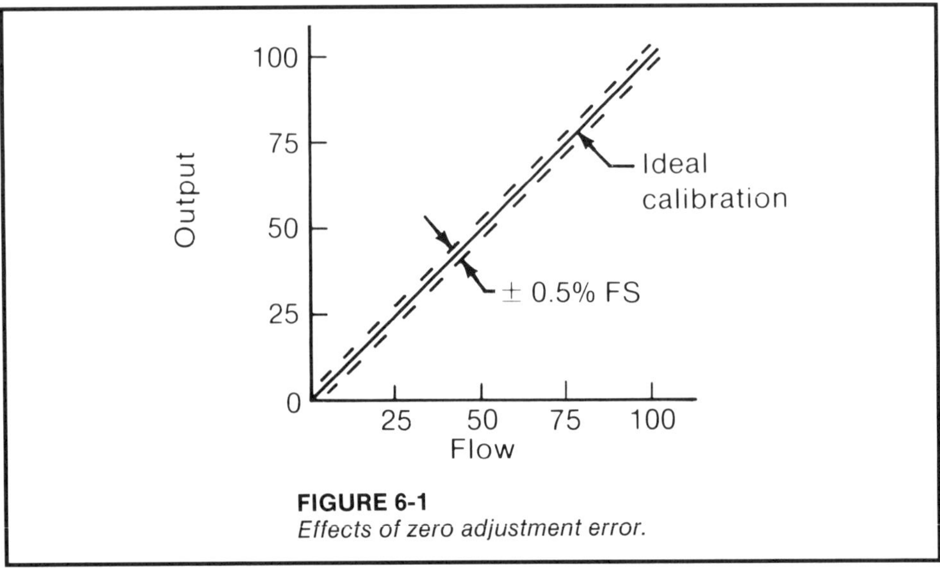

FIGURE 6-1
Effects of zero adjustment error.

Some transmitters, either by the nature of the flowmeter design or due to electronic techniques, do not require zero adjustment since the zero is well defined or zero shifts are automatically compensated for and generally no error is associated with their zero adjustment.

EXAMPLE 6-2

Problem: A flowmeter has an automatic zero feature that continually adjusts zero to compensate for any zero shifts that may occur during flowmeter operation. Calculate the resultant zero adjustment error.

Solution: There is no error associated with the zero adjustment.

Span Adjustment

The span adjustment of the transmitter is used to calibrate the output of the flowmeter to a signal that is equivalent to the full scale flow. This adjustment is performed by simulating an input to the transmitter that corresponds to a full scale flow condition at operating conditions and adjusting the span adjustment. Span adjustment is critical to the proper operation of the flowmeter because inaccuracies in setting span will cause an error in the flowmeter output that diminishes linearly with flow and can be expressed as an error that is a function of actual flow. While not preferred, the span can be adjusted at other than a full scale flow condition; however, the error associated with the span adjustment will generally be greater than that of calibration at full scale flow conditions.

Some transmitters that have frequency outputs do not require span adjustment, as the output frequency is the same as the input frequency from the flowmeter and generally no additional error is associated with the electronic span adjustment.

EXAMPLE 6-3

Problem: Determine the error attributable to span adjustment and drift when the span of a linear flowmeter can be adjusted and maintained to within ±0.5 percent of full scale.

Solution: As the span changes the slope of the calibration curve, any error in setting the span adjustment will cause a shift that will be a function of flow. Therefore, the span adjustment error is ±0.5 percent of rate, the effects of which are illustrated in Figure 6-2.

Scaling Factor Calibration Scaling factors are used to force the output of a flowmeter to read directly in engineering units and are usually associated with flowmeters that generate frequency outputs. Scaling factors are usually set with thumbwheel switches or potentiometers per manufacturer specifications.

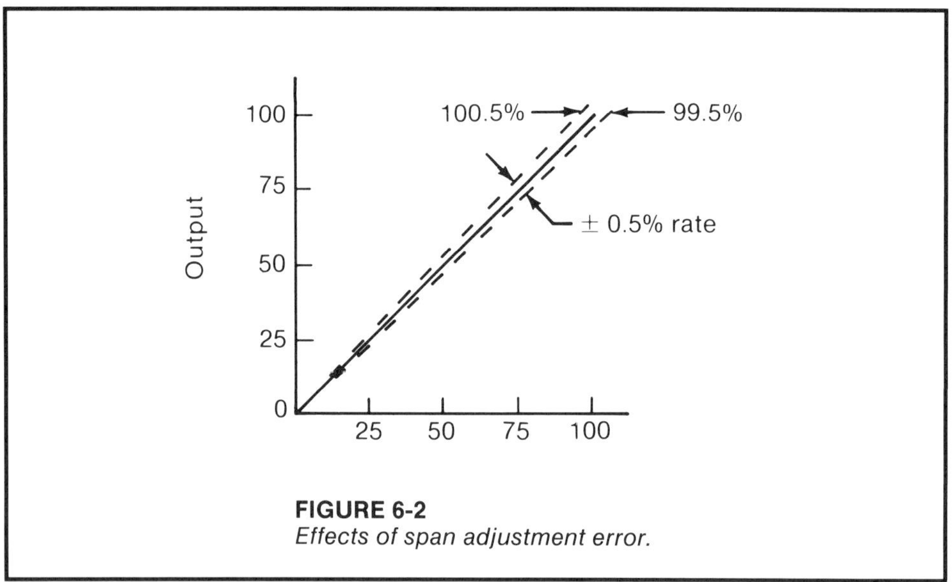

FIGURE 6-2
Effects of span adjustment error.

EXAMPLE 6-4

Problem: A digital flowmeter outputs 4 pulses per gallon of liquid that passes through the flowmeter. Calculate the scaling factor necessary to develop an output in which each pulse would represent 1000 gallons.

Solution:

$$(4 \text{ pulses/gallon}) \times 1000 \text{ gallons} = 4000 \text{ pulses}$$

From the above calculation, a scaling factor of 4000 is necessary such that one output pulse is generated for every 4000 pulses generated by the flowmeter.

Analog Output Calibration Calibration of an analog output is often required in addition to the flowmeter calibration, most often when the flowmeter inherently produces a frequency output that must be converted to an analog output. The transmitter can be thought of as having two stages, one dedicated to the primary flowmeter element and another in series dedicated to generating an analog output. The analog output introduces additional error due to the nonlinearities of the conversion and uncertainties involved in setting the zero and span. As a result, flowmeters may have up to 4 to 6 adjustments, depending upon design.

EXAMPLE 6-5

Problem: How many adjustments are there in a digital flowmeter that has an automatic zero feature with an analog output? An analog flowmeter with an analog output? A digital flowmeter without an automatic zero?

Solution: The primary flow measurement of a digital flowmeter with an automatic zero feature requires a span adjustment, while the frequency to current converter has both zero and span adjustments, for a total of 3 adjustments. Similarly, the analog flowmeter has a zero and span adjustment, while the digital flowmeter without an automatic zero feature has both zero and span adjustments for both the primary flow measurement and the frequency to current converter, resulting in a total of 4 adjustments.

Verification of Operation

It is often desired to verify the operation of a flowmeter system before putting it in service, after having calibrated all of the components and connecting the system. This is possible with some technologies by simulating flow conditions at a point as close as possible to the flowmeter sensing system, taking care not to damage the flowmeter during simulation. This can often be performed by flowing a compatible fluid through the flowmeter and verifying that the flowmeter responds.

EXAMPLE 6-6

Problem: What are the disadvantages of verifying flowmeter operation by operating a flowmeter designed for liquid service with compressed gas?

Solution: This procedure can cause damage to some flowmeters.

EXERCISES

6.1 Is a flow calibration facility traceable to the National Institute of Standards and Technology?

6.2 Why is liquid flow measurement more accurate than gas flow measurement?

6.3 What Reynolds numbers can be simulated in a flow laboratory?

6.4 What type of calibration errors are introduced by zero calibration errors and zero drift? Span calibration errors and span drift?

7

Measurement of Flowmeter Performance

Introduction One of the factors upon which flowmeter selection is based is flowmeter performance. As in any industry, manufacturers tend to represent their products in a manner that best presents that product. An understanding of the measures of flowmeter performance is necessary to properly interpret and evaluate flowmeter performance specifications.

Applicable Range Flowmeters measure accurately over a range of flows, but in most applications the *accurate* measurement range is not the same as the *desired* measurement range. Turndown is a measure of the range over which a flowmeter will perform accurately; it will vary with flowmeter technology, manufacture, and application. For example, the turndown of a linear flowmeter that measures accurately from 25 to 100 percent of full scale output is 4:1, though the flowmeter may measure lower flows without stated accuracy. Note also that a flowmeter that generates an accurate squared output from 25 to 100 percent of full scale will only have a 2:1 turndown after the square root is taken. Finally, a flowmeter with a 12:1 turndown based upon the flowmeter capacity will result in only 3:1 turndown when full scale flow is set at 25 percent of meter capacity.

As can be seen from the above, meaningful turndown specifications are in terms of flowmeter full scale flow so as to represent the accurate measurement range of a flowmeter to the *application at hand.* Turndown must be examined carefully to determine over what range flow measurement is accurately performed in relation to the application, instead of in relation to extreme operating conditions that are stated by manufacturers but are seldom encountered.

Flowmeter Composite Accuracy The composite accuracy of a flowmeter includes the effects of accuracy, linearity, and repeatability of the flowmeter at reference conditions and represents the ability of the output of a calibrated flowmeter to correspond to the flow of a fluid through the flowmeter.

The type of flowmeter accuracy statement is typically dependent upon calibration errors that can be expressed in general as

$$\text{calibration error} = \frac{\text{zero error}}{\text{span}} + \frac{\text{span error}}{\text{span}}$$

$$= \%\ \text{full scale} + \%\ \text{rate}$$

Flowmeters that have both zero and span errors are typically specified with a full scale accuracy statement that accurately describes flowmeter performance.

When the zero flow condition is well defined, the zero error is zero and the first term in the above equation drops out, resulting in a percentage of rate accuracy statement. This is more desirable than the full scale statement described above.

Accuracy is stated at reference temperature and ideal flow conditions, which are usually not duplicated in actual industrial applications at nominal operating conditions. Corrections can be made in flowmeter calibration to compensate for inaccuracies introduced as a result of operation at nominal operating conditions, as opposed to reference operating conditions.

Often flowmeter accuracy is specified in terms of probabilities, although manufacturer specifications rarely state this. This is most common in flowmeters that are not wet calibrated to determine their true operating characteristics. An example of this is a flowmeter that will perform with ± 1.5 percent rate accuracy with a 95 percent confidence level. Therefore, the performance of approximately 5 percent of these flowmeters will not be within the stated accuracy.

Flowmeter accuracy is also affected by fluctuations in operating conditions and tolerances in flowmeter construction. These parameters are considered in the following subsections as part of calculations of the overall flowmeter system accuracy. When a composite accuracy statement is not available, accuracy, linearity, and repeatability can be considered individually in calculating overall flowmeter system accuracy.

Transmitter Accuracy Transmitters serve the function of converting signals produced by the primary flowmeter element into standard signals that can be transmitted to standard instrumentation.

Transmitters that convert analog signals usually have a zero adjustment, the net result being that transmitter specifications are typically expressed as a percent of full scale. Digital flowmeter transmitters will also introduce some percentage of full scale error if an analog output signal is desired; however, when the output pulses can be used directly, the transmitter will have no zero adjustment and, hence, a percent of rate accuracy specification.

Microprocessor-based transmitters offer improved stability of the zero and span adjustments, potentially offering superior long-term accuracy in addition to built-in diagnostics. "Smart" microprocessor-based transmitters offer remote digital recalibration of the transmitter, as well as other convenient features.

EXAMPLE 7-1

Problem: What are typical flowmeter transmitter accuracy specifications for an analog output signal?

Solution: Analog and digital flowmeter transmitter specifications are typically between 0.2 and 0.5 percent full scale and between 0.1 and 0.2 percent full scale, respectively.

Linearization Accuracy Analog computing devices used to linearize signals have associated accuracy statements that are usually expressed as a percentage of full scale since the zero and span are adjustable. While the percentages may appear to be small, the errors can be significant at low flows. These devices are subject to drift with time and varying operating conditions, and it should be noted that inaccuracies due to power supply voltage fluctuations and temperature can be greater than the calibration accuracy of the device. Also, most manufacturers fail to specify the long-term stability of the device. In short, it can often be difficult to quantify the inaccuracies of analog conversion devices over the long term.

Squared output meters are linearized by taking the square root of the output signal. Without linearization, for example, controlling flow over a large flow range requires careful controller tuning, since percentage changes in flowmeter signal produce different magnitudes of correction depending on whether the flow is in the upper or lower portion of the flow range.

If the raw transmitter signal requires linearization, additional errors introduced by the linearizing device can be significant. It is important to recognize this when considering the overall measurement accuracy for a specific application. Many digital flowmeters are sufficiently linear to require little or no linearization of the transmitted signal when the flowmeter is properly designed and operated within the applicable range.

Digital Conversion Accuracy When computer manufacturers boast of the inherent accuracies of digital systems, they usually refer to the "exact" calculations that are performed internally. The errors encountered in the conversion of an analog or frequency signal to a series of binary bits, which can be used internally to interface a computer or microprocessor-based device, can be significant. The length of this string of bits is not to be confused with the length of the words used internally to the computer. For example, a 16-bit computer may use an 8-bit analog-to-digital converter to interface to the flowmeter signal, or vice versa.

Devices that are used to perform the above conversions have practical as well as theoretical limitations. In addition to drift errors that can occur, the number of binary bits in the conversion is limited, and an error results because the series of binary bits produced has a fixed resolution due to its digital nature. For a linear flowmeter, this can be represented as:

$$\text{conversion accuracy} = \text{converter accuracy}/2^n$$

where n is the number of bits in the conversion. The error is a percent of full scale as it is the same for all flows.

EXAMPLE 7-2

Problem: For a linear flowmeter, find the conversion error of an 8-bit analog-to-digital converter that has an accuracy of ±1 bit.

Solution:

$$\text{conversion accuracy} = \pm 1\ \text{bit}/2^8$$
$$= \pm 0.39\%\ \text{FS}$$

It can be seen that this conversion results in significant errors. If, however, a 12-bit conversion were performed, the conversion error would be 1/16 of the above, or ±0.024 percent FS.

Nonlinear flowmeter outputs that are fed directly into an analog-to-digital converter have the same absolute conversion error calculated above. This error, however, refers not to the flow, as above, but to the output of the flowmeter. The effect of linearization on the flow can vary with the severity of the nonlinearity at the measured flow and can be determined by calculation.

EXAMPLE 7-3

Problem: Determine the effect of a resolution error of ±0.4 percent FS on a squared input signal if the square root extraction is performed with negligible error.

Solution: Results of calculations are summarized in the following table.

Input (Q^2)	Resolution Error	Flow (Q)	Resolution Error
100.00%	0.4%	100.20%	0.20% FS (0.20% rate)
56.25%	0.4%	75.27%	0.27% FS (0.36% rate)
25.00%	0.4%	50.40%	0.40% FS (0.80% rate)
6.25%	0.4%	25.79%	0.79% FS (3.16% rate)
1.00%	0.4%	11.83%	1.83% FS (18.3% rate)

Note: The resolution errors at 50 percent flow approach almost one percent of rate and increase significantly at lower flowrates. A 12-bit conversion would reduce these errors by a factor of 16.

Indicator Accuracy Indicators, depending on the technology employed, have accuracies that typically vary from 0.5 to 5 percent of full scale and are therefore significant in the overall measurement of flow. If a nonlinear signal is to be indicated directly on an indicator with a nonlinear scale, the flow error can be tabulated similarly to the methods used for conversion errors that were calculated in the table in Example 7-4. Computers and microprocessor-based devices that have a numerical representation of flow introduce no error in addition to the digital conversion error, unless the flow is rounded off.

EXAMPLE 7-4

Problem: Calculate the rate error associated with an accurate indicator with a squared scale and an accuracy of ±2 percent full scale.

Solution: The rate error associated with a ±0.5 percent full scale error is summarized in the following table.

Input	Indication Error	Indication	Rate Error
100.00%	0.50%	100.5%	0.5%
56.25%	0.67%	75.7%	0.9%
25.00%	1.00%	51.0%	2.0%
6.25%	2.00%	27.0%	8.0%
1.00%	5.00%	15.0%	50.0%

Note: The indicator introduces significant error to the measurement at low flows. If a less accurate indicator with an accuracy of ±2 percent full scale were used, the resultant rate error would increase the above error fourfold.

Totalization Accuracy Totalizer accuracy is usually expressed as a percent of full scale since totalizers typically have zero adjustments; however, no error is introduced when a digital totalizer counts pulses generated by a digital flowmeter. Selection of a digital flowmeter and counter totalization system can improve the overall accuracy of flow totalization by eliminating the approximately 0.5 percent full scale error caused by a typical analog totalizer.

Overall Flowmeter System Accuracy The accuracy of the overall measurement is a function of the errors previously presented. The maximum error that can result in a measurement can be obtained by summing each error in the system, but realistically this will not occur often due to the low probability of all of the components being in error by the maximum amount in the same direction. The problem of overall system accuracy then becomes one of determining an accuracy within which the overall system will perform an acceptable percentage of the time. One method of estimating the overall system accuracy is to calculate the root mean square average as follows:

$$\text{system accuracy} = \pm [\Sigma (X_i \times \text{accuracy}_i)^2]^{1/2}$$

where X_i is the sensitivity coefficient associated with each source of error in the system. The sensitivity coefficient expresses the relationship of how the error generated by a device affects the flow measurement, a few of which are tabulated below for the instruments in a typical compensated gas flow measurement.

Device	Linear Flowmeter	Squared Output Flowmeter
Transmitter	1	1/2
Multiplier/divider	1	1/2
Linearizer	—	1
Indicator	1	1
Totalizer	1	1

The system accuracy calculated above does not take into account bias errors that are not compensated for, such as Reynolds number effects, temperature and pressure effects in noncompensated systems, and the like.

EXAMPLE 7-5

Problem: Estimate the expected accuracy that will be achieved given the following parameters of a linear and a squared output flowmeter that are operated accurately over a 10:1 flow range in liquid service.

Device	Linear	Square Output
Flowmeter element	±0.75% rate	±0.75% rate
Transmitter	±0.1% FS	0.2% FS
Linearizer	NA	±0.25% FS
Indicator	±0.5% FS	0.5% FS

Solution: The accuracy of the linear flowmeter is tabulated as follows:

Instrument	Accuracy (Rate)	X	$(X \times \text{Accuracy})^2$
Flowmeter	±0.75%	1	0.5625
Transmitter	$\pm 0.1\%\ Q_{FS}/Q$	1	$0.01\ (Q_{FS}/Q)^2$
Indicator	$\pm 0.5\%\ Q_{FS}/Q$	1	$0.25\ (Q_{FS}/Q)^2$
Total			$0.5625 + 0.26\ (Q_{FS}/Q)^2$

The accuracy of the linear flowmeter system is

$$\pm\,[0.5625 + 0.26\ (Q_{FS}/Q)^2]^{1/2}$$

The accuracy of the squared output flowmeter is tabulated as follows:

Instrument	Accuracy (Rate)	X	$(X \times \text{Accuracy})^2$
Flowmeter	±0.75%	1	0.5625
Transmitter	$\pm 0.2\%\ Q_{FS}/Q$	1/2	$0.01\ (Q_{FS}/Q)^2$
Square root extractor	$\pm 0.25\%\ Q_{FS}/Q$	1	$0.0625\ (Q_{FS}/Q)^2$

Indicator	$\pm 0.5\%\ Q_{FS}/Q$	1	$0.25\ (Q_{FS}/Q)^2$
Total			$0.5625 + 0.3225\ (Q_{FS}/Q)^2$

The accuracy of the squared output flowmeter system is

$$\pm\ [0.5625 + 0.3225\ (Q_{FS}/Q)^2]^{1/2}$$

The rate accuracies associated with each system are tabulated as follows:

Flow	Linear Flowmeter	Squared Output Flowmeter
10%	±5.15%	±5.81%
25%	±2.17%	±2.43%
50%	±1.27%	±1.38%
75%	±1.01%	±1.07%
100%	±0.91%	±0.95%

The accuracy of the overall system can be significantly poorer than that of the flowmeter primary. Note that even though flowmeter accuracy is ±0.75 percent rate, other errors have dramatic effects at low flows. Linear flowmeter accuracy can be seen to be slightly superior to that of an equal squared output flowmeter. System error can be reduced by using a computer or microprocessor-based control system, which would reduce the 0.25 percent FS and 0.5 percent linearizer and indicator errors, respectively, to an estimated 0.1 percent FS conversion error. It should be noted that uncertainties associated with the process data can easily add ±5 percent or more to the overall system uncertainty.

A more complete analysis of flowmeter system accuracy for a pressure- and temperature-compensated squared output flowmeter in gas service is tabulated below to illustrate the complexity to which flowmeter accuracy can be analyzed.

Influence Factors	Accuracy (% Rate)	X	$(X \times \text{Accuracy})^2$
Flowmeter coefficient	±0.75	1	0.5625
Pipe diameter (β = 0.75)	±0.20	$\frac{-2\ \beta^4}{1 - \beta^4}$	0.0343
Bore diameter (β = 0.75)	±0.07	$\frac{2}{1 - \beta^4}$	0.0419
Gas expansion factor	±0.07	1	0.0049
Differential pressure	±0.60	1/2	0.0900
Flowing density	±0.15	1/2	0.0056
Pressure (absolute)	±0.50	1/2	0.0625

Influence Factors	Accuracy (% Rate)	X	$(X \times \text{Accuracy})^2$
Temperature (absolute)	± 0.25	1/2	0.0156
Compressibility	± 0.20	1/2	0.0100
Transmitter	$\pm 0.2\ Q_{FS}/Q$	1/2	$0.01\ Q_{FS}/Q$
Square root extractor	$\pm 0.25\ Q_{FS}/Q$	1	$0.0625\ Q_{FS}/Q$
Indicator	$\pm 0.5\ Q_{FS}/Q$	1	$0.25 Q_{FS}/Q$
Total			$0.8273 +$ $0.3225\ Q_{FS}/Q$

The accuracy of this flowmeter system is

$$\pm\,[0.8273 + 0.3225\ Q_{FS}/Q]^{1/2}$$

which does not take into account inaccuracies due to operational conditions that do not correspond to reference operating conditions and uncertainties associated with process data.

EXERCISES

7.1 Under what conditions is the accuracy of a device stated?

7.2 What is the approximate additional error that is introduced when an analog output is required from a digital flowmeter?

7.3 Why do analog flowmeters usually have accuracy statements stated as a function of full scale?

7.4 Why do digital flowmeters usually have accuracy statements expressed as a function of rate?

7.5 Why is the use of a linearizer conceptually undesirable?

7.6 Calculate the conversion accuracy of a 10-bit analog-to-digital converter.

7.7 Compare the accuracies of digital and analog flow totalization systems given the following:

Instrument	Linear	Squared
Flowmeter	±1% rate	±1% rate
Transmitter	NA	±0.2% FS
Totalizer with linearizer	NA	±0.5% FS
Counter	±1 digit	NA

8 Miscellaneous Considerations

Introduction There are a number of important aspects of flowmeter selection that go beyond the specific flowmeter technology. Safety is often paramount in the selection of the proper flowmeter for an application. Improper materials of construction can result in high maintenance and replacement costs, loss of calibration, etc. Thermal expansion that is unaccounted for and improper upstream pipe hydraulics can also result in measurement error. All of these factors must be taken into account in the flowmeter selection and installation and in assessing the installed uncertainty.

Materials of Construction Necessity dictates that a flowmeter be constructed of materials that are compatible with the fluid being measured at the operating conditions at which the fluid is flowing. Depending on the service, flowmeters are susceptible to corrosion, abrasion, contamination, and failures due to excessive pressure or temperature. Gasket and seal materials as well as metals must be considered in the analysis.

Corrosion Care must be taken to ensure that flowmeter calibration will not change due to corrosion. In an extreme case, corrosion may result in complete loss of service or in a spill or leak. Materials specified must satisfy the piping specifications as a minimum. However, because metering elements are more sensitive to dimension changes, often the wetted parts of the flowmeter element, the transmitter, should be specified with even more corrosion-resistant materials.

Figure 8-1 represents the corrosion rate of a metal due to attack by an acid at various concentrations and at various temperatures. Note that the metal exhibits good resistance to the acid at both high and low concentrations; however, between these concentrations, the corrosion rate is clearly unacceptable. Note also that the corrosion rate typically doubles with each 10°C increase in temperature.

Material compatibility of commonly used substances can be determined from tables and graphs provided by flowmeter manufacturers and in chemical

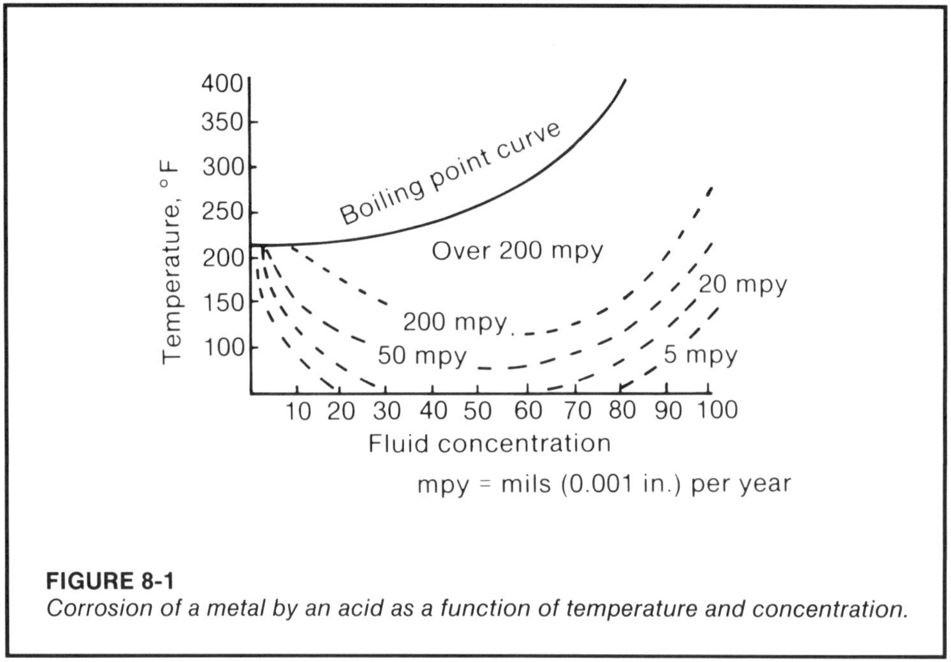

FIGURE 8-1
Corrosion of a metal by an acid as a function of temperature and concentration.

handbooks and by experience. Extreme care should be taken to ensure that the tables or graphs apply to the application at hand, as they are often condensed and contain information from which compatibility is inferred. If there is any doubt, further investigation should be done to avoid potential problems and hazards. If necessary, a corrosion test should be performed.

It should be noted that not all grades of a given material may be applicable to a given application. For example, some grades of stainless steel may exhibit low corrosion in a given service, while another grade may exhibit an unsatisfactorily high corrosion rate in the same service. Information regarding material compatibility of the fluid with specific material grades should be solicited from knowledgeable sources so as to avoid potential problems. For example, the desired information is not whether a fluid is compatible with stainless steel, but rather whether the fluid is compatible with specific grades or perhaps all grades of stainless steel.

Determining whether a fluid is compatible with the materials of construction is generally the responsibility of the person specifying the flowmeter. Get as much operating and maintenance history from other instruments in the same service as possible. Corrosion, especially in primary measuring elements, is difficult to assess because rates as low as 0.002 in./yr may be detrimental to the measured accuracy. In some cases an experimental or trial and error basis must be considered.

All wetted parts, including lubricants, gaskets, seals, fittings, bearings, and the body, must be considered for material compatibility and as a minimum should conform to the material specification for piping and valves.

Abrasion Abrasion results when the fluid is abrasive or contains solids that contact flowmeter components and erode them due to mechanical contact between the fluid and the flowmeter. When the fluid is abrasive, hardened materials of construction may be selected to minimize abrasion. An obstructionless flowmeter design that has no restrictions may be used so as to minimize fluid movement against flowmeter components and hence minimize abrasion.

Pressure and Temperature The flowmeter must be capable of functioning under the operating pressures and temperatures of the fluid. This applies not only to the flowmeter body and all wetted parts but also to any attached components, such as sensors, which are not directly in the flow stream but may have pressure or temperature limitations. The flowmeter can only be used until its weakest component fails, so consideration must be given to all of the above parameters.

Flange Ratings When the fluid exceeds the pressure and temperature ratings of the flanges, the flange connection will probably blow a gasket before the flowmeter leaks, cracks, or explodes, thereby causing a hazardous condition.

The pressure-temperature relationships of ANSI flanges rated at 150, 300, and 600 psi are illustrated in Figure 8-2. Note that pressures can exceed the nominal flange pressure ratings if the temperature is low.

EXAMPLE 8-1

Problem: The maximum pressure and temperature of a fluid are 200 psi and 150°F, respectively. Does this application require 150, 300, or 600 psi rated flanges?

Solution: An ANSI flange rated at 150 psi can handle the 200 psi pressure at over 300°F. Purchasing a flowmeter with flanges rated at 150 psi instead of 300 psi is technically correct for this service and can reduce the cost of the flowmeter substantially.

Contamination In certain processes, the fluid passing through the flowmeter can be contaminated by the materials contained in the wetted parts of the flowmeter. Individuals familiar with the process should be able to determine if this is the case and suggest alternate materials of construction, if necessary. This is particularly important in food grade services.

EXAMPLE 8-2

Problem: Carbon leaches out of carbon steel, contaminating the flowing fluid and adversely affecting the quality of the product. What can be done to avoid this problem if a flowmeter is to be installed in the pipe?

Solution: From the above information, low carbon materials should be used in the flowmeter. This includes not only the flowmeter body but also the bearings, gaskets, seals, and the like. Some grades of stainless steel may be compatible with the fluid, but it is usually necessary to have a materials and corrosion specialist identify the permissible options.

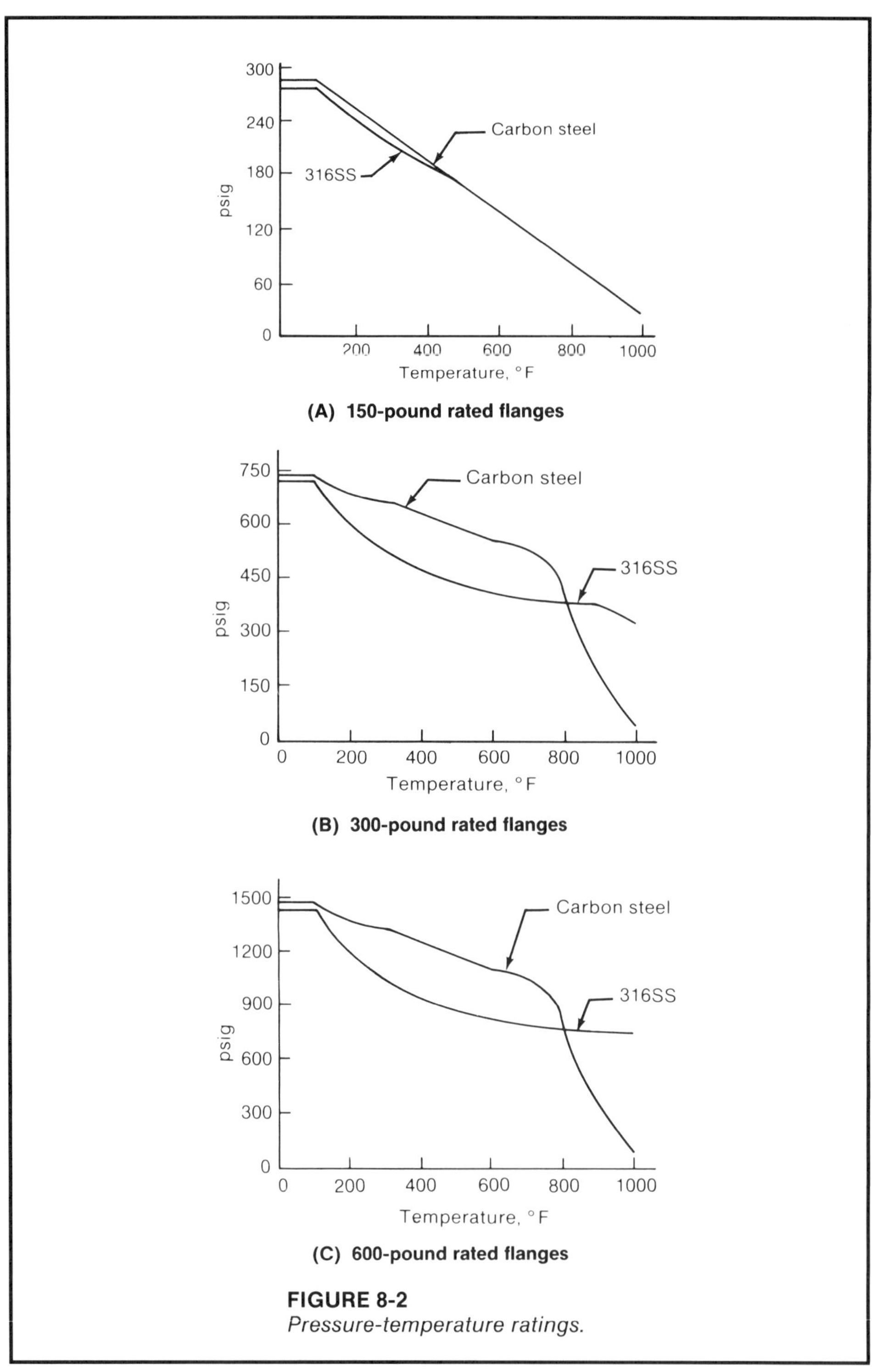

(A) 150-pound rated flanges

(B) 300-pound rated flanges

(C) 600-pound rated flanges

FIGURE 8-2
Pressure-temperature ratings.

Piping Considerations Various piping considerations applicable to flowmeters are virtually independent of the technology employed to perform the measurement. Flowmeter performance is stated under essentially ideal reference conditions, and failure to reproduce those conditions compromises flowmeter performance. In some flowmeter designs, high quality machined pipe should be used due to manufacturing tolerances of standard pipe, which can cause variances in the inside diameter of the pipe (and hence the cross-sectional area) and can result in significant flow measurement uncertainty.

Pipe Hydraulics Many flowmeters are sensitive to the flow profile in the pipe, which necessitates having a predictable flow profile at the flowmeter inlet. This is accomplished by positioning straight runs of pipe upstream and downstream of the flowmeter to provide straight pipe without obstructions or bends so as to set up flow profiles that are predictable and not distorted, thereby minimizing flow profile effects. When a flowmeter cannot be located with sufficient straight run, a flow conditioner can be used to reduce upstream straight run requirements by reducing swirl and distortion, as shown in Figures 8-3, 8-4, 8-5, and 8-6.

It should be noted that pipe hydraulics can be altered by other factors such as jetting through obstructions such as valves, foreign objects in the pipe, build-up of solids in the pipe, and protrusions into the pipe by such things as misaligned gaskets.

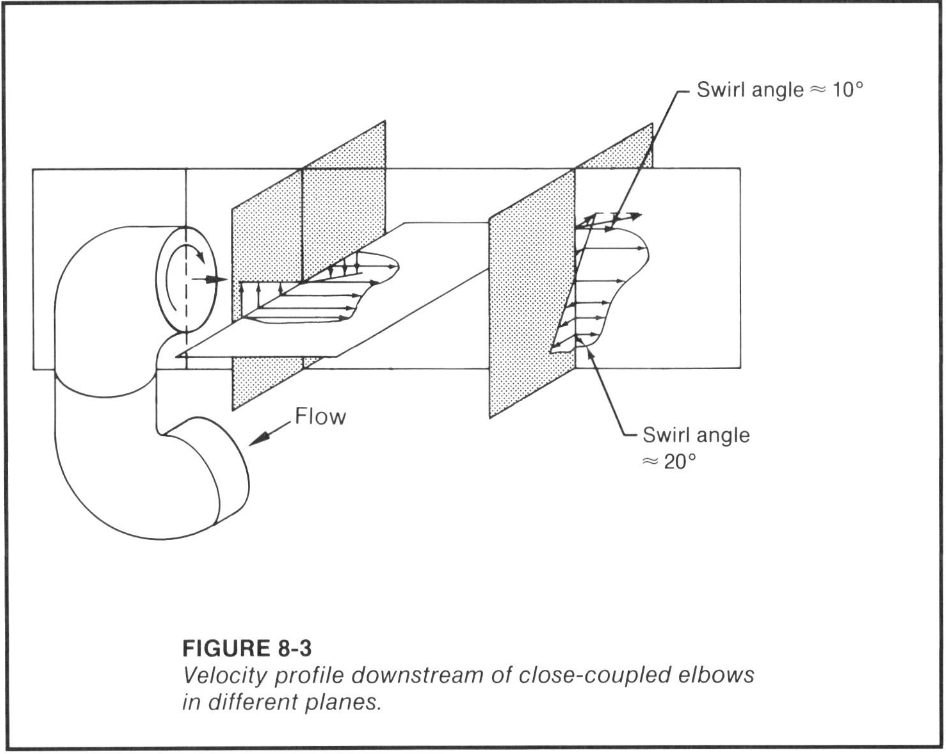

FIGURE 8-3
Velocity profile downstream of close-coupled elbows in different planes.

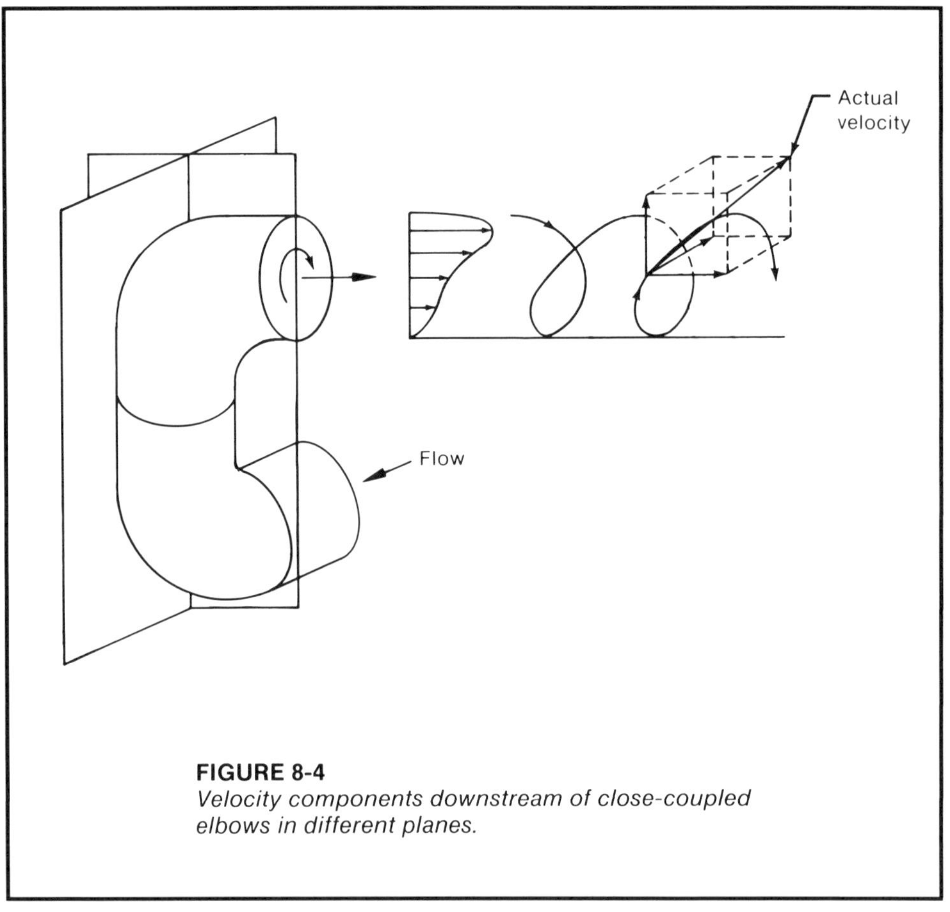

FIGURE 8-4
Velocity components downstream of close-coupled elbows in different planes.

Orientation Liquid flowmeter installations should be in an orientation that ensures that the flowmeter remains full of liquid when a measurement is desired (see Figure 8-7). This can be accomplished by locating the flowmeter in a submerged leg of the piping system or in a portion of the pipe in which the fluid is rising. Mounting a flowmeter in a pipe flowing downwards can cause significant measurement error in most applications and should be avoided.

Gas or vapors present in the liquid will adversely affect the flow measurement and possibly damage the flowmeter. Air eliminators are available and can be installed upstream of the flowmeter to provide some degree of immunity to the problem of entrained gases.

Care should be taken in liquid applications where the pipe can drain the flowmeter empty when a pump or similar device is turned off. As there is no liquid downstream of the flowmeter, the initial flow of liquid that reaches the flowmeter when the pump is turned on can have sufficient momentum to damage the flowmeter, not to mention the damage to piping. This can be avoided by ensuring

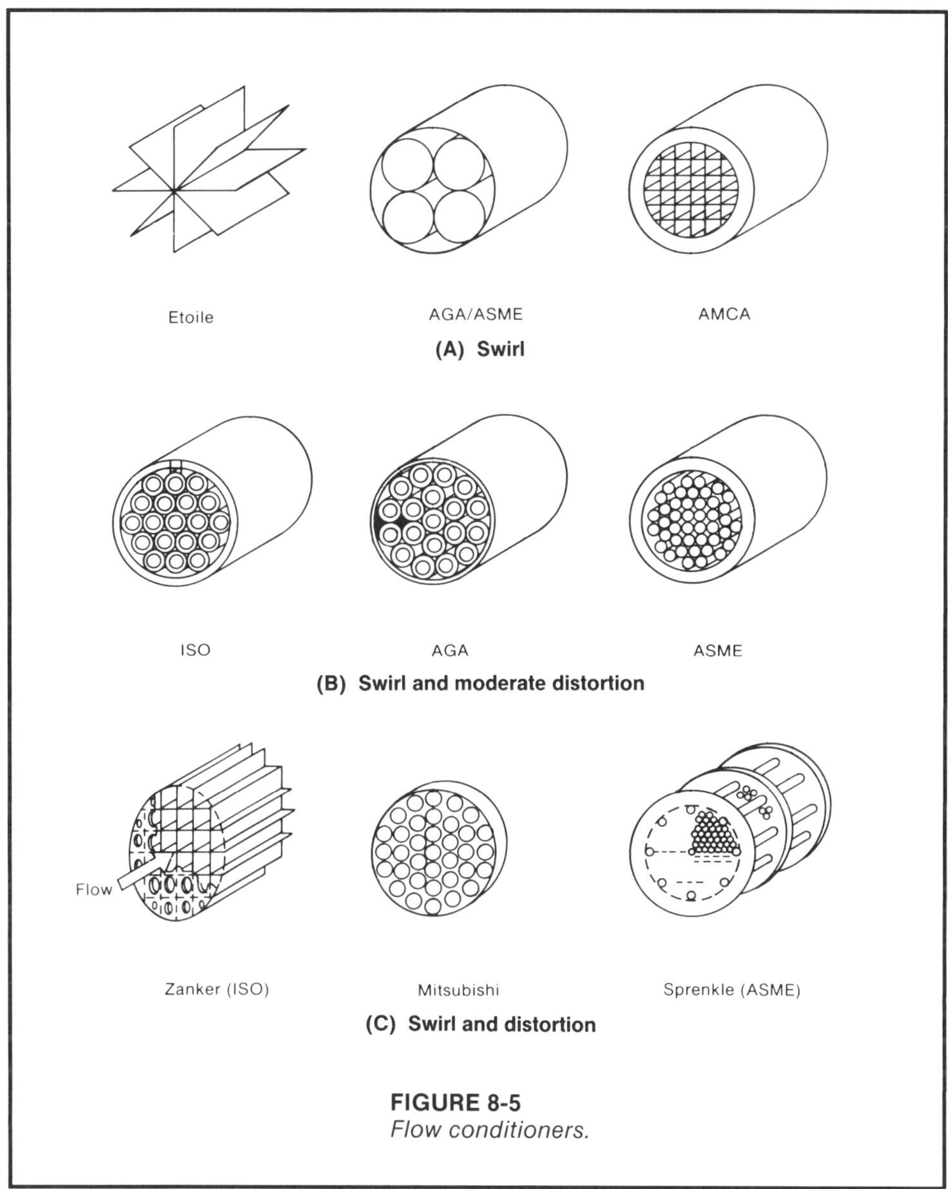

FIGURE 8-5
Flow conditioners.

that the flowmeter is always filled with liquid, or that the initial flow that fills the pipe is turned on slowly. These requirements must be met by proper startup procedures, documentation, and piping design.

Most flowmeters are unidirectional, so the flowmeter must be mounted with the proper orientation in the pipe. Typically, there is a flow direction arrow either cast into or attached to the flowmeter body to indicate the proper flow direction through the flowmeter.

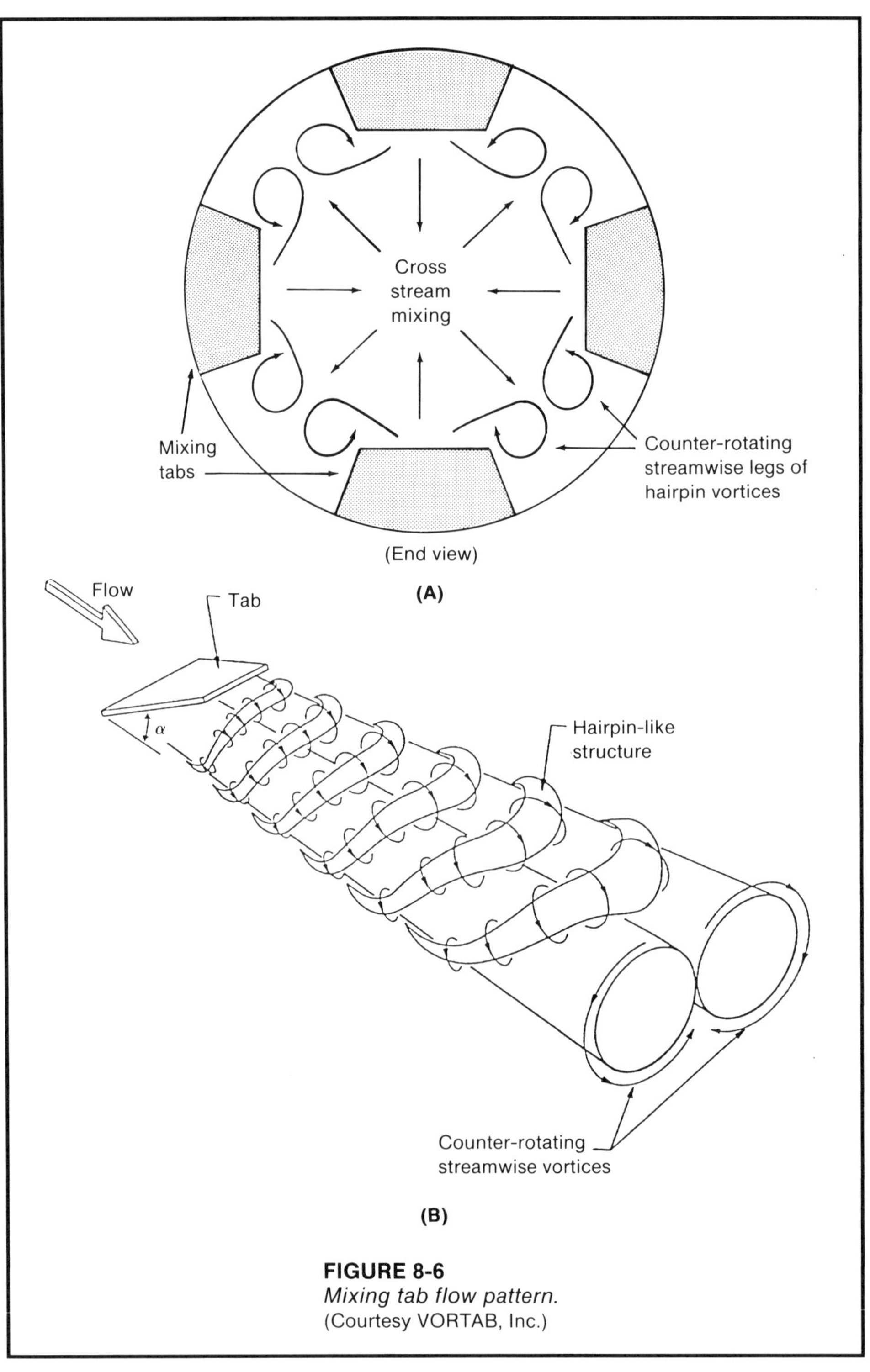

FIGURE 8-6
Mixing tab flow pattern.
(Courtesy VORTAB, Inc.)

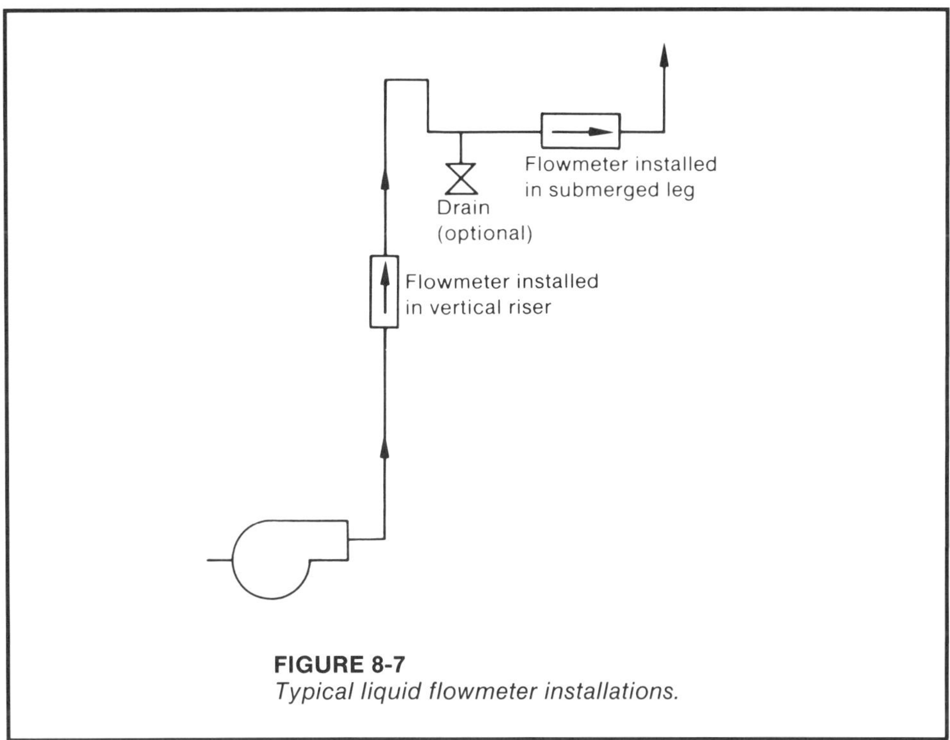

FIGURE 8-7
Typical liquid flowmeter installations.

Bypass Piping Bypass piping may be desirable when the flowmeter cannot be taken out of service in the case of failure or for maintenance (see Figure 8-8). It should be noted that the need for bypass valving is determined not only by the construction of the flowmeter but also by the nature of the process itself.

Hydrotest Considerations Often, portions of a flowmeter are not rated, nor are they compatible with the pressures, materials, or methods used to hydrotest a pipe. If this is the case, the flowmeter should be either valved out of service or removed from the line during hydrotesting to avoid damage.

Dirt Dirt that is present in the process fluid can damage or plug some flowmeters that are designed with high tolerance moving parts or small passageways. Strainers or filters can be located upstream of the flowmeter to protect it against this possibility.

Coating Coating, a phenomenon that can be present in liquid or gaseous streams, can result in changes in flowmeter geometry that can affect measurement accuracy. When it is suspected that a fluid can cause coating, a flowmeter that minimizes coating effects should be selected. Although coating can cause some flowmeters to cease to operate, in most cases coating causes measurement errors that are not apparent unless the wetted parts of the flowmeter are examined.

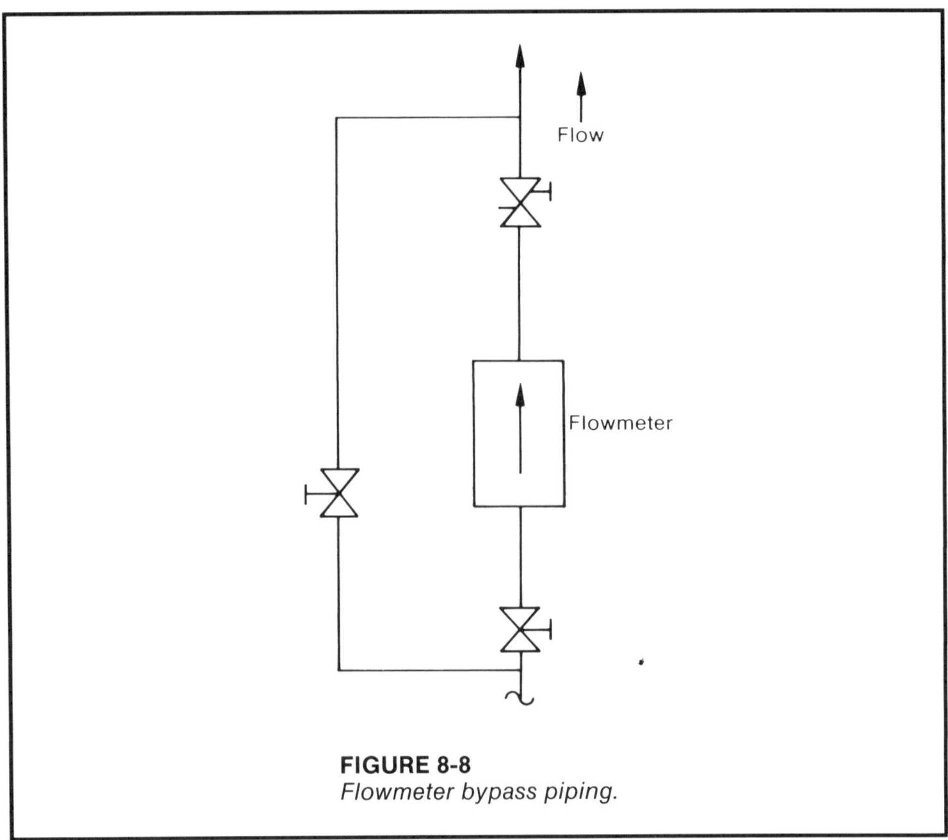

FIGURE 8-8
Flowmeter bypass piping.

Safety Safety must be considered of paramount importance in selecting a flowmeter. Even flowmeters on safe fluids can be dangerous to personnel and property under certain operating pressures and temperatures.

The flowmeter body and flanges must be sufficiently rated for the service and any failure modes. Flowmeters that do not require gaskets or seals, such as completely welded assemblies, usually are preferred where safety may pose a problem. Where safety is a concern, the flowmeter should be so designed that two seals or gaskets must fail before fluid is allowed to escape. The safety of a flowmeter is checked by examining requirements for various modes of failure, such as leakage.

Leakage The possibility of leakage poses a serious problem when the fluid to be measured is hazardous or lethal. Materials of construction must be examined with extreme care, and they must exhibit low corrosion and erosion under operating conditions. A design with no seals or gaskets is conceptually preferable, but most designs utilize redundant seals or gaskets, both of which must fail before the fluid can escape. Often, specially machined flanges are required to properly seal the flowmeter with no leaks.

It should be noted that not all leakage problems are safety-related. Leakage of a flowmeter element in virtually any service can result in failure of the transmitter that is exposed to the fluid.

Area Electrical Classification All instruments and electrical devices must be designed and installed to meet any hazardous area classifications that may apply in regard to fire safety. Information is generally available from flowmeter manufacturers regarding device suitability in hazardous areas and how the device meets this requirement.

Lubricants and Contamination Some applications require special handling and cleaning of all parts in contact with the fluid. The most notable such fluid is oxygen, and oxidants in general, where a fire or an explosion can result if residual cutting oils, etc., from manufacturing come in contact with the fluid. Flowmeters for this service must use special lubricants and must be specially cleaned. Some materials are even capable of reacting with fluorocarbons, so caution is advised. The flowmeter flanges should be sealed, backfilled with inert dry gas, and not opened until installation, at which time the flowmeter should be inspected (preferably with ultraviolet light) for dirt, fingerprints, or contamination. If there is any sign of contamination during inspection, the flowmeter should be properly cleaned with a compatible solvent before installation.

Thermal Expansion

Most flowmeters measure or infer fluid velocity, from which the flow is calculated based upon the area of the flowmeter through which the fluid passes. Due to the cubical coefficient of expansion, materials from which a flowmeter is constructed will expand and contract with varying temperature. Therefore, the effective area through which the fluid passes can vary with temperature, although usually in a predictable manner. Errors are introduced as the temperature varies from the flowmeter reference temperature.

The flowmeter can be scaled to correct the measurement for the nominal operating temperature in order to minimize error. This is usually sufficient to correct the majority of applications, but when the operating temperature varies significantly from the nominal temperature, compensation may be required.

EXAMPLE 8-3

Problem: Calculate the correction required when operating a flowmeter with a temperature coefficient of 0.3%/100° F, at nominal temperature of 165° F.

Solution: Assuming a 75° F reference temperature, the correction is:

$$\begin{aligned} \text{correction} &= (165°\text{F} - 75°\text{F}) \times 0.3\%/100°\text{F} \\ &= 0.27\% \end{aligned}$$

Wiring

Flowmeter field wiring is typically of a 2-wire, 3-wire, or 4-wire design, as shown in Figure 8-9.

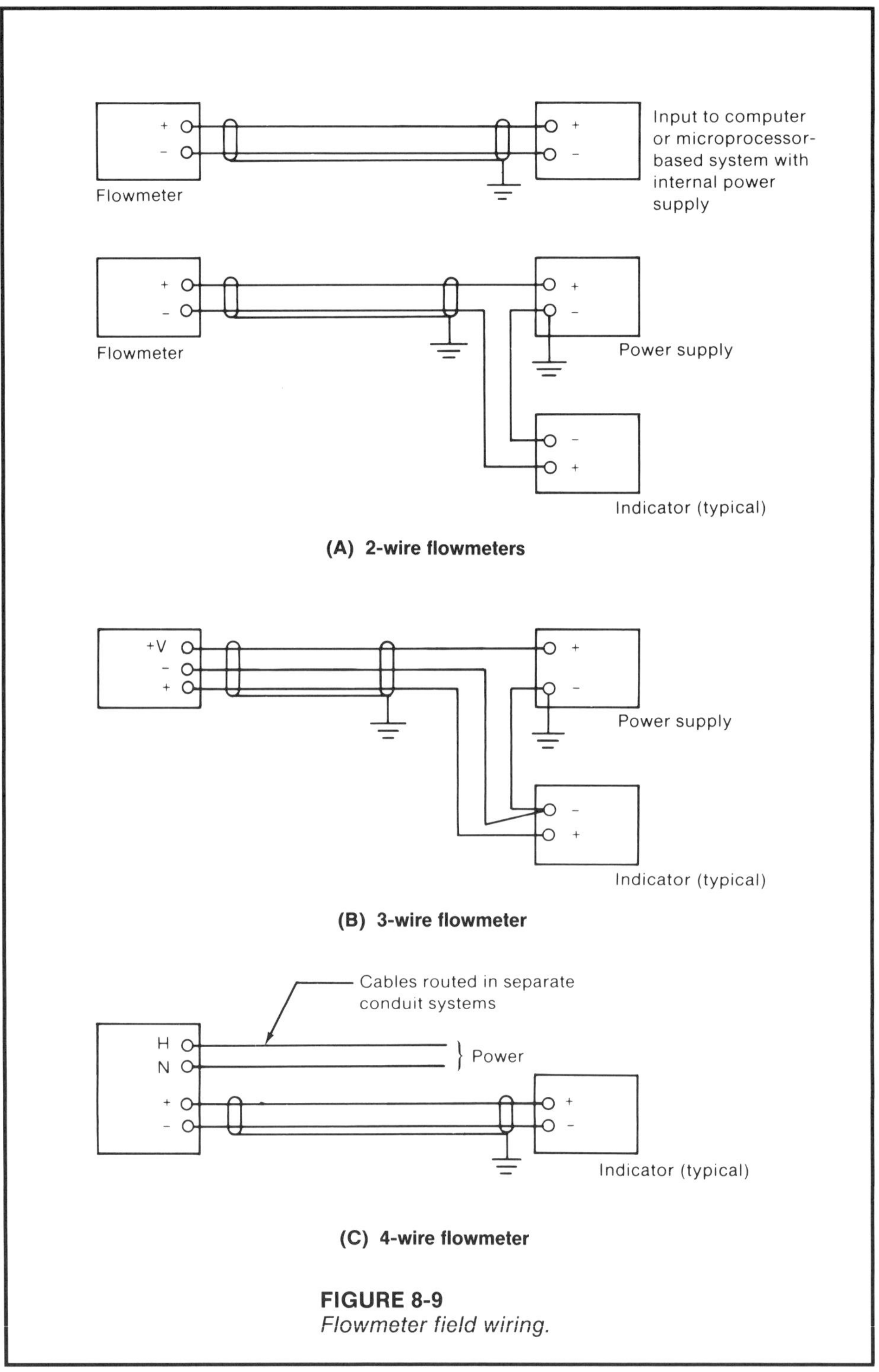

FIGURE 8-9
Flowmeter field wiring.

The 2-wire design is conceptually preferable but not always technically possible. It is more economical to install, as only a single shielded pair is required to be field run to the flowmeter. In some computer- and microprocessor-based control systems, a separate power supply is not necessary because it is built into the system, which reduces the amount of interconnecting wiring. Therefore, a typical installation may require as few as four signal terminations.

Three-wire designs may require a single shielded triplet in conjunction with a power supply, while 4-wire designs typically require power and signal cables that should be installed in separate conduit systems.

EXERCISES

8.1 Can a fluid at 50 percent concentration be more corrosive than the pure fluid? Explain.

8.2 Which parts of a flowmeter must be compatible with the fluid?

8.3 Define abrasion.

8.4 Can a 300-pound ANSI rated flange be used on a flowmeter that is exposed to a fluid at 350 psi and 250°F? Explain.

8.5 Why are flow conditioners used?

8.7 Why are hydrocarbons not to be used to lubricate flowmeters in oxygen service?

8.8 Are the effects of thermal expansion predictable? Why or why not?

8.9 Is a 2-wire, a 3-wire, or a 4-wire design flowmeter preferable? Why?

9

Introduction to Flowmeters

Introduction Interestingly enough, up to this point attention has been focused on fluid properties and measures of flowmeter performance without having defined what a flowmeter is, how it is used to measure flow, or why. Flowmeters can be divided into various classifications and types to aid in understanding their attributes.

Flowmeter Categories Flowmeters, which use many varied principles to measure flow, can be grouped into general categories, some of which may overlap one another but nonetheless are useful in describing some of the factors involved in flowmeter selection. These categories are:

I. Flowmeters with wetted moving parts
II. Flowmeters with no wetted moving parts
III. Obstructionless flowmeters
IV. Flowmeters with sensors mounted external to the pipe

Class I flowmeters by their nature require moving parts to operate. These flowmeters, such as positive displacement and turbine flowmeters, utilize high tolerance machined moving parts upon which the operation and performance of the flowmeter depend. These moving parts are subject to wear and damage, which can result in partial and catastrophic flowmeter failure. Any alteration of geometry or wear will increase the uncertainty associated with the flow measurement. While these flowmeters may not appear to be conceptually appealing, long-term accuracy of some designs has proven to be excellent when the flowmeter is properly applied, calibrated, and installed, although lower reliability due to susceptibility to sudden catastrophic failure is not a desirable feature. These flowmeters are usually not applicable to other than clean fluids.

Flowmeters that have no moving parts, categorized as Class II, such as orifice plate flowmeters and vortex shedding flowmeters, are conceptually more appealing than those that do not; however, wear of machined surfaces of the flowmeter that exceeds tolerances may cause added uncertainty in the flow measurement. The lack of moving parts results in fewer catastrophic failures, although other problems such as plugging of impulse tubing or excessive pressure drop begin to crop up, depending upon the design. Fluids other than clean fluids can be handled by this class of flowmeters, but very dirty fluids and very abrasive fluids may pose long-term wear problems.

Class III or obstructionless flowmeters are considered separately but are usually a subset of flowmeters with no moving parts. These are flowmeters that allow the fluid to pass through the flowmeter undisturbed, such as magnetic flowmeters. One advantage of Class III flowmeters is that while the fluid may be dirty and abrasive, the flowmeter will still maintain a reasonable service life if properly applied and installed.

Flowmeters that have sensors located external to the pipe are considered as Class IV flowmeters and are usually a subset of obstructionless flowmeters. This classification typically has the advantage of not only being obstructionless but also of having no wetted parts, such as an ultrasonic flowmeter with externally mounted transducers. This eliminates the requirement of ensuring that the wetted parts of the flowmeter are compatible with the fluid.

From this analysis it would seem that flowmeters should be specified in order of the preference of their classification, but it should be noted that while Class IV flowmeters offer considerable promise in effecting flow measurement, only limited success has been achieved in applying them. There may be other overriding technical and economic factors influencing flowmeter selection that would prohibit flowmeter selection on the basis of classification; therefore, successful flow measurement is a blend of trade-offs.

Category I	**Category II**
Positive Displacement Hydraulic Wheatstone Bridge Turbine Variable Area	Differential Pressure Oscillatory Target Thermal
Category III	**Category IV**
Coriolis Mass Magnetic Ultrasonic	Clamp-on Ultrasonic

FIGURE 9-1
Flowmeter categories.

Flowmeter Types Flowmeters can be grouped into general types that are useful in describing some of the factors involved in flowmeter selection, as follows:

A. Volumetric
B. Velocity
C. Inferential
D. Mass

Type A volumetric flowmeters, such as positive displacement flowmeters, measure flow by measuring volume directly. Volumetric flow-measuring devices usually use high tolerance machined parts to physically trap precisely known quantities of fluid as they rotate.

Velocity flow measurements, which are effected using Type B flowmeters, are those in which the velocity of the flow is measured and multiplied by the area through which the fluid flows to determine the total flow. Various principles can be used to measure velocity as illustrated by the number of different designs, examples of which include turbine, vortex shedding, and Doppler ultrasonic flowmeters.

Type C inferential flowmeters measure flow by inferring the flow through a pipe from some physical phenomenon. An example of this is to infer flow from the differential pressure across a restriction in a pipe as is the case of an orifice plate flowmeter. This measurement does not measure volume, nor velocity, but rather flow is inferred from the measured differential pressure and accepted experimental correlations.

Type A	Type B
Positive Displacement	Magnetic Oscillatory Turbine Ultrasonic
Type C	**Type D**
Differential Pressure Target Variable Area	Coriolis Mass Hydraulic Wheatstone Bridge Thermal

FIGURE 9-2
Flowmeter types.

Type D mass flowmeters measure mass directly. An example of this is a Coriolis mass flowmeter that measures mass directly as a function of the force that the mass produces as it accelerates in a curved pipe.

While flow measurements can be used for a variety of industrial applications, the paradox of how flow is measured by different types of flowmeters can best be explained in terms of a chemical reaction involving liquid ingredients. The reaction requires that each liquid be in the exact mole proportion to each other in order to completely react, as illustrated at the top of Figure 9-3.

Therefore, each reactant must be in the proper mass proportion to each other. As most flowmeters are not Type D and do not measure mass per unit time but rather measure or infer volume per unit time, the required volumetric flow must be calculated from the mass where the pressure, temperature, density, viscosity, and the like, are the nominal operating conditions of the liquid. Once the required volume is calculated, it can be measured by a flowmeter at process conditions, where the measurement, if exactly equal to the calculated volume, represents the volume required at assumed operating conditions, measured at real process conditions, with no compensation made for differences between the assumed operating conditions and the real process conditions. If calculations are

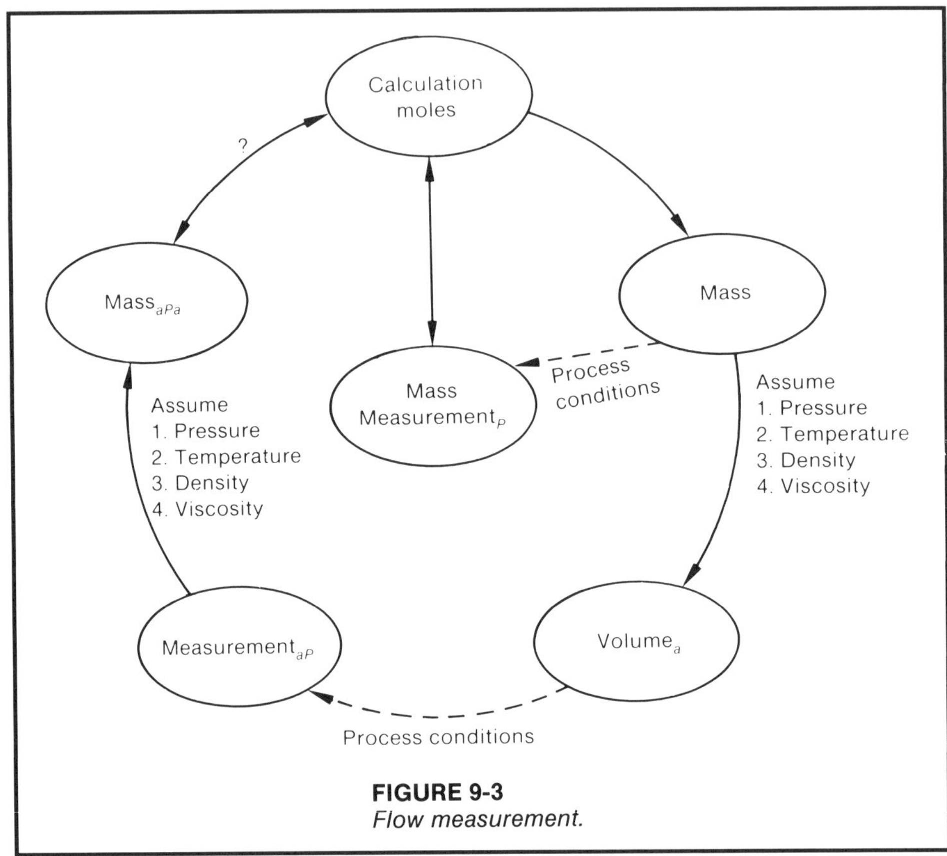

FIGURE 9-3
Flow measurement.

performed to convert the measured volume to mass using nominal operating conditions and then to moles, then it is very unlikely that it will be in agreement with the original desired number of moles.

This means that although the volume under nominal conditions may be accurately calculated, the process conditions and fluid properties can vary sufficiently such that it is questionable if the measured volume represents the required quantity of liquid. In real terms, this means that too much or too little liquid could be added to the reaction due to differences between the nominal operating conditions and the actual process conditions.

This example illustrates the paradox of flow measurement, that is, that the quantity that is typically measured is not the measurement that is desired, but rather what is available and can be economically implemented.

A mass flow measurement avoids the pitfalls of volumetric flow measurement by measuring mass directly. Therefore, the errors associated with mass measurement are those associated with the measurement and not the process. Disregarding the measurement accuracy of the instrument, as was done above, the net result is the measurement of the mass of the liquid, which corresponds directly to the number of moles required. This will result in the addition of the proper quantity of liquid to the reaction. However, measurement of mass may not be desirable for other applications, such as the filling of a tank.

Introduction to Flowmeter Technology Sections In subsequent sections, flowmeters that utilize various technologies will be discussed with the intent of imparting a working knowledge of the operation, performance, installation, and maintenance considerations of available flowmeter technologies.

The basic principle of operation of each flowmeter will be discussed in sufficient detail to apply the technology. Detailed design equations and derivations that do not add to this are omitted.

Operating constraints of a technology, such as type of fluid, pressure, temperature, viscosity, flow range, size, Reynolds number, and the like, as well as materials and types of flowmeter construction, will form the basis for a flowmeter selection procedure.

Expected flowmeter performance will be examined as it relates to flowmeter uncertainty, deviations in operating conditions, and flowmeter pressure losses. When possible, accuracy statements relating to the uncertainties of the primary element only will be presented. In most cases, this allows the user to analyze flowmeter primary and secondary instrumentation independently.

Typical applications and sizing requirements for each technology are discussed to illustrate the limits of flowmeter performance and typical application. Equations, charts, or empirical information are used as required to impart a working knowledge of the subject matter. It should be noted that some flowmeters can be sized by the user, while others require that sizing be performed by the manufacturer due to the complexity of the calculations necessary to select the applicable flowmeter size and calibration. Some flowmeters can be sized by the user using rough calculations, but precise calculations that take all variables into account to determine the final dimensions or adjustments of the flowmeter should be performed by the manufacturer.

Piping requirements, tap locations, flowmeter orientation, transmitter location, and the like are discussed to effect technically correct flowmeter installation. Maintenance requirements and possible problems that may occur after the flowmeter is installed are examined, including discussions of long-term wear, serviceability, effects of repair on performance, and expected flowmeter reliability.

EXERCISES

9.1 What are the four flowmeter catetories? What are the advantages of each?

9.2 Which classifications of flowmeters are applicable when the following fluids are to be measured? Why?
   a. Liquid containing solids
   b. Abrasive liquid
   c. Corrosive liquid
   d. Lubricative liquid

9.3 What are the four types of flowmeters? What are the advantages of each type? Disadvantages?

9.4 Which types of flowmeters are applicable when the following flow measurement goals and parameters are set? Why?
   a. Multiple products with different densities
   b. Control gas volume through equipment
   c. Fill a drum to a predetermined level
   d. Operating conditions are tightly controlled.

10

Differential Pressure Flowmeters

Introduction Differential pressure or head-type flowmeters represent one of the most commonly used flowmeter technologies. Their versatility, cost, and simplicity make them attractive for many applications. Differential pressure producers can be applied to virtually all low viscosity liquid flow measurement applications, as well as to most gas applications.

Differential pressure producing flow elements utilize empirical correlations to quantify the relationship between the produced differential pressure and the volumetric flow through a carefully specified restriction in a pipe. Neither the mass, velocity, nor volume are measured directly, but rather the flow is inferred from the hydraulic similarity to flowmeters that have been carefully tested under laboratory conditions.

Orifice Plate Flowmeters

Orifice plate technology represents one of the most accepted and versatile methods for measuring flow. Its simplicity is attractive from both maintenance and application perspectives. However, to achieve the full performance of orifice plate technology, a considerable amount of detail must be attended to.

Principle of Operation Head producing flowmeters are described by Bernoulli's equation, which states that the sum of the static energy (pressure head), the kinetic energy (velocity head), and the potential energy (elevation head) of the fluid are approximately conserved in the flow across a constriction in a pipe and by continuity. Bernoulli's equation at each flow cross section is given by:

$$\frac{P}{\rho \times g} + \frac{v^2}{2g} + y = \text{constant}$$

where g is the acceleration of gravity and y is the elevation head of the fluid.

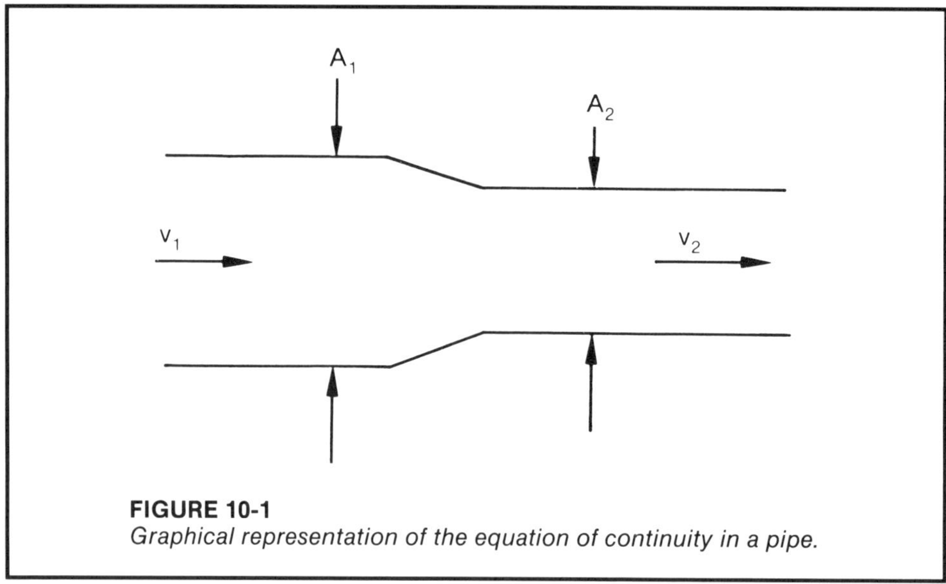

FIGURE 10-1
Graphical representation of the equation of continuity in a pipe.

The equation of continuity provides a relation between the velocity and the fluid flow rate for incompressible fluids. In a pipe this may be represented as:

$$Q = A_1 \times v_1 = A_2 \times v_2$$

The continuity relationship requires that the velocity of the fluid increase when the cross-sectional area of the pipe is reduced.

EXAMPLE 10-1

Problem: Calculate the velocity of a liquid in a 3 inch schedule 40 pipe if the liquid has a velocity of 10.0 feet per second in a 2 inch schedule 40 pipe.

Solution:

$$Q = A_1 \times v_1 = A_2 \times v_2$$

$$= 1/4\ \pi \times (2.067 \text{ in.})^2 \times (10 \text{ ft/sec}) = 1/4\ \pi \times (3.068 \text{ in.})^2 \times (v_2)$$

Solving for v_2,

$$v_2 = (2.067 \text{ in.}/3.068 \text{ in.})^2 \times (10 \text{ ft/sec})$$

$$= 4.54 \text{ feet per second}$$

Applying Bernoulli's equation to the upstream and downstream locations of an orifice plate or other flow element results in:

$$P_1 + 1/2\ \rho \times v_1^2 = P_2 + 1/2\ \rho \times v_2^2$$

The difference in elevation head drops out of the equation, if the flow is horizontal. Combining this result with the equation of continuity and rearranging terms yields

$$P_1 - P_2 = 1/2\,\rho\,[v_2^2 - v_1^2]$$
$$= 1/2\,\rho\,[(A_1/A_2)^2 - 1] \times v_1^2$$
$$= 1/2\,\rho\,[(D/d)^4 - 1]^2 \times v_1^2$$
$$= 1/2\,\rho\,[(D/d)^4 - 1]^2 \times Q^2/A_1^2$$

This shows that the differential pressure generated across an orifice is proportional to the square of the flow through the orifice plate. This relation is valid with some modification for compressible fluids. It should be noted that the differential pressure across a device is termed the *dynamic* pressure, while the pressure present in the pipe is termed the *static* pressure.

Using the idealized result just developed, the flow through an orifice plate can be represented empirically by:

$$Q = \text{constant} \times (\Delta P/\rho)^{1/2}$$

The constant adjusts for the dimensional units, non-ideal fluid losses and behavior, discharge coefficients, pressure tap location, operating conditions, gas expansion factor, Reynolds number and the like. These variations are accounted for empirically by flow testing.

EXAMPLE 10-2

Problem: Flow through an orifice plate flowmeter is controlled at 100 gpm. Estimate the effects of a change in specific gravity from the nominal 1.060 used for calculation purposes to 1.076 at operating conditions.

Solution: The effect can be calculated by forming a ratio of the two flow conditions:

$$Q\,/\,Q_0 = (SG_0\,/\,SG)^{1/2} = (1.060\,/1.076)^{1/2} = 0.99254$$

For estimating purposes, a 1 percent change in density causes a -1/2 percent change in the flow measurement. In this example, the estimated effect would be -1/2 (1.076/1.06) or -0.75 percent, which agrees closely with the calculated result.

Construction

Orifice Plate

The orifice plate is usually constructed of metal, into which an opening of a predetermined size and shape is machined to tight tolerances. It is installed between two flanges in the pipe in such a manner as to effectively form a restriction in the flow through the pipe.

Orifice plates generally are unidirectional. As the direction of the orifice plate cannot be determined once it is installed in the pipe, standard industry practice is to stamp or affix key dimensional information on the upstream side of the orifice plate handle. The handle gives the orifice a paddle-like appearance.

The thin, *concentric*, square-edged orifice plate is the most commonly applied type of orifice plate. The machined opening is circular and is in a position such

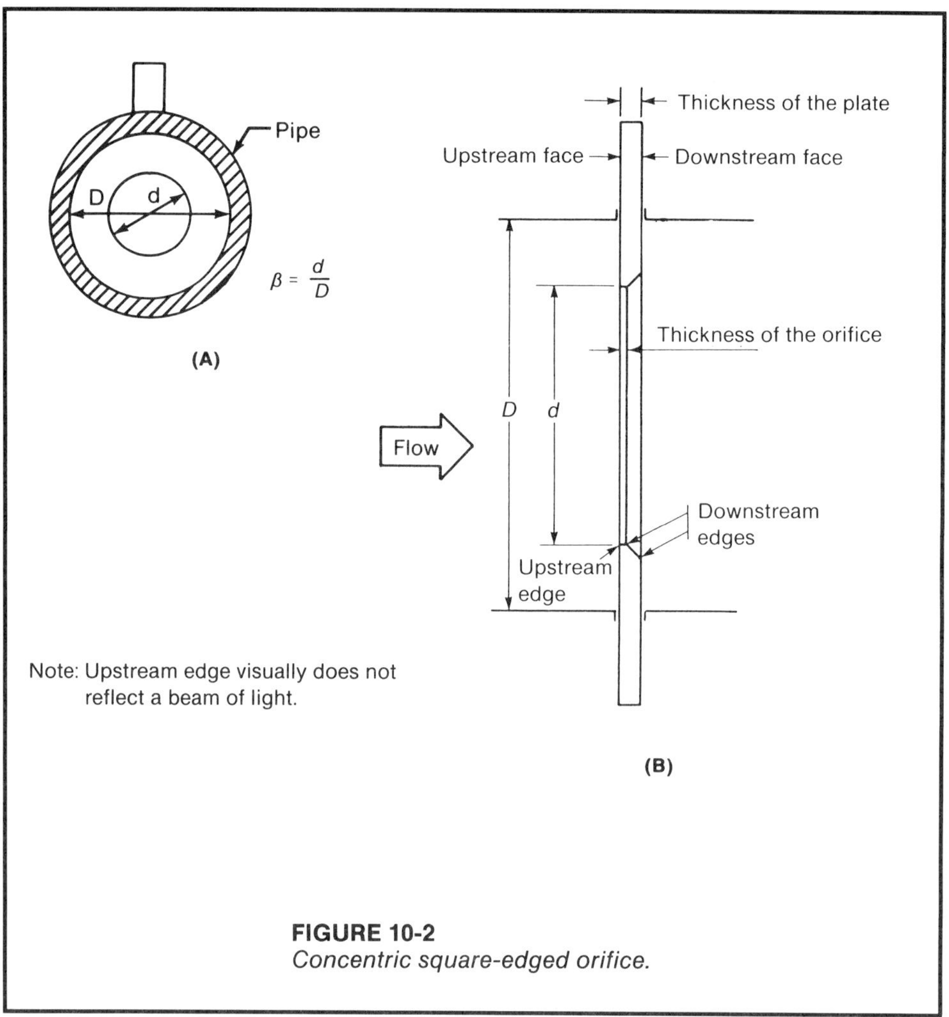

FIGURE 10-2
Concentric square-edged orifice.

that upon installation the circle will theoretically be positioned in the center of the pipe. As the plate is thick in relation to the diameter of the pipe, the back of the orifice is usually beveled or counter-bored to make the orifice plate effectively thinner and performance more predictable. The diameter ratio of the orifice to the pipe ID (termed the beta ratio) can be used to characterize the orifice plate.

Conical orifice plates are not frequently applied since the lack of generally accepted coefficient data limits its usefulness.

Eccentric orifice plates, shown in Figure 10-4, have a circular opening machined in the same manner as a concentric orifice plate, but located nearly tangent to the top of the pipe for liquids and tangent to the bottom of the pipe for gases. This type of orifice plate can be used to allow entrained gases or liquid in two-phase flows to flow through the orifice plate instead of building up in front of it affecting the accuracy of the flowmeter. The actual discharge coefficient is

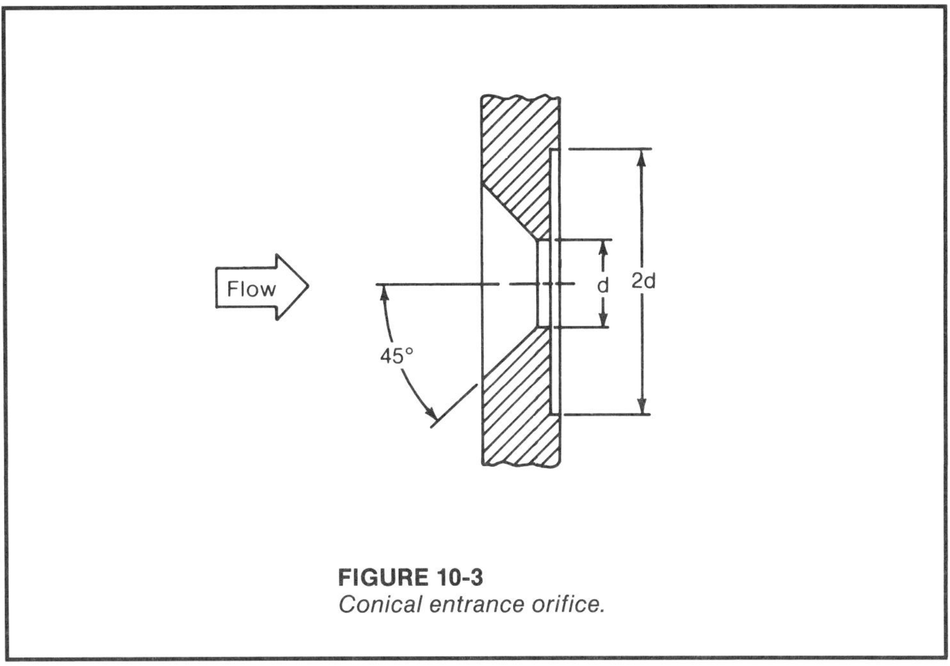

FIGURE 10-3
Conical entrance orifice.

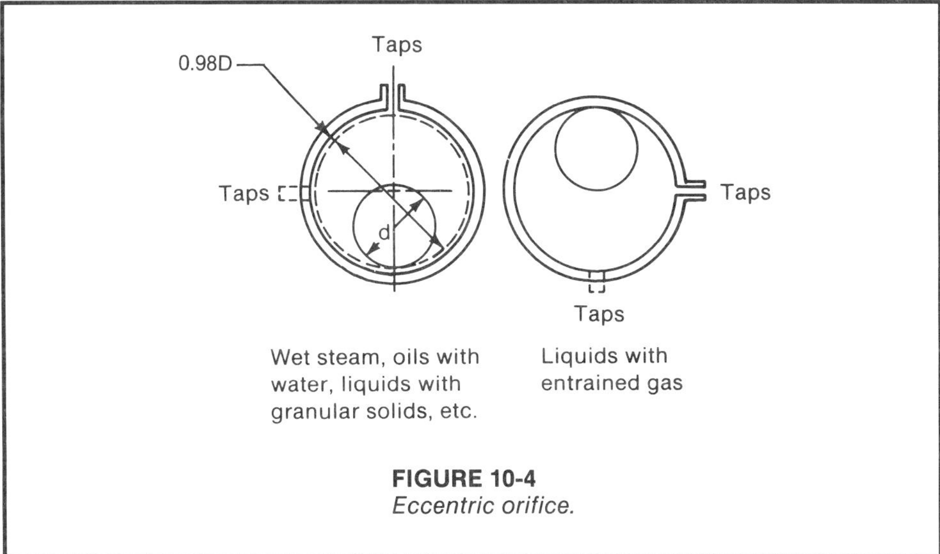

FIGURE 10-4
Eccentric orifice.

dependent upon whether the taps are diametrically opposite taps or at 90° to one another and varies as a function of Reynolds number in a manner similar to a concentric orifice plate.

An *integral* orifice plate is a machined concentric orifice assembly that is mounted inside or directly attached to the transmitter. Integral orifice flowmeters are applied to small flows, typically in the 1/2 to 1-1/2 inch pipe size.

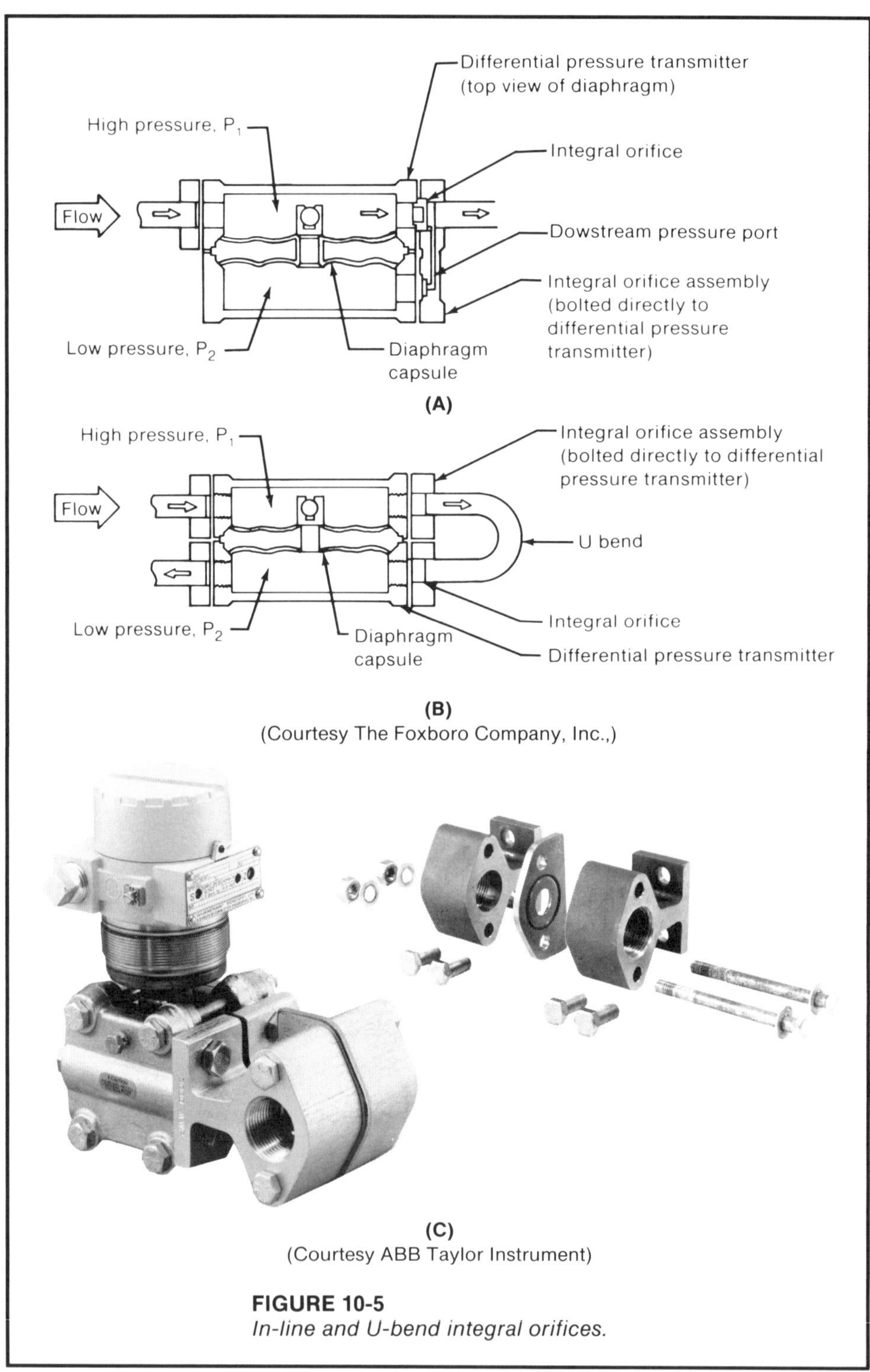

(Courtesy The Foxboro Company, Inc.,)

(C)

(Courtesy ABB Taylor Instrument)

FIGURE 10-5
In-line and U-bend integral orifices.

The *quadrant* orifice plate has a rounded upstream orifice edge, as illustrated in Figure 10-6, such that the orifice plate is linear at low Reynolds numbers.

Segmental orifice plates, as shown in Figure 10-7, have a segmental opening that is machined in the same manner as a concentric orifice plate, but located tangent to the top of the pipe for liquids and tangent to the bottom of the pipe for gases. This type of orifice plate can be used to allow entrained air, liquid, or particulate matter to flow through the orifice plate instead of building up in front of it and affecting the accuracy of the flowmeter. The discharge coefficient varies as a function of Reynolds number, as with a concentric orifice plate, but is not as predictable.

EXAMPLE 10-3

Problem: Select the restriction that the fluid is subject to in a pipe for each of the following:

1. Concentric
2. Conical
3. Eccentric
4. Integral
5. Quadrant
6. Segmental

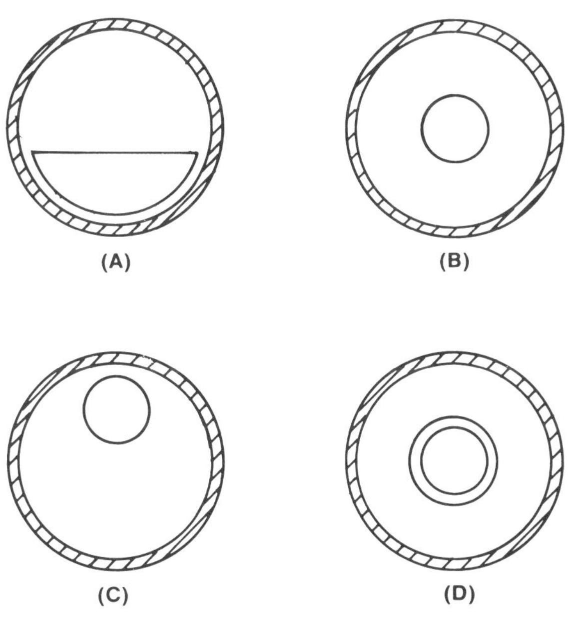

Solution:

1-B, 2-D, 3-C, 4-B, 5-B, 6-A

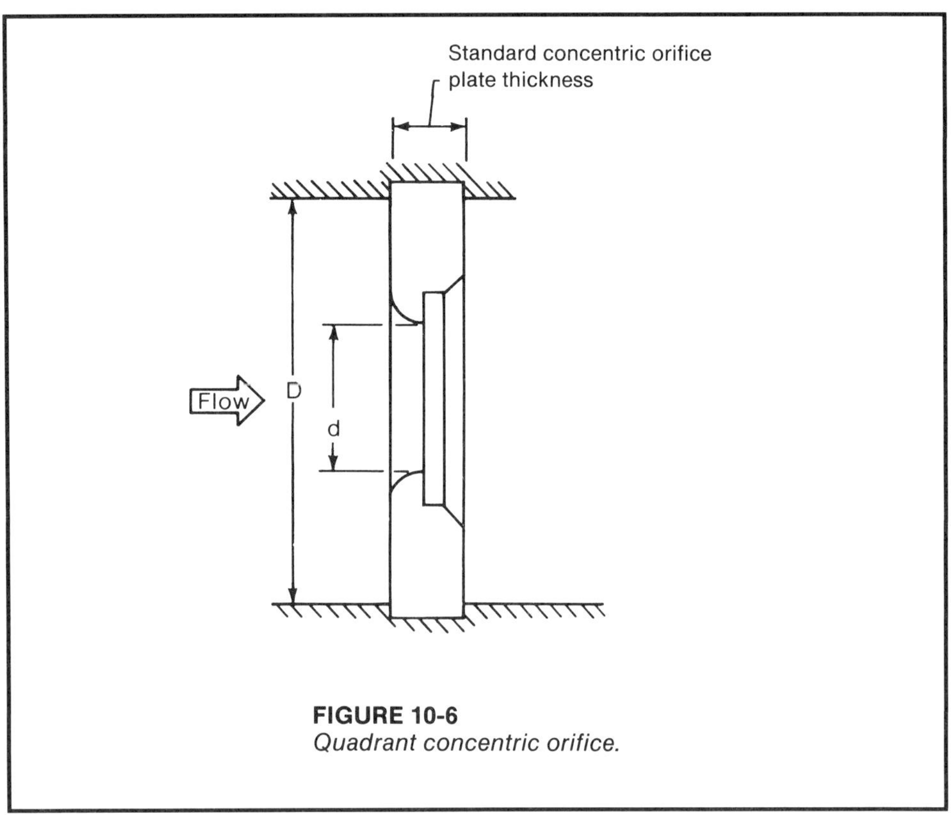

FIGURE 10-6
Quadrant concentric orifice.

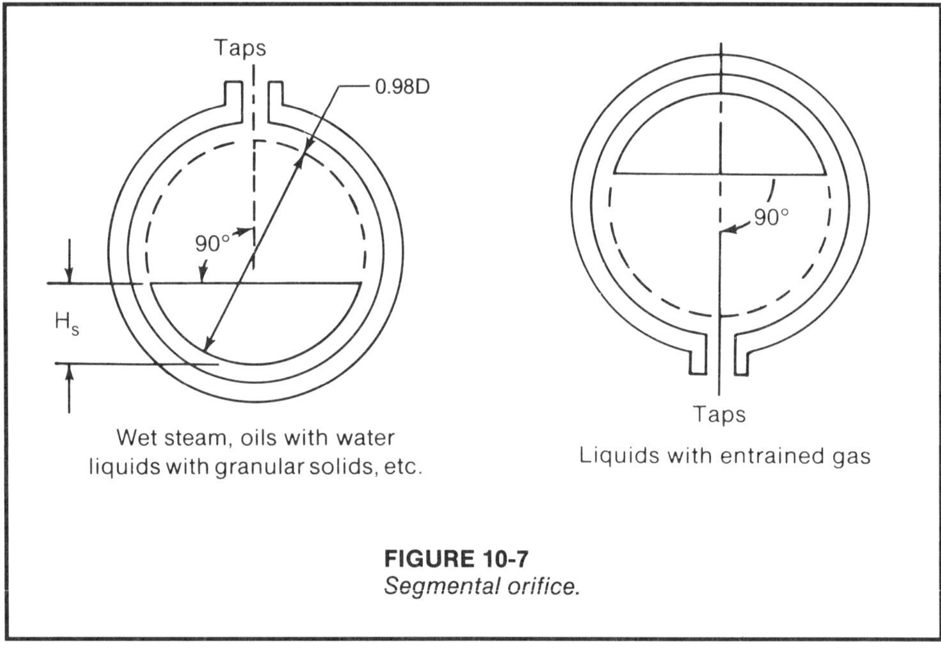

FIGURE 10-7
Segmental orifice.

Vent and Weep Holes

Orifice plates may be specified with either a vent or a weep hole for liquid or gas service, respectively. The vent hole allows gas that may accumulate upstream of the orifice plate at the top of the pipe to pass through the flowmeter. Accumulation of gas at the flowmeter inlet can affect the discharge coefficient of the flowmeter element and decrease the accuracy of the measurement. Weep holes are commonly used at the bottom of the pipe to allow condensation that may form to pass through the flowmeter without affecting the discharge coefficient of the meter.

When the fluid being measured is not clean, the weep or vent hole can plug. This can result in the settling of solids or the accumulation of gas upstream of the orifice plate, affecting the accuracy of the flow measurement. As the vent and weep holes represent an area through which fluid can flow in addition to the bore of the orifice plate, the cross-sectional area introduced by a vent or weep hole may be used in the bore calculation. However, this is not always required since the weep hole correction is generally significantly less than a fractional percentage of the total flow.

Taps

Pressure taps are located upstream and downstream of the orifice plate to allow measurement of the developed pressure differential. Some taps are welded in the pipe, while other arrangements require that the taps be integral to prefabricated flanges. Taps located in the flanges are often preferred on smaller pipes to eliminate the possibility of positioning errors that may occur when performing field welds. Taps located on the pipe are usually more economical and more easily implemented on larger pipes. The tap location is also a factor in orifice calculations.

Corner taps are located within the flanges such that the pressures sensed are indicative of the pressures at the upstream and downstream faces of the orifice plate, as illustrated in Figure 10-8.

Flange taps are located within the flanges in such a manner that the pressures sensed are indicative of the pressures at the distance of 1 inch upstream of the upstream face and 1 inch downstream of the downstream face of the orifice plate (see Figure 10-9).

Full flow taps are located on the pipe in such a manner that the pressures sensed are indicative of the pressures $2.5D$ upstream of the upstream face and $8D$ downstream of the downstream orifice face (see Figure 10-9).

Radius taps, which are most commonly applied in larger pipe sizes, are located in such a manner that the pressures sensed are indicative of the pressures at the distance of $1D$ upstream of the upstream face and $0.5D$ downstream of the upstream face of the orifice plate.

Vena contracta taps are located in such a manner that the pressures sensed are indicative of the pressure $1D$ upstream of the upstream face and the vena contracta, which is the point of lowest local pressure downstream of the orifice plate, in order to develop the maximum possible differential pressure, as shown in Figure 10-10.

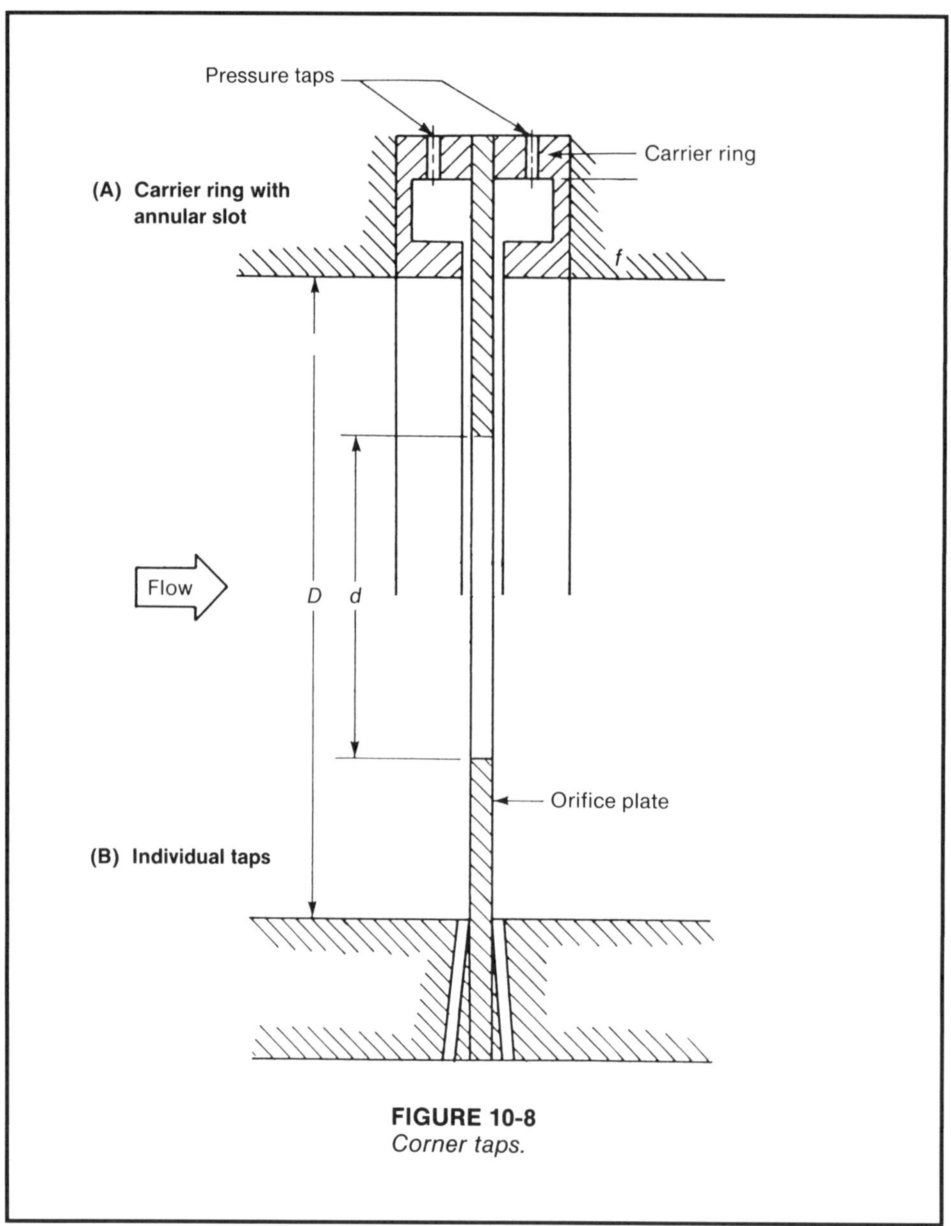

FIGURE 10-8
Corner taps.

The location of the vena contracta is a function of the beta ratio of the orifice plate. Therefore, relocation of the downstream tap is required when the beta ratio is changed.

Flange taps are the dominant configuration in the United States for small pipe sizes, while radius taps are preferred over vena contracta and full flow taps for larger pipe sizes.

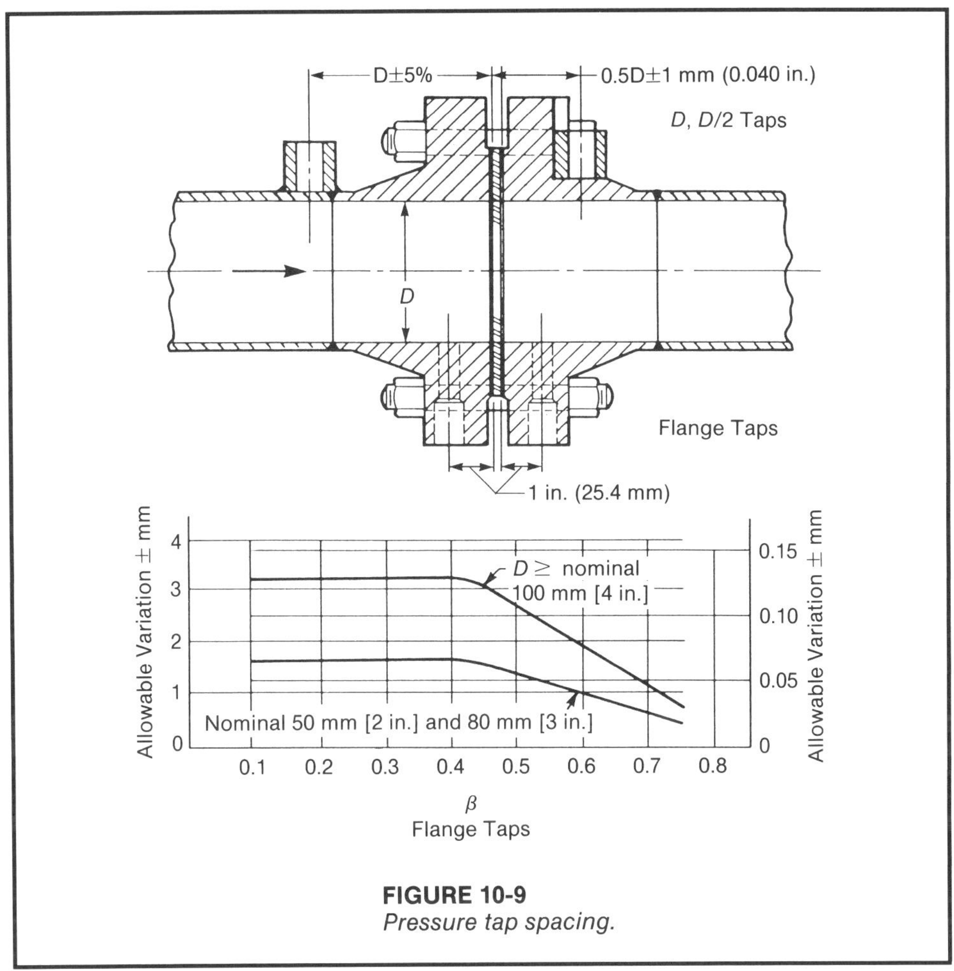

FIGURE 10-9
Pressure tap spacing.

EXAMPLE 10-4

Problem: Match the attribute that best describes each type of tap.

1. Corner	A. Applicable to large pipes for all beta ratios.
2. Flange	B. Detects maximum differential generated by the orifice, but the tap must be moved when an orifice with a different beta ratio is installed.
3. Full flow	C. Detects differential pressure at the orifice faces and applicable to small line sizes.
4. Radius	D. Preferred in North America for moderately small pipe sizes.
5. Vena contracta	E. Preferred for large pipe sizes.

Solution:

1-C, 2-D, 3-A, 4-E, 5-B

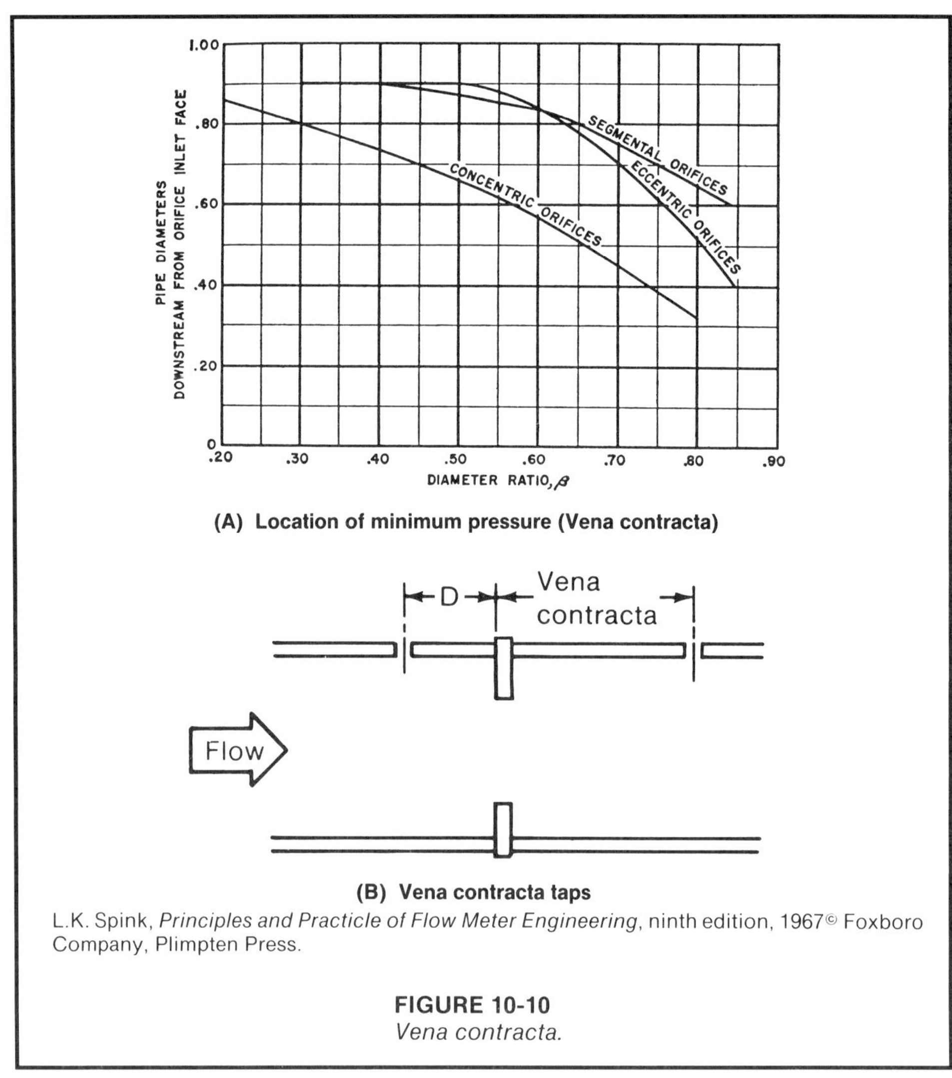

(A) Location of minimum pressure (Vena contracta)

(B) Vena contracta taps

L.K. Spink, *Principles and Practicle of Flow Meter Engineering*, ninth edition, 1967© Foxboro Company, Plimpten Press.

FIGURE 10-10
Vena contracta.

Impulse Tubing

Impulse tubing is used to transmit the pressures generated at the taps to the transmitter. A shut-off valve is usually located at each tap to allow the impulse tubing and transmitter to be taken out of service without affecting the flow in the pipe. As this valve separates the process from the instrument, it should be rated to safely shut off the process fluid and be compatible with the piping specifications.

Sensing Systems

The pressure generated by the flowmeter is sensed by a differential pressure transmitter, typically with block and bypass valves or a 3-valve manifold, which is used to take the transmitter in and out of service.

EXAMPLE 10-5

Problem: Draw a schematic of an orifice plate, impulse tubing, and transmitter that has a bypass valve.

Solution:

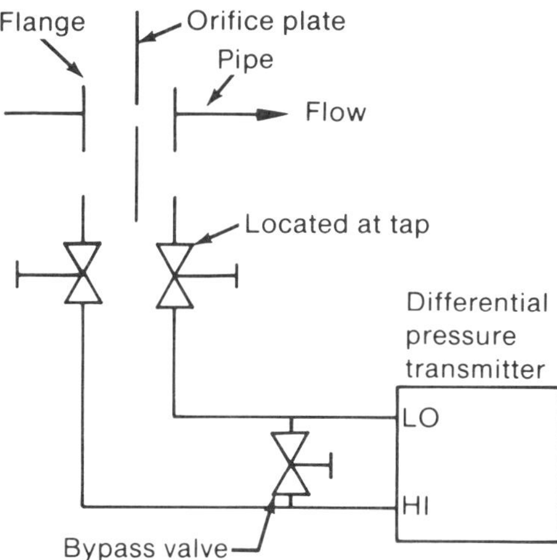

Wetted Parts

The wetted parts of an orifice plate flowmeter include the orifice plate, orifice flanges, impulse tubing, shut-off valves, bypass or manifold valve, and the wetted parts of the differential pressure transmitter.

The orifice plate must be compatible with the process fluid, as wear or corrosion of the edges of the plate will affect the discharge coefficient of the meter and thus its calibrated accuracy. Stainless steel is often used as a material of construction, although the orifice plate may be fabricated from virtually any machinable material. The valves, impulse tubing, and manifold must also be compatible with the process fluid and satisfy any applicable piping codes that usually apply to the first shut-off or block valve.

All wetted parts of the differential pressure transmitter must be compatible with the process fluid or, if used, the seal liquid. It should be noted that the diaphragm of the differential pressure transmitter is very thin and not much corrosion is necessary to result in transmitter failure. Care must be taken to ensure that the diaphragm is corrosion resistant or at least protected from the process fluid.

Specification of impulse tubing need not only take into account the process fluid inside the impulse tubing, but also the environment in which it is to be operated, the workability, and the cost.

Operating Constraints Orifice plate flowmeter secondary devices are temperature- and pressure-limited by the flange ratings of the pipe, tubing, and valves, as well as by the ability of the process fluid to be cooled sufficiently so that the differential pressure transmitter will operate within its temperature specifications.

Approximate Reynolds' number constraints are as follows:

	R_D Constraint
Concentric	
(under 2 in.)	1000+
(2 in. and over)	5000 *d*+
Conical	$250\ \beta < R_D < 200{,}000\ \beta$
Eccentric	10,000 - 1,000,000
Integral	1000 *d*/*D*
Quadrant	$250 - 3200 < R_D < 60{,}000 - 280{,}000$
Segmental	10,000 - 1,000,000

Rangeability of the differential pressure transmitter is limited to approximately 10:1, while the flow turndown is usually limited to 3.5:1 due to the nature of a squared output. Differential pressure transmitters with different calibrated ranges can be "stacked" or installed in parallel across the orifice plate to achieve flow turndowns of 10:1 or better, when the orifice plate is operated within its operating constraints. However, care must be taken in measuring fractional inches of water column in developed differential.

EXAMPLE 10-6

Problem: It is desired to increase the flow range of an orifice plate flowmeter with a transmitter that has a full scale calibration range of 0 to 5 in. WC to 0 to 30 in. WC. If the transmitter were calibrated at 20 in. WC, can the same transmitter be recalibrated to increase the full scale flow range by 25 percent?

Solution: As the differential pressure generated by the orifice plate is proportional to the square of the flow

$$\frac{\Delta P_2}{\Delta P_1} = \frac{Q_2{}^2}{Q_1{}^2} = \frac{(1.25)^2}{(1.00)^2} = 1.5625$$

$$\begin{aligned} \Delta P_2 &= 1.5625\ \Delta P_1 \\ &= 1.5625 \times 20 \text{ in. WC} \\ &= 31.25 \text{ in. WC} \end{aligned}$$

Even though it may be possible to calibrate the differential pressure transmitter to the calculated calibration differential, the calculated calibration differential pressure exceeds the specifications of the differential pressure transmitter, and a differential pressure transmitter with the proper calibration range should be used.

It should be noted that the manufacturing parameters of the orifice plate are often calculated for each application. The flowmeter will operate over a wider range; however, the maximum differential pressure must be within the calibration range of the transmitter.

Performance The inherent accuracy of a properly installed square-edge orifice plate is on the order of ±0.6 percent rate. However, the flowmeter accuracy is typically poorer due to other influence factors and bias errors, even if instrument inaccuracy is excluded. Other orifice types are characterized by inherent uncertainties as shown in Table 10-1.

When the orifice plate is measuring flow at the lower portion of the range, the total measurement error can increase dramatically. This is due to the fact that the developed differential decreases with the square of the flow. As a result, for a flow turndown of 3.5:1 the measured differential must cover a 12:1 turndown. Process uncertainties such as pressure, temperature, compressibility, density, and expansion factor effects may further add to the inaccuracies of the flow measurement. Generally these factors are the dominant sources of uncertainty.

EXAMPLE 10-7

Problem: What is the coefficient accuracy that can be expected from a concentric orifice plate with a beta ratio of 0.6 when it is installed with flange taps? With pipe taps?

Solution: The coefficient accuracies are ±0.6 percent and ±1.6 percent of rate for flange and full flow taps, respectively, which illustrates increased uncertainty when using full flow taps.

EXAMPLE 10-8

Problem: For which type of orifice plate can the flow coefficient be most accurately determined?

Solution: Examination of Table 10-1 shows that the flow coefficient of the concentric orifice plate can be most accurately determined and is hence preferred.

Applications Orifice plate technology is versatile and can be applied to virtually all gases and low viscosity liquids, as the flowmeter is individually sized to satisfy Reynolds number and differential pressure constraints.

EXAMPLE 10-9

Problem: Is a concentric orifice plate applicable, given the following data?

$$Q = 50 \text{ gpm}$$
$$SG = 1.13$$
$$\mu_{cP} = 10 \text{ cP}$$
$$d = 1.033 \text{ inches}$$
$$D = 2.067 \text{ inches}$$

Solution: Determine if Reynolds number constraints are satisfied by calculating operating Reynolds number and the required Reynolds number, as follows:

$$\begin{aligned} R_D &= (3160\ Q\ \text{gpm} \times \text{SG})/(\mu_{cP} \times D) \\ &= (3160 \times 50\ \text{gpm} \times 1.13)/(10\,\text{cP} \times 2.067\ \text{in.}) \\ &= 8638 \end{aligned}$$

Reynolds number must be in excess of

$$\min R_D = 5000\,d = 5000 \times 1.033\ \text{in.} = 5165$$

for accurate operation. This means that for an accurate turndown of 3.5:1, Reynolds number at full scale flow must be in excess of 3.5 times min R_D or 18078.

Reynolds number at full scale flow of 8638 is above the minimum required Reynolds number of 5165, which means that the flowmeter will operate accurately over an estimated turndown of 8638/5165, or 1.67:1, which means that a square-edge orifice plate transmitter will not perform accurately over the entire desired flow range.

If the viscosity were 4 cP, Reynolds number at full scale would be 21594, which would fully satisfy the R_D constraints of a square-edge orifice plate.

EXAMPLE 10-10

Problem: Is a square-edge orifice plate applicable, given the following data?

$$\begin{aligned} Q &= 50\ \text{acfm} \\ \rho &= 1.0\ \text{lb/ft}^3 \\ \mu_{cP} &= 0.017\,\text{cP} \\ d &= 1.033\ \text{inches} \\ D &= 2.067\ \text{inches} \end{aligned}$$

Solution: Determine if Reynolds number constraints are satisfied by calculating operating R_D and the required R_D as follows.

$$\begin{aligned} R_D &= (379\ Q\ \text{acfm} \times \rho)/(\mu_{cP} \times D) \\ &= (379 \times 50\ \text{acfm} \times 1.0\ \text{lb/ft}^3)/(0.017\,\text{cP} \times 2.067\ \text{in.}) \\ &= 557{,}353 \end{aligned}$$

Reynolds number must be in excess of

$$\min R_D = 5000\,d = 5000 \times 1.033\ \text{in.} = 5165$$

for accurate operation. The operating full scale Reynolds number is in excess of 3.5 times min R_D so the square-edge orifice plate is applicable. A square-edge orifice plate would be applicable to most gas applications due to the low viscosities and hence large Reynolds numbers associated with gases.

TABLE 10-1
Recommended Accuracy and Restrictions

Primary device	*Nominal pipe diameter D, in. (mm)*	*Beta ratio* β	*Pipe Reynolds number* (R_D) *range*	*Coefficient accuracy, %*
Venturi				
Machined inlet	2–10 (50–250)	0.4–0.75	2×10^5 to 10^6	± 1
Rough cast	4–32 (100–800)	0.3–0.75	2×10^5 to 10^6	± 0.7
Rough-welded sheet-iron inlet	8–48 (200–1500)	0.4–0.7	2×10^5 to 10^6	± 1.5
Universal Venturi Tube[1]	≥ 3 (≥ 75)	0.2–0.75	$>7.5 \times 10^4$	± 0.5
Lo-Loss[1]	3–120 (75–3000)	0.35–0.85	1.25×10^5 to 3.5×10^6	± 1
Nozzle				
ASME	2–16 (50–400)	0.25–0.75	10^4 to 10^7	± 2.0
ISA	2–20 (5–500)	0.3–0.6	10^5 to 10^6	± 0.8
		0.6–0.75	2×10^5 to 10^7	$2\beta - 0.4$
Venturi nozzle	3–20 (75–500)	0.3–0.75	2×10^5 to 2×10^6	$\pm 1.2 \pm 1.5\beta^4$
Orifice				
Corner, flange, *D* and *D*/2	2–36 (50–900)[2]	0.2–0.6	10^4 to 10^7	± 0.6
		0.6–0.75	10^4 to 10^7	$\pm \beta$
		0.2–0.75	2×10^3 to 10^4	$\pm 0.6 \pm \beta$
2½*D* and 8*D* (Pipetaps)	2–36 (50–900)	0.2–0.5	10^4 to 10^7	± 0.8
		0.51–0.7		± 1.6
Quadrant-edged	1–30 (25–750)	0.24–0.6	$R_{D,\min} < R_D < 10^5\beta$	± 2
Flanged and corner			$R_{D,\min} = 1000\beta + 9.4(\beta - 0.24)^8$	

(continued)

TABLE 10-1
Recommended Accuracy and Restrictions
(continued)

Primary device	*Nominal pipe diameter D, in. (mm)*	*Beta ratio β*	*Pipe Reynolds number (R_D) range*	*Coefficient accuracy, %*
Conical entrance	>1 (25)	0.1–0.316	$250\beta \leq R_D \leq 5000\beta$ for C = 0.730	±2
Corner			$5000\beta \leq R_D \leq 2 \times 10^5\beta$ for C = 0.734	±2
Honed meter runs	½–1½ (12–40)	0.1–0.8	>1000	±0.75
Flange, corner				
Integral flow orifice assembly (IFOA) (Foxboro)		Std sizes	$1500/\beta \leq R_D/\beta$	±0.75
IFOA (Foxboro)				
Eccentric				
Flange and vena contracta	4 (100)	0.3–0.75	10^4 to 10^6	±2
	6–14 (150–350)	0.3–0.75	10^4 to 10^6	±1.5
Segmental				
Flange and vena contracta	4–14 (150–350)	0.35–0.75	10^4 to 10^6	±2

ISO 5167 (1980) and ASME *Fluid Meters* (1971) show slightly different values for some devices.
1. The manufacturer should be consulted for recommendations.
2. For $½ \leq D \leq 1½$ in. ($12 \leq D$* 40 mm) use honed-orifice meter run or Foxboro IFOA.

(From Miller, *Flow Measurement Engineering Handbook*, © 1989, McGraw-Hill Book Company. Used with permission.)

Sizing Many factors are considered in the sizing of differential pressure producers. Precise calculations are usually performed on a computer due to interaction of flowmeter parameters and the numerical tedium of precise calculations. The method for sizing orifice plates and other differential producing devices is based upon ASME/ISO discharge coefficients and uses equations presented in ASME MFC-3M-1989 and *Flow Measurement Engineering Handbook* by R. W. Miller.

Plant calculations are performed by calculating a sizing factor and estimating the approximate beta ratio of the flowmeter. Precise calculations are performed by iteration. Calculations are presented in terms of mass flow. It should be noted that orifice sizing slide rules are available to estimate the approximate beta ratio.

Step 1

Select a full scale flow, design flow, and differential pressure for the flowmeter.

In gas applications, the differential pressure should be selected such that the expansion factor variation is kept to less than 1 percent, i.e.,

$$\Delta p/P \text{ less than or equal to } 0.04$$

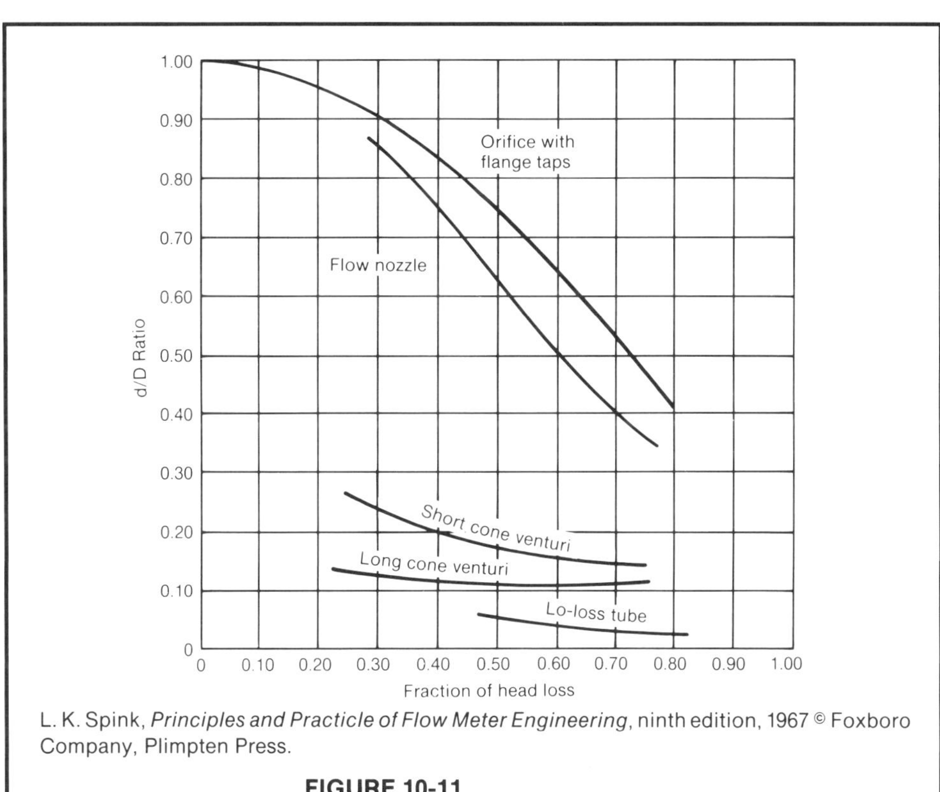

L. K. Spink, *Principles and Practicle of Flow Meter Engineering*, ninth edition, 1967 © Foxboro Company, Plimpten Press.

FIGURE 10-11
Head loss versus diameter ratio.

In commonly used units of inches of water column and psia, the relationship is approximated by

$$\Delta P_{\text{in. WC}} / P_{\text{psia}} \text{ less than or equal to } 1.0$$

When the design flow is not known, assume that the design flow is 80 percent of the full scale flow. A differential pressure of 100 inches of water column (25 kPa) is assumed when this parameter is not otherwise specified. The design flow and the differential pressure at design flow are used in all subsequent calculations to evaluate the flowmeter coefficient at design conditions, thereby minimizing the averaged flow error.

Step 2
Calculate Reynolds number at design flow and operating conditions to ensure that it is greater than the minimum values tabulated below.

	Liquid	*Gas (vapor)*
Reynolds number		
Orifice	$R_D \geq 10{,}000$	$R_D \geq 10{,}000$
Venturi nozzle	$R_D \geq 100{,}000$	$R_D \geq 10{,}000$
Lo-Loss	$R_D \geq 100{,}000$	$R_D \geq 10{,}000$
Expansion factor	$Y_1 = 1.0$	$\Delta P_{\text{in. WC}} / P_{1\text{ psia}} \leq 0.5$
	$Y_2 = 1.0$	$\Delta P_{\text{in. WC}} / P_{2\text{ psia}} \leq 1.0$

Step 3
Calculate the sizing factor at design flow and operating conditions

$$S_M = \frac{Q_{\text{lb/hr}}}{358.9268\, F_a \times D^2 \times F_p{}^{1/2} \times \rho_{\text{upstream}}{}^{1/2} \times \Delta P_{\text{in. WC}}{}^{1/2}}$$

where ρ is in pounds per cubic feet and D is in inches. F_a is the thermal expansion factor, which is defined as follows: when the coefficients of linear expansion, often called the thermal expansion coefficients, of the primary element and pipe are approximately the same.

$$F_a = 1 + 2\,\alpha \times (T_{°\text{F}} - 68)$$

Figure 10-12 shows the relationship between the expansion factor and temperature for most commonly used orifice plate materials.

F_p is the liquid compressibility factor correction, which is 1.0 for gases and most liquid applications. F_p can be estimated by using the graph in Figure 10-13.

Step 4

Calculate the approximate beta for the appropriate primary device using S_M calculated above and the equations in Table 10-2.

Beta is typically limited to between 0.12 and 0.75. When beta is outside these limits, or when beta is desired to be different from the calculated beta, return to Step 1 and select a higher or lower differential pressure, or a larger or smaller pipe size to decrease or increase beta, respectively.

For an iterative solution, continue with the next step; however, when approximate bore that yields an accuracy of approximately 2 percent is desired, proceed to Step 9.

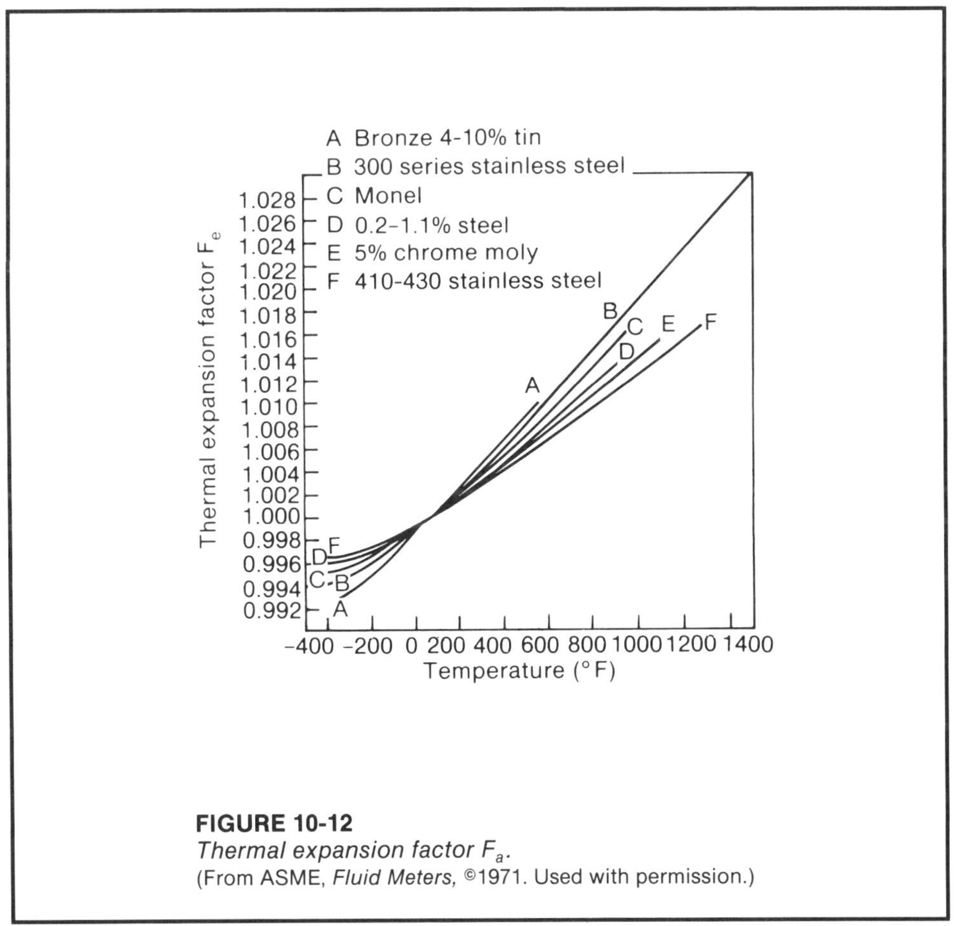

FIGURE 10-12
Thermal expansion factor F_a.
(From ASME, *Fluid Meters,* ©1971. Used with permission.)

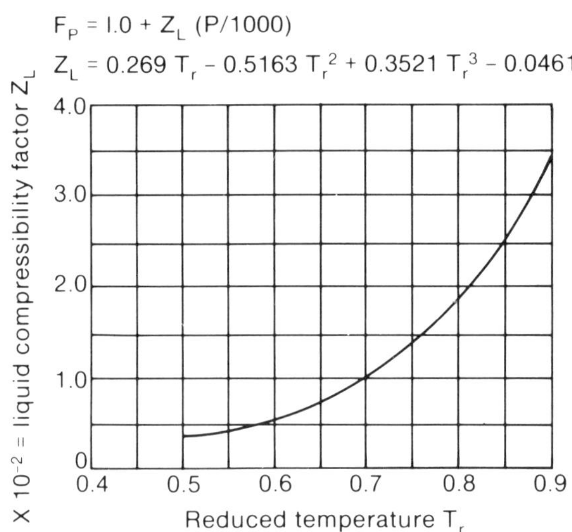

FIGURE 10-13
Generalized liquid compressibility factor.
(From Miller, *Flow Measurement Engineering Handbook*, © 1989, McGraw-Hill Book Company. Used with permission.)

Step 5
Using beta and Table 10-3, calculate the discharge coefficient which has the form:

$$C = C_{\text{infinity}} + (b/\mathrm{R_D}^n)$$

Step 6
For liquids, set Y_1 equal to 1.0. For gases and vapors, calculate the gas expansion factor upstream of the flowmeter, Y_1, using equations in Table 10-4. Subscripts 1 and 2 represent conditions upstream and downstream of the flowmeter, respectively, and k is the isentropic exponent for an ideal gas:

$$k = C_p/C_v$$

where C_p and C_v are the specific heats at constant pressure and volume, respectively.

Step 7
Calculate the next estimate for beta as

$$\beta = \left[1 + \left(\frac{C \times Y_1}{S_M}\right)^2\right]^{-1/4}$$

TABLE 10-2
β_0 Approximate Sizing Equations

Type	*Equation*
Venturi	
Machined inlet	$\beta_0 = \left[1 + \left(\frac{0.995}{S_M}\right)^2\right]^{-1/4}$
Rough-cast inlet	$\beta_0 = \left[1 + \left(\frac{0.984}{S_M}\right)^2\right]^{-1/4}$
Rough-welded sheet iron	$\beta_0 = \left[1 + \left(\frac{0.985}{S_M}\right)^2\right]^{-1/4}$
Universal Venturi Tube†	$\beta_0 = \left[1 + \left(\frac{0.9797}{S_M}\right)^2\right]^{-1/4}$
Lo-Loss tube‡	$\beta_0 = \left[1 + \left(\frac{0.92}{S_M} - 0.31\right)^2\right]^{-1/4}$
Nozzle	
ASME long radius	$\beta_0 = \left[1 + \left(\frac{0.9975}{S_M}\right)^2\right]^{-1/4}$
ISA	$\beta_0 = \left[1 + \left(\frac{0.9944}{S_M} - 0.118\right)^2\right]^{-1/4}$
Venturi nozzle (ISA inlet)	$\beta_0 = \left[1 + \left(\frac{0.989}{S_M} - 0.09\right)^2\right]^{-1/4}$
Orifice	
Corner, flange, *D*-and-*D*/2 taps	
$R_D < 200{,}000$	$\beta_0 = \left[1 + \left(\frac{0.6}{S_M} + 0.06\right)^2\right]^{-1/4}$
$R_D > 200{,}000$	$\beta_0 = \left[1 + \left(\frac{0.6}{S_M}\right)^2\right]^{-1/4}$
2½*D* and 8*D* taps	$\beta_0 = \left[1 + \left(\frac{0.61}{S_M} + 0.55\right)^2\right]^{-1/4}$
Eccentric, all taps	$\beta_0 = \left[1 + \left(\frac{0.607}{S_M} + 0.088\right)^2\right]^{-1/4}$
Segmental, all taps	$\beta_0 = \left[1 + \left(\frac{0.634}{S_M} - 0.062\right)^2\right]^{-1/4}$
Quadrant ($\beta \leq 0.6$)	$\beta_0 = \left[1 + \left(\frac{0.76}{S_M} + 0.26\right)^2\right]^{-1/4}$
Conic, corner ($\beta \leq 0.3$)	$\beta_0 = \left[1 + \left(\frac{0.734}{S_M}\right)^2\right]^{-1/4}$

†From BIF CALC 440/441; the manufacturer should be consulted for exact coefficient information.
‡Derived from Badger Meter, Inc. Lo-Loss flow-tube coefficient curve.
(From Miller, *Flow Measurement Engineering Handbook*, ©1989, McGraw-Hill Book Company. Used with permission.)

TABLE 10-3
Equations and Values for C_∞, b, and n

Primary device	*Discharge coefficient C_∞ at infinite Reynolds number*	*Reynolds number term*	
		Coefficient b	*Exponent n*
Venturi			
Machined inlet	0.995	0	0
Routh cast inlet	0.984	0	0
Rough welded sheet-iron inlet	0.985	0	0
Universal Venturi Tube[b]	0.9797	0	0
Lo-Loss tube[c]	$1.005 - 0.471\beta + 0.564\beta^2 - 0.514\beta^3$	0	0
Nozzle:			
ASME long radius	0.9975	$-6.53\beta^{0.5}$	0.5
ISA	$0.9900 - 0.2262\beta^{4.1}$	$(-0.00175\beta^2 + 0.0033\beta^{4.15})\,10^{6.9}$	1.15
Venturi nozzle (ISA inlet)	$0.9558 - 0.196\beta^{4.5}$	0	0
Orifice:			
Corner taps	$0.5959 + 0.0312\beta^{2.1} - 0.184\beta^8$	$91.71\beta^{2.5}$	0.75
Flange taps (D in inches)			
$D \geq 2.3$	$0.5959 + 0.0312\beta^{2.1} - 0.184\beta^8 + 0.09\dfrac{\beta^4}{D(1-\beta^4)} - 0.0337\dfrac{\beta^3}{D}$	$91.71\beta^{2.5}$	0.75
$2 \leq D \leq 2.3$[d]	$0.5959 + 0.0312\beta^{2.1} - 0.184\beta^8 + 0.039\dfrac{\beta^4}{1-\beta^4} - 0.0337\dfrac{\beta^3}{D}$	$91.71\beta^{2.5}$	0.75

(continued)

TABLE 10-3
Equations and Values for C_∞, b, and n[a] (continued)

Primary device	Discharge coefficient C_∞ at infinite Reynolds number	Reynolds number term: Coefficient b	Reynolds number term: Exponent n
Flange taps (D^* in millimeters)			
$D^* \geq 58.4$	$0.5959 + 0.0312\beta^{2.1} - 0.184\beta^8 + 2.286 \dfrac{\beta^4}{D^*(1-\beta^4)} - 0.856 \dfrac{\beta^3}{D^*}$	$91.71\beta^{2.5}$	0.75
$50.8 \leq D^* \leq 58.4$[d]	$0.5959 + 0.0312\beta^{2.1} - 0.184\beta^8 + 0.039 \dfrac{\beta^4}{1-\beta^4} - 0.856 \dfrac{\beta^3}{D^*}$	$91.71\beta^{2.5}$	0.75
D and $D/2$ taps	$0.5959 + 0.0312\beta^{2.1} - 0.184\beta^8 + 0.039 \dfrac{\beta^4}{1-\beta^4} - 0.0158\beta^3$	$91.71\beta^{2.5}$	0.75
2½D and 8D taps[e]	$0.5959 + 0.461\beta^{2.1} - 0.48\beta^8 + 0.039 \dfrac{\beta^4}{1-\beta^4}$	$91.71\beta^{2.5}$	0.75
Eccentric orifice (flange taps)[f]			
180° taps			
$D \leq 4$ (100 mm)	$0.5875 + 0.3813\beta^{2.1} + 0.6898\beta^8 - 0.1963 \dfrac{\beta^4}{1-\beta^4} - 0.3366\beta^3$	$7.3 - 15.7\beta + 170.8\beta^2 - 399.7\beta^3 + 332.2\beta^4$	0.75
$D > 4$ (100 mm)	$0.5949 + 0.4078\beta^{2.1} + 0.0547\beta^8 + 0.0955 \dfrac{\beta^4}{1-\beta^4} - 0.5608\beta^3$	$-139.7 + 1328.8\beta - 4228.2\beta^2 + 5691.9\beta^3 - 2710.4\beta^4$	0.75
90° taps			
$D \leq 4$ (100 mm)	$0.6284 + 0.1462\beta^{2.1} - 0.8464\beta^8 + 0.2603 \dfrac{\beta^4}{1-\beta^4} - 0.2886\beta^3$	$69.1 - 469.4\beta + 1245.6\beta^2 - 1287.5\beta^3 + 486.2\beta^4$	0.75
$D > 4$ (100 mm)	$0.6276 + 0.0828\beta^{2.1} + 0.2739\beta^8 - 0.0934 \dfrac{\beta^4}{1-\beta^4} - 0.1132\beta^3$	$-103.2 + 898.3\beta - 2557.3\beta^2 + 2977\beta^3 - 1131.3\beta^4$	0.75

(continued)

TABLE 10-3
Equations and Values for C_∞, *b*, and *n* (continued)

Primary device	Discharge coefficient C_∞ at infinite Reynolds number	Reynolds number term: Coefficient *b*	Reynolds number term: Exponent *n*
Vena contracta taps[f]			
180° taps			
$D \leq 4$ (100 mm)	$0.6261 + 0.1851\beta^{2.1} - 0.2879\beta^8 + 0.1170 \frac{\beta^4}{1-\beta^4} - 0.2845\beta^3$	$23.3 - 207\beta + 821.5\beta^3 - 1388.6\beta^3 + 900.3\beta^4$	0.75
$D > 4$ (100 mm)	$0.6276 + 0.0828\beta^{2.1} + 0.2739\beta^8 - 0.0934 \frac{\beta^4}{1-\beta^4} - 0.1132\beta^3$	$55.7 - 471.4\beta + 1721.8\beta^2 - 2722.6\beta^3 + 1569.4\beta^4$	0.75
90° taps			
$D \leq 4$ (100 mm)	$0.5917 + 0.3061\beta^{2.1} + 0.3406\beta^8 - 0.1019 \frac{\beta^4}{1-\beta^4} - 0.2715\beta^3$	$-69.3 + 556.9\beta - 1332.2\beta^2 + 1303.7\beta^3 - 394.8\beta^4$	0.75
$D > 4$ (100 mm)	$0.6016 + 0.3312\beta^{2.1} - 1.5581\beta^8 + 0.6510 \frac{\beta^4}{1-\beta^4} - 0.7308\beta^3$	$52.8 - 434.2\beta + 1571.2\beta^2 - 2460.9\beta^3 + 1420.2\beta^4$	0.75
Segmental[f]			
Flange taps			
$D \leq 4$ (100 mm)	$0.5866 + 0.3917\beta^{2.1} + 0.7586\beta^8 - 0.2273 \frac{\beta^4}{1-\beta^4} - 0.3343\beta^3$	0	0
$D > 4$ (100 mm)	$0.6037 + 0.1598\beta^{2.1} - 0.2918\beta^8 + 0.0244 \frac{\beta^4}{1-\beta^4} - 0.0790\beta^3$	0	0

(continued)

TABLE 10-3
Equations and Values for C_∞, b, and n (continued)

Primary device	*Discharge coefficient C_∞ at infinite Reynolds number*	*Reynolds number term*	
		Coefficient b	*Exponent n*
Vena contracta			
$D \leq 4$ (100 mm)	$0.5925 + 0.3380\beta^{2.1} + 0.4016\beta^{8} - 0.1046 \dfrac{\beta^4}{1-\beta^4} - 0.3212\beta^3$	0	0
$D > 4$ (100 mm)	$0.5922 + 0.3932\beta^{2.1} + 0.3412\beta^{8} - 0.0569 \dfrac{\beta^4}{1-\beta^4} - 0.4628\beta^3$	0	0
Quadrant (corner and flange taps)[f]			
$D \geq 1.5$ (40 mm)	$0.7746 - 0.1334\beta^{2.1} + 0.4098\beta^{8} + 0.0675 \dfrac{\beta^4}{1-\beta^4} + 0.3865\beta^3$	0	0
Conic orifice	$250\beta \leq R_D \leq 500\beta$ 0.734		
$D \geq 1$ (25 mm)	$5000\beta < R_D \leq 200{,}000\beta$ 0.730	0	0
Honed orifice meter runs			
Flange taps	$[0.5980 + 0.468(\beta^4 + 10\beta^{12})](1 - \beta^4)^{0.5}$	$(0.87 + 8.1\beta^4)(1 - \beta^4)^{0.5}$	0.5
$\frac{1}{2} \leq D \leq 1\frac{1}{2}$			
Corner taps (D in inches) $\frac{1}{2} \leq D \leq 1\frac{1}{2}$	$\left[0.5991 + \dfrac{0.0044}{D} + \left(0.3155 + \dfrac{0.0175}{D}\right)(\beta^4 + 2\beta^{16})\right](1 - \beta^4)^{0.5}$	$\left[\dfrac{0.52}{D} - 0.192 + \left(16.48 \dfrac{1.16}{D}\right)(\beta^4 + 4\beta^{16})\right](1 - \beta^4)^{0.5}$	0.5

(continued)

TABLE 10-3
Equations and Values for C_∞, b, and n (continued)

Primary device	*Discharge coefficient C_∞ at infinite Reynolds number*	*Reynolds number term*	
		Coefficient b	*Exponent n*
Corner taps (D^* in millimeters)	$\left[0.5991 + \frac{1.1176}{D^*} + \left(0.3155 + \frac{0.4445}{D^*}\right)(\beta^4 + 2\beta^{16})\right](1 - \beta^4)^{0.5}$	$\left[\frac{13.2}{D^*} - 0.192 + \left(16.48 \frac{29.46}{D^*}\right)(\beta^4 + 4\beta^{16})\right](1 - \beta^4)^{0.5}$	0.5
$\frac{1}{2} \leq D \leq 1\frac{1}{2}$ Corner taps (D^* in millimeters)	$\left[0.5991 + \frac{1.1176}{D^*} + \left(0.3155 + \frac{0.4445}{D^*}\right)(\beta^4 + 2\beta^{16})\right](1 - \beta^4)^{0.5}$	$\left[\frac{13.2}{D^*} - 0.192 + \left(16.48 \frac{29.46}{D^*}\right)(\beta^4 + 4\beta^{16})\right](1 - \beta^4)^{0.5}$	0.5
Integral Flow Orifice Assembly Foxboro (IFOA)			
Quadrant-edged $D = \frac{1}{2}$ (12.5 mm)	$1.1126 - 99.13\beta^2 + 8006\beta^4 - 26900\beta^8$	$-10.72\beta^{1/2} + 3823\beta^{5/2} - 309{,}300\beta^{9/2}$	0.5
Square-edged $D = \frac{1}{2}$ (12.5 mm)	$0.6479 - 0.3505\beta^2 + 0.3853\beta^4 + 4.645\beta^8$	$-0.4356\beta^{1/2} + 33.49\beta^{5/2} - 88.33\beta^{9/2}$	0.5
Square-edged $D = 1$ (12.5 mm)	$0.6050 - 0.1837\beta^2 + 0.6615\beta^4 - 1.094\beta^8$	$1.646\beta^{1/2} + 2.394\beta^{5/2} - 4.899\beta^{9/2}$	0.5
Square-edged $D = 1\frac{1}{2}$ (40 mm)	$0.6122 - 0.1076\beta^2 + 0.3416\beta^4 - 0.684\beta^8$	$0.2368\beta^{1/2} + 14.3\beta^{5/2} - 12.86\beta^{9/2}$	0.5

[a] Detailed Reynolds-number, line-size, beta-ratio, and other limitations are given in Table 10-1.
[b] From BIF CALC-440/441; the manufacturer should be consulted for exact coefficient information.
[c] Derived from the Badger Meter, Inc. Lo-Loss tube coefficient curve; the manufacturer should be consulted for exact coefficient information.
[d] For $\frac{1}{2} \leq D \leq 1\frac{1}{2}$ in ($12 \leq D^* \leq 40$ mm) use flow coefficient equation with $C = \sqrt{1 - \beta^4}\,K$.
[e] SOURCE: Stolz (1978).
[f] SOURCE: Freeman (1988).

(From Miller, *Flow Measurement Engineering Handbook*, © 1989, McGraw-Hill Book Company. Used with permission.)

TABLE 10-4
Summary of Gas (Vapor) Expansion-Factor Equations

	Equation	
	CONTOURED PRIMARY ELEMENTS (NOZZLE, VENTURI, VENTURI NOZZLE, ETC.)	
Upstream measurements	$Y_1 = \left\{ \dfrac{(1-\beta^4)\,[k/(k-1)]\,(P_2/P_1)^{2/k}\,[1-(P_2/P_1)^{(k-1)/k}]}{[1-\beta^4 (P_2/P_1)^{2/k}]\,[1-(P_2/P_1)]} \right\}^{1/2}$	$\dfrac{P_2}{P_1} = 1 - x_1$
Downstream measurements	$Y_2 = Y_1 \sqrt{1 + x_2}$	$\dfrac{P_2}{P_1} = \dfrac{1}{1 + x_2}$
	CONCENTRIC ORIFICE	
Corner, flange, *D* and *D*/2 taps		
Upstream measurements	$Y_1 = 1 - (0.41 + 0.35\beta^4)\,\dfrac{x_1}{k}$	
Downstream measurements	$Y_2 = \sqrt{1 + x_2} - (0.41) + 0.35\beta^4)\,\dfrac{x_2}{k\sqrt{1 + x_2}}$	$x_1 = \dfrac{\Delta p_{\text{in. WC}}}{27.73 P_{1\,\text{psia}}}$
2½*D* and 8*D*		
Upstream measurements	$Y_1 = 1 - [0.333 + 1.145\,(\beta^2 + 0.7\beta^5 + 12\beta^{13})]\,\dfrac{x_1}{k}$	
Downstream measurements	$Y_2 = \sqrt{1 + x_2} = [0.333 + 1.145\,(\beta^2 + 0.7\beta^5 + 12\beta^{13})]\,\dfrac{x_2}{k\sqrt{1 + x_2}}$	$x_2 = \dfrac{\Delta p_{\text{in. WC}}}{27.73 P_{1\,\text{psia}}}$
	ECCENTRIC ORIFICE	
Upstream measurements	$Y_1 = 1 - (0.1926 + 0.574\beta + 0.9675\beta^2 - 4.24\beta^3 + 3.62\beta^4)\,\dfrac{x_1}{k}$	
Downstream measurements	$Y_2 = \sqrt{1 + x_2} - (0.1926 + 0.574\beta + 0.9675\beta^2 - 4.24\beta^3 + 3.62\beta^4)\,\dfrac{x_2}{k\sqrt{1 + x_2}}$	

Step 8
Repeat steps 5, 6, and 7 until two consecutive iterations of beta differ by less than 0.0001.

Step 9
Calculate the bore of the flowmeter using

$$d = \beta \times D$$

Several additional factors exist that can be introduced into the denominator of the expression for S_M to compensate for special measuring situations, such as to correct for steam quality (gas-liquid flows), drain or vent holes, water vapor in a gas, and the differential pressure transmitter located substantially above or below the primary element.

EXAMPLE 10-11

Problem: Size the bore of a flange tapped concentric orifice plate required to measure 0 to 100 gpm of a liquid flowing in a 2-inch schedule 80 pipe at 50 psi and 80° F with a viscosity of 0.63 cP at flowing conditions, and base and flowing specific gravities of 0.736 and 0.726, respectively. Base conditions are at 60°F and the liquid has a critical temperature of 795°R. The design flow is 80 gpm.

Solution:

Step 1
Select basic flowmeter parameters.

Full scale flow	100 gpm
Maximum differential pressure	100 in. WC
Design flow	80 gpm
Differential pressure @ design flow ($0.8^2 \times$ FS)	64 in. WC

Step 2
Calculate Reynolds number at design conditions.

Q_{gpm} = 80 gpm

SG = 0.726

μ_{cP} = 0.63 cP

D = 1.939 inches

R_D = (3160 × 80 gpm × 0.726)/(0.63 cP × 1.939 in.) = 150,243

Reynolds number is sufficiently high.

Step 3
Calculate the sizing factor S_M.

$$Q_{lbs/hr} = 80 \text{ gpm} \times 8.334 \text{ lbs/gal} \times 0.726 \times 60 \text{ min/hr}$$
$$= 29{,}442 \text{ lbs/hr}$$
$$F_a = 1 + 2 \times (9.6 \times 10^{-6}) \times (80°F - 68°F) = 1.00023$$
$$T_R = (460°F + 80°F)/795°R = 0.68 \text{ such that } Z_L = 0.009$$
$$F_p = 1 + 0.009/[(14.7 \text{ psi} + 50 \text{ psi})/1000] = 1.0000582$$
$$D^2 = (1.939 \text{ inches})^2 \times 3.7597 \text{ inches}^2$$
$$\rho_{upstream} = 0.726 \times 62.3363 \text{ lb/ft}^3 = 45.256 \text{ lb/ft}^3$$
$$\Delta P_{in.\,WC} = (0.8^2 \times 100) = 64 \text{ in. WC}$$
$$S_M = 29{,}442/(358.9268 \times 1.0002 \times 3.7597 \text{ in.}^2 \times (45.256 \text{ lb/ft}^3)^{1/2} \times (64 \text{ in. WC})^{1/2} \times 1.0000582^{1/2})$$
$$= 0.40530$$

Step 4
Calculate β_0.

From Table 10-2:

$$\beta_0 = [1 + [(0.6/S_M) + 0.06]^2]^{-1/4}$$

Note that β_0 can be calculated on a calculator with a square root function as follows:

$$\beta_0 = 1/[(1 + [(0.6/0.40530) + 0.06]^2)^{1/2}]^{1/2}$$
$$= 1/[(3.3728)^{1/2}]^{1/2}$$
$$= 1/1.8365^{1/2} = 1/1.35518 = 0.73791$$

Calculate the approximate bore from Step 9 as

$$d = 0.73791 \times 1.939 = 1.431 \text{ inches}$$

Continue to Step 5 for precise calculations.

Step 5
Calculate the discharge coefficient C.

Substituting 0.73791 for β into the equation for a 2-inch orifice plate with flange taps yields:

$$C_{infinity} = 0.60556$$
$$b = 42.897$$
$$n = 0.75$$

such that

$$C_0 = 0.60556 + (42.897/150{,}240^{0.75}) = 0.61118$$

Step 6
Calculate Y.

$$Y_1 = 1 \text{ for liquids}$$

Step 7
Calculate β_1.

$$\beta_1 = (1 + [(0.61118 \times 1)/0.40530]^2)^{-1/4} = 0.74341$$

Step 8
Repeat Steps 5, 6 and 7.

	Iteration n	
	1	2
β_{n-1}	0.73791	0.74341
C_{n-1}	0.61118	0.61121
β_n	0.74341	0.74340

Step 9
Calculate bore.

The bore is calculated as:

$$d = 0.74340 \times 1.939 = 1.441 \text{ inches}$$

EXAMPLE 10-12

Problem: Size the bore of a flange tapped eccentric orifice plate required to measure 0 to 200 scfm of an ideal gas flowing in a 2-inch schedule 40 pipe at 15 psi and 90°F with a viscosity of 0.011 cP at flowing conditions and a specific gravity of 0.63 at standard conditions. The gas has an isentropic constant of 1.25.

Solution:

Step 1
Select basic flowmeter parameters.

Full scale flow	200 scfm
Maximum differential pressure	20 in. WC
Design flow (assume 80% FS)	160 scfm
Differential pressure @ design flow ($0.8^2 \times$ FS)	12.8 in. WC

Step 2
Calculate Reynolds number and $\Delta p/P$ at design conditions.

Q_{acfm} = 160 acfm × [(460°F + 90°F)/(460°F + 59°F)] × [14.7 psi/(14.7 psi + 15 psi)] = 83.92 acfm

ρ = 0.63 × 0.0764 lb/ft^3 air × [(460°F + 59°F)/(460°F + 90°F)] × [(14.7 psi + 15 psi)/14.7 psi]

= 0.91765 lb/ft^3

μ_{cP} = 0.011 cP

D = 2.067 inches

R_D = (379 × 83.92 × 0.091765)/(0.011 × 2.067) = 128,366

Reynolds number is sufficiently high.

$\Delta p/P$ = 12.8 in. WC/(14.7 psi + 15 psi) = 0.43

This ratio satisfies the maximum of 0.5 required for calculating an approximate beta, as well as 1.0 for precise calculations.

Step 3
Calculate the sizing factor S_M.

$Q_{lb/hr}$ = 160 scfm × 0.0764 lb/ft^3 × 0.63 × 60 min/hr

= 0.091765 lb/ft^3 × 83.92 acfm

= 462.1 lb/hr

F_a = 1 + 2 × (9.6 × 10^{-6}) × (90 – 68) = 1.00042

F_p = 1.0

D^2 = (2.067 inches)^2 = 4.2725 inches^2

$\rho_{upstream}$ = 0.091765 lb/ft^3 from above

$\Delta P_{in.\ WC}$ = $(0.8)^2$ × 20 in. WC = 12.8 in. WC

S_M = 462.1 lb/hr / (358.9268 × 1.00042 × 4.2725 in.^{2} × (0.091765 lb/ft^3)$^{1/2}$ × (12.8 in. WC)$^{1/2}$ × 1.0$^{1/2}$)

= 0.27792

Step 4
Calculate β_0.

From Table 10-2:

$$\beta_0 = [1 + [(0.6/0.27792) + 0.088]^2]^{-1/4} = 0.63766$$

Calculate the approximate bore from Step 9 as

$$d = 0.63766 \times 2.067 \text{ in.} = 1.318 \text{ inches}$$

Continue to Step 5 for precise calculations.

Step 5
Calculate the discharge coefficient C.

Substituting 0.63766 for beta into the equation for a 2-inch orifice plate with flange taps yields:

$$C_{\text{infinity}} = 0.60650$$
$$b = 29.78$$
$$n = 0.75$$

such that

$$C_0 = 0.60650 + (29.78/128{,}366^{0.75}) = 0.61089$$

Step 6

Calculate Y.

$$X_1 = 12.8/[27.73 \times (14.7 + 15)]\ 0.01554$$
$$\beta_0 = 0.63766$$
$$k = 1.25$$
$$(Y_1)_0 = [1 - (0.41 + 0.35 \times (0.63766)^4) \times (0.01554/1.25)] = 0.99418$$

Step 7
Calculate β_1.

$$\beta_1 = [1 + [(0.61089 \times 0.99418)/0.27792]^2]^{-1/4}$$
$$= 0.64506$$

Step 8
Repeat Steps 5, 6 and 7.

	Iteration n	
	1	2
β_{n-1}	0.63766	0.64506
C_{n-1}	0.61089	0.61102
$(Y_1)_{n-1}$	0.99418	0.99415
β_n	0.64506	0.64501

Step 9
Calculate bore.

The bore is calculated as

$$d = 0.6450 \times 2.067 = 1.333 \text{ inches}$$

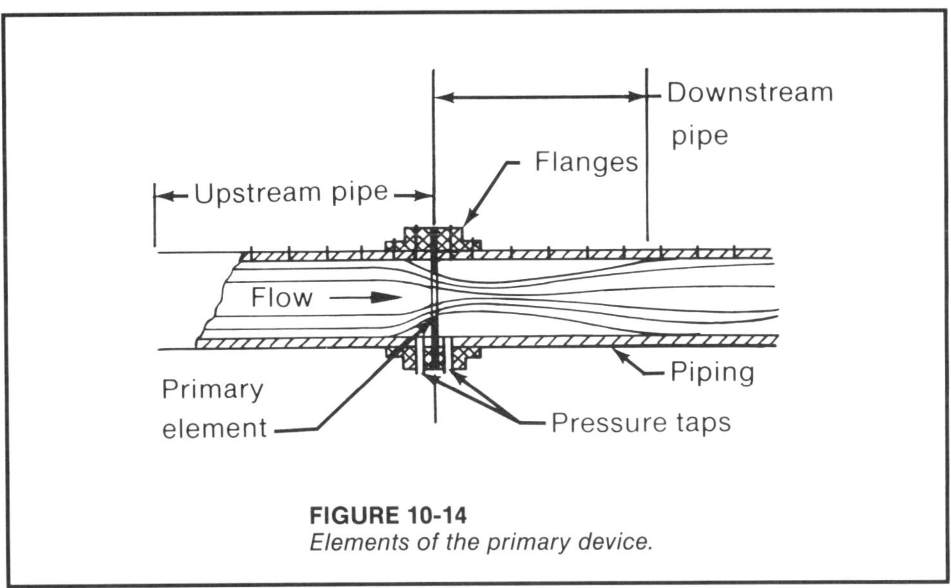

FIGURE 10-14
Elements of the primary device.

Installation Orifice plate technology is relatively simple in concept; however, there are a number of installation guidelines that must be followed so that expected accuracy is achieved.

Hydraulic Requirements

Orifice plates are sensitive to the velocity profile of the fluid entering the flowmeter. As a result, there are a number of requirements that must be satisfied to effect a satisfactory installation.

Required upstream and downstream straight run (that is, the distance required upstream and downstream of the orifice plate between the orifice plate and the nearest pipe fitting, which should include no pipe connections other than those required for the temperature and pressure taps required for the flowmeter) is dependent upon the beta ratio of the orifice plate. In general, the higher the beta ratio, the higher the requirements, which typically vary from 6 to $38D/2$ to $4D$ (or higher for certain piping configurations) for beta ratios between 0.12 to 0.72.

ASME requires that piping $4D$ upstream and $2D$ downstream of the orifice plate have a 350 finish free of mill scale, holes, bumps, and the like, while the pipe diameter should not depart from the average by more than 0.33 percent. These requirements exceed ASTM pipe specifications of commonly used pipe and the quality of field welds that would typically be made at the inlet and outlet flanges.

AGA requirements are more stringent in that they apply to the entire upstream and downstream straight run. Seamless pipe with a 300 finish may be used; any grooves, scoring, pits, ridges resulting from seams, distortion caused by welding, and the like, that exceed the tolerances in Figure 10-16 must be corrected by filling in, grinding, or filing.

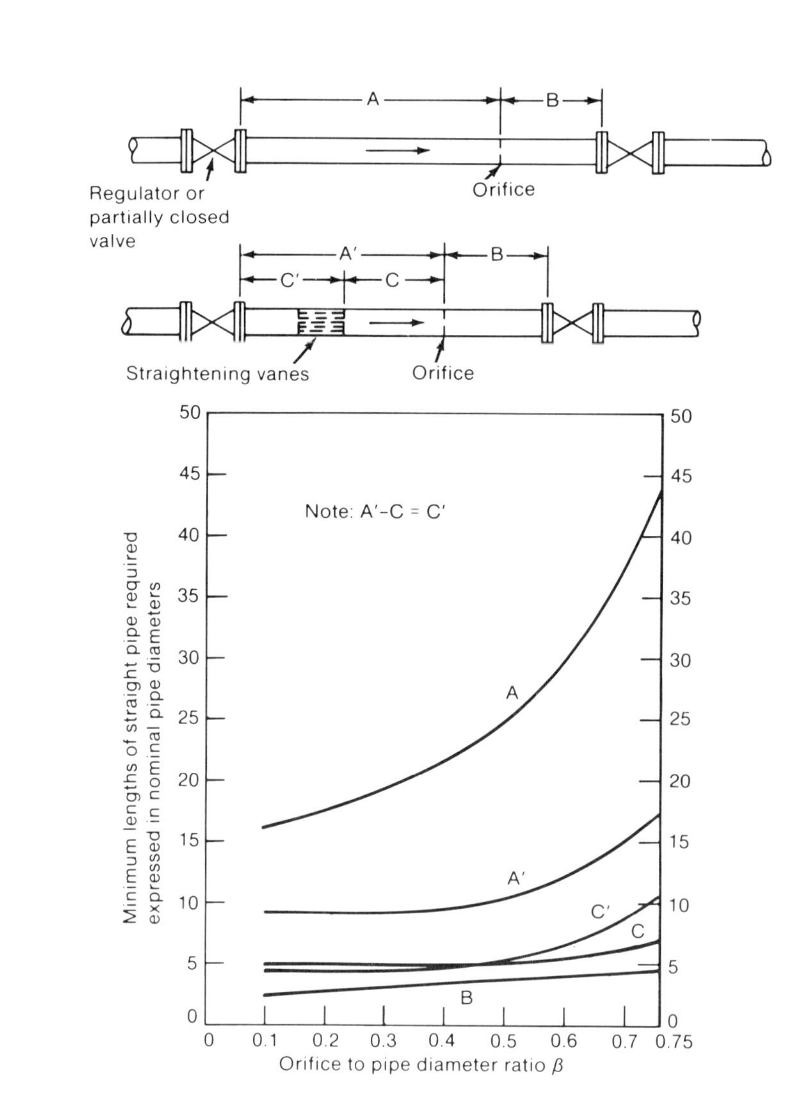

NOTES:

1. When "pipe taps" are used, lengths A, A', and C shall be increased by 2 pipe diameters, and B by 8 pipe diameters.
2. When the diameter of the orifice may require changing to meet different conditions, the lengths of straight pipe should be those required for the maximum orifice to pipe diameter ratio that may be used.

FIGURE 10-15A

Requirements for orifice metering of natural gas.

(From ANSI/API 2530, ©1978, American Gas Association. Used with permission.)

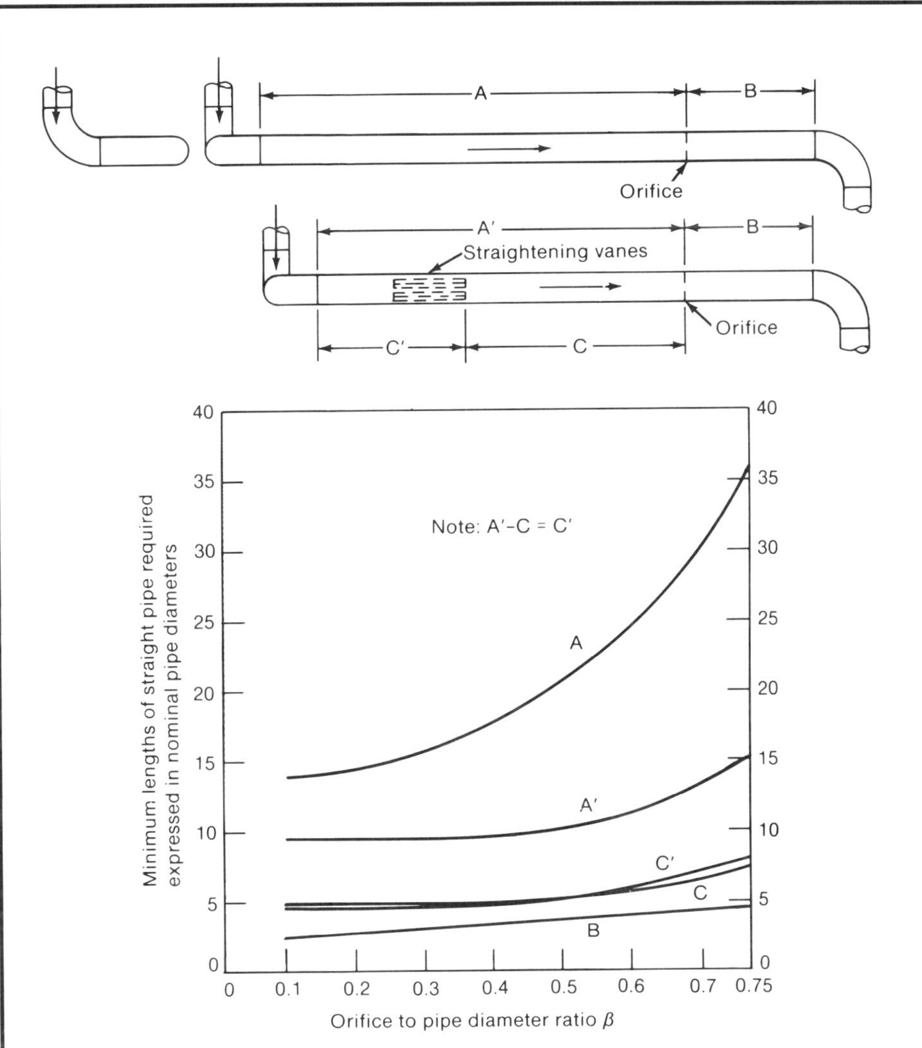

NOTES:

1. When "pipe taps" are used, lengths A, A′, and C shall be increased by 2 pipe diameters, and B by 8 pipe diameters.
2. When the diameter of the orifice may require changing to meet different conditions, the lengths of straight pipe should be those required for the maximum orifice to pipe diameter ratio that may be used.
3. When the 2 ells shown in the above sketches are closely (less than [3D]) preceded by a third which is not in the same plane as the middle or second ell, the piping requirements shown by A should be doubled.

FIGURE 10-15B
Requirements for orifice metering of natural gas.
(From ANSI/API 2530, ©1978, American Gas Association. Used with permission.)

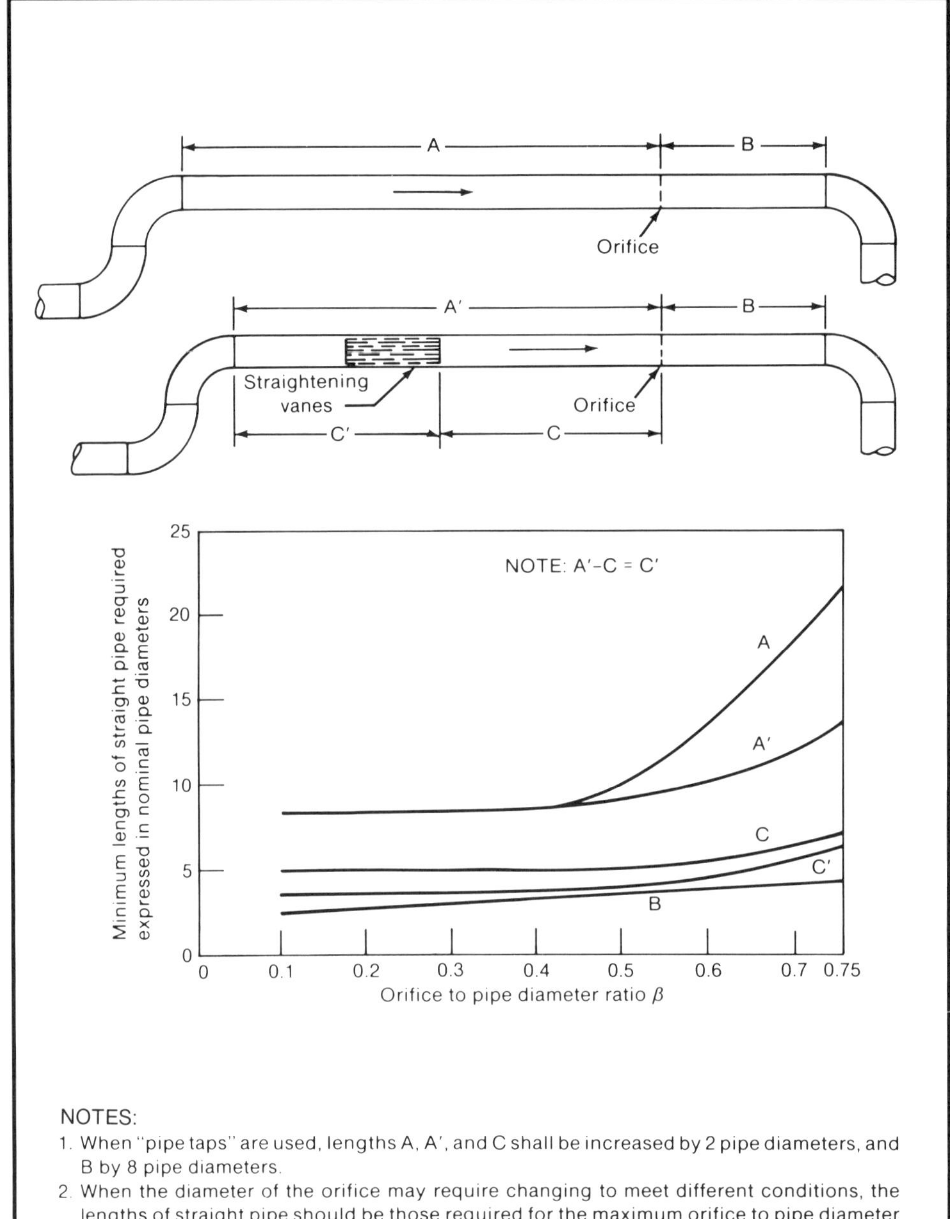

NOTES:

1. When "pipe taps" are used, lengths A, A′, and C shall be increased by 2 pipe diameters, and B by 8 pipe diameters.
2. When the diameter of the orifice may require changing to meet different conditions, the lengths of straight pipe should be those required for the maximum orifice to pipe diameter ratio that may be used.

FIGURE 10-15C

Requirements for orifice metering of natural gas.

(From ANSI/API 2530, ©1978, American Gas Association. Used with permission.)

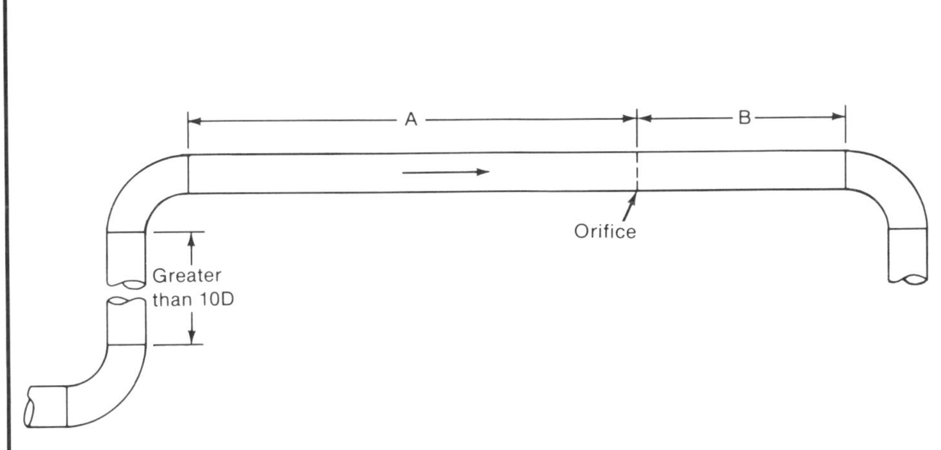

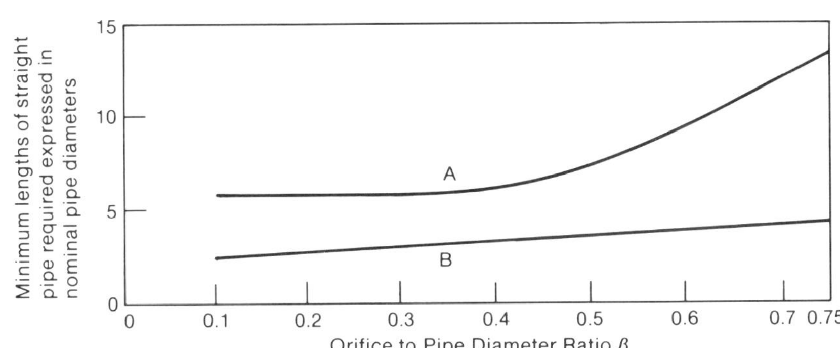

NOTES:

1. When "pipe taps" are used, length A shall be increased by 2 pipe diameters, and B by 8 pipe diameters.
2. When the diameter of the orifice may require changing to meet different conditions, the lengths of straight pipe should be those required for the maximum orifice to pipe diameter ratio that may be used.
3. The straight run of pipe between the elbows must be at least 10 diameters in length. If this length is less than 10 diameters, Figure 3C shall be applicable.

FIGURE 10-15D
Requirements for orifice metering of natural gas.
(From ANSI/API 2530, ©1978, American Gas Association. Used with permission.)

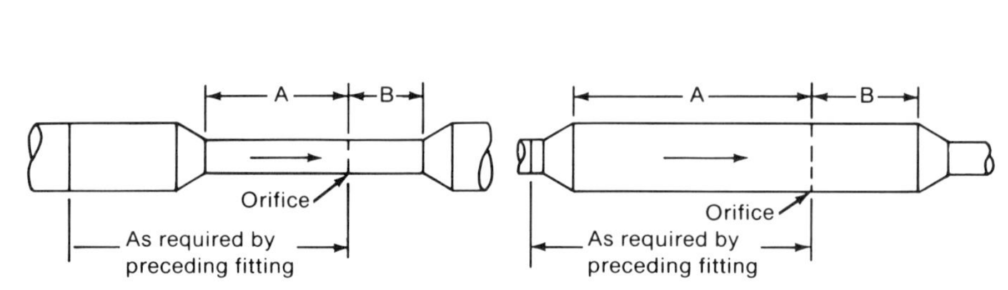

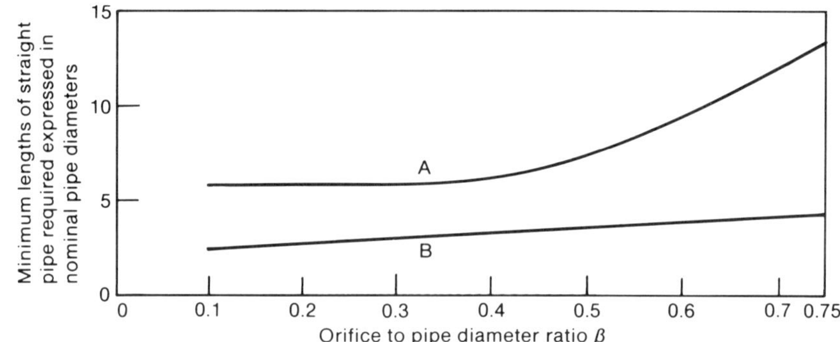

NOTES:

1. When "pipe taps" are used, length A shall be increased by 2 pipe diameters, and B by 8 pipe diameters.
2. When the diameter of the orifice may require changing to meet different conditions, the lengths of straight pipe should be those required for the maximum orifice to pipe diameter ratio that may be used.
3. Straightening vanes will not reduce required lengths of straight pipe A. Straightening vanes are not required because of the reducers. They may be required because of other fittings which precede the reducer. Length A is to be increased by an amount equal to the length of the straightening vanes whenever they are used.

FIGURE 10-15E

Requirements for orifice metering of natural gas.

(From ANSI/API 2530, ©1978, American Gas Association. Used with permission.)

	Upstream (Inlet) of the Primary Device											Downstream
β	Single 90° Bend or Tee (Flow From One Branch Only)	Two or More 90° Bends in the Same Plane [Note (1)]	*C′*	*C*	Two or More 90° Bends in Different Planes [Note (1)]	*C′*	*C*	Reducer (2*D* to *D* *Over a* Length of 1.5*D* to 3*D*)	Expander (0.5*D* to *D*Over a Length of 1*D* to 2*D*)	Glove Valve Fully Open [Note (2)]	Full Bore Ball or Gate Valve Fully Open	All Fittings Included in This Table
0.20	6	7	3.5	5	17	4.5	5	5[3)]	8	6	6	2
0.25	6	7	3.5	5	17	4.5	5	5[3)]	8	6	6	2
0.30	6	8	3.5	5	17	4.5	5	5[3)]	8	6	6	2.5
0.35	6	8	3.5	5	18	4.5	5	5[3)]	8	6	6	2.5
0.40	7	9	3.5	5	18	4.5	5	5[3)]	8	10	6	3
0.45	7	9	4	5	19	5	5	5[3)]	9	12	6	3
0.50	7	10	4	5	20	5	5	5	9	15	6	3
0.55	8	11	4	5	22	5.5	5.5	5	10	18	7	3
0.60	9	13	4.5	5.5	24	6	5.5	5	11	22	7	3.5
0.65	11	16	5	6	27	6.5	6	6	13	25	8	3.5
0.70	14	18	5.5	6.5	31	7	6.5	7	15	25	10	3.5
0.75	18	21	6	7	35	8	7	11	19	25	12	4

	Fittings	Minimum Upstream Straight Length Required
For all β values	Abrupt symmetrical reduction having a diameter ratio ≥ 0.5	15

GENERAL NOTES:

(a) Columns *C* and *C′* are for flow conditioners.

(b) No additional uncertainty is indicated when the straight lengths are longer than twice the value.

(c) All straight lengths are expressed as multiples of the diameter *D*. They shall be measured from the upstream face of the primary device.

(d) Interpolation for intermediate β values is to be used.

(e) The pipe roughness, at least over the length indicated in this Table shall not exceed maximum roughness of 9 μm [350 μin.].

NOTES:

(1) The insertion of 5*D* to 10*D* straight lengths between the two bends is sufficient to make the combined effect the same as the single bends in the left column.

(2) These lengths require no additional uncertainty.

(3) These lengths require no additional uncertainty, but the uncertainties for shorter lengths are not well enough known to be given.

FIGURE 10-15F *Recommended straight lengths for nozzles and orifice plates for 0.5% additional uncertainty (from ASME MFC—3M—1989).*

These piping requirements are rarely adhered to in practice due to cost constraints, lack of general knowledge of the requirements, and the lack of definitive data describing the effects of not maintaining the proper pipe interior surface finish.

EXAMPLE 10-13

Problem: What are the upstream and downstream pipe requirements for the orifice installations shown in the figure?

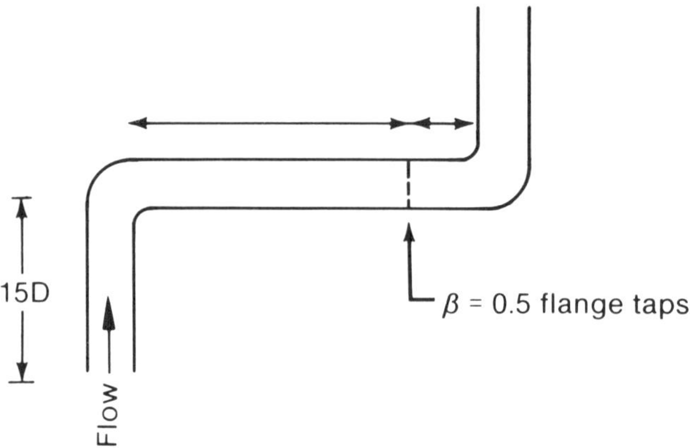

(1)

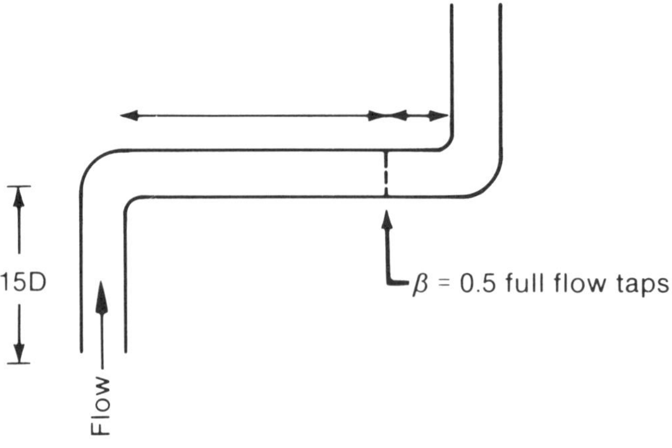

(2)

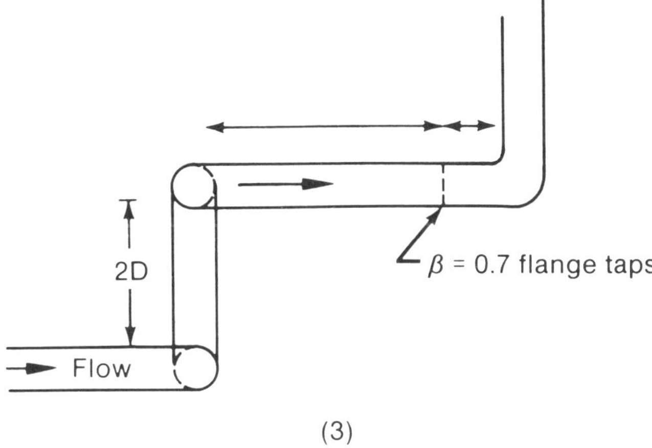

(3)

Solution:

(1) $7.5D$ / $3.5D$ from Figure 10-15D
(2) $9.5D$ / $11.5D$ from Figure 10-15D, Note 1
(3) $62D$ / $8D$ from Figure 10-15B, Note 3

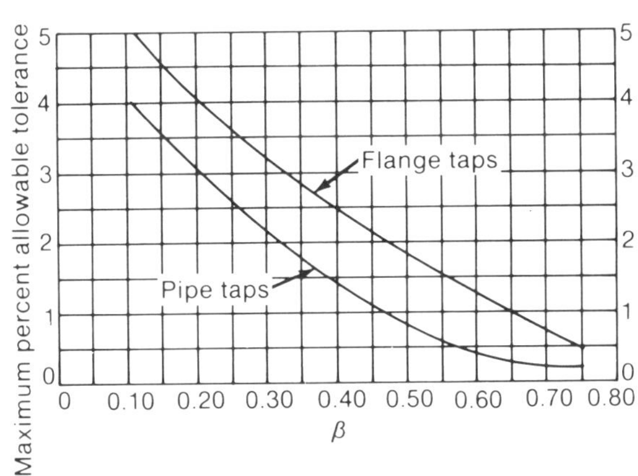

Note: It is recommended that, for new installations in which the orifice to pipe diameter ratio is likely to be changed, the tolerance permitted for variations in pipe size be the same as that given for the maximum orifice to pipe diameter ratio β=0.75.

FIGURE 10-16
Maximum percent allowable tolerance between measured upstream inside diameter and the published inside diameter, referred to the published inside diameter.
(From ANSI/API 2530, ©1978, American Gas Association. Used with permission.)

Piping

AGA recommends that the orifice plate be concentrically located in the pipe within 3 percent of the inside pipe diameter. Orifice plate alignment is more critical in small diameter runs, which is reflected by ASME's more stringent recommendation that the eccentricity toward the measuring taps be less than

$$0.0025\,D/(0.1 + 2.3 \times \beta^4)$$

for $D \geq 4$ in. and 0.030 inch for $D \leq 3$ in. and 1.5% D in other planes.

Orifice plates must be kept clean at all times and free from accumulations of any type, either on the orifice plate or within the upstream and downstream straight runs. Such accumulations will affect the velocity profile of the fluid and hence the accuracy of the measurement. When the inside diameter is unmeasured, uncertainty is introduced due to pipe fabrication tolerances.

Tap Orientation

Taps for liquid service should be located on the side of the pipe; for dry gas service, taps can be located on the side or the top of the pipe. When the gas is condensable and the impulse tubing is designed for liquid filling, the impulse tubing must allow the same condensate leg to be present on each side of the differential pressure transmitter diaphragm, independent of whether the taps are on the side or the top of the pipe.

Taps should not be located on the bottom of the pipe since dirt that may be flowing in the fluid along the bottom of the pipe may plug the taps. However, some tap configurations of eccentric and segmental orifice plate installations may have taps located on the bottom of the pipe.

Impulse Tubing

Significant errors can result from incorrect installation of the impulse tubing between the taps and the transmitter. In principle, liquids require that the impulse tubing be completely full of liquid and void of any pockets of gas. This requires that the impulse tubing be constantly sloped downward towards the transmitter from the tap located on the side of the pipe (see Figure 10-17).

Gas service usually requires that the impulse line be self-draining so that condensate or impurities cannot accumulate on one side of the differential pressure transmitter. Such accumulation can cause an error in the sensed differential pressure that is equivalent to the height of the condensate. Exceptions to this include some condensable vapors such as steam. In such cases, the condensate protects the transmitter from the hot condensing vapors.

For condensible gas service where the condensate is used to transmit the pressure from the taps to the transmitter, the impulse tubing should be void of any gas pockets and should be installed to maintain the condensate legs of equal height, which is usually more simply implemented with the tap being on the side

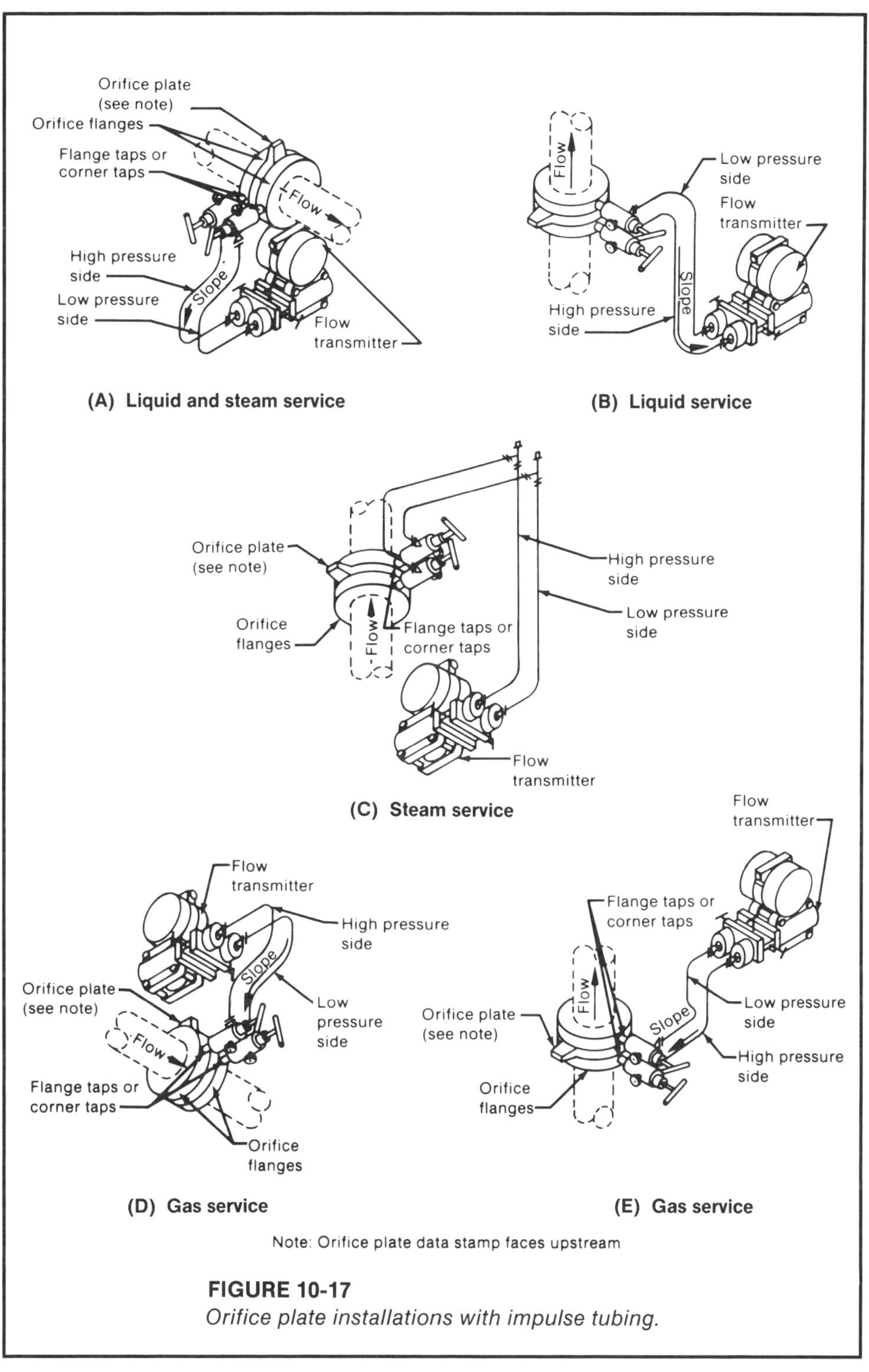

FIGURE 10-17
Orifice plate installations with impulse tubing.

of the pipe. For horizontal pipe installations, the design of the impulse tubing can be identical to that for liquid service. However, in vertical installations where one tap is above the other, the impulse tubing from the lower tap can be modified so that the condensate reaches the same height as the condensate from the upper tap. The zero calibration of the differential pressure transmitter may be modified to compensate for a difference in static head due to unequal accumulation of condensate in the two lines.

Close-coupled installation of the transmitter can simplify many of the impulse tubing-related considerations when flange taps are used. Most differential pressure transmitters are constructed in such a manner that the dimension between the high and low pressure transmitter taps is equal to the distance between flange taps, since the taps are located at fixed distances upstream and downstream of the front edge of the orifice plate, independent of pipe size, as illustrated in Figure 10-18.

Situations in which fluids contain quantities of dirt sufficient to plug the impulse tubing can be remedied by continuously purging the impulse tubing with a clean compatible fluid. The purge maintains flow of the clean fluid from the transmitter towards the taps, thereby not allowing any foreign matter from the fluid to enter the impulse tubing. Flow through the purge system should be sufficient to keep all foreign matter out of the impulse tubing. This flow may be pressure- or flow-regulated to prevent flow imbalance manifested as pressure imbalance. The purges can be balanced by adjusting the purge flows at zero flow conditions for a null reading on the transmitter.

EXAMPLE 10-14

Problem: How can incorrect impulse tubing installation cause error in liquid flow measurement? In gas flow measurement?

Solution: In liquid service, air in impulse tubes can be compressed as the pressure in the impulse tubes changes, usually causing a flow measurement that is bouncy and subject to error. In gas service, when condensate does not accumulate equally in each impulse tube, the difference of the forces exerted by the condensate on the differential pressure transmitter will cause errors.

Insulation and Heat Tracing

As impulse tubing is relatively small and is normally at no flow conditions, insulation and/or heat tracing may be required for process considerations, climatic conditions, and personnel protection. Insulation and heat tracing may be required to keep liquids from solidifying and gases from condensing. Otherwise, the result might be measurement error or failure of the measurement system. Ambient temperature fluctuations may freeze impulse tubes, while impulse tubes that contain fluids at high temperature that can be accessed by personnel from walkways and the like may present a safety hazard.

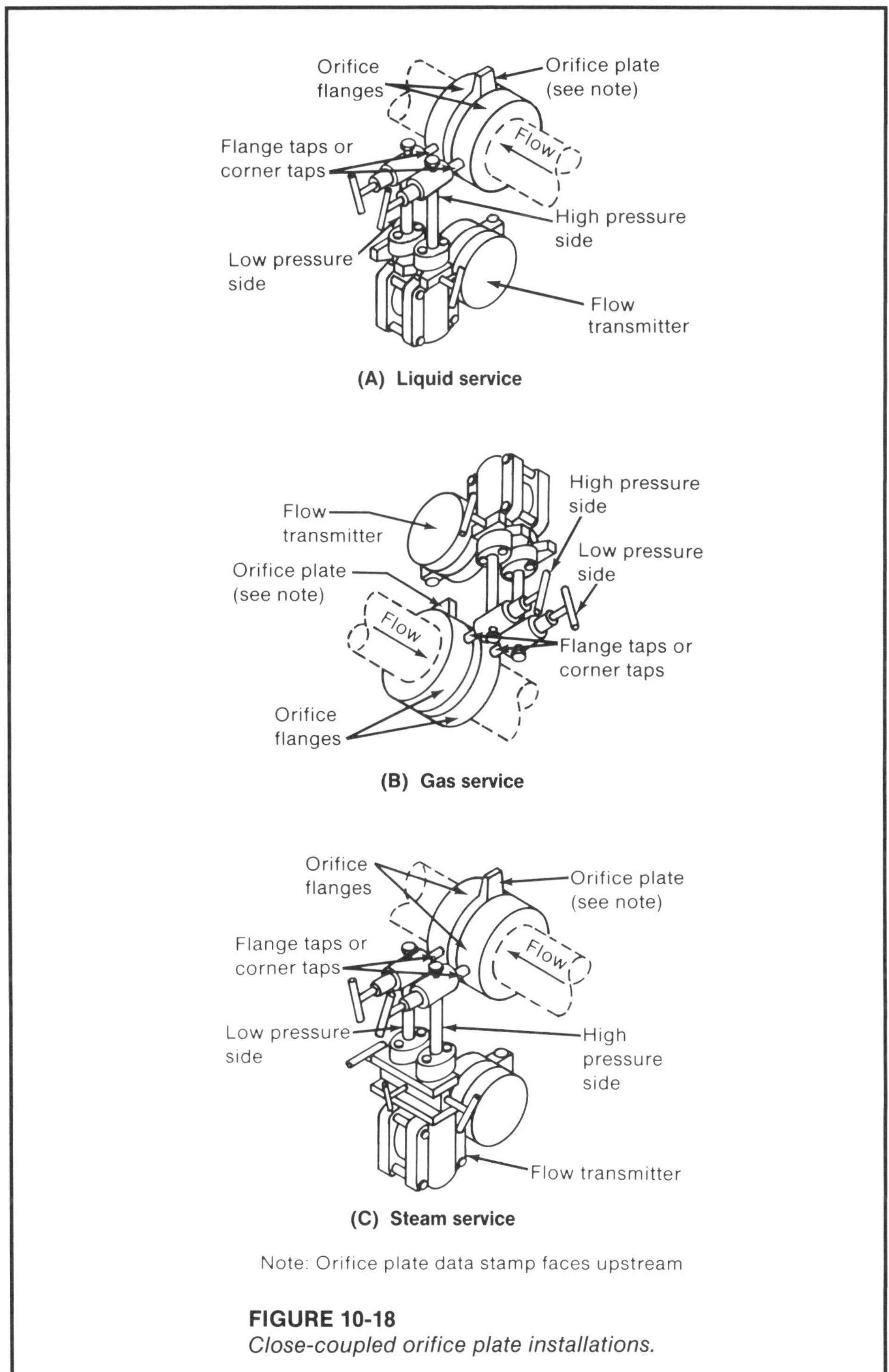

FIGURE 10-18
Close-coupled orifice plate installations.

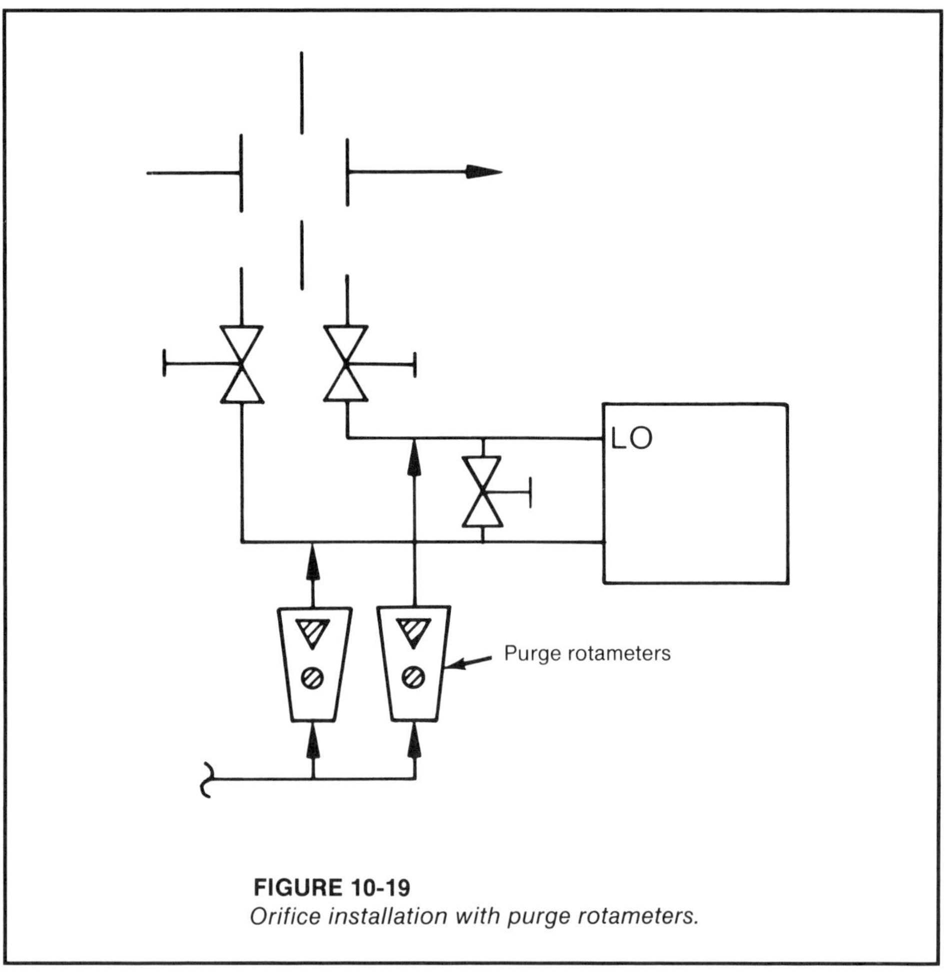

FIGURE 10-19
Orifice installation with purge rotameters.

Cabling

Differential pressure transmitters are typically 2-wire devices, although 3-wire and 4-wire transmitters are available.

Maintenance Orifice plate flowmeters require no routine maintenance other than routine calibration checks. However, problems such as wear, build-up, tap and impulse tubing pluggage, leaks, and transmitter failures can occur.

Orifice Plate Bore Wear

The majority of orifice plate applications utilize sharp edge orifice plates counterbored on the downstream side to make the orifice plate thin in relation to the inside diameter of the pipe. If the sharp edge of an orifice plate becomes dulled, it may start to approach the operation of a quadrant orifice plate, which has coefficients that can vary significantly from those of a sharp edge orifice plate. The net result of long-term wear is a measurement drift in which the flowmeter measurement will be lower than the flow in the pipe, as less differential pressure is developed across the larger orifice plate bore.

Orifice plate wear or corrosion may be a problem. As the velocity at the edge of the orifice plate is higher than that in the pipe, both effects may be accelerated.

EXAMPLE 10-15

Problem: The sensitivity coefficient of a concentric square edge orifice plate to changes in bore is $-2 / (1-\beta^4)$. Calculate the bias error caused by 0.005 inch of uniform wear on a 2-inch orifice plate flowmeter with a bore of 1.033 inches.

Solution:

$$X_d = 2 / (1-[1.033 \text{ in.}/2.067 \text{ in.}]^4) = 2.133$$

The bias error associated with 0.01 inch of wear is

$$\begin{aligned} \text{Bias error} &= X_d \times \Delta d \\ &= 2.133 \times (0.01 \text{ in.}/1.033 \text{ in.}) \\ &= 0.021 \end{aligned}$$

Therefore, the resultant bias error is approximately 2.1 percent. It should be noted that error due to orifice plate wear is more significant in smaller pipe sizes than in larger pipe sizes.

Build-Up

Build-up of foreign material that will affect the velocity profile of the fluid can occur upstream and downstream of the orifice plate as well as on the orifice plate. This phenomenon will adversely affect the measurement accuracy of the flowmeter. Build-up can often be avoided through selection of an orifice plate geometry that will alleviate this problem in applications where build-up is suspected.

EXAMPLE 10-16

Problem: The sensitivity coefficient of a concentric orifice plate to changes in pipe wear or coating is $-2 \times \beta^4 / (1-\beta^4)$. Calculate the bias error caused by 0.02-inch uniform pipe coating on a 2-inch orifice plate flowmeter with a bore of 1.033 inches.

Solution:

$$X_D = -2 \times (1.033 \text{ in.}/2.067 \text{ in.})^4 / (1-[1.033 \text{ in.}/2.067 \text{ in.}]^4)$$

The bias error associated with coating causing a 0.04-inch change in pipe diameter is

$$\text{Bias error} = X_D \times \beta D$$
$$= -0.133 \times (0.04 \text{ in.}/2.067 \text{ in.})$$
$$= -0.0026$$

Therefore, the resultant bias error is approximately -0.26 percent. It should be noted that error due to pipe wear and coating is more significant in smaller pipe sizes than in larger pipe sizes. The effects of pipe coating and wear are less than an equivalent amount of orifice plate wear.

Removing the Transmitter from Service

Differential pressure transmitters must be removed from service with care so as not to damage the transmitter or shift calibration by putting full pipe pressure on one side of the diaphragm. This is accomplished by first opening the bypass valve at the transmitter while it is still in service, exposing both sides of the differential pressure transmitter are exposed to the same pressure. Other valves, such as valving to the taps or transmitter drain or vent valves, can then be opened without damaging the transmitter.

Placing the Transmitter in Service

Differential pressure transmitters must be placed in service with care so as not to damage the transmitter or shift calibration by putting full pipe pressure on one side of the diaphragm. This is accomplished by first opening the bypass valve at the transmitter with the flowmeter system not in service; that is, with both valves at the taps closed, exposing both sides of the differential pressure transmitter to the same pressure. The valves to the taps are opened, which allows fluid to flow through the impulse tubing when the pipe is in service due to the differential pressure developed across the orifice plate. The bypass valve is closed to place the transmitter in service.

Tap and Impulse Tubing Pluggage

Dirt and foreign matter in the fluid can plug the tap or impulse tubing. When this is suspected, the bypass valve can be opened with flow in the pipe in an attempt to free the contamination by allowing the differential pressure developed across the orifice plate to push the contaminant out of the system and back into the pipe. If this does not work and the pluggage cannot be opened by applying an inert fluid to one point of the system with the bypass open, some dismantling of the impulse tubing may be required to remove the pluggage.

Leakage

Any piping connection in the impulse tubing and valving is subject to leakage, which will bleed off pressure and affect the differential pressure sensed by the transmitter. Normal piping techniques can be used to correct the leaks.

Transmitter Failure

Transmitter failure is usually remedied by removing and replacing the transmitter. Suspected failure can be verified with the transmitter in place by verifying calibration.

Spare Parts

Spare parts are limited to commodity items such as impulse tubing, fittings, manifolds, valves and the like, with the exception of the differential pressure transmitter.

Calibration

The orifice plate is an inferential flowmeter, so the physical condition of the orifice plate and the bore should be checked. Calibration of the differential pressure transmitter can be performed by simulating the calculated inputs to the differential pressure transmitter and making the required zero and span adjustments.

Other Technologies

Other forms of differential pressure flowmeters are available. They have many of the attributes and require many of the same considerations that are applicable to orifice plates. Sizing information for some of these technologies is included with orifice plate sizing information.

Critical Flow Elements When a gas accelerates through a nozzle, its density decreases. The mass flow can be progressively increased until a maximum flow is reached where the velocity is sonic, and further decreasing the downstream pressure will not increase the mass flow rate. This is referred to as *choked* or *critical flow.*

In liquid service, when the pressure of the liquid is reduced below the vapor pressure in the flowmeter, a flashing/cavitation zone will form and restrict the flow. In this condition, decreasing the downstream pressure will not increase the flow through the flowmeter.

The velocity of the fluid reaches sonic velocity at a critical pressure ratio of the absolute downstream pressure to the absolute upstream pressure. Operation of the flowmeter above the critical pressure ratio results in subsonic flow. Operation of the flowmeter below the critical pressure ratio results in critical flow conditions. In critical flow operation, the mass flow can be increased only by increasing the upstream pressure.

Critical flow elements using the principles of venturi flow measurement are generally more predictable than those using specially designed orifice plates. Critical flow nozzles are often used as secondary standards to test and calibrate flowmeters at flow testing facilities.

Elbow Flowmeters Elbow flowmeters operate on the principle that when a fluid moves around the curved path of an elbow, the angular acceleration of the fluid results in a centrifugal force that creates a differential pressure between the inner and outer radii. The differential pressure that is produced is proportional to

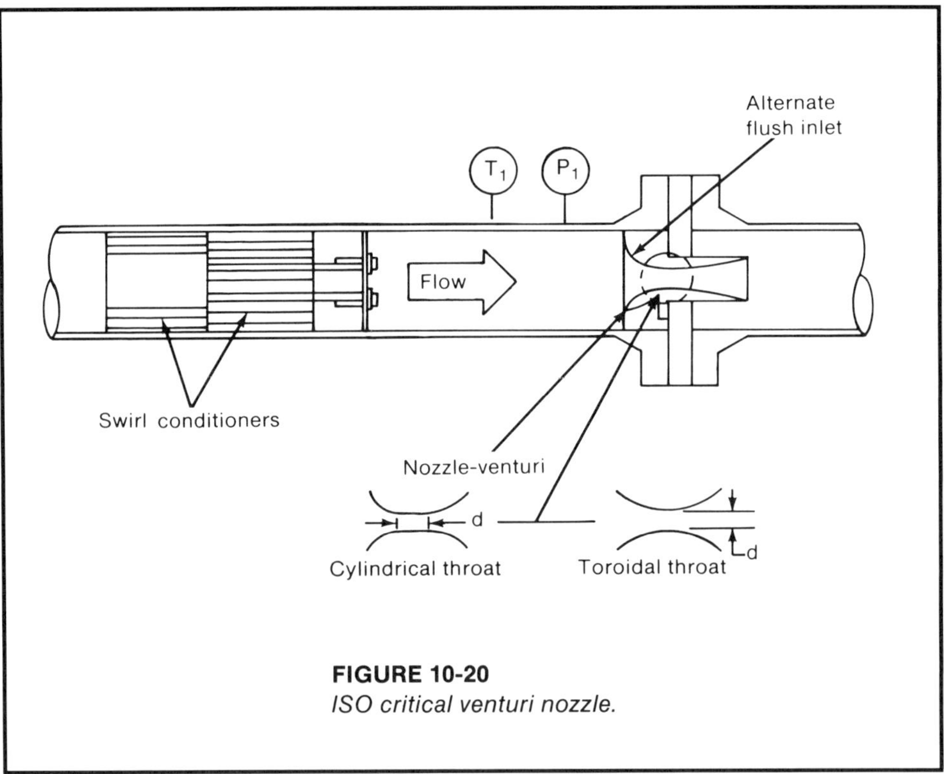

FIGURE 10-20
ISO critical venturi nozzle.

the square of the flow through the flowmeter. Inaccuracy of the flow coefficient can be expected to exceed ±4 percent of rate.

Flow Nozzles Flow nozzles are generally selected for high temperature and high velocity applications. The discharge coefficients are better documented at high Reynolds number conditions than are the orifice plate designs and are less sensitive to high flow and velocities than are orifice plates.

Laminar Flow Elements A laminar flow element is a flowmeter in which a differential pressure proportional to flow is generated when a fluid flows through the flowmeter in the laminar flow regime. The relationship between the differential pressure and flow is defined by the Hagan-Poiseuille Law for flow in a horizontal pipe with fully developed laminar flow:

$$Q = \text{constant} \times \Delta p \,/\, (\mu_{\text{cP}} \times L)$$

The flow measurement is linearly proportional to the differential pressure. However, the measurement is also inversely proportional to the viscosity, which is a distinct disadvantage. Viscosity changes can be minimized by controlling the temperature of the fluid through immersing the flowmeter in a controlled temperature bath. Various proprietary configurations of laminar flowmeters are available. Most utilize capillary or tube bundles, as shown in Figure 10-23.

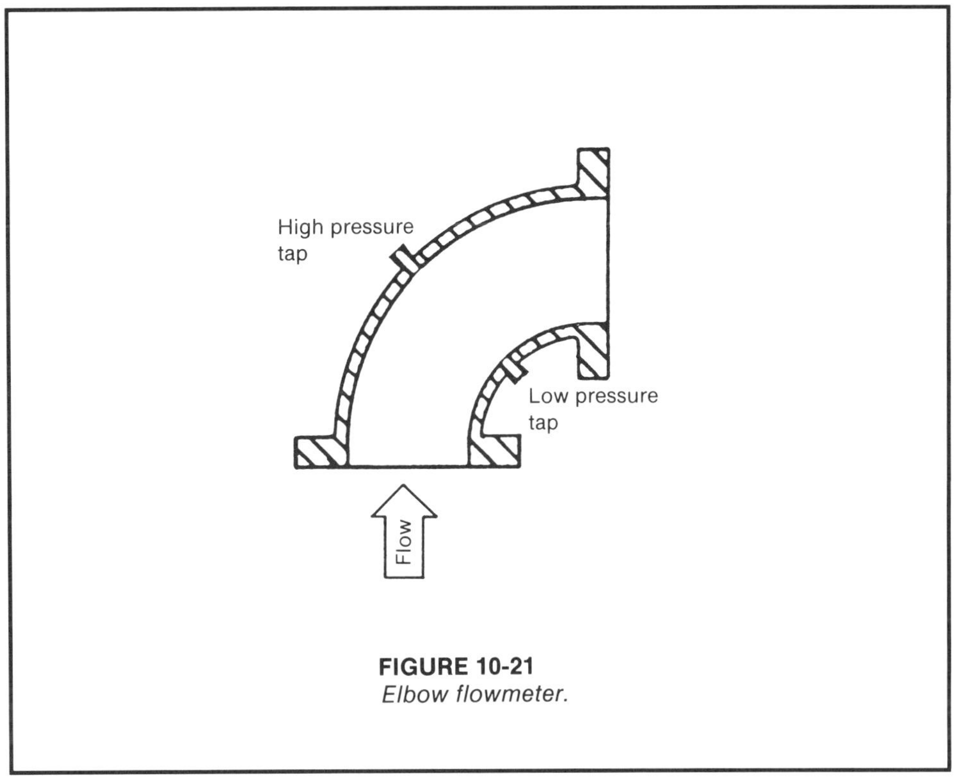

FIGURE 10-21
Elbow flowmeter.

Low Loss Flow Tubes Low loss flow tubes are available that produce high differential pressures at lower permanent pressure loss than other differential producing devices. Most designs are proprietary and information concerning these devices must be obtained from the manufacturer.

Segmental Wedge The segmental wedge is similar to a segmental orifice plate, but with a contoured wedge substituted for the relatively thin segmental orifice plate. The wedge is illustrated in Figure 10-25. The use of a wedge allows the flowmeter to accurately measure fluids operating at Reynolds numbers greater than 500.

Installation requirements are similar to those for an orifice plate. However, chemical seals are available that allow differential pressure transmitters with diaphragm seals to be installed directly in the pipe, minimizing the possibility of plugging taps or impulse tubing.

Venturi The classical or Herschel venturi is typically available for sizes 2 in. and larger. Flow through the venturi is relatively streamlined and thereby exhibits a relatively low permanent pressure loss. Several propriety designs are also available. Typical coefficient accuracies are in the range of ± 0.5 to 1.5 percent, depending upon flowmeter design and construction and whether or not the flowmeters are flow calibrated.

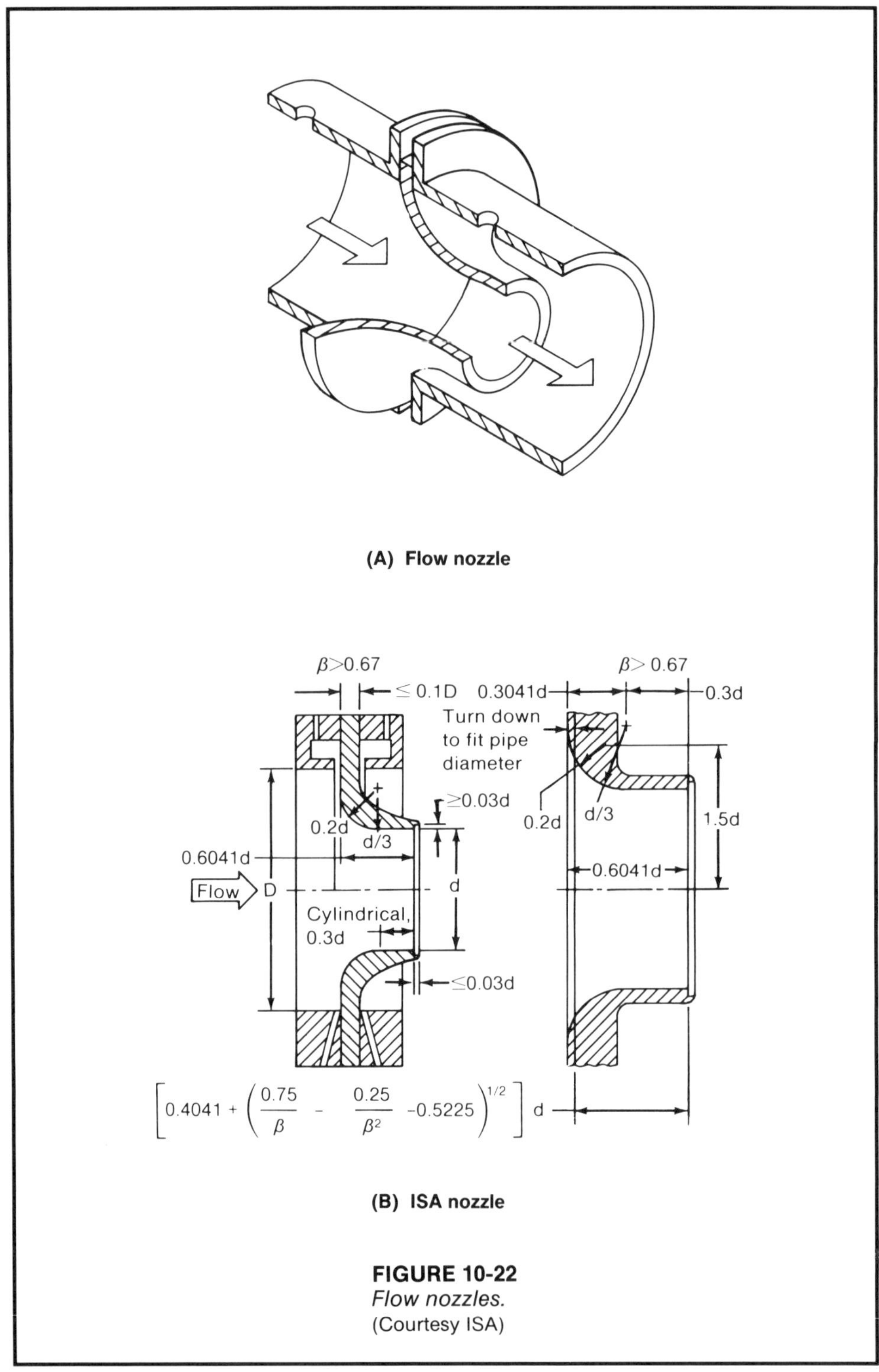

FIGURE 10-22
Flow nozzles.
(Courtesy ISA)

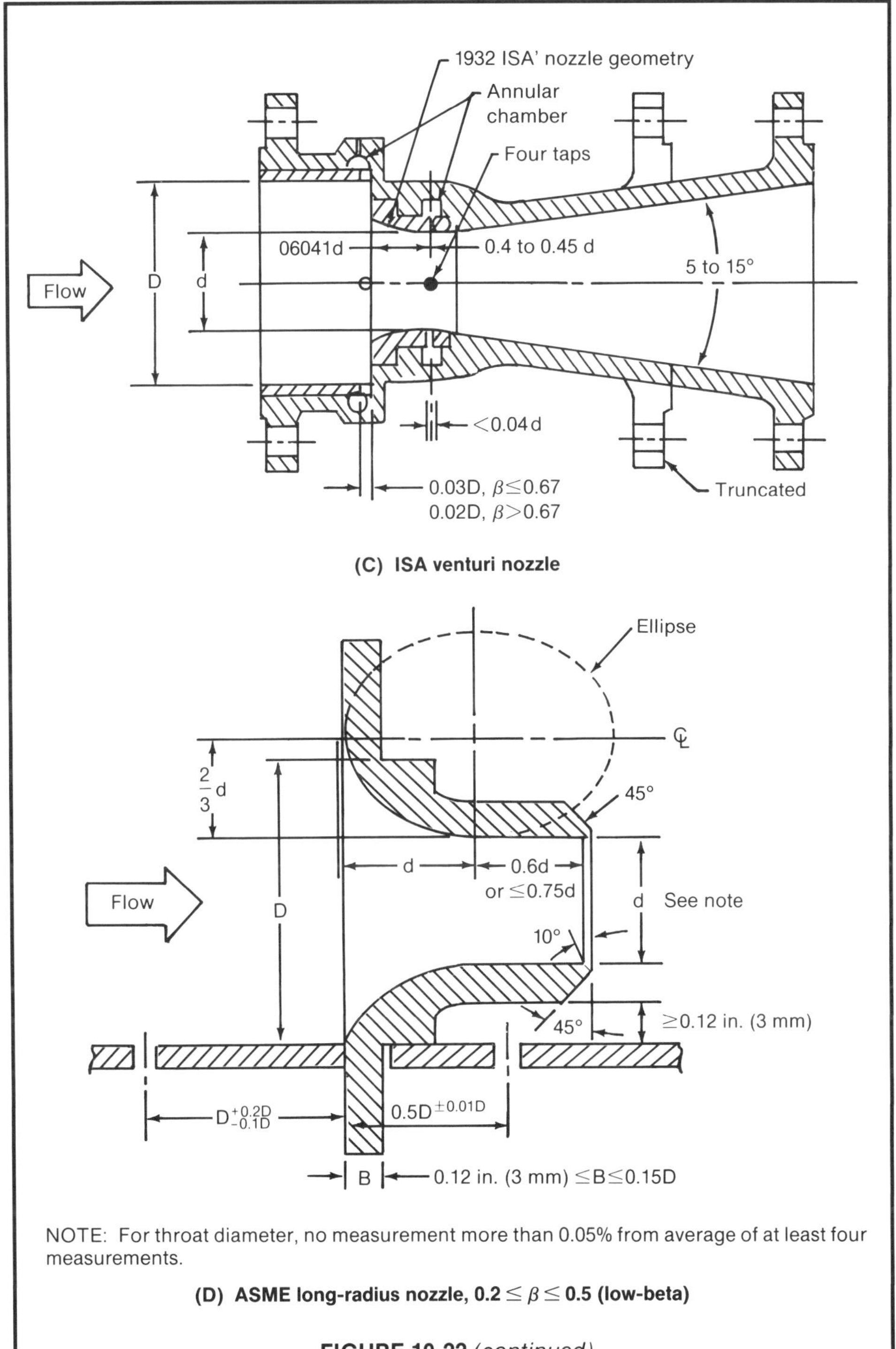

(C) ISA venturi nozzle

NOTE: For throat diameter, no measurement more than 0.05% from average of at least four measurements.

(D) ASME long-radius nozzle, $0.2 \leq \beta \leq 0.5$ (low-beta)

FIGURE 10-22 *(continued)*

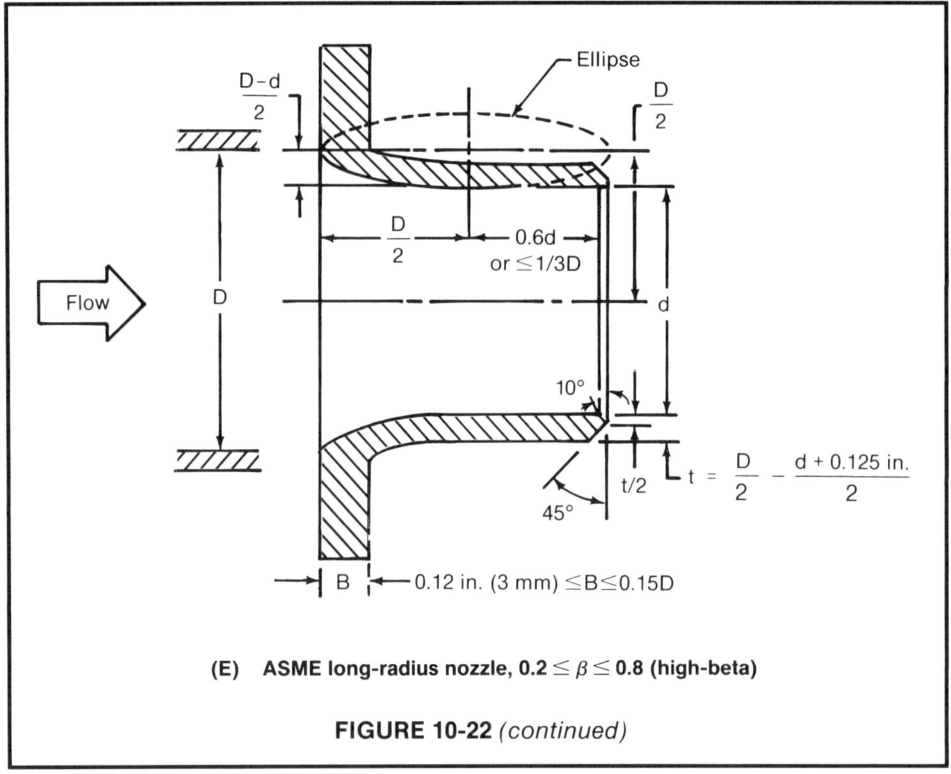

(E) ASME long-radius nozzle, 0.2 ≤ β ≤ 0.8 (high-beta)

FIGURE 10-22 *(continued)*

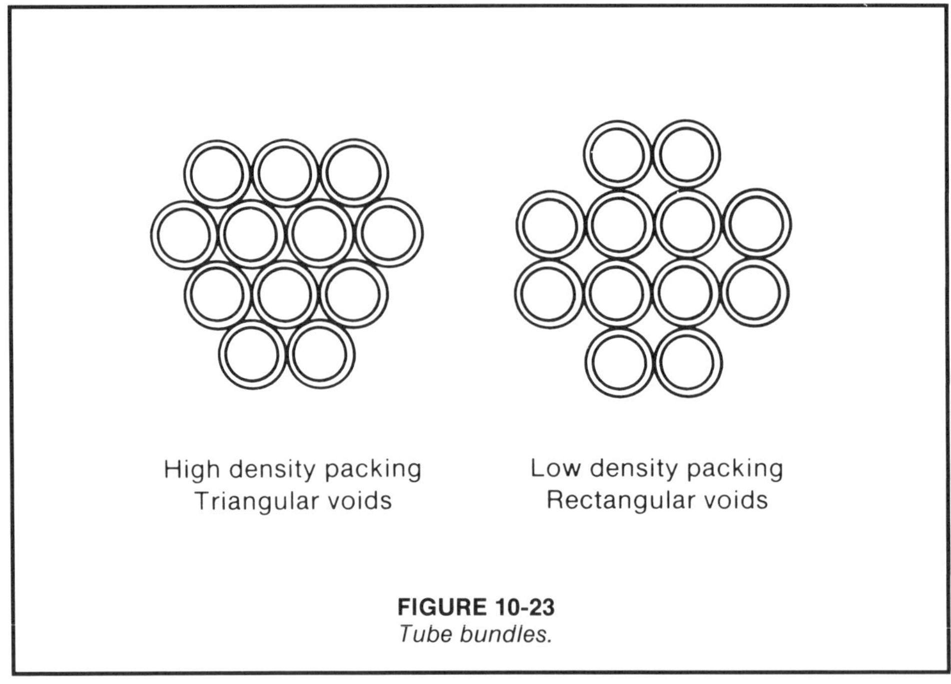

FIGURE 10-23
Tube bundles.

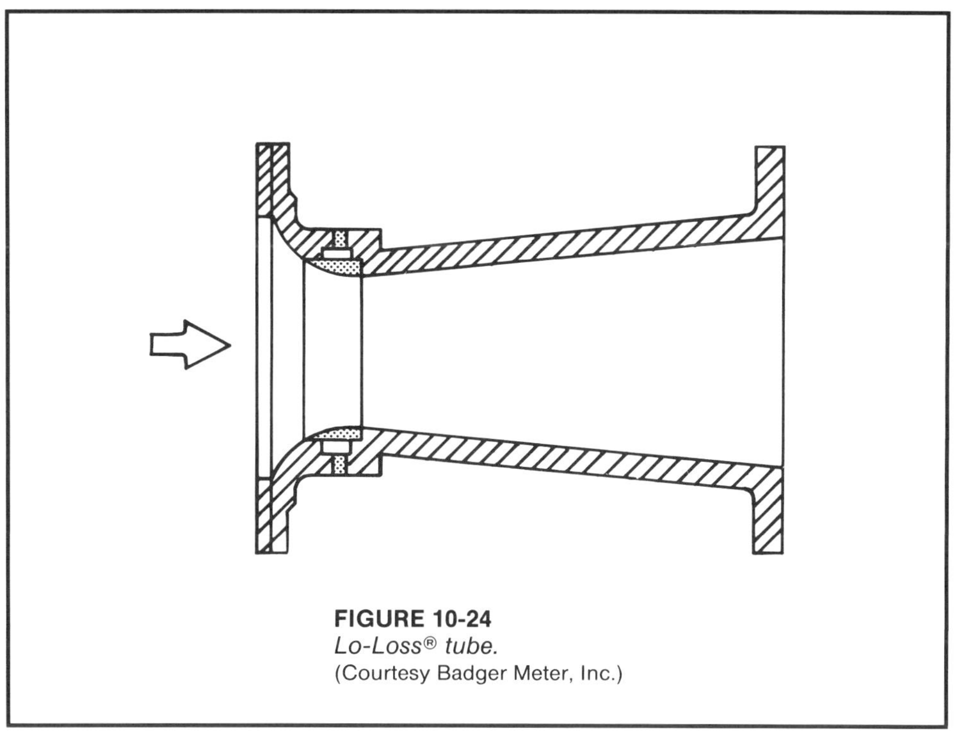

FIGURE 10-24
Lo-Loss® tube.
(Courtesy Badger Meter, Inc.)

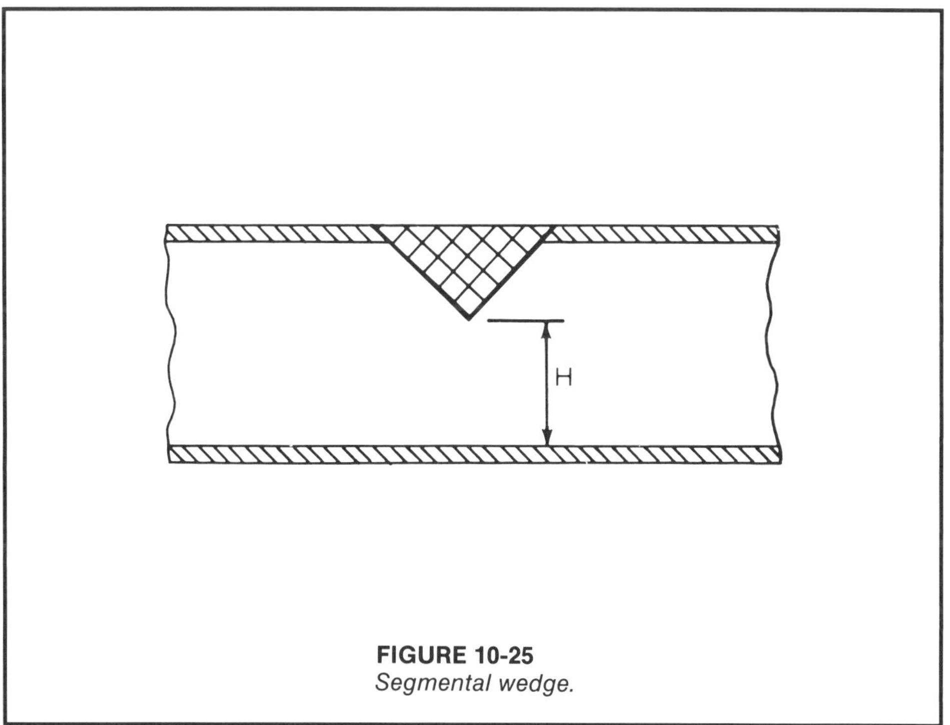

FIGURE 10-25
Segmental wedge.

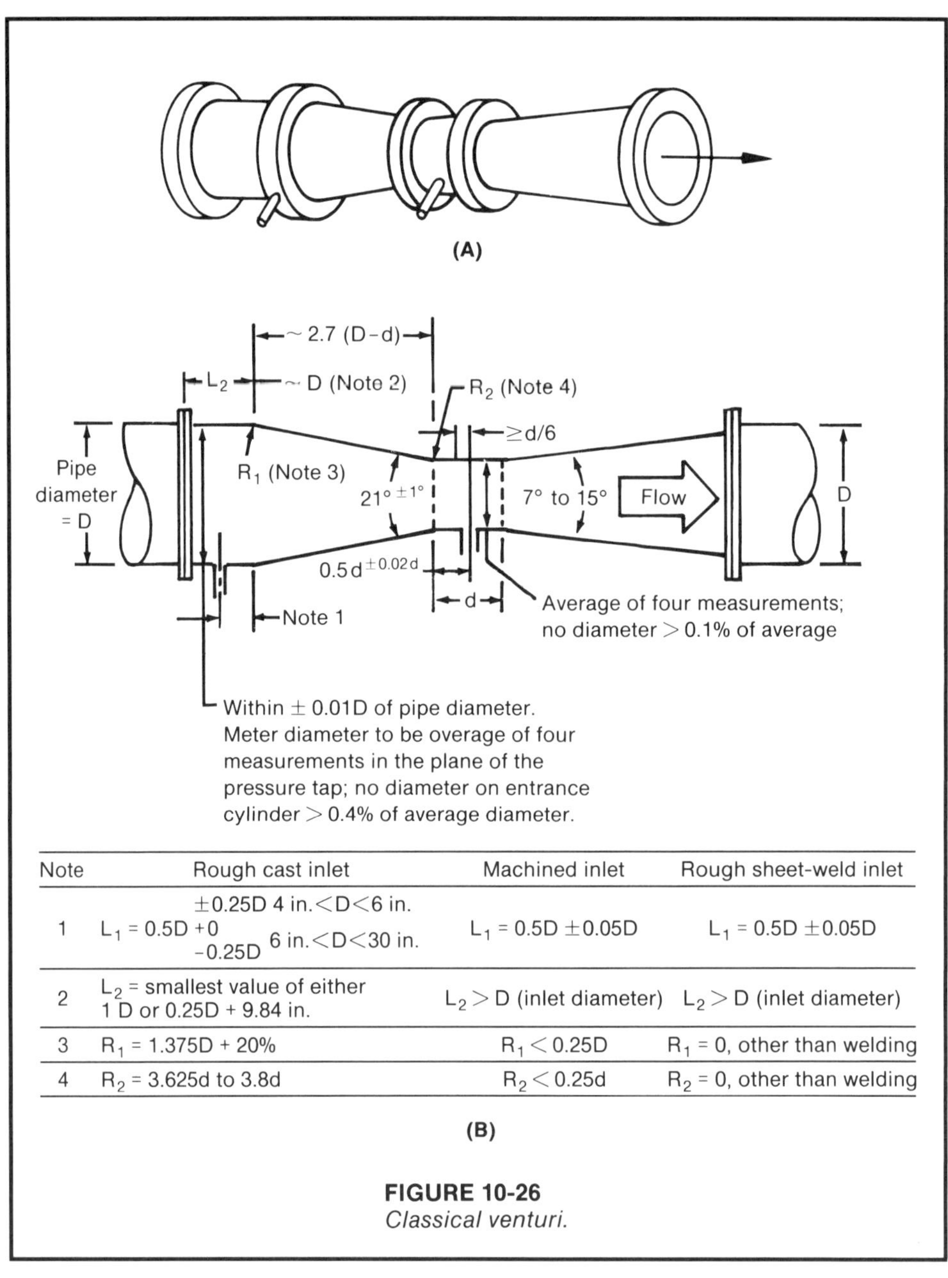

Note	Rough cast inlet	Machined inlet	Rough sheet-weld inlet
1	$L_1 = 0.5D \begin{smallmatrix}\pm 0.25D\\ +0\\ -0.25D\end{smallmatrix}$ 4 in. < D < 6 in.; 6 in. < D < 30 in.	$L_1 = 0.5D \pm 0.05D$	$L_1 = 0.5D \pm 0.05D$
2	L_2 = smallest value of either 1 D or 0.25D + 9.84 in.	$L_2 > D$ (inlet diameter)	$L_2 > D$ (inlet diameter)
3	$R_1 = 1.375D + 20\%$	$R_1 < 0.25D$	$R_1 = 0$, other than welding
4	$R_2 = 3.625d$ to $3.8d$	$R_2 < 0.25d$	$R_2 = 0$, other than welding

(B)

FIGURE 10-26
Classical venturi.

V-Cone The V-cone is a proprietary device where a differential pressure is generated as the fluid passes through an annular opening in the flowmeter. The V-cone can be applied to fluid flows operating at Reynolds Numbers greater than approximately 10,000. Installation requirements are similar to those of an orifice plate. (See Figure 10-27.)

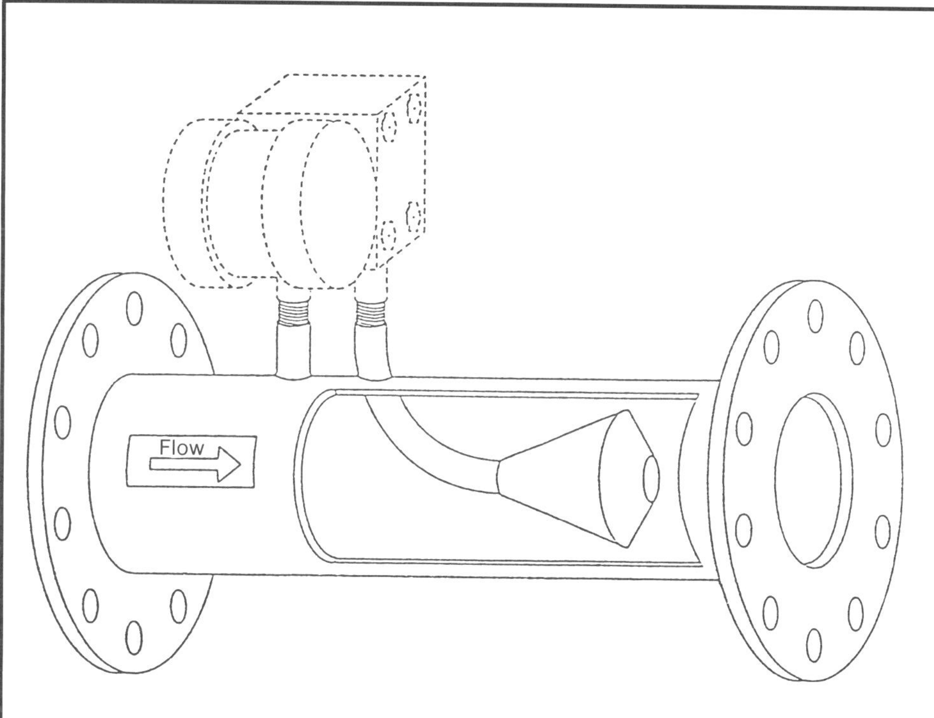

FIGURE 10-27
V-cone.
(Courtesy KETEMA/McCrometer Division)

EXERCISES

10.1 Draw the schematic of an orifice plate flowmeter, impulse tubing and transmitter that has a 3-valve manifold.

10.2 What are the preferred types of taps for small pipes? For large pipes? Why?

10.3 Calculate the velocity of a gas in a 6-inch pipe if its velocity in an 8-inch pipe is 32 feet per second. How does the reduced line size affect orifice sizing for a given mass flow?

10.4 Calculate the effects on the flowmeter measurement of an uncertainty in process pressure of 50 psi ± 5 psi.

10.5 Why are concentric orifice plates used more often?

10.6 Why are vena contracta taps uncommon?

10.7 Calculate the new calibration range of a differential pressure flow transmitter originally calibrated at 0 to 100 in. WC for a flow of 0 to 50 gpm, if the new flow range is 0 to 40 gpm.

10.8 Why are upstream and downstream straight runs required for orifice plate flowmeter installations?

10.9 How should impulse tubing be installed for liquid service?

10.10 Which is more significant in causing flow measurement errors, orifice wear or pipe wear or process uncertainties?

10.11 Size the bore of a flange tapped concentric orifice plate required to measure 0 to 600 gpm of a liquid flowing in a 6-inch schedule 40 pipe at 100 psi and 120°F with a viscosity of 1.36 cP at flowing conditions. Base and flowing specific gravities are 1.065 and 1.093, respectively. Base conditions are at 60°F and the liquid has a critical temperature of 965°R.

10.12 Size the bore of a flange tapped concentric orifice plate required to measure 0 to 2500 scfm of an ideal gas flowing in a 6-inch schedule 40 pipe at 100 psi and 70°F with a viscosity of 0.014 cP at flowing conditions and a specific gravity of 1.10 at standard conditions. The gas has an isentropic constant of 1.36.

10.13 Size the bore of a flange tapped concentric 316 stainless steel orifice plate required to measure 0 to 10,000 lb/hr of steam flowing in a 4-inch schedule 80 pipe at 225 psi and 397°F with a density of 0.5213 lb/ft^3 and a viscosity of 0.0193 cP at flowing conditions. The isentropic constant of steam at operating conditions is 1.29.

11

Magnetic Flowmeters

Introduction While magnetic flow measurement techniques have been applied for decades, recent technological refinements have resulted in instruments that are relatively easy to apply and install as well as being more economical than previous designs. Flow is obstructed only if the flowmeter is sized less than line size. Magnetic flowmeters exhibit true unobstructed flow characteristics as they have no protrusions into the flow stream.

The trend in magnetic flowmeters is toward increased application of miniature dc design due to the desirability of dc design features, reduced size and weight, standard design that handles most applications, less cabling requirements, and lower cost. Estimates of the applicability of miniature dc magnetic flowmeters range from 70 to 90 percent of total applications.

Principle of Operation

Faraday's Law From classical physics, Faraday's Law of Electromagnetic Induction is the underlying principle of operation of many electrical devices. This law states that the magnitude of the voltage induced in a conductive medium moving through a magnetic field and at a right angle to the field is directly proportional to the product of the strength of the magnetic flux density (B), the velocity of the medium(v), and the path length (L) between the probes.

$$E = \text{constant} \times B \times L \times v$$

This result is completely analogous to the voltage induced in a wire caused by its movement at right angles to an applied magnetic field. The faster the wire is passed through the magnetic field, the more voltage will be induced.

This principle can be applied most notably to electrical power generation in magneto-hydrodynamics where translational energy provided by high temperature combustion processes is converted into electrical energy.

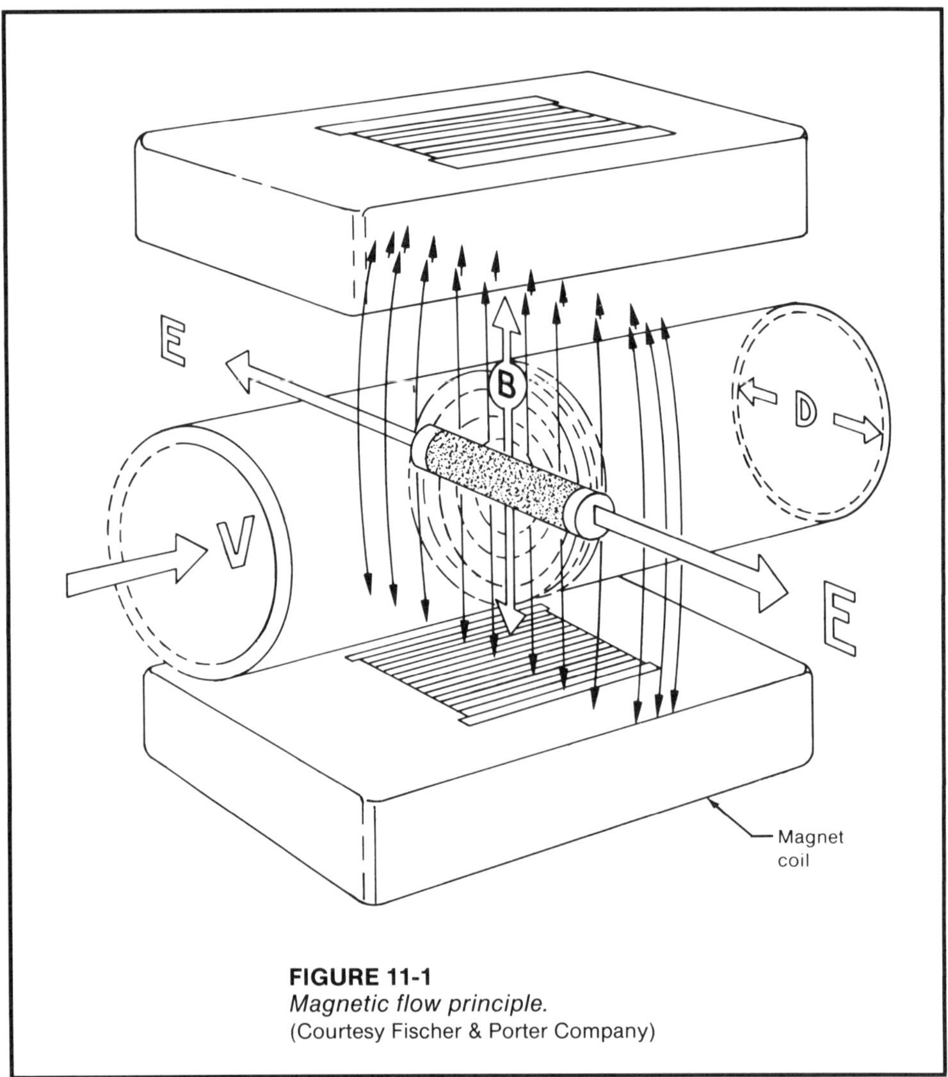

FIGURE 11-1
Magnetic flow principle.
(Courtesy Fischer & Porter Company)

Magnetic flowmeters apply Faraday's Law in the following way: when a conductive liquid passes through a homogeneous field, a voltage is generated along a path between two electrodes positioned within the magnetic field on opposite sides of the pipe. The path length is the distance between the electrodes. From Faraday's Law as applied to magnetic flowmeters, the induced voltage becomes:

$$E = \text{constant} \times B \times D \times v$$

Therefore, if the magnetic field is constant and the distance between the electrodes is fixed, the induced voltage is directly proportional to the velocity of the liquid.

Since the volumetric flow is related to the average fluid velocity,

$$Q = A \times v$$

with

$$A = \pi \times D^2/4$$

In pipes the induced voltage can be expressed as

$$E = (\text{constant} \times B \times 4/\pi \times D) \times Q$$

All of the terms within the parenthesis are held constant in a well designed flowmeter, resulting in an induced voltage output that is linearly proportional to the liquid flow.

AC Magnetic Flowmeters Alternating current (ac) magnetic flowmeters excite the flowing liquid with an ac electromagnetic field.

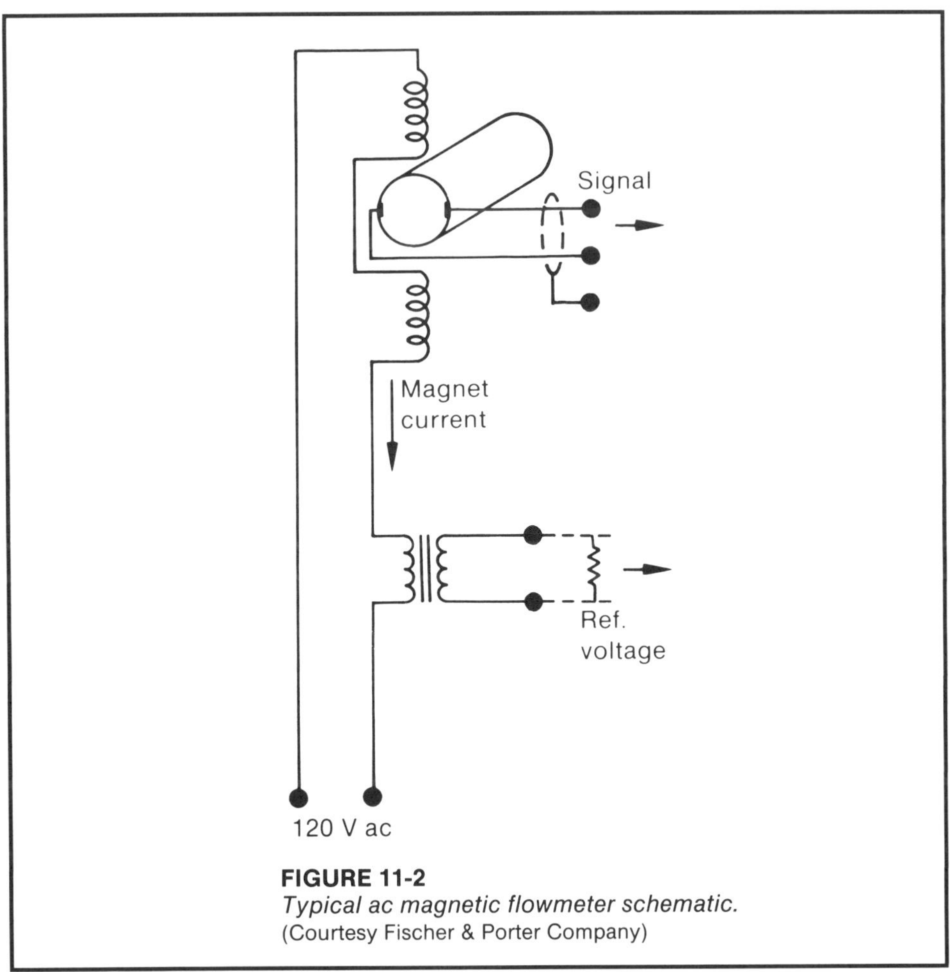

FIGURE 11-2
Typical ac magnetic flowmeter schematic.
(Courtesy Fischer & Porter Company)

One of the greatest difficulties in magnetic flowmeter design is that the amplitude of the voltage across the electrodes is in the order of a few millivolts and is relatively small when compared to extraneous voltages and noise that may be present in real process applications. Noise sources include:

- Stray voltages present in the process liquid
- Capacitive coupling between signal and power circuits of the flowmeter
- Capacitive coupling and lead losses in interconnecting wiring
- Electrochemical emf produced as a result of electrolytic interaction between the electrode and the process fluid
- Inductive coupling of the magnets within the flowmeter

Calibration requires that a zero adjustment be made to compensate for noise that may be present. The flowmeter should be full (of process fluid at zero flow conditions) to properly perform the zero adjustment. Zero adjustments performed with other than the process fluid can result in calibration error if the liquids have differing conductivities.

Similarly, if the electrodes should become coated with an insulating substance, the effective conductivity that the electrodes sense will be altered, causing the calibration to shift. This results in additional inaccuracies. If the coating changes with time, the flowmeter will continually require calibration and will not be repeatable.

If it is assumed that the current to the magnet is constant, the magnetic field will be constant, and the amplitude of the voltage generated at the electrodes will be linearly proportional to the flow through the flowmeter. In practice, the current to the magnet may vary slightly due to line voltage and frequency variations. One scheme to minimize this effect is to use a reference voltage proportional to the strength of the magnetic field to compensate for variations in the magnetic field.

Special cabling practices specified by the manufacturer must be followed to ensure that noise is not introduced to the flowmeter system. Typically, manufacturer recommendations include the use of two conduits, one each to handle the power and signal cables. A maximum length between the primary flowmeter and electronics is imposed to minimize or eliminate noise and sensitivity problems.

DC Magnetic Flowmeters Unlike ac magnetic flowmeters, direct current (dc) or pulsed magnetic flowmeters excite the flowing liquid with a dc electromagnetic field as shown in Figure 11-3.

Instead of the ac waveforms that are generated in a conventional magnetic flowmeter, the dc or pulsed magnetic flowmeter excites the magnet with a pulsed dc current. The current to the magnet is turned on, and a dc voltage is induced at the electrodes that represents the sum of the flow signal and the noise that is present. The current to the magnet is then turned off, and the voltage induced at the electrodes represents the noise that is present. Subtracting the measurement of the flowmeter when no current flows through the magnet from the measurement when current does flow through the magnet, effectively cancels out the effects of noise.

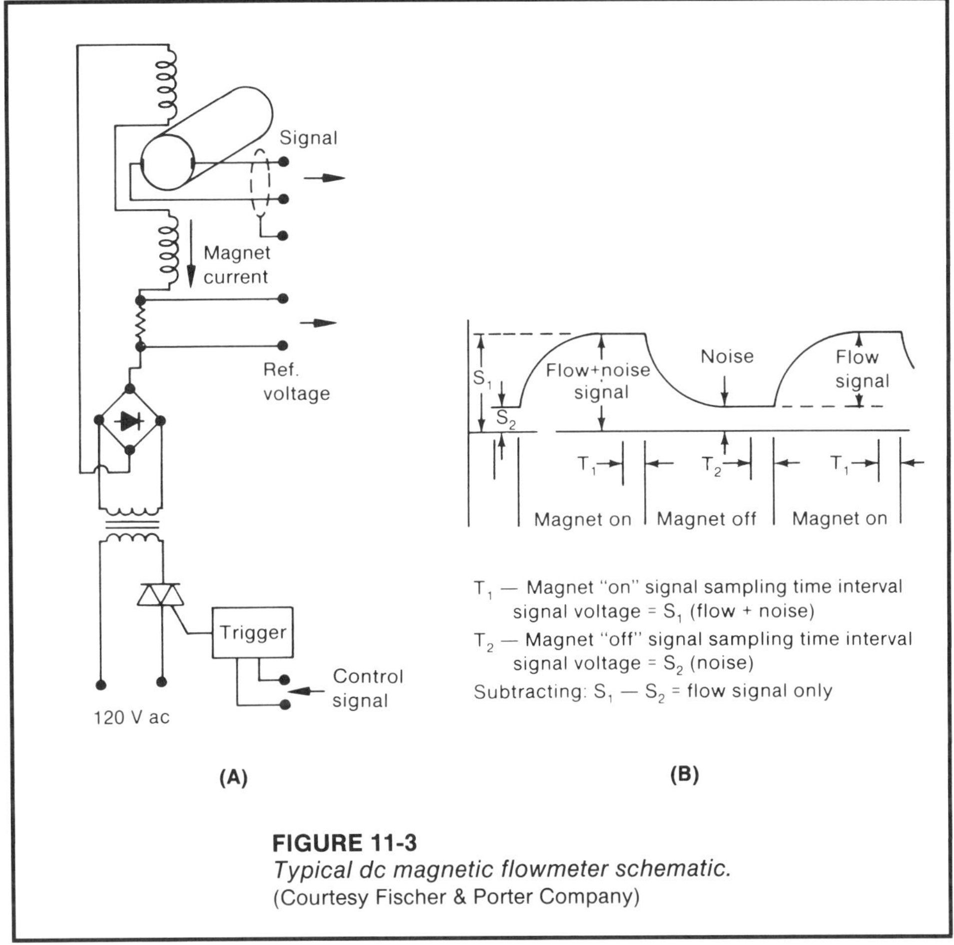

FIGURE 11-3
Typical dc magnetic flowmeter schematic.
(Courtesy Fischer & Porter Company)

The zero compensation inherent in the dc design eliminates the necessity of a zero adjustment, as the flow signal is extracted regardless of the zero shifts that may occur due to noise or electrode coating. The flowmeter need not be full of liquid at zero flow conditions to calibrate the zero, as is required in the conventional design. Insulating electrode coating can shift the effective conductivity without affecting flowmeter performance, as long as the effective conductivity remains high enough for the flowmeter to operate. In contrast, similar changes of effective conductivity with an ac design would result in significant zero shifts. Therefore, the dc design is less susceptible to drift, electrode coating, and changing process conditions than is a conventional ac magnetic flowmeter. To avoid electrolytic polarization of the electrodes, however, bi-polar pulsed dc meters are also available.

DC magnetic flowmeters do not exhibit good response times due to the pulsed rather than continuous nature of the design. However, zero to full scale response times of a few seconds do not create problems in the great majority of applications. As the dc design energizes the magnet only part of the time, power requirements are correspondingly reduced.

Assuming that the dc current to the magnet is constant, the magnetic field will be constant, and the difference of the amplitudes of the dc voltages generated at the electrodes will be linearly proportional to the flow through the flowmeter. In practice, the current to the magnet may vary slightly due to line voltage and frequency variations. One scheme to minimize this effect is illustrated in Figure 11-3A, where a reference voltage, which is proportional to the strength of the magnetic field is used to compensate for variations in the magnetic field.

Typically, only one conduit is required between the primary flowmeter and the electronics since the noise introduced into the signal cable by the dc power to the magnet is negligible, and any noise that is constantly present is effectively cancelled. Cabling limitations between the primary flowmeter and the transmitting electronics are typically not as stringent as those for a conventional ac flowmeter.

Miniature DC Magnetic Flowmeters Developments in dc magnetic flowmeter technology have resulted in the miniature dc magnetic flowmeter of wafer design, which is smaller, has reduced weight and power requirements, and is lower in price. Materials of construction and other optional features are extremely limited. Teflon® or ceramic liners and exotic metal electrodes are typically offered as standard in order to handle a wide variety of applications with standard production flowmeters.

EXAMPLE 11-1

Problem: What type of magnetic flowmeter is preferred for service in which 10 gallons of liquid is to be added to a reactor at 20 gpm?

Solution: As dc magnetic flowmeters exhibit response times of a few seconds, an ac magnetic flowmeter would be preferred to minimize errors that would occur when the flow is initially turned on and when the flow is turned off.

Electrodeless Magnetic Flowmeters Electrodeless magnetic flowmeters are miniature dc magnetic flowmeters with non-wetted electrodes, making these magnetic flowmeters suitable for applications where electrode coating presents a potential problem. Materials of construction are limited to ceramic liners.

Construction

The construction of a magnetic flowmeter primary is shown in Figure 11-4.

The magnetic coils create a magnetic field that passes through the flowtube and into the process liquid. When a conductive liquid flows through the flowmeter, a voltage is induced between the electrodes, which are in contact with the process liquid and isolated electrically from the pipe walls by a nonconductive liner to prevent a short circuit of the electrode signal voltage. The liner also serves to

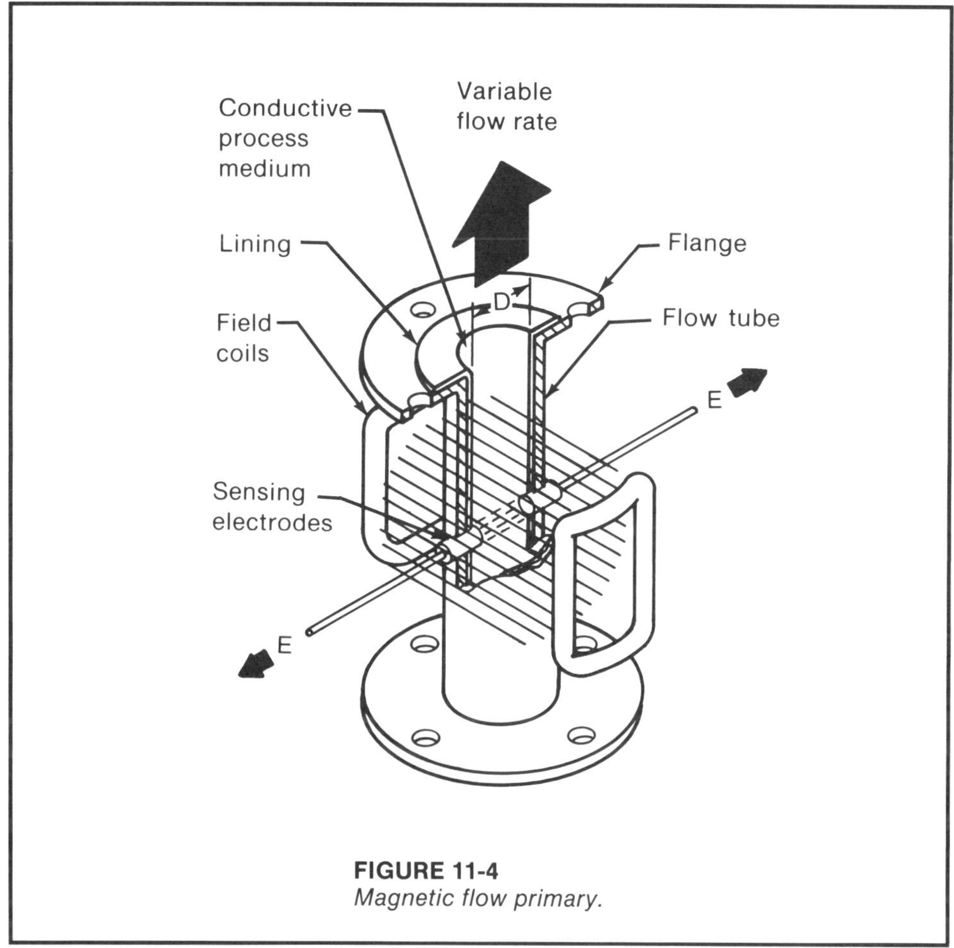

FIGURE 11-4
Magnetic flow primary.

protect the flowtube. Each electrode is held in place by an electrode holder, which is also in contact with the process so as to form a seal around the electrode.

A conventional magnetic flowmeter design is shown in Figure 11-5. This is a replaceable tube type where the field coils are located external to the flow tube. Face-to-face dimensions of these flowmeters are relatively large, as the mating flanges, which can be constructed of carbon steel, must be far enough removed from the coils and electrodes so as not to affect the measurement.

In some dc designs, the field coils are located closer to the liquid, within the flowtube, so that carbon steel flanges could be located closer to the magnet and electrodes without affecting the measurement, thereby reducing the face-to-face dimension somewhat (see Figure 11-6).

The miniature dc and electrodeless magnetic flowmeters are so compact that their face-to-face dimensions are short enough that they can be installed between two flanges. As a result, a dramatic weight and size reduction is achieved with the wafer design.

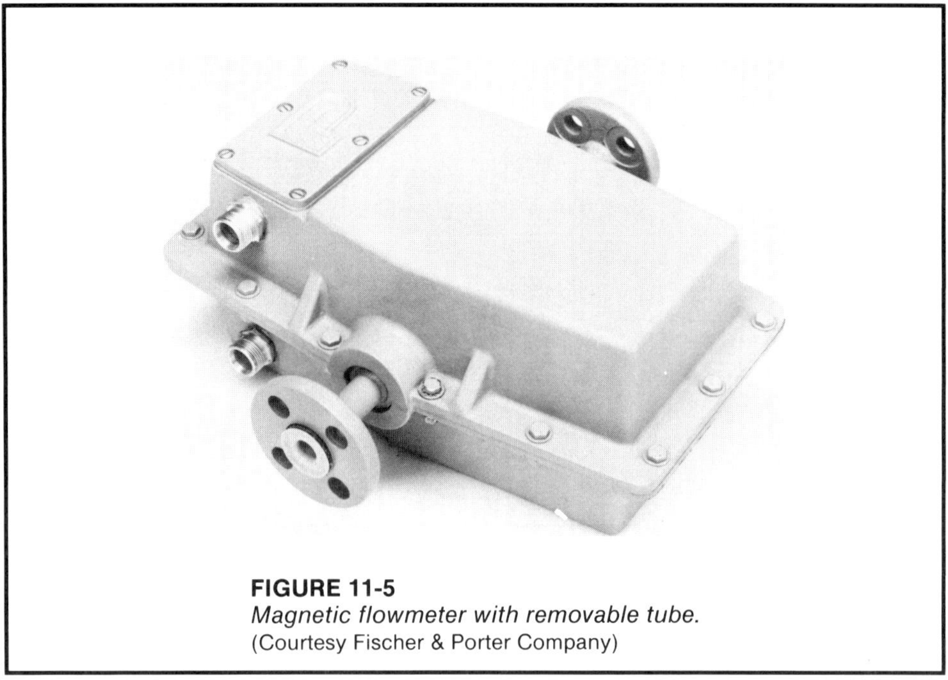

FIGURE 11-5
Magnetic flowmeter with removable tube.
(Courtesy Fischer & Porter Company)

Devices that attempt to clean the electrodes continuously or periodically by ultrasonic or electrical means are available from various manufacturers for both conventional ac and dc magnetic flowmeters. Cleaners are specified more often for conventional ac magnetic flowmeters as these flowmeters are affected by relatively small amounts of insulating coating. Ultrasonic cleaners are specified for dc magnetic flowmeters when an insulating coating that will cause the flowmeter to cease to operate is anticipated, but they are not yet offered in miniature dc magnetic flowmeters.

Wetted parts of a magnetic flowmeter include the liner, electrodes, and electrode holder. Conventional ac and dc magnetic flowmeters offer many materials of construction to suit process corrosivity and temperature constraints, such as rubber, Teflon®, polyurethane, and polyethylene liners; and stainless steel, tantalum, titanium, platinum, Monel®, Alloy 20, and Hastelloy® electrodes and electrode holders. The liner should be chosen to withstand the abrasive and corrosive properties of the liquid, while the electrode must, in addition, withstand abrasion and corrosion and not become coated with insulating deposits.

Miniature dc and electrodeless magnetic flowmeters have very limited or no choice of materials of construction. However, the materials that are offered are sufficiently exotic that the flowmeters can be used in the majority of magnetic flowmeter applications. Typical standard materials of construction are Teflon® or ceramic liners, which are suitable for most applications, tantalum or zirconium electrode holders for acid and base applications, respectively, and tungsten carbide or exotic metal electrodes. Electrodeless magnetic flowmeters do not have wetted electrodes.

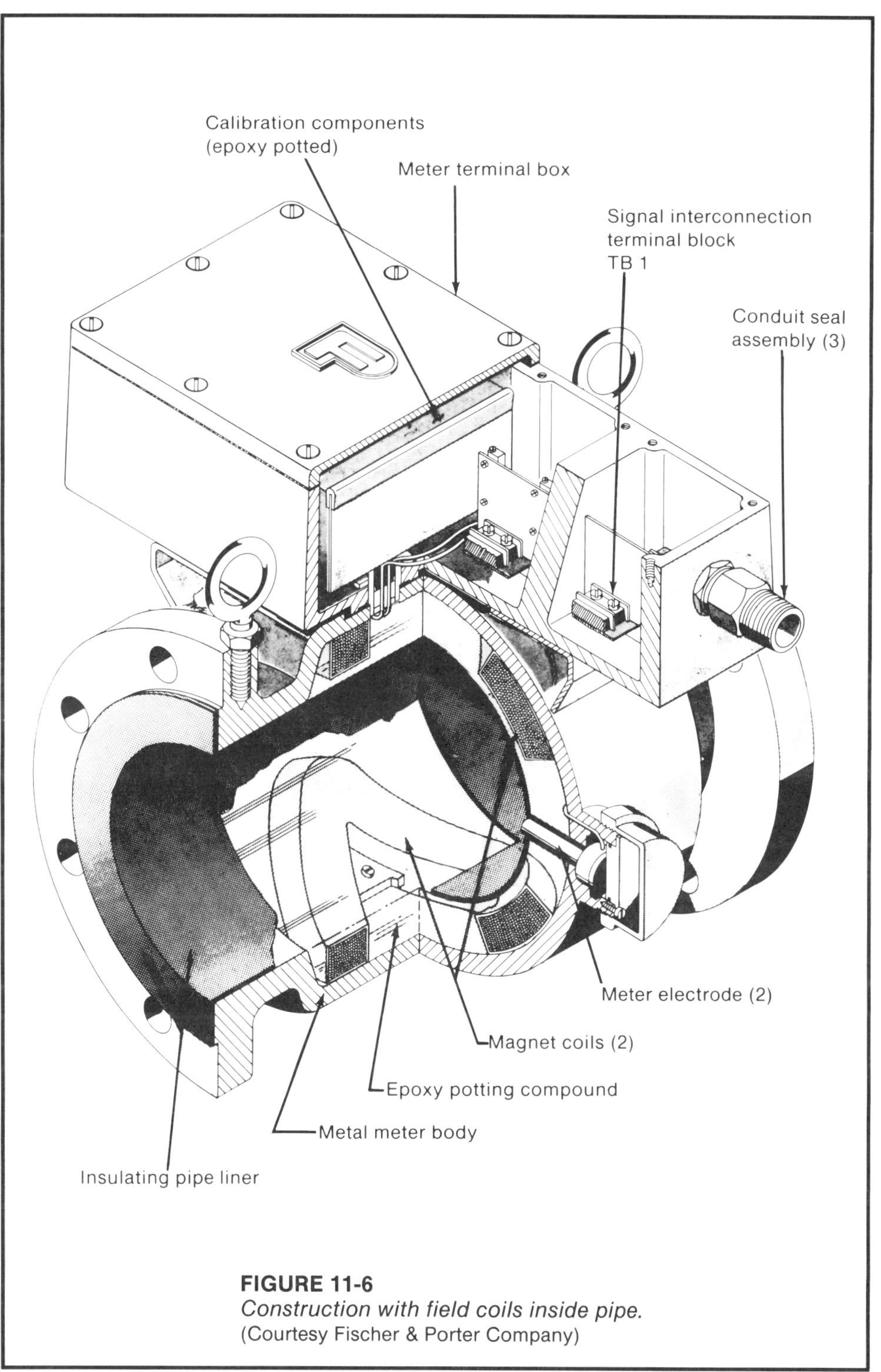

FIGURE 11-6
Construction with field coils inside pipe.
(Courtesy Fischer & Porter Company)

FIGURE 11-7
Mini Mag X™ minature dc magnetic flowmeter.
(Courtesy Fischer & Porter Company)

Flowtubes, which must pass the magnetic field created by the magnet, are usually constructed of stainless steel, which has essentially no magnetic properties and adds strength to the flowmeter.

As flanges are protected by the liner and are not wetted, they need not be compatible with the process liquid.

Operating Constraints Operation of a magnetic flowmeter is generally limited by the operating limits of the liner and the pressure rating of the flanges, which is typically 150 or 300 pounds. Care should be taken when designing for vacuum service, as some liners are not structurally strong enough and can be collapsed. Maximum temperatures will vary up to approximately 200°C depending upon the liner material selected. Ceramic liners are subject to cracking due to sudden (relatively large) changes in process fluid temperature.

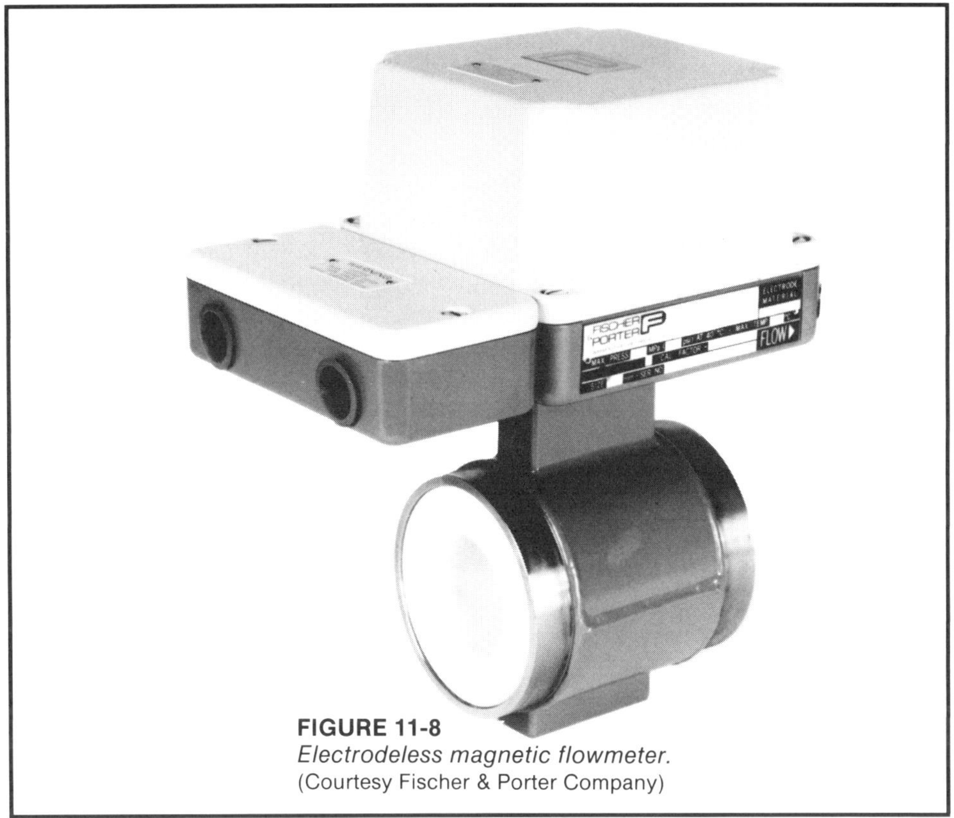

FIGURE 11-8
Electrodeless magnetic flowmeter.
(Courtesy Fischer & Porter Company)

The liquid to be measured should have a minimum conductivity of 1 to 5 microsiemens per centimeter (μS/cm). Electrodeless magnetic flowmeters and other special designs are available that will measure down to approximately 0.05 μS/cm. As most common applications involve liquids whose conductivity is greater than 5 μS/cm there is little difference between standard flowmeters in this respect. However, the minimum conductivity of a magnetic flowmeter can be affected by the distance from the flowmeter tube to the transmitter electronics.

The full scale velocity of the flowmeter is typically 3 to 30 feet per second. Some flowmeters can be adjusted down to 1 foot per second full scale with decreased accuracy.

EXAMPLE 11-2

Problem: As most liquids are operated at velocities of less than 10 feet per second so as to reduce pressure losses and pipe wear, why might it be desirable to operate a magnetic flowmeter at higher velocities?

Solution: Higher velocities may be desirable to aid in keeping the electrodes clean to prevent loss of signal and to produce a larger electrode voltage.

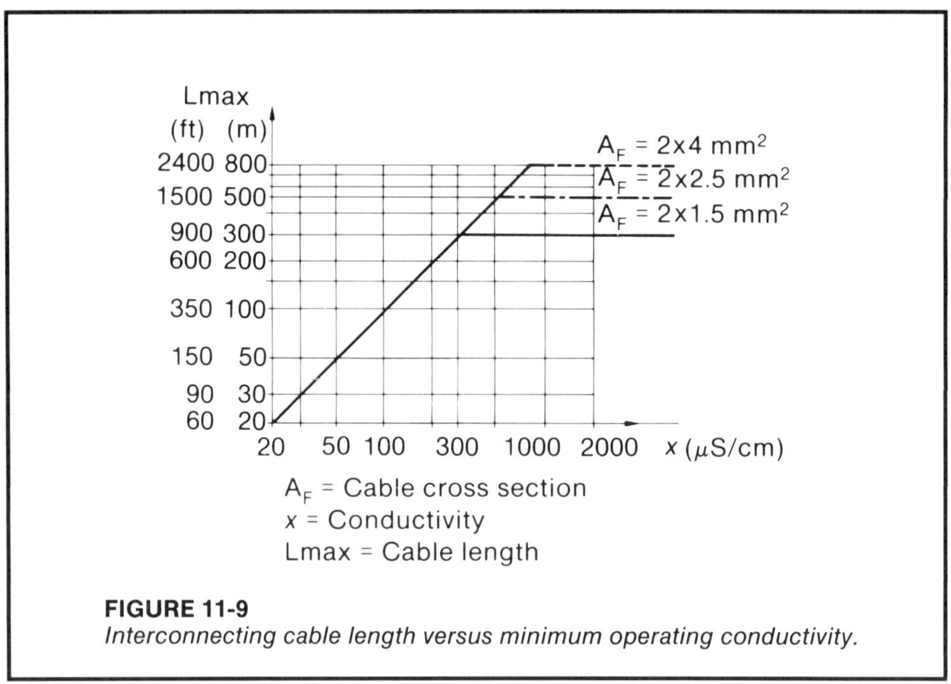

FIGURE 11-9
Interconnecting cable length versus minimum operating conductivity.

Performance Conventional ac magnetic flowmeters have accuracy statements that are expressed as a function of full scale, typically 0.5 to 1 percent FS. Some dc magnetic flowmeters have similar accuracy statements. The majority of dc magnetic flowmeters have a well defined zero due to the automatic zeroing nature of the electronic circuits, and hence have percentage of rate accuracy statements that are typically 0.5 to 2 percent rate.

Care should be taken when determining what the performance will be and over which range it applies. Some manufacturers state accuracy ranges over which a given turndown for any full scale calibration applies, while others state accuracy for specific velocity ranges.

Velocity constraints of magnetic flowmeters should be examined in detail in order to determine the expected range of accurate performance. These constraints may be absolute (typically from 1 to 30 fps or 3 to 30 fps) or a function of the calibration (typically from 10 to 100 percent of the calibrated range for any range from 3 to 30 fps). Design velocities above 10 to 15 ft/sec are not recommended except in special applications, as accelerated pipe erosion and liner damage can occur. At these velocities, excessive pressure drop and cavitation can result in extreme cases. Below the minimum velocities, flowmeter accuracy deteriorates to the value of the error at the minimum velocity, resulting in a full scale error. As pipes are sized to operate at a typical velocity of 7 fps or less, the range of accurate performance of a flowmeter with absolute constraints can be 2.3 to 7:1 or less. The flowmeter with a 10:1 range of accurate performance reference to is clearly more desirable in this respect.

Applications Magnetic flowmeters can be applied only to conductive liquids such as acids, bases, slurries, waste water, foods, dyes, polymers, emulsions, sludge, and mixtures that have conductivities greater than the minimum conductivity requirement. Magnetic flowmeter manufacturers have compiled extensive conductivity data on the more common magnetic flowmeter applications. However, this is not complete in light of the many varying and often proprietary liquids that must be measured under varying conditions. One rule of thumb used to determine if a liquid is sufficiently conductive is that if the process liquid is composed of more than 10 percent of a sufficiently conductive liquid, the mixture is sufficiently conductive to apply a magnetic flowmeter.

Part of the difficulty in applying magnetic flowmeters is the limited data available on the conductivities of liquids other than those commonly encountered. If there is any doubt as to the conductivity of a liquid under operating conditions, most manufacturers will measure a sample to accurately determine the conductivity and hence the applicability of this technology.

Liquids on which magnetic flowmeters will not operate generally include organics and hydrocarbons. Their conductivities are often orders of magnitude less than those required for magnetic flowmeter operation.

The lack of any direct Reynolds number constraints and the obstructionless straight-through design of the magnetic flowmeter make it practical for applications that involve conductive liquids that would plug other flowmeters or that have high viscosity. Materials of construction generally do not enter into the elimination of the applicability of magnetic flowmeters due to the wide variety of corrosion resistant materials available.

Magnetic flowmeters are available that measure bidirectional flow, that is, the flow in either direction.

EXAMPLE 11-3

Problem: Magnetic flowmeters are generally not applicable to hydrocarbons. Explain.

Solution: The conductivity of hydrocarbons is usually very low and does not satisfy the minimum conductivity requirements for magnetic flowmeter technology.

Sizing Magnetic flowmeters range in size from 1/10 in. to 96 in. and larger, while miniature magnetic flowmeters are available through approximately 4 inches. Sizing a magnetic flowmeter is a matter of selecting the size in which the liquid velocities of interest are reasonable and within the measurement range of the flowmeter. This can be determined by calculating the desired liquid velocities through the selected flowmeter and comparing them with the range of velocities that the flowmeter can measure, which is a function of the ability of the flowmeter to accurately measure the millivolt level electrode voltage. This information is often graphed in terms of flow rather than velocity for smaller, more commonly applied sizes.

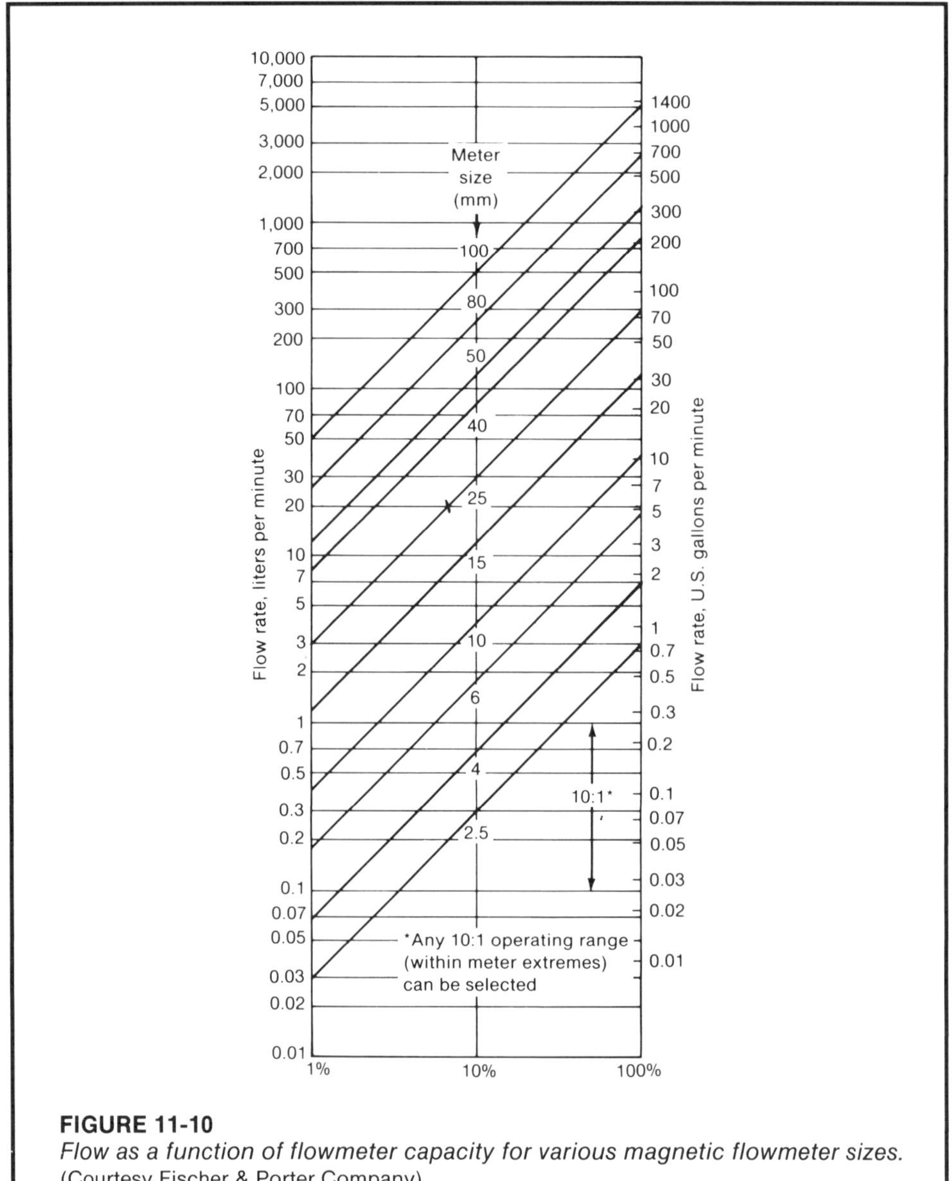

FIGURE 11-10
Flow as a function of flowmeter capacity for various magnetic flowmeter sizes.
(Courtesy Fischer & Porter Company)

EXAMPLE 11-4

Problem: Size a magnetic flowmeter for the following applications based upon the following process data:

A. 15 percent water, 80 percent acid, and 5 percent alcohol flowing at 100 gpm

B. Liquid with a conductivity of 10 to 300 μS/cm flowing at 10 gpm
C. No 2 fuel oil flowing at 1000 gpm

Solution: Both liquids A and B have sufficient conductivity to apply magnetic flowmeter technology. Liquid C does not, and another flowmeter technology should be investigated.

Assuming that the magnetic flowmeter has a 10:1 turndown for all ranges of 3 to 30 feet per second full scale, the velocities of liquids A and B in various size pipes are

Pipe Size	Velocity A	Velocity B
1/2 in.	—	10.56 fps
1 in.	37.1 fps	3.71 fps
2 in.	9.56 fps	0.956 fps
3 in.	4.34 fps	—

For liquid A, a 2-inch or 3-inch magnetic flowmeter can be applied, but the 2-inch flowmeter is more economical. Liquid B could be measured by either a 1/2-inch or a 1-inch flowmeter. The decision of which flowmeter to specify is dependent upon the user's present and future requirements.

It should be noted that if the magnetic flowmeter were accurate over the range of 3 to 30 feet per second, liquids A and B could be measured accurately with turndowns of 3.2:1 and 3.5:1 with 2-inch and 1/2-inch flowmeters, respectively, which reduces performance and flexibility of the flowmeter. In selecting the range, caution must be exercised at fluid velocities exceeding 10 to 15 fps because of excessive pressure drops, pipe erosion, and possible cavitation.

Installation

Installation of a magnetic flowmeter, while relatively straightforward, is a multi-faceted task. Hydraulic, electrical, and orientation requirements must be satisfied for the flowmeter to function properly.

Hydraulic Requirements

Despite myths to the contrary, magnetic flowmeters are somewhat sensitive to flow profiles that are not symmetrical in the pipe. Upstream and downstream straight run requirements vary with manufacturer from 3 to $5D/2D$ from the electrode to $5D/5D$ from the faces of the flowmeter. Straight run requirements for flowmeters that require minimum distances from the electrodes are typically already included within the face-to-face dimensions of conventional ac and dc magnetic flowmeters in small sizes. It can be seen that sufficient straight run from the electrode is provided within a 2-in. flowmeter, as the face-to-face dimension is approximately 15 in., excluding at least 2 in. of additional straight run that would be required for flanges before a bend could be put in the pipe.

Straight run requirements are not satisfied within larger magnetic flowmeters nor within miniature magnetic flowmeters. The approximately 4-in. face-to-face dimension of a 2-in. miniature magnetic flowmeter will dictate some upstream and downstream piping requirements external to the flowmeter, while large flowmeters are constructed with as small a face-to-face dimension as possible for economy.

Upstream and downstream piping requirements should be determined and adhered to in order to maintain flowmeter accuracy. It should be noted that they may vary with manufacture, as some designs utilize characterized coils that distribute the magnetic field in the pipe or electronic means to minimize piping effects.

EXAMPLE 11-5

Problem: Determine the straight run requirements of a 2-inch magnetic flowmeter with a 4-inch face-to-face dimension.

Solution: Assuming that there is a $5D/2D$ requirement from the electrodes and that the electrodes are centered in the magnetic flowmeter body, 10 in./4 in. are required from the electrode. As approximately 4 in. of each requirement is satisfied within the flowmeter body, gasket, and flanges, the actual requirements are 6 in./0 from the faces of the flowmeter.

Piping Conventional ac and dc magnetic flowmeters have flanges at the inlet and outlet that must be bolted to flanges on the pipe. Each manufacturer's face-to-face dimensions are different, with the net result that once a flowmeter is installed, piping changes are required if a flowmeter is replaced with one of different manufacture.

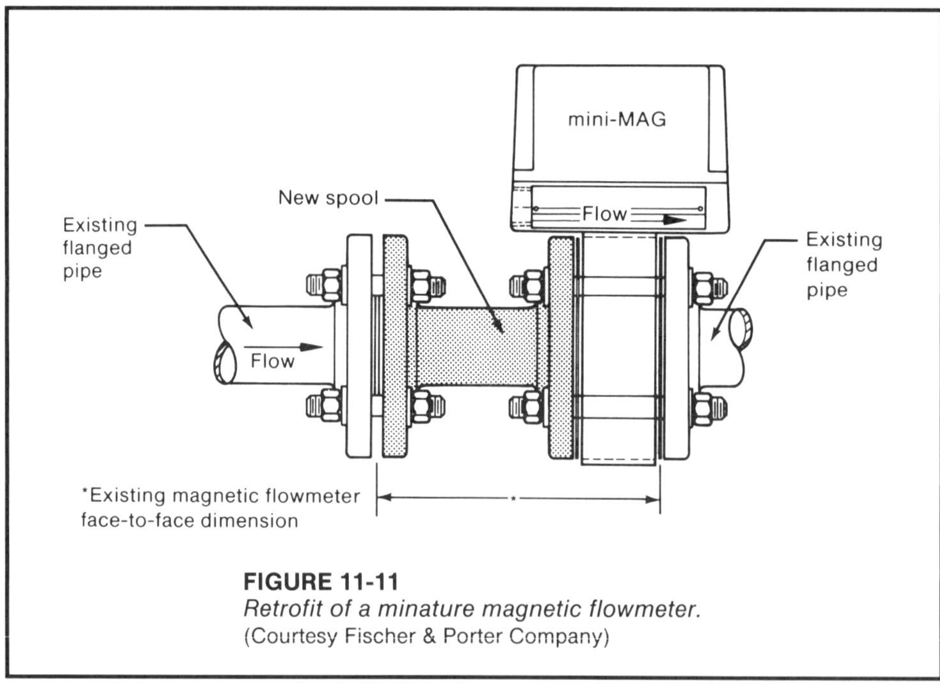

FIGURE 11-11
Retrofit of a minature magnetic flowmeter.
(Courtesy Fischer & Porter Company)

Miniature dc and electrodeless magnetic flowmeters are of a wafer design and are pancaked between pipe flanges using special length bolts. Due to the relatively short face-to-face dimension of this design, miniature dc magnetic flowmeters can be used to retrofit existing magnetic flowmeter installations by inserting a spool piece to take up the pipe length that the miniature magnetic flowmeter does not occupy.

Piping Orientation Magnetic flowmeters must be kept full of liquid at all times for accurate measurement. The flowmeter measures velocity as sensed by the electrodes. This velocity, when multiplied by the area that the liquid flows through, yields the flow. When the electrodes are covered with liquid but the pipe is not full, the measured velocity will be multiplied by the area of the flowmeter rather than by the area through which the liquid is flowing. This results in a substantial error. In the case where the liquid does not contact the electrodes, there is no measurement.

The electrodes should be orientated in the horizontal plane to avoid problems created by air bubbles that may be present in the liquid. If the electrodes were orientated in the vertical plane as shown in Figure 11-13B, air bubbles could cause a lack of liquid contact with the electrodes, resulting in sporadic or complete loss of signal.

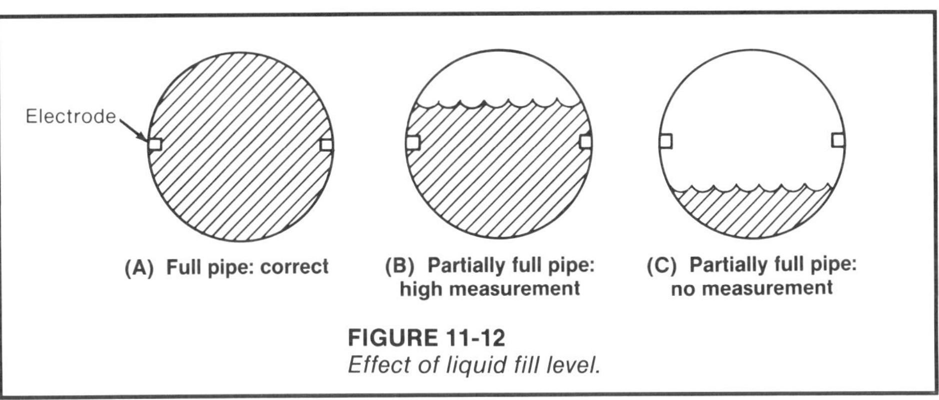

FIGURE 11-12
Effect of liquid fill level.

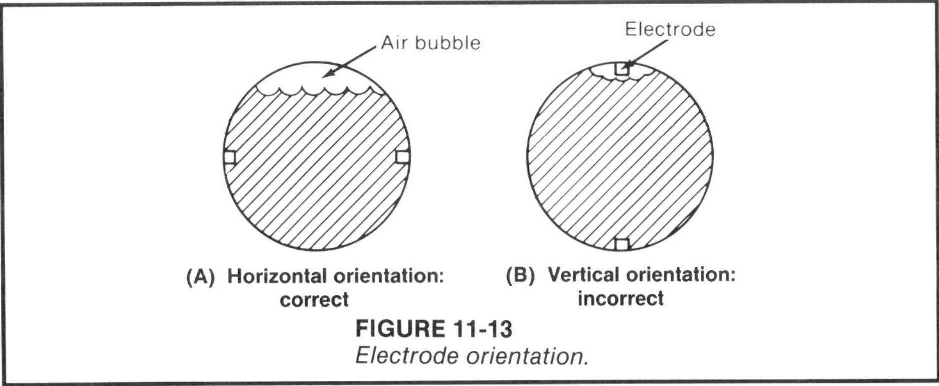

FIGURE 11-13
Electrode orientation.

Installation of the flowmeter in the pipe must be performed in accordance with the flow direction arrow affixed to the flowmeter.

Grounding is required for magnetic flowmeters to shield or isolate the relatively low voltage signal that is measured at the electrodes from relatively high common-mode potentials that may be present in the liquid. If the pipe is conductive and contacts the process liquid, the flowmeter should be grounded to the pipe both upstream and downstream of the flowmeter. If the pipe is constructed from a nonconductive material such as plastic or a conductive material that is insulated from the process liquid such as plastic-lined steel pipe, grounding rings should be installed in contact with the liquid (see Figure 11-14).

If the flowmeter is not grounded relative to the potential of the liquid in the pipe, then the flowmeter electrodes may be exposed to excessive common-mode noise. This limits the turndown of the meter and, in an extreme case, may prevent operation of the electronics or damage them, as the least resistance path to ground for any stray voltages would be via the electrodes and electronics.

Cabling Magnetic flowmeters are 4-wire devices that require an external source of power for operation. Plant wiring practice usually dictates that the power and the analog signal output be fed to the magnetic flowmeter system in separate conduits. Power is usually wired directly to the primary flowmeter since the majority of the power is required by the field coils. Then it is fed to the transmitter where the output signal is generated. Connections are also required from the electrodes at the flowmeter to the transmitter (Figure 11-15).

Conventional ac magnetic flowmeters require individual conduits for electrode signal and power between the primary flowmeter and the transmitter to avoid the effects of capacitive coupling that can occur between the millivolt level electrode signal cable and the power cable.

The power and electrode signal can be run in one conduit for dc magnetic flowmeters, as the frequency of excitation of the magnet is low enough and the sampling period, long enough that the effects of any induced voltage from the power cable are cancelled out. Special cable may be required by the manufacturer, but cabling costs are less than for conventional systems.

Miniature magnetic flowmeters usually have the electronics mounted integral to the flowmeter. This arrangement represents the most economical cabling method since no conduit is required between the flowmeter and the transmitter.

Maintenance

Magnetic flowmeters require no routine maintenance other than periodic calibration checks. However, problems such as electrode coating, liner damage, and electronic failures can occur. Conventional ac magnetic flowmeters and dc magnetic flowmeters often offer some degree of field repairability. The miniature magnetic flowmeters do not, and damaged assemblies should be replaced when failure occurs. It should be noted that some accuracy is lost if the wetted parts are removed or if their position with respect to one another is tampered with. Higher accuracy systems are often implemented by matching the primary flowmeter to a transmitter in the factory. Any modification of either can result in some loss of accuracy.

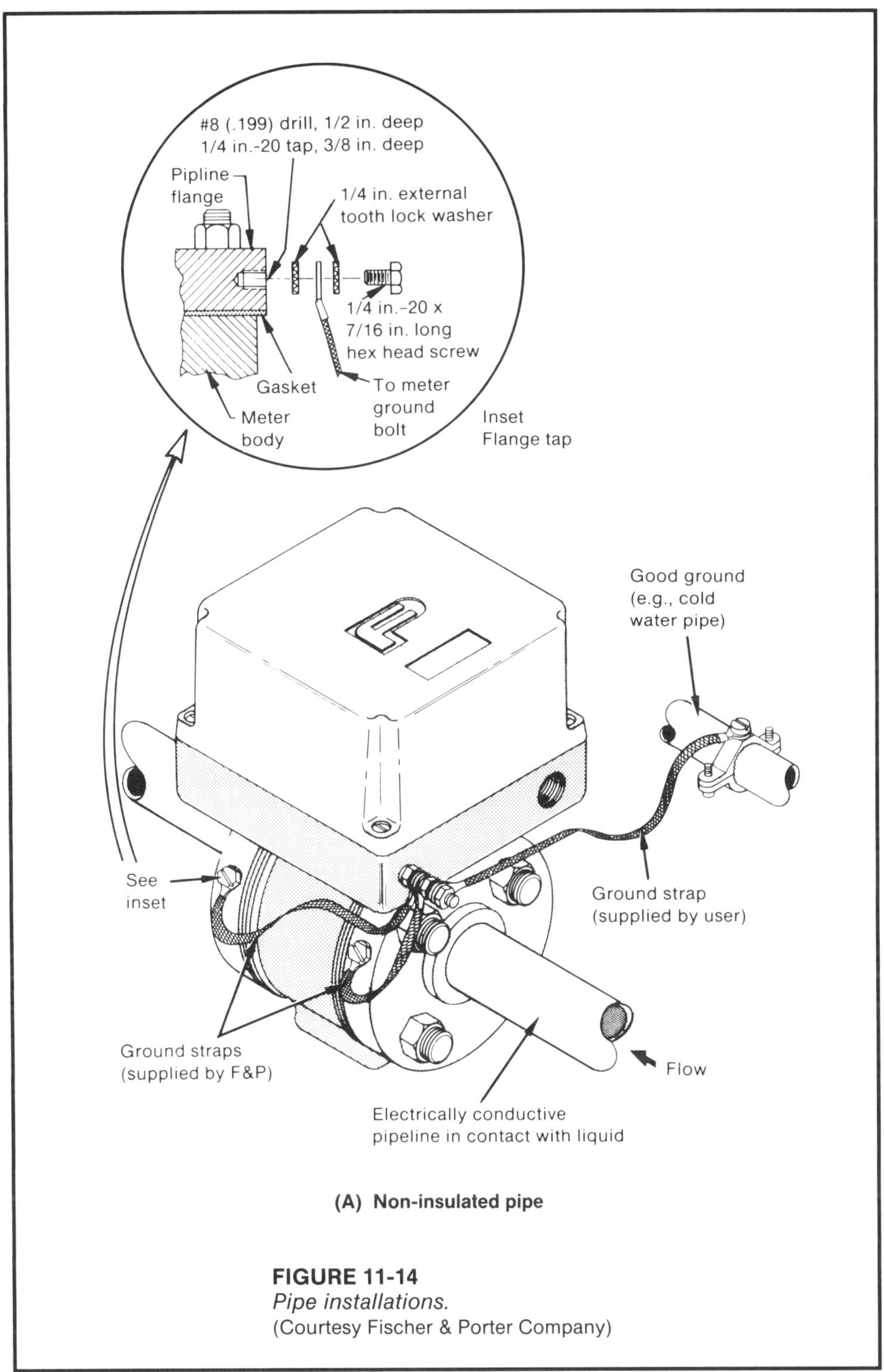

(A) Non-insulated pipe

FIGURE 11-14
Pipe installations.
(Courtesy Fischer & Porter Company)

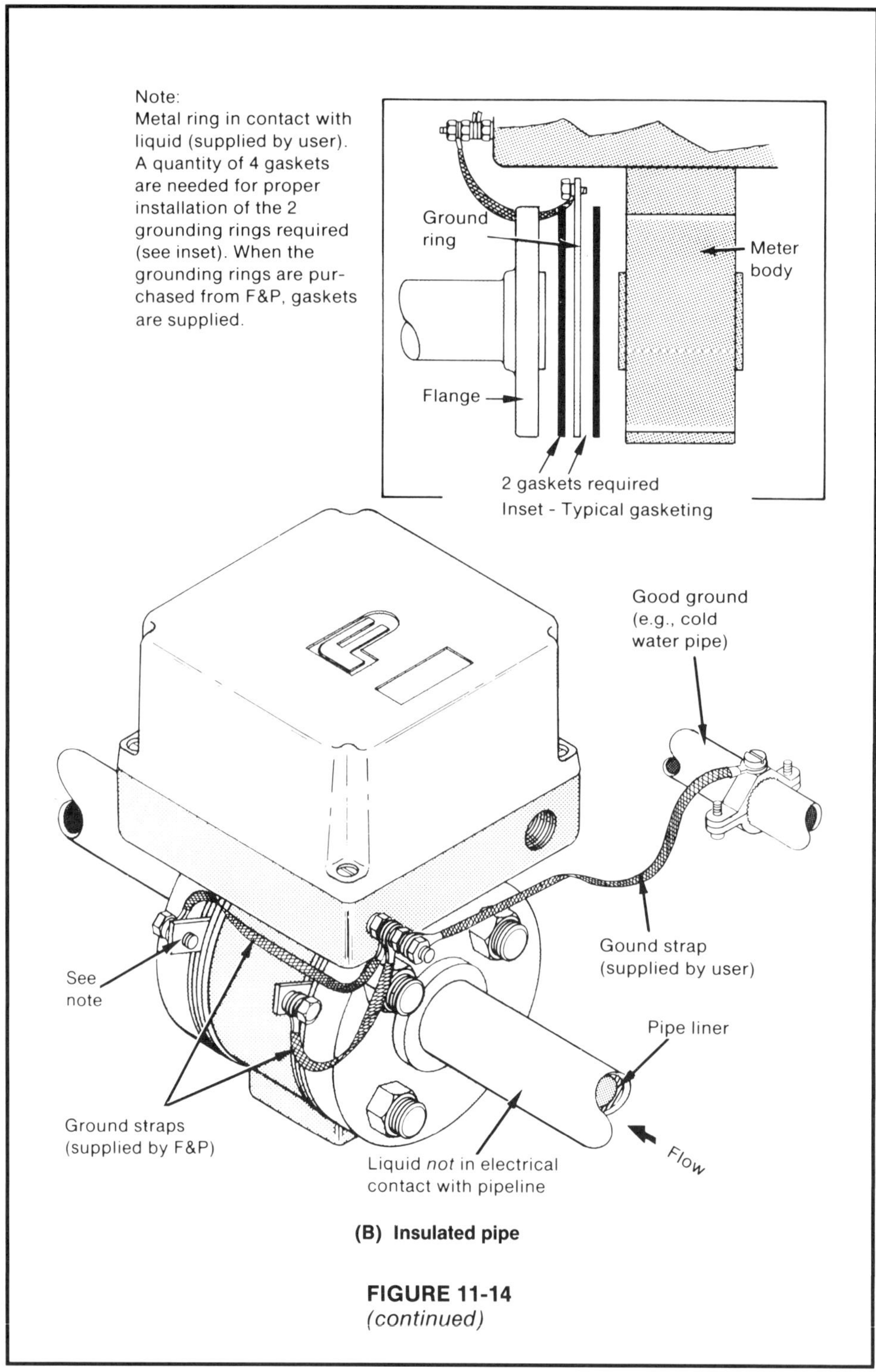

(B) Insulated pipe

FIGURE 11-14
(continued)

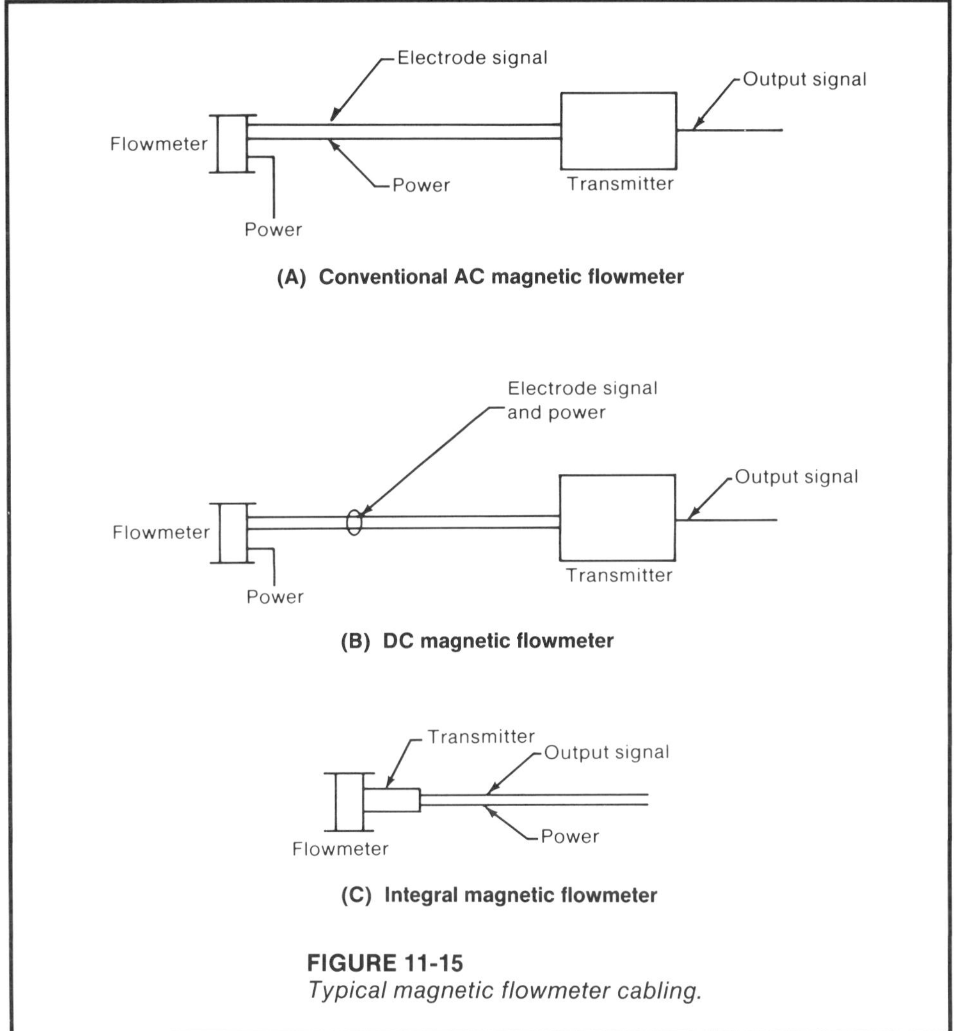

FIGURE 11-15
Typical magnetic flowmeter cabling.

Electrode Coating AC flowmeters are susceptible to nonconductive coating, as the coating will cause a calibration shift due to the effective change in conductivity that the electrodes sense. DC and miniature dc magnetic flowmeters are affected by a nonconductive coating only when the coating becomes so extreme that the effective conductivity between the electrodes is less than the minimum required for the flowmeter to operate.

While some flowmeter designs allow removal and replacement of the electrode, manufacturers should be consulted about methods for cleaning electrodes that do not require removal or replacement of the flowmeter. If this is a continual problem, addition of an ultrasonic cleaner or replacement of a conventional ac magnetic flowmeter with a dc magnetic flowmeter could be considered. An electrodeless design should be considered where possible.

Liner Damage Magnetic flowmeter liners can be damaged by foreign materials (such as debris or solids which are present in the line), wear due to abrasion, liquids that are incompatible with the liner material, excess temperature, or by using improper methods to remove the flowmeter from the pipe. Some flowmeter designs allow field replacement of the liner, but most require that the flowmeter be replaced or returned to the manufacturer for overhaul. Corrosion of the electrodes may contribute indirectly to liner damage if the process fluid leaks around the probes and behind the liner. Ceramic liners are subject to cracking when exposed to large temperature variations.

Shorted Electrode Should one electrode become shorted to the fluid, the magnetic flowmeter will read approximately 50% of the true flow.

Electronic Failure Most dc, miniature dc, and electrodeless magnetic flowmeters have some type of reference signal that can be injected into the circuit by switching jumpers. This will check the operation of approximately 90 percent of the circuits and is useful when a malfunction is suspected.

Spare Parts Spare parts requirements for conventional ac and dc magnetic flowmeters can include various electrodes, liners, flowtubes for each size, and the electronics. Miniature dc magnetic flowmeters require that a complete flowmeter primary for each size and the electronics be stocked. It should be noted that miniature dc magnetic flowmeters are standard items and have few options, so the number of inventoried spare parts is effectively reduced. Failure of a miniature dc magnetic flowmeter usually results in the salvage of either the primary flowmeter or the electronics.

Calibration Calibration of the electronics can be accomplished with a magnetic flowmeter calibrator or by electronic means. Magnetic flowmeter calibrators are precision instruments that inject the output signal of the primary flowmeter into the transmitter, which effectively checks and calibrates all of the electronic circuits. An alternate calibration method, although not as accurate, is to make adjustments based upon test signals injected into the transmitter, circuit test measurements, or thumbwheel switches per manufacturer specifications.

Calibration of conventional ac magnetic flowmeters should be performed at zero flow with the flowmeter full of the process liquid. It should be noted that dc and miniature dc magnetic flowmeters have an automatic zero adjustment and require a span adjustment only for the primary flowmeter, but they may require additional zero and span adjustments for an analog output.

EXERCISES

11.1 What are the advantages of a dc magnetic flowmeter versus an ac magnetic flowmeter?

11.2 What are the advantages of a miniature dc magnetic flowmeter versus a dc magnetic flowmeter? Disadvantages?

11.3 What are the typical velocity and conductivity constraints of a magnetic flowmeter?

11.4 Assuming that the liquid is sufficiently conductive to apply magnetic flowmeter technology, size magnetic flowmeters for 30 gpm, 50 gpm, and 1000 gpm.

11.5 What is the preferred magnetic flowmeter technology for applications where electrode coating poses a problem?

12

Mass Flowmeters

Introduction Flowmeters that measure mass directly, as opposed to indirectly as a function of other physical properties, have recently been developed into practical flowmeter designs. Applications of one design are a function of mass flow, which makes it virtually independent of the physical properties of the fluid, as long as the fluid can be put through the flowmeter. Another design has ranges of operating conditions over which the flowmeter will operate as a mass flowmeter. The net result is a true mass measurement and the ability to valve various products through a common flowmeter and to effect cost savings in some applications.

Coriolis Mass Flowmeters The Coriolis mass flowmeter is a true mass flowmeter with many applications. It is relatively easy to apply and size. Due to the versatility of the flowmeter and the advantages of mass flow measurement, the temptation is often present to apply this technology to all applications. The relatively high cost of Coriolis mass flowmeters tends to restrict the use of this technology. However, the additional cost may be justifiable in applications such as pilot plants, where these flowmeters can be used for different fluids and reused on future projects, or when fluid properties are properly defined or varying.

Principle of Operation Coriolis mass flowmeters are based on the conservation of angular momentum as it applies to the Coriolis acceleration of the fluid. For the purposes of discussion, an illustration of the existence and meaning of Coriolis acceleration is more desirable than a detailed derivation. Consider a man standing on a rotating turntable as shown in Figure 12-1A. Since the man is

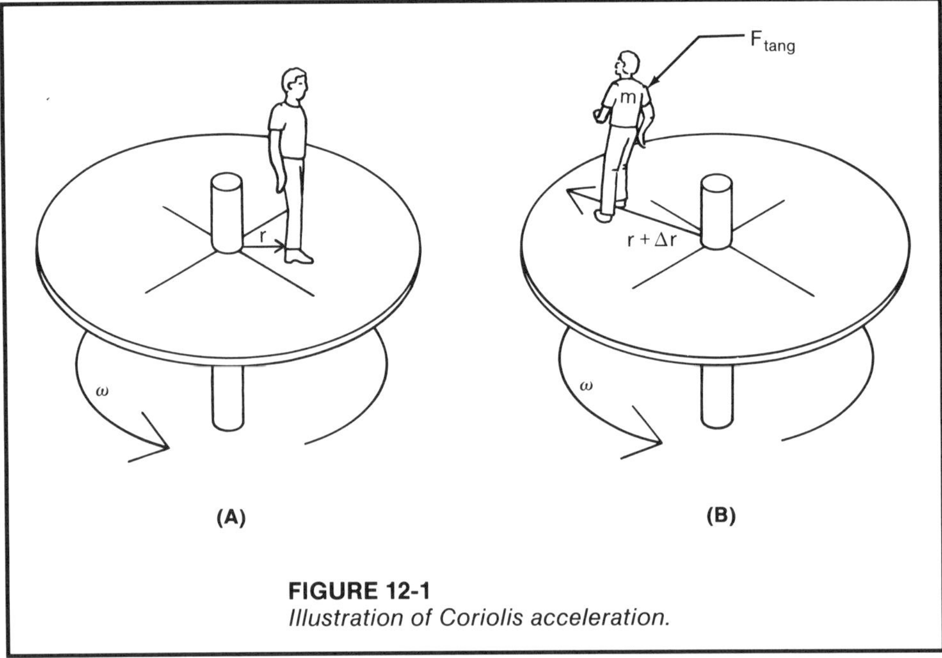

FIGURE 12-1
Illustration of Coriolis acceleration.

standing in the same place on the turntable and the turntable is rotating at a constant speed, the forces acting against the man in the plane tangential to the turntable are:

$$F_{tang} = m \times a_{tang}$$
$$= m \times \frac{\Delta v_{tang}}{\Delta t}$$

where the change in velocity per unit time is, by definition, the acceleration. The tangential velocity of the man is given as

$$v_{tang} = r \times \omega$$

and is a constant; the man is stationary, his distance from the center of rotation is constant, and the rotational speed of the turntable is constant. Therefore, the tangential velocity of the man does not change, Δv_{tang} is zero, and the force exerted on the man in this plane is also zero.

If the man were to walk away from the center of rotation, as depicted in Figure 12-1B, a nonzero force is exerted on the man in the tangential plane as the distance between the man and the center of rotation is changing. In this case,

$$v_{tang} = (r + \Delta r) \times \omega$$
$$= (r \times \omega) + (\Delta r \times \omega)$$

The first term in the above equation has been shown not to result in any forces that act on the man in this plane. The second term reflects the effects of the changing position of the man in relation to the center of rotation. The nonconstant nature of this term results in a change of tangential velocity that is nonzero, and hence a force acts on the man in the tangential plane. The force is created as a result of the Coriolis acceleration acting on the man as he changes position in relation to the center of rotation of the turntable.

Coriolis mass flowmeters exploit the existence of the force exerted by the Coriolis acceleration of a fluid. The flowmeter consists of a vibrating tube, in which the Coriolis acceleration is created and measured, as shown in Figure 12-2.

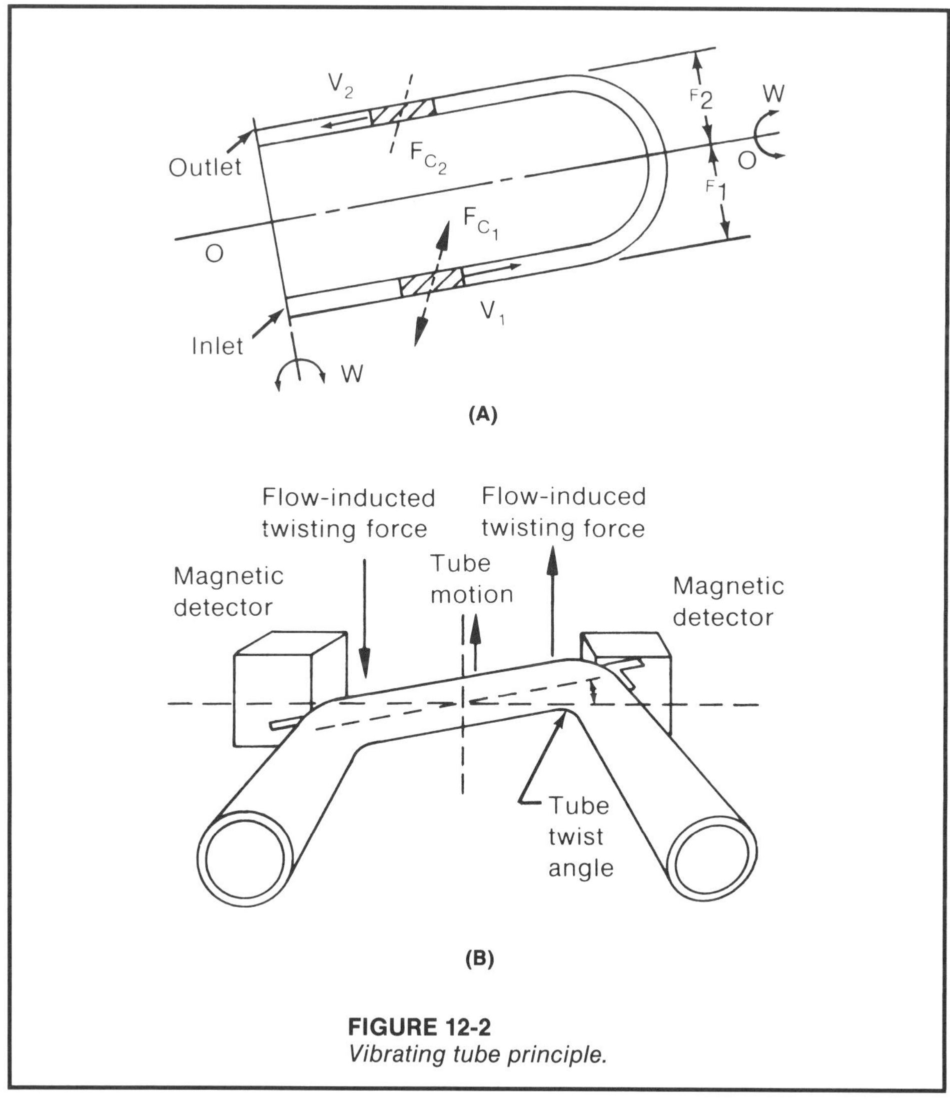

FIGURE 12-2
Vibrating tube principle.

In place of the rotational motion described in the illustration of Coriolis acceleration, the inlet and outlet are held fixed while the tube is vibrated sinusoidally about an axis formed between the inlet and outlet. In one half of the tube, fluid flows away from the axis of rotation while in the other half, the same amount of fluid flows towards the axis of rotation. At a given instant, the fluid in each half of the tube has an associated Coriolis acceleration that acts in opposite directions due to the opposite directions of fluid flow in relation to the axis of rotation. Coriolis accelerations in opposite directions result in forces in opposite directions, which tend to twist the tube. The twist is directly proportional to the mass flow through the tube.

Construction The basic construction of various Coriolis mass flowmeters is shown in Figure 12-3. The tubes, which are constructed to have predictable vibratory characteristics, are vibrated by the drive assembly. The twist of the tube is sensed by the detector system.

The dual tube design, where the fluid flows through two parallel tubes, effectively cancels the effects of vibration, as the flowmeter maintains physical relationships between components when vibration is present that could not be maintained in the single tube design. The flow is split in some designs, which can limit the ability to effectively clear a plugged flowmeter.

The only wetted part of the Coriolis mass flowmeter is the tube itself, which is typically constructed of stainless steel. Other corrosion resistant metals, such as Hastelloy®, are also used in tube construction. Teflon®-lined flowmeters may prove to be more economical than flowmeters made of exotic metals in many applications.

Designs are available with standard or thin-wall tubes. The thin-wall design is more sensitive and is applied to gas and low velocity liquid flow applications. It should be noted that the rate at which the fluid attacks the metal or liner should be examined closely. Although the flowmeter is fully pressure rated, its construction is such that the wall thickness of the tube may be thinner than that of associated piping. This means that if the fluid attacks the flowmeter tube, its characteristics can change over time, which could affect the accuracy of the instrument. If the attack is allowed to continue, the tube may lose its strength as it loses it walls, which could cause a leak.

Operating Constraints Coriolis mass flowmeters operate when the mass flow is within the accurate measurement range of the instrument and neither the pressure nor the temperature constraints of the instrument and flanges are exceeded. High temperature sensors can withstand temperatures of up to approximately 425°C, while flanges are available through 600 lb ratings (see Figure 12-4).

A more practical application constraint, but not a flowmeter performance constraint, is that the pressure drop across the flowmeter cannot exceed the maximum allowable pressure drop that the total system will accept. When this happens, the desired flow cannot be made to flow through the system due to the overall pressure drops in the piping system, of which the flowmeter is but a part. This should be examined closely, especially if the flowmeter is to measure high

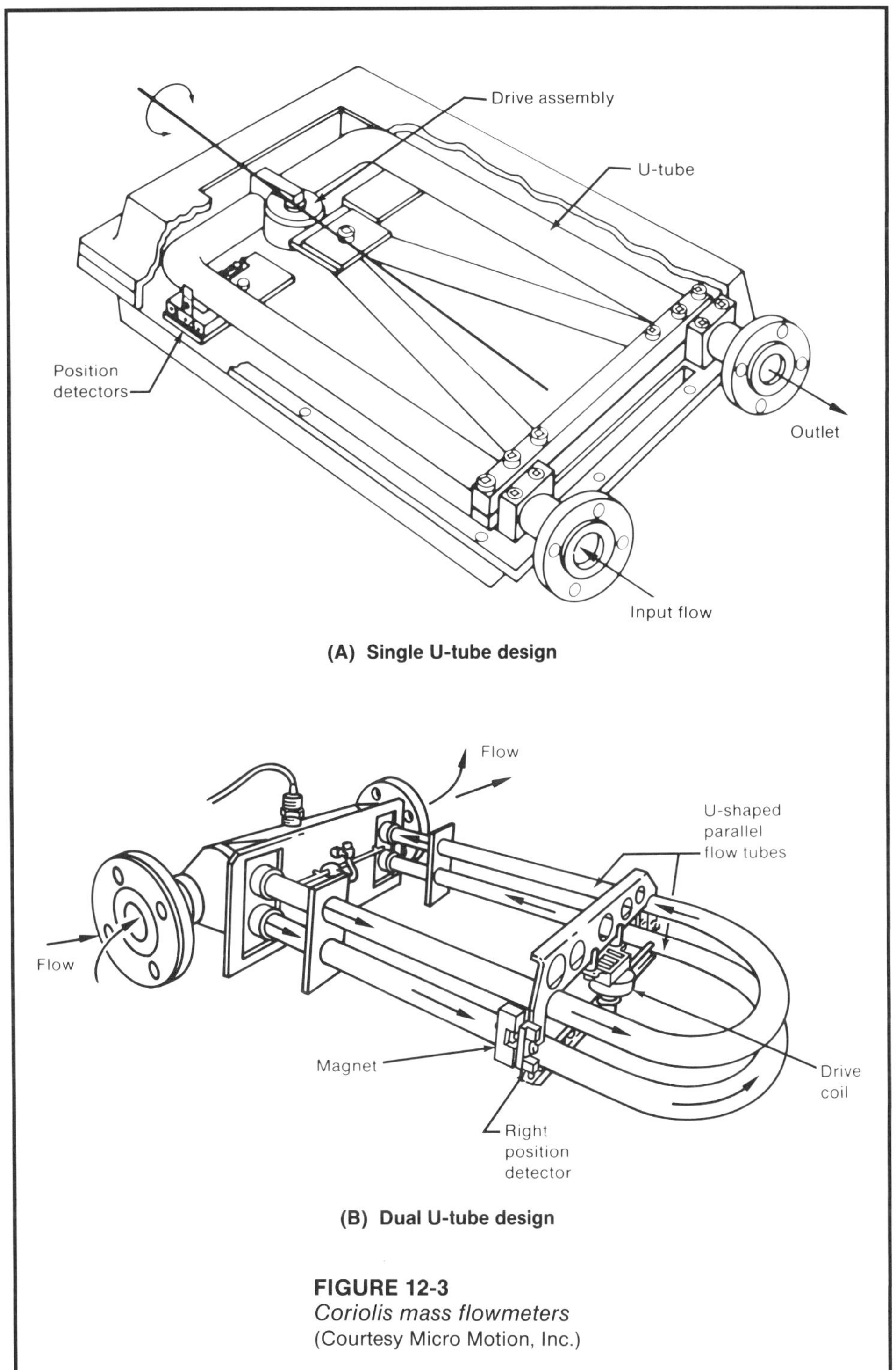

(A) Single U-tube design

(B) Dual U-tube design

FIGURE 12-3
Coriolis mass flowmeters
(Courtesy Micro Motion, Inc.)

(C) Twinloop design
(Courtesy EXAC Corporation)

(D) Straight flowtube design
(Courtesy Endress + Hauser)

FIGURE 12-3 (continued)
Coriolis mass flowmeters.

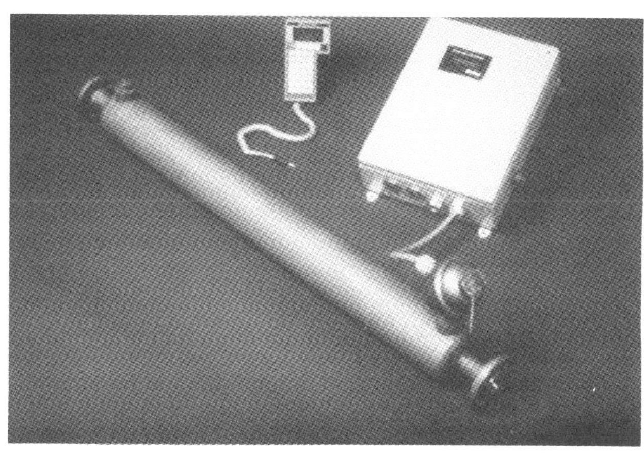

(E) Straight flowtube design
(Courtesy Bailey Controls Company)

(F) Double helix design
(Courtesy Fischer & Porter Company)

FIGURE 12-3 (continued)

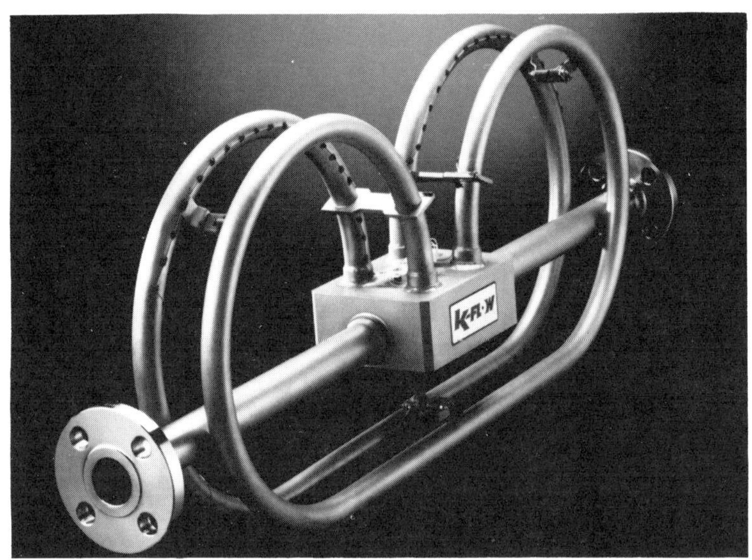

(G) B-tube design
(Courtesy K-Flow Corporation)

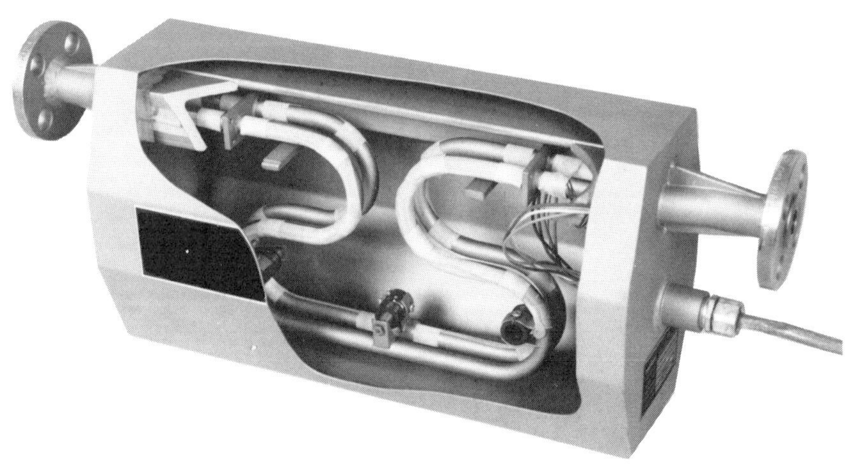

(H) Omega tube design
(Courtesy Schlumberger Industries, Measurement Division)

FIGURE 12-3 (continued)

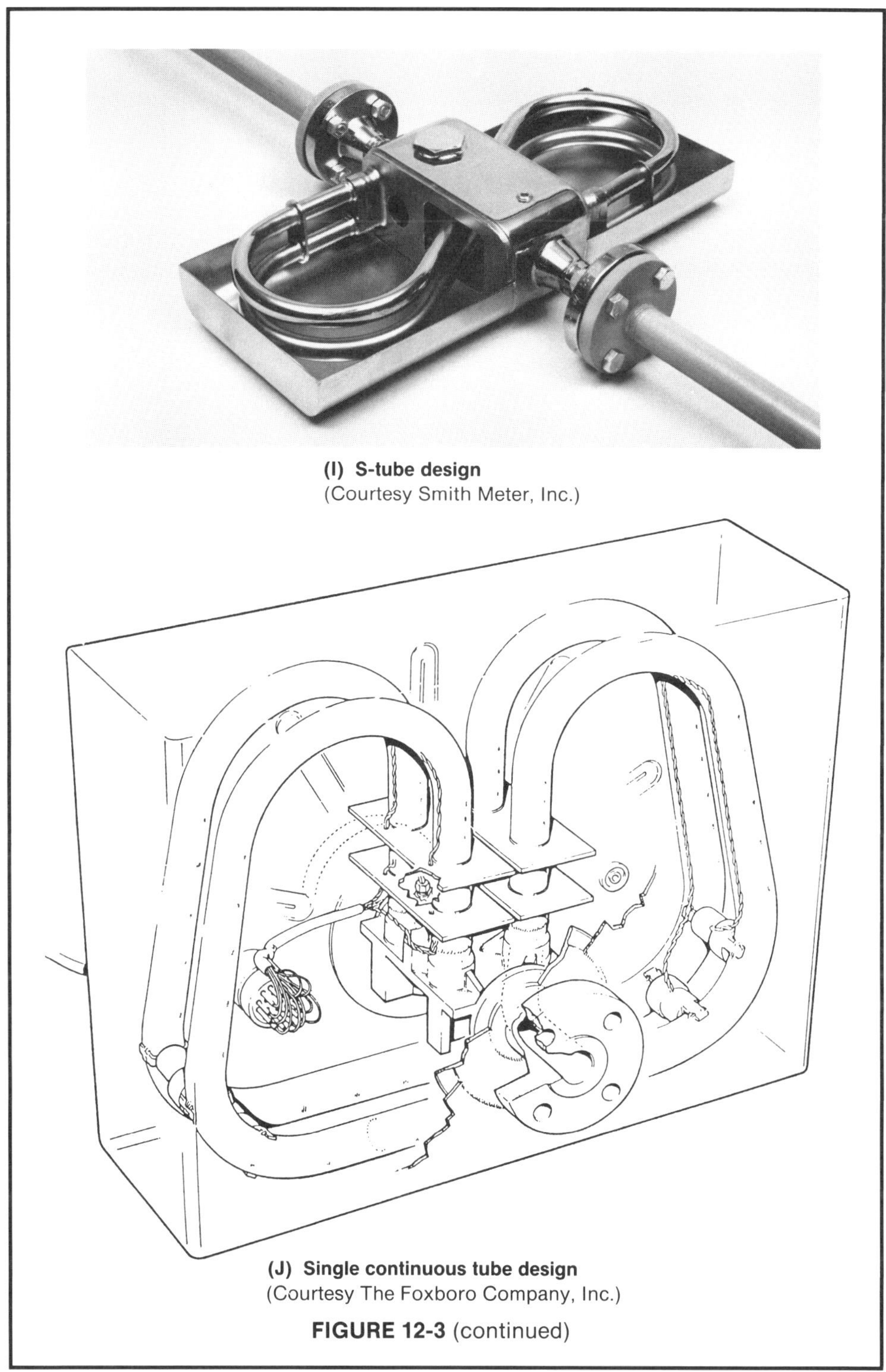

(I) S-tube design
(Courtesy Smith Meter, Inc.)

(J) Single continuous tube design
(Courtesy The Foxboro Company, Inc.)

FIGURE 12-3 (continued)

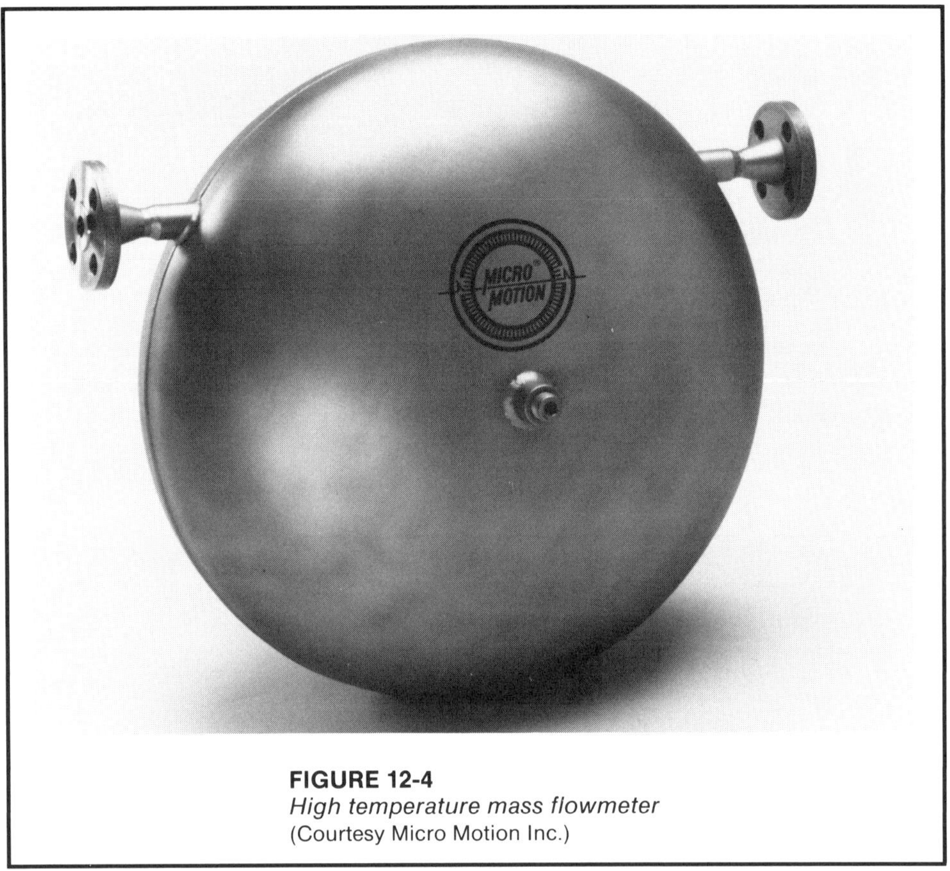

FIGURE 12-4
High temperature mass flowmeter
(Courtesy Micro Motion Inc.)

flows or if the fluid is highly viscous. The pressure drop at a given set of operating conditions will be different for different designs.

Performance The accuracy of the Coriolis mass flowmeter is claimed to be 0.2–0.4 percent rate over its accurate measurement range. However, there are various stipulations associated with this statement. When the flow is less than the minimum accurate flow, the error is constant and equal to the error at the minimum accurate flow. This results in a full scale accuracy statement at low flows. When the process temperature varies more than $\pm 11\,^{\circ}C$ from the temperature at which the zero was performed on the flowmeter, a 0.01 percent nominal meter capacity error is added to the accuracy statement.

Applications Coriolis mass flowmeters, which have no Reynolds number constraints, can be applied to virtually any liquid or gas flowing at a sufficient mass flow to operate the flowmeter.

Typical liquid applications include harsh chemicals, low to medium viscosity liquids, foods, slurries, and blending systems. While the installed cost of a

Coriolis mass flowmeter is relatively high, the ability to use one flowmeter to measure various fluids, eliminating individual flowmeters for each fluid, may be justifiable.

Applications of this technology to gases is somewhat limited in that the density of low pressure gases is usually too low to accurately operate the flowmeter. The advantage of mass measurement in gas applications is that, when applicable, it can eliminate the need for pressure and temperature compensation and the hardware necessary to implement these functions, which may justify the added cost for this type of flowmeter in some applications.

As the frequency of the vibration of the tube varies with the density of the fluid within the tube, a density signal is inherently generated. This signal may be used to infer composition of some fluids.

Sizing Coriolis mass flowmeters range in size from 1/16 in. to 6 in. and larger, which corresponds to accurate mass flows as low as 1 pound per hour. Sizing a Coriolis mass flowmeter involves calculating the desired mass flow measurement range, selecting the flowmeter size that will measure that range accurately, and then verifying that the pressure drop across the flowmeter is less than the maximum pressure drop the flowmeter can take in the piping system. This can be done by using graphs and procedures developed by the manufacturer or by obtaining the equivalent lengths of pipe for each size flowmeter and calculating the friction losses. A graph typical of the pressure loss associated with this type of flowmeter is shown in Figure 12-5 for a liquid application with a specific gravity of 1.0. Proprietary equations are available from the manufacturer to allow Figure 12-5 to be used for gas applications.

EXAMPLE 12-1

Problem: Size two Coriolis mass flowmeters for a full scale flow of 100 pounds per minute of a liquid with a specific gravity of 1.0 if one flowmeter is operated at a viscosity of 25 cP and the other is operated at 250 cP. The maximum acceptable pressure drop across each flowmeter is 10 psi.

Solution: A 1-in. flowmeter exhibits a 3-psid pressure drop at 25 cP and a 28-psid pressure drop at 250 cP. The 1-in. flowmeter is acceptable at 25 cP, but a larger flowmeter would be required at 250 cP.

Installation There are no straight run requirements for this type of flowmeter.

Piping

Recommendations for the installation of the single-tube Coriolis mass flowmeter design vary greatly depending on the application. The flowmeter should be securely supported, so that pipe vibration and/or self-induced vibration cannot cause the flowmeter body to twist, thereby affecting the measurement. This often necessitates installation of a rather substantial support.

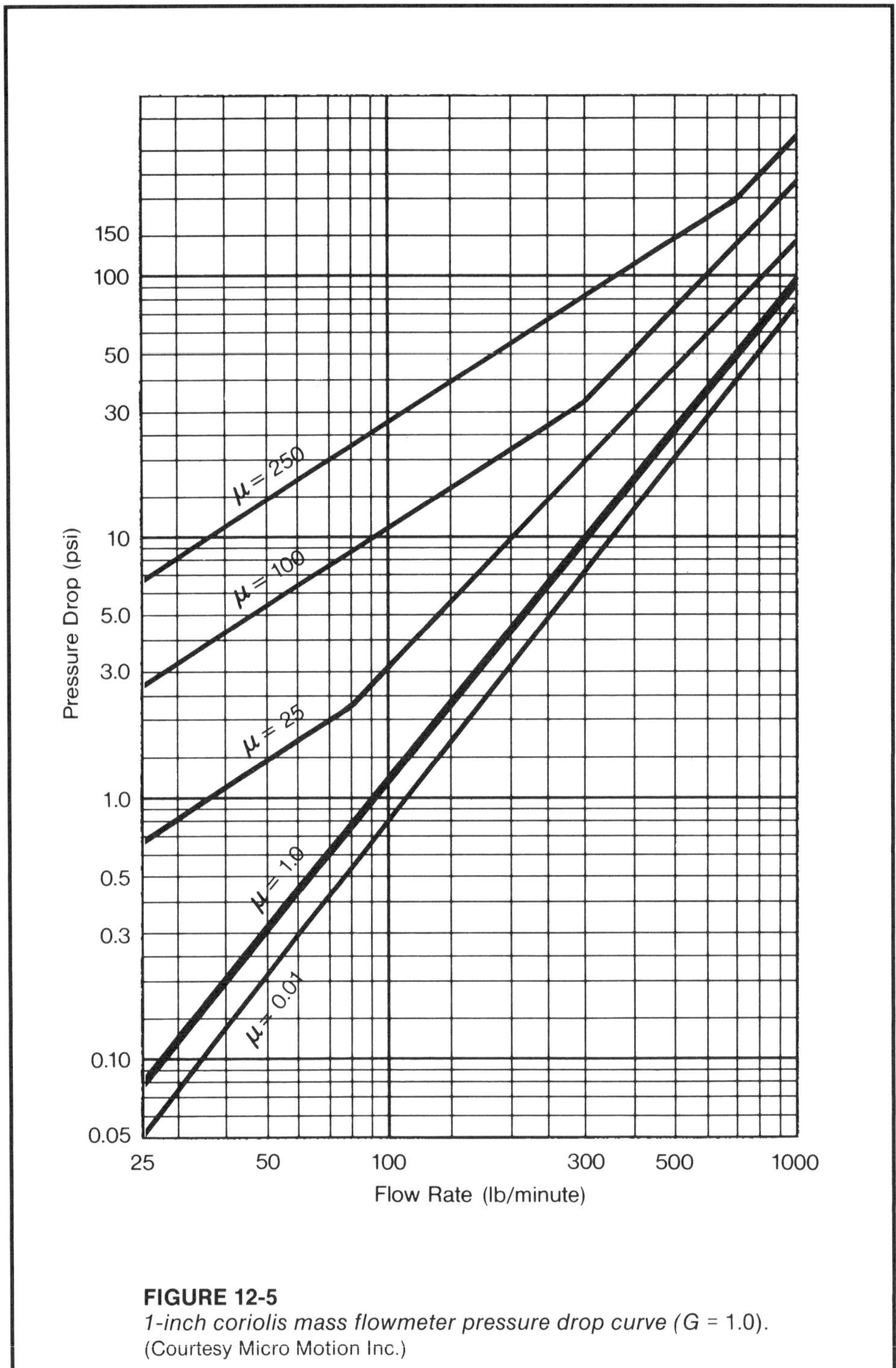

FIGURE 12-5
1-inch coriolis mass flowmeter pressure drop curve (G = 1.0).
(Courtesy Micro Motion Inc.)

The dual tube design, where the fluid flows through the tubes so that the effects of vibration are cancelled (as the flowmeter maintains relationships between its internal components when vibration is present), presents considerably fewer mounting constraints. This design can be installed directly in the pipe with no supporting structures other than those required to properly support the piping system. The reduced installation cost inherent with simplified piping and support requirements in new and retrofit applications makes this design clearly preferable.

Piping Orientation

For liquid applications, the Coriolis mass flowmeter must be oriented so that the meter is completely full of liquid at all times and no bubbles of gas can accumulate in the flowmeter. Therefore, the flowmeter should not be located at the high point of the piping system.

For gas applications, the flowmeter must be oriented so that no condensate or liquid present in the pipe can collect in the flowmeter. Therefore, the flowmeter should not be located in a low point of the piping system.

Cabling

Coriolis mass flowmeters are usually specified as 4-wire devices that require an external source of power to operate. Three-wire models are available that require an additional power supply to generate low voltage dc power to operate the flowmeter.

Maintenance Coriolis mass flowmeters require no routine maintenance other than periodic zero calibration checks. However, problems such as wear and coating of the tube or electronic failures can occur.

Tube Wear/Coating/Corrosion

As the measurement is a direct function of the force due to the motion of mass, coating of the inside of the tubes will not affect the operation of the flowmeter. Excessive coating conditions can cause the tubes to be constricted in such a manner that the piping system cannot supply sufficient mass flow for the flowmeter to be operated within its accurate measurement range. This will result in a loss of range and a loss of accuracy if the flow is less than the minimum accurately measurable flow of the flowmeter.

Wear and corrosion of the tubes can cause a long-term shift in the accuracy of the measurement due to a gradual change in the effective mechanical characteristics of the tubes.

Electronic Failure

Electronic failures are usually handled by board replacement.

Spare Parts

Spare parts are typically limited to electronic circuit boards and sensors.

Calibration

Zero and span calibration of most flowmeters is performed digitally under zero flow conditions at operating temperature. Variations of more than $\pm 11°C$ from the temperature at which the zero adjustment was performed result in reduced flowmeter accuracy.

Hydraulic Wheatstone Bridge

This flowmeter technology is a true mass flowmeter with some constraints on the operating conditions of the liquid. The use of a pump proves to be an asset in achieving low pressure losses required for process reasons.

Principle of Operation This technology is, in principle, the hydraulic equivalent of the electrical Wheatstone bridge (see Figure 12-6).

Construction Typical construction of a mass flowmeter utilizing the Wheatstone bridge principle is shown in Figure 12-7. A pump and motor arrangement is used to provide a constant flow for the bridge assembly, which houses the precision orifices. A differential pressure transmitter is then used to sense the flow signal. The pump, bridge assembly, and differential pressure transmitter are the only wetted parts and can be of stainless steel construction.

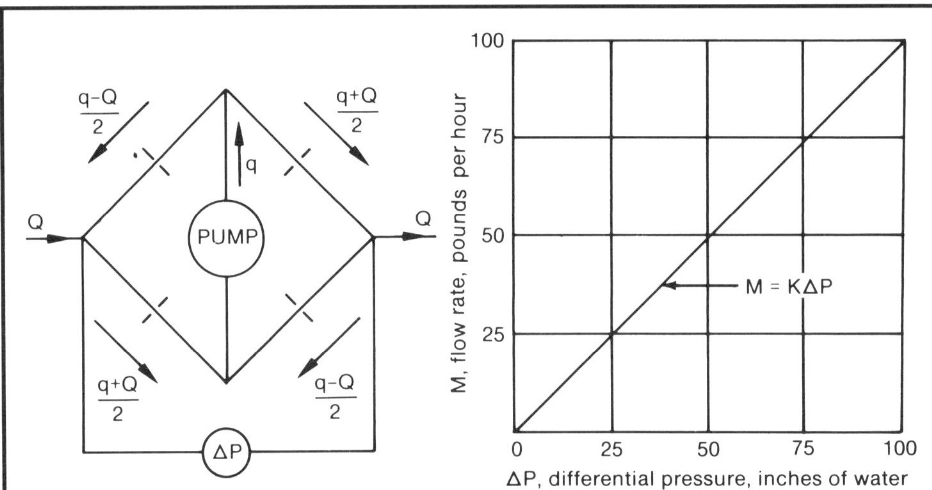

Four matched orifices make up the bridge, and an integral constant flow recirculating pump establishes the internal reference flow. Sensing a situation of unbalance generated by external flow through the meter, the hydraulic bridge produces an output of differential pressure, which is both linear and proportional to the true mass liquid flow.

FIGURE 12-6
Principle of operation.
(Courtesy FLO-TRON, Inc.)

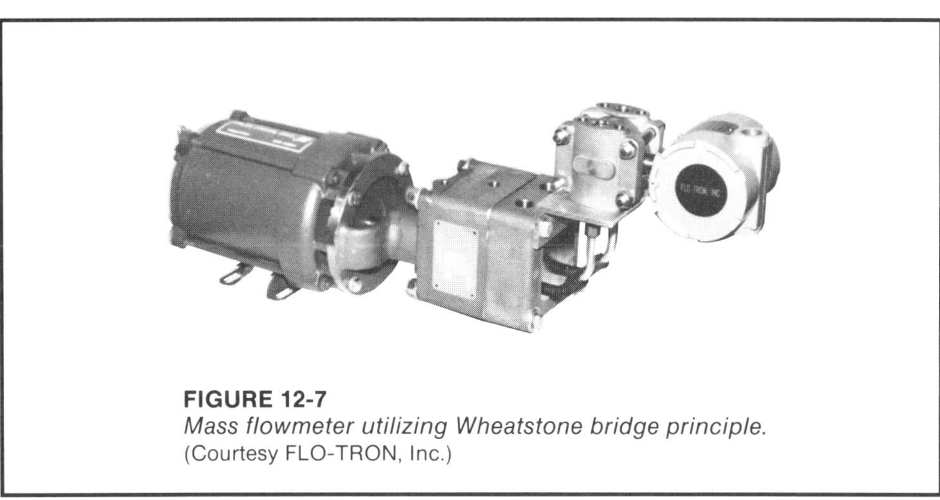

FIGURE 12-7
Mass flowmeter utilizing Wheatstone bridge principle.
(Courtesy FLO-TRON, Inc.)

Operating Constraints Operation is limited to clean liquids whose pressures and temperatures are typically limited to 1000 psig and approximately -30 to 150°C, respectively. To achieve mass flows less than 1 pound per hour, the temperature must be controlled within ±1°C. There are no Reynolds number constraints, but the pump effectively limits applications to those in which the viscosity is less than 50 cP. Large viscosity variations can cause shifts in accuracy.

Performance Accuracy statements are typically the sum of 0.5 percent rate and 0.01 to 0.03 percent full scale, although higher accuracy flowmeters are available. Repeatability statements are typically 0.25 percent rate plus 0.01 to 0.03 percent full scale. Accurately measurable flows range from approximately 0.1 pound per hour to 600 pounds per minute.

The flowmeter bridge assembly can operate over ranges of 50 to 100:1 or more, but the differential pressure transmitter loses considerable accuracy at the low end of its scale due to its full scale accuracy statement. This can be remedied by paralleling more than one transmitter, each with a different calibration range, to achieve accurate measurement of low flows.

Applications Applications are limited to clean liquids with viscosities less than 50 cP.

Sizing Each mass flowmeter is custom designed for each application; therefore, calculations and sizing are performed by the manufacturer.

Installation The motor, pump, and bridge assembly should be mounted on a solid base, as should any piece of rotating equipment. There are no upstream or downstream piping requirements. Power requirements of the motor and the signal requirements of the differential pressure transmitter necessitate power and analog cables that are usually installed in separate conduits.

Since small orifices are utilized, it is recommended that a filter be installed upstream of the flowmeter to avoid pluggage.

Maintenance Routine maintenance is generally limited to lubrication of rotating equipment. However, since the orifices in the bridge assembly are small, any dirt that may be present in the liquid may cause plugging, which would necessitate a thorough cleaning of the bridge assembly.

Spare Parts

Spare parts should be maintained for the differential pressure transmitter, motor, and pump.

Calibration

Calibration of the differential pressure transmitter can be performed by adjusting the zero and span with zero and the full scale differential across the transmitter, respectively.

EXERCISES

12.1 Is a 1-in. Coriolis mass flowmeter applicable for a maximum mass flow of 100 pounds per minute when the liquid has a viscosity of 25 cP, given that the maximum acceptable pressure drop is 5 psi?

12.2 Is a 1-in. Coriolis mass flowmeter applicable for a maximum mass flow of 50 pounds per minute when the liquid has a viscosity of 250 cP, given that the maximum acceptable pressure drop is 5 psi?

12.3 Why is the hydraulic Wheatstone bridge technology limited to clean liquids?

12.4 Why is the hydraulic Wheatstone bridge technology limited to liquids with viscosities of less than 50 cP?

13

Open Channel Flowmeters

Introduction The majority of industrial liquid flows are carried in closed conduits that are operated full of fluid. This is not always the case, however, for high volume flows of water in irrigation systems, water works, or in sanitation and storm drain systems. The flows in these processes are typically classified as open channel flows and are characterized by low system heads and high volumetric flow rates.

Open channel flow rates are commonly inferred from established flow characteristics of carefully constructed restrictions to the flow. The geometry of such a restriction establishes a fixed relationship between flow and the liquid level near or within the element. The two most common restrictions used are the weir and the flume.

Weirs

Principle of Operation A *weir* is a restriction in an open channel in which a crested dam or plate is placed across the flowing stream. The weir is designed so that the liquid springs clear of the sharp-edged crest of the dam, resulting in a sheet of fluid. This sheet is referred to as the nappe and should be freely ventilated underneath so as not to affect the flow over the weir. The depth of the liquid above the crest of the weir is used to infer the flow. The profile of a weir and various weir geometries are illustrated in Figure 13-1.

The rectangular and Cipolletti weirs have approximately a 3/2 power relationship between level and flow, while the V-notch weir has an approximate 5/2 power relationship. The V-notch weir has no crest but the measured differential is referenced to the bottom of the "V".

Construction A weir may consist of a dam of metal (thin plate) or concrete (broad crested) with specified openings. A level-sensing device upstream of the weir is used to sense the liquid head. A variety of materials may be used depending

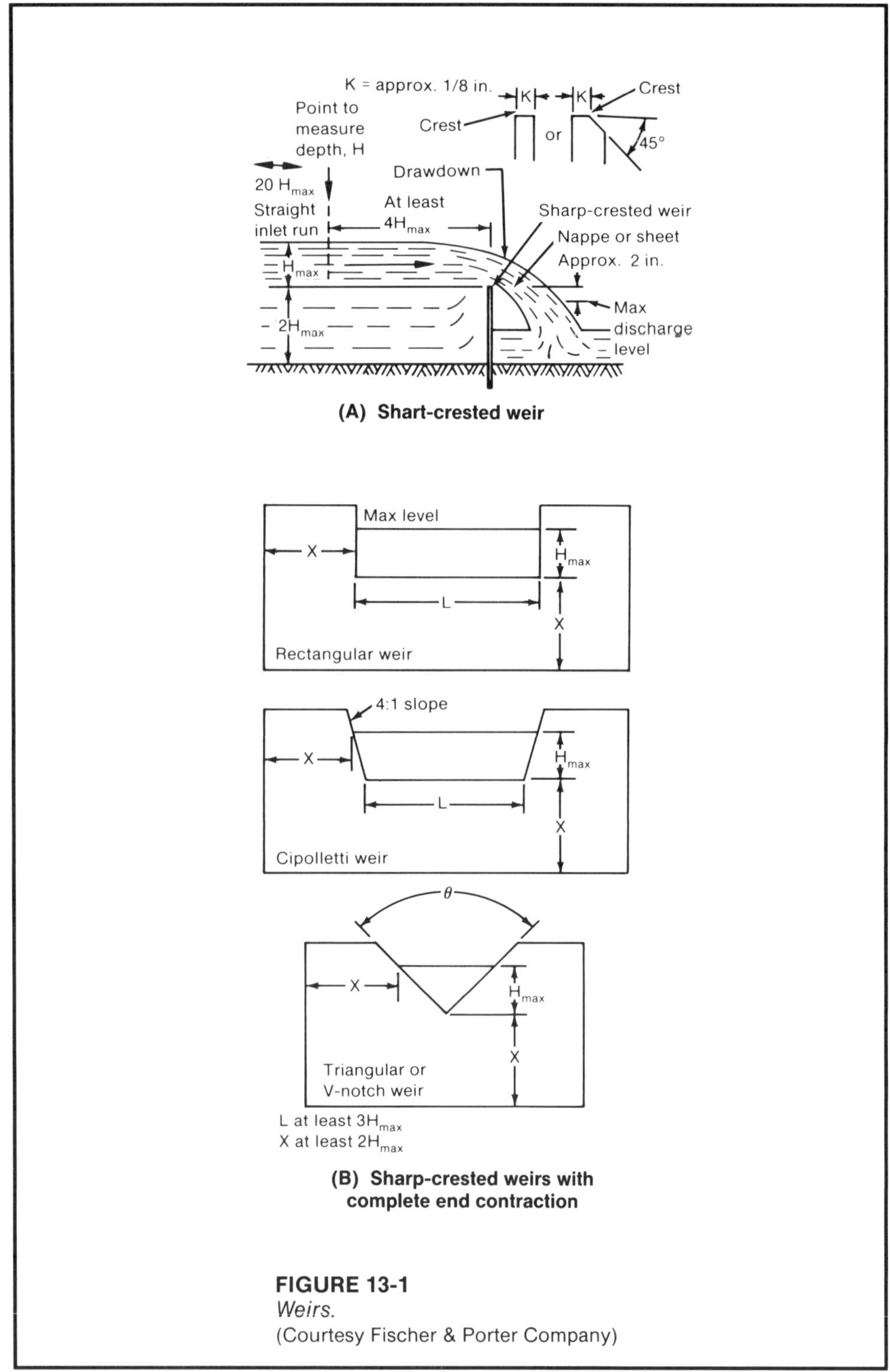

FIGURE 13-1
Weirs.
(Courtesy Fischer & Porter Company)

on the nature of the flowing fluid. Typically, fiberglass construction with metal crest, concrete with metal crest, or all metal construction are available.

Operating Constraints Weirs, aside from being operated within their flow limits, must also be operated within the available system head. Typically, most applications are gravity fed and the amount of permanent head loss (that is, the difference in level before and after the flowmeter) may be limited by physical requirements on the elevation of the inlet and outlet.

Operation of the weir is sensitive to the approach velocity of the liquid, often necessitating a stilling basin or pond upstream of the weir. Such a basin reduces the fluid velocity and provides a place for debris to settle out. Accumulation of foreign material and debris adjacent to the flowmeter will affect the operation of the flowmeter. Self-cleaning bar screens well upstream of the flowmeter may be considered if debris is a continual problem.

Performance Weirs can achieve accuracies of 2 to 5 percent of rate and turndowns of as high as 25:1. However, the reduced accuracy of the level transmitter may become significant in the lower portion of the flow range. The V-notch weir has a very good turndown and its coefficient does not vary excessively over a wide range of flow.

Applications Weirs are typically applied to liquid flow measurements in which relatively large head is available to establish the free-flow conditions over the weir. V-notch weirs may be applied in low flow situations or in applications that require large turndown.

Sizing Weir size may be estimated by using the graphs of the relationship between flow and the liquid head upstream of the flowmeter as shown in Figure 13-2.

EXAMPLE 13-1

Problem: Size a weir for the measurement of 0 to 5000 gallons per minute of water that is flowing in an open channel, assuming that the difference in upstream and downstream elevations is adequate.

Solution: Reading directly from Figure 13-2, a 90° V-notch weir could be used; however, this would entail a liquid head measurement of approximately 22 inches, which would exceed the available head. Rectangular and Cipolletti weirs in sizes ranging from 4 to 10 feet develop heads of approximately 6 to 13 inches, depending on size.

Installation Installation of the weir and the level transmitter should be in accordance with manufacturer recommendations. A stilling pond or basin may be required to reduce velocity and the effects of flow turbulence of the liquid upstream of the flowmeter. Careful leveling of the crest is also required.

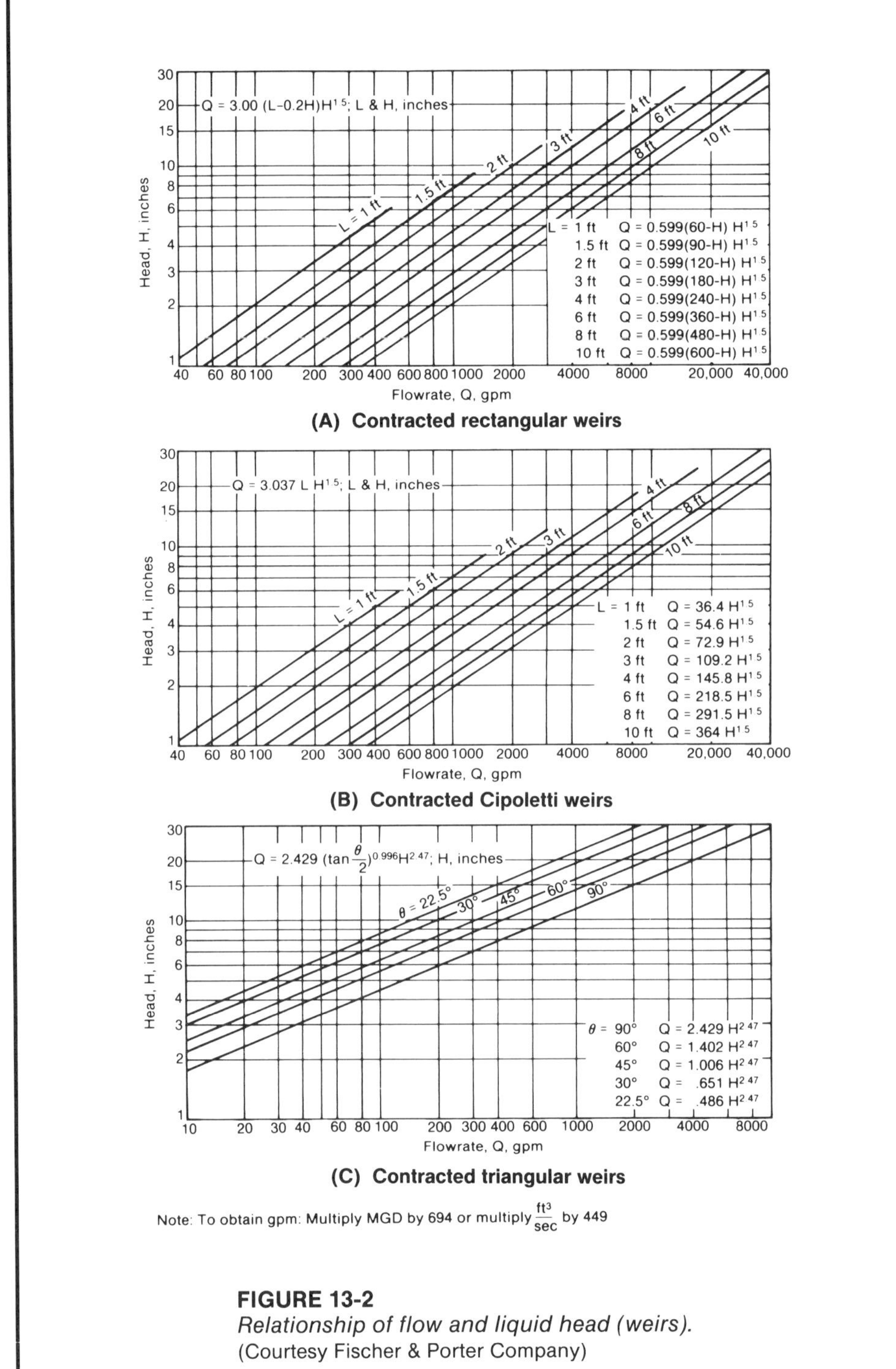

FIGURE 13-2
Relationship of flow and liquid head (weirs).
(Courtesy Fischer & Porter Company)

Level transmitters are typically 2-wire devices, although 3-wire and 4-wire devices are also available. Level transmitters must be calibrated to compensate for any hydrostatic heads resulting from the elevation of the level transmitter relative to the crest or the bottom of the weir notch. Non-contact measurements must also be made to account for the elevation of the transducer above the crest level.

Maintenance Weir flowmeters require routine maintenance in addition to periodic calibration checks of the level transmitter.

Accumulation of Debris

The operation of the weir is sensitive to any foreign material or debris that may be present upstream of the flowmeter. Therefore, any such debris should be removed, which usually necessitates periodic cleaning of the area upstream of the flowmeter to maintain accuracy.

Electronic Failures

Electronic failures are usually handled by board replacement in the level transmitter.

Spare Parts

Spare parts are typically limited to the electronic circuit boards and sensors of the level transmitter.

Calibration

Adjustment of zero and the span of the level transmitter should be performed per the manufacturer's instructions, which will vary significantly with manufacturer, as well as with the technology that is employed to measure level. It should be noted that the level sensed by the level transmitter should be compensated to account for differences in sensor elevation relative to the bottom of the notch or the weir crest.

Parshall Flumes

Principle of Operation Parshall flumes operate on the principle that a converging section of channel restricts the flow from the sides and causes a change in the depth of the liquid as flow varies. The diverging section assures that the downstream level is less than 50 to 80 percent of the level in the converging section, depending on size. The shape of a Parshall flume and a profile of the flow through the flume are shown in Figure 13-3.

Parshall flumes have an approximate 3/2 power relationship between level and flow.

Construction A Parshall flume consists of a fabricated section of open channel that consists of a converging section, a throat, and a diverging section designed to increase velocity at the throat of the flume. This results in a corresponding change in level of the fluid as the flow changes. In typical applications,

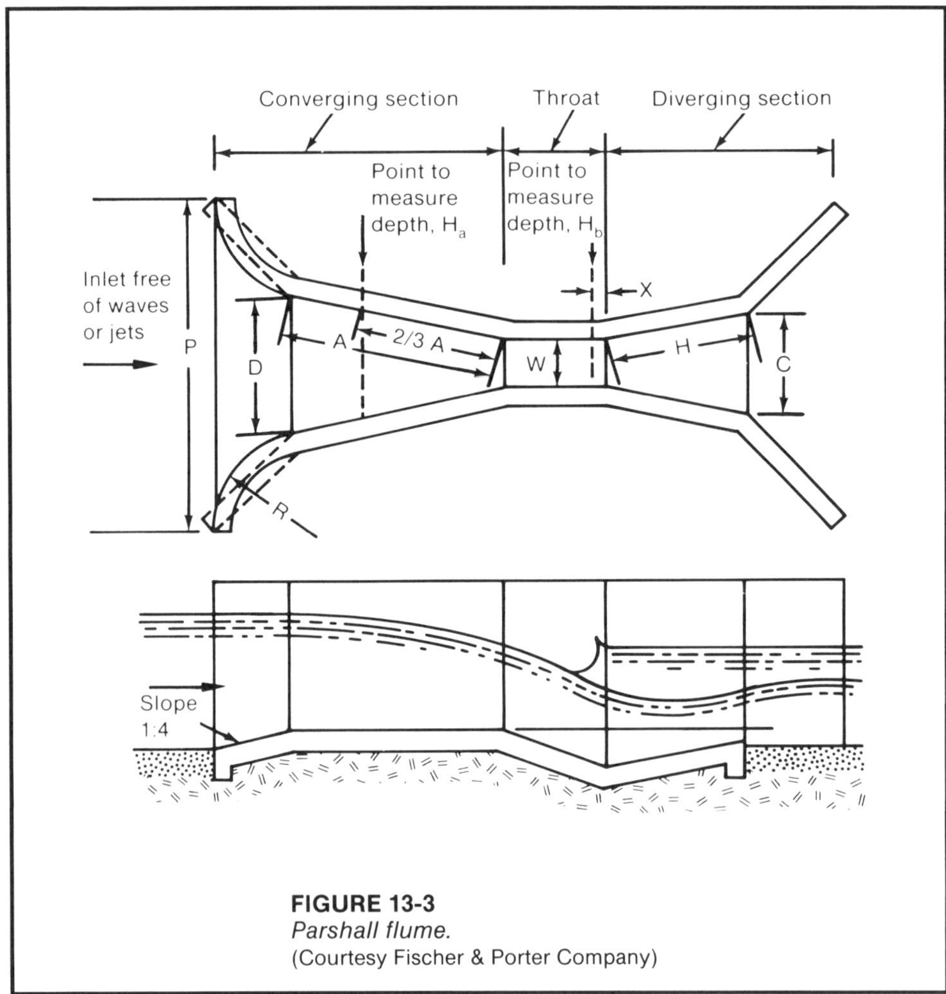

FIGURE 13-3
Parshall flume.
(Courtesy Fischer & Porter Company)

a level measurement device (which comprises the flow transmitter) is required only upstream of the flowmeter. Some applications, however, require level transmitters both upstream and downstream of the flume. Materials of construction must be compatible with the flowing fluid and typically include carbon steel, fiberglass reinforced plastic (FRP), and concrete.

Operating Constraints Flumes, aside from being operated within their flow limits, must also be operated within the level restrictions of the flowmeter. Parshall flumes are usually operated under free-flow conditions in which the discharge liquid level is low enough that it exerts no backpressure on the high velocity jet that is present at the throat of the flume. The Parshall flume will also operate under submerged flow conditions; however, two level measurements (at points H_a and H_b) are required.

Flume size "W"	Limit of free flow
1 in to 3 in.	$H_b/H_a < 0.5$
6 in. to 9 in.	$H_b/H_a < 0.6$
1 ft to 8 ft	$H_b/H_a < 0.7$
10 ft to 50 ft	$H_b/H_a < 0.8$

FIGURE 13-4
Limits of free flow.

The permanent head loss of a Parshall flume (that is, the difference in level between the level before and after the flowmeter) is relatively small; therefore, flumes may be applied where the elevation gradient of the channel is small.

Operation of the weir is sensitive to any waves or jets that may be present upstream of the flowmeter, which can necessitate a section of straight channel upstream of the flume.

Performance Flumes can achieve accuracies of 3 to 10 percent of rate and turndowns of as high as 40:1; however, the reduced accuracy of the level transmitter may become significant in the lower portion of the flow range.

Applications Flumes are applied to open channel liquid flow measurements in which relatively small permanent head loss can be tolerated, such as for irrigation systems. Because of their streamline design, they are also applied to liquids containing debris and foreign matter as these materials tend to be swept through the flowmeter. This minimizes the need for cleaning.

Sizing Flume sizing can be estimated by using the graphs in Figure 13-5, which show the relationship between flow and the liquid head that is generated upstream of the flowmeter for different size flumes.

EXAMPLE 13-2

Problem: Size a flume for the measurement of 0 to 5000 gallons per minute of water that is flowing in an open channel where the generated head is limited to 10 inches.

Solution: Reading directly from Figure 13-5, a 4-, 6-, or 8-foot flume could be used. Barring other constraints, the 4-foot flume should be applied to the flow, as it would be more economical than either the 6- or 8-foot flume. Maximum flow through the 4-foot flume would generate a liquid head of approximately 9.5 inches.

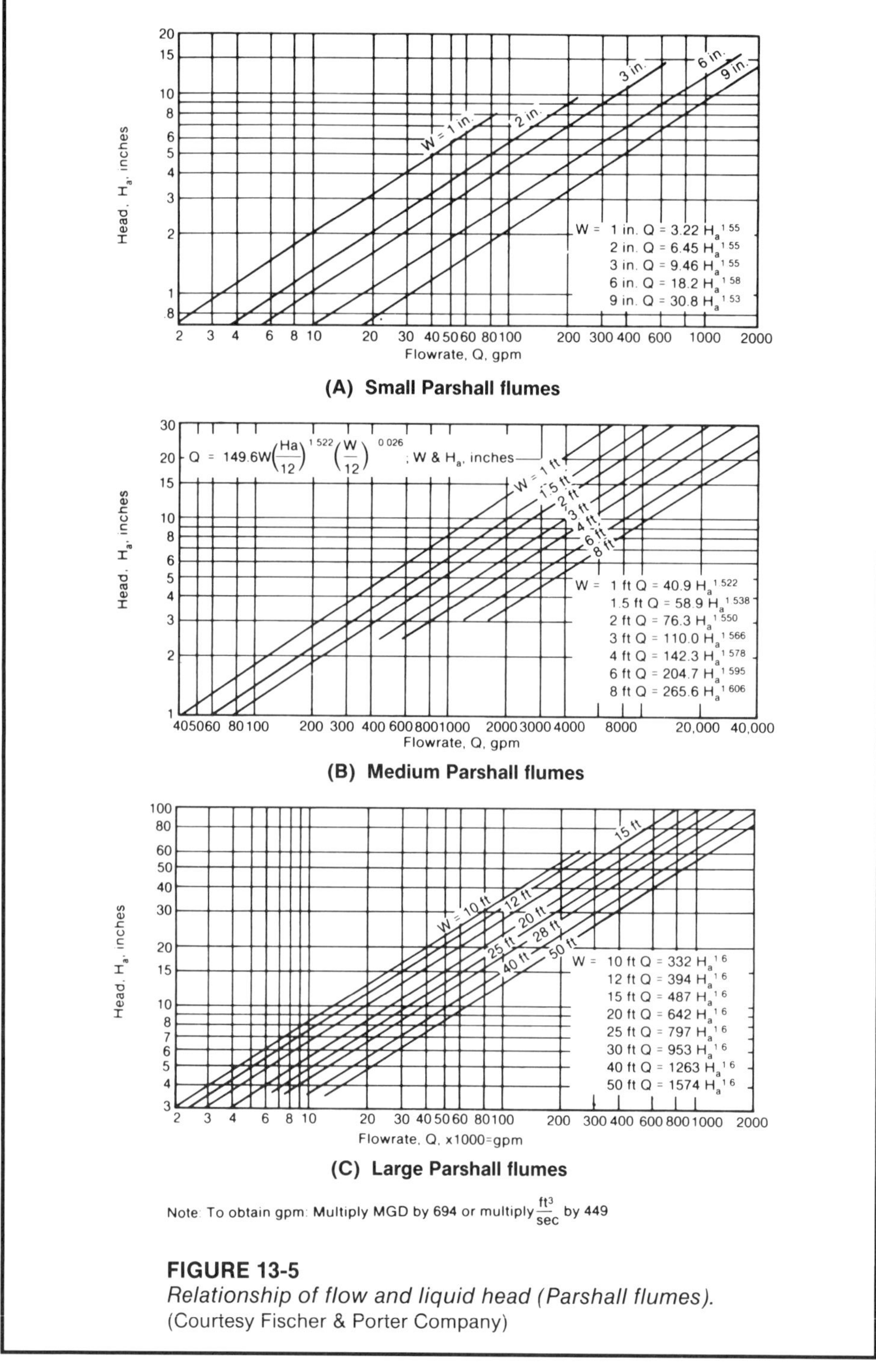

FIGURE 13-5
Relationship of flow and liquid head (Parshall flumes).
(Courtesy Fischer & Porter Company)

Installation Installation of the flume and the level transmitter should be in accordance with the manufacturer's recommendations. A straight section of channel may be required upstream of the flume to reduce any waves or jetting of the liquid upstream of the flowmeter.

Level transmitters are typically 2-wire devices, but 3-wire and 4-wire devices are also available. Level transmitters must be calibrated to compensate for the elevation of the level transmitter relative to the zero flow or reference elevation.

Maintenance Flumes typically require no particular maintenance other than periodic calibration checks of the level transmitter.

Accumulation of Debris

Due to the acceleration of the liquid and the nature of the operation of the flume, virtually any foreign material or debris that may be present upstream of the flowmeter is washed downstream. This is advantageous if the stream has high sand or silt content.

Electronic Failures

Spare parts are typically limited to the electronic circuit boards and sensors of the level transmitter.

Spare Parts

Spare parts are typically limited to the electronic circuit boards and sensors of the level transmitter.

Calibration

Adjustment of zero and span of the level transmitter should be performed per manufacturer's instructions, which will vary significantly with manufacturer as well as with the technology that is employed to measure level. It should be noted that the level sensed by the level transmitter should be zeroed relative to the zero flow condition.

EXERCISES

13.1 Size a weir for the measurement of 0 to 500 gallons per minute of water that is flowing in an open channel, assuming that the difference in upstream and downstream elevations is sufficient to operate the flowmeter.

13.2 Size a weir for the measurement of 0 to 500 gallons per minute of water that is flowing in an open channel where the generated head is limited to 10 inches.

14 Oscillatory Flowmeters

Introduction Oscillatory flowmeters employ physical phenomena that inherently cause discrete changes in some parameter that is a function of the flow through the flowmeter. Some of these flowmeters are applied independent of the fluid state (liquid or vapor).

The lower installed cost and better performance of some oscillatory flowmeters, as compared to more traditional technologies, has precipitated a shift towards oscillatory flowmeters in many applications.

Fluidic Flowmeters

Fluidic flowmeter technology represents a method of measuring low viscosity liquids with a large turndown and reasonable accuracy.

Principle of Operation Fluidic flowmeters are based on the Coanda Effect, which causes a liquid to attach itself to a surface, and fluidics, which is typified by feedback action of the liquid on itself.

A portion of the liquid flows through the bottom feedback passage as shown in Figure 14-1. The physical construction of the fluidic flowmeter is such that the liquid attaches itself to one side of the flowmeter by means of the Coanda Effect. A small portion of the main flow is diverted back through a control port. The feedback flow acts on the main flow so as to divert the main flow to the opposite wall. The feedback action is repeated on the opposite wall, resulting in a continuous self-induced oscillation. The frequency of oscillation is directly proportional to the velocity of the liquid and hence the volumetric flow.

Construction Basic construction of the fluidic flowmeter is shown in Figure 14-2. Oscillations created by the geometry of the flowmeter are detected by changes in the effective cooling of a heated thermistor or by the motion of a deflection sensor installed in one of the feedback passages. Flow through the feedback passage causes a greater cooling effect on the thermistor than does the

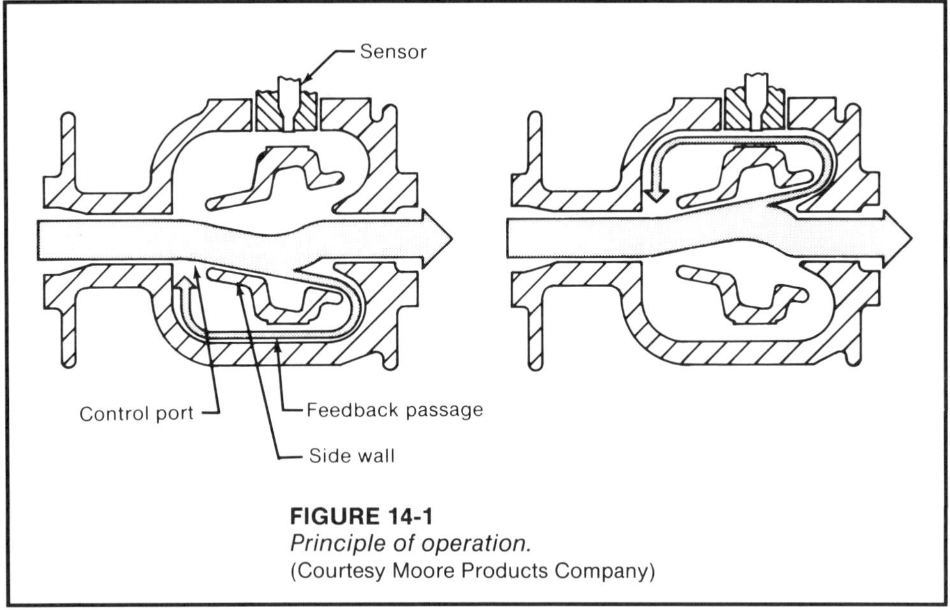

FIGURE 14-1
Principle of operation.
(Courtesy Moore Products Company)

feedback passage under no flow conditions. This temperature difference is detected and conditioned by the transmitter. In the deflection sensor design, flow through the feedback passage causes sensor motion with respect to the no flow sensor position. The sensor signal is conditioned by the transmitter.

Wetted parts of a fluidic flowmeter include the body, which is constructed of stainless steel, and the sensor assembly, which is constructed of stainless steel and Teflon®.

Operating Constraints Operation of the fluidic flowmeter is limited to liquids with less than 2 percent solids that are flowing through the flowmeter with a pipe Reynolds number of greater than 500 to 3000, depending on the design. Accuracy is degraded at lower Reynolds numbers before the flowmeter reaches its minimum operating Reynolds number, typically between 200 and 3000, depending on the design, when the oscillations cease and the flowmeter turns off.

Pressure and temperature are limited by the flange rating of the body and approximately -40 to 175°C, respectively. Cavitation within the flowmeter can usually be avoided by maintaining sufficient inlet pressure.

EXAMPLE 14-1

Problem: Determine the turndown that can be expected when measuring a liquid with a specific gravity of 1.19 and a viscosity of 3 cP with a 2-inch fluidic flowmeter that has a full scale flow of 100 gpm.

Solution: Reynolds number at full scale flow is calculated to be

$$R_D = (3160 \times 100 \times 1.19)/(3 \times 2.067) = 60642$$

Assuming that the fluidic flowmeter is linear at Reynolds numbers as low as 3000, the expected turndown is 60642/3000, approximately 20:1, barring any other constraints.

Performance Fluidic flowmeters have accuracy statements that range from ±1.25 to 2.00% of rate plus ±0.1% of full scale.

The accurate measurement range of fluidic flowmeters in low viscosity service is typically 20:1, although it can be as high as 50:1 in some applications. This is due to the high velocities at which the flowmeter can be operated, coupled with a relatively low Reynolds number constraint.

Applications Fluidic flowmeters can be applied to liquids such as acids, bases, water, fuel oils, chemicals, and the like, provided that the pipe Reynolds number is greater than the minimum for flowmeter operation and that the materials of construction are compatible.

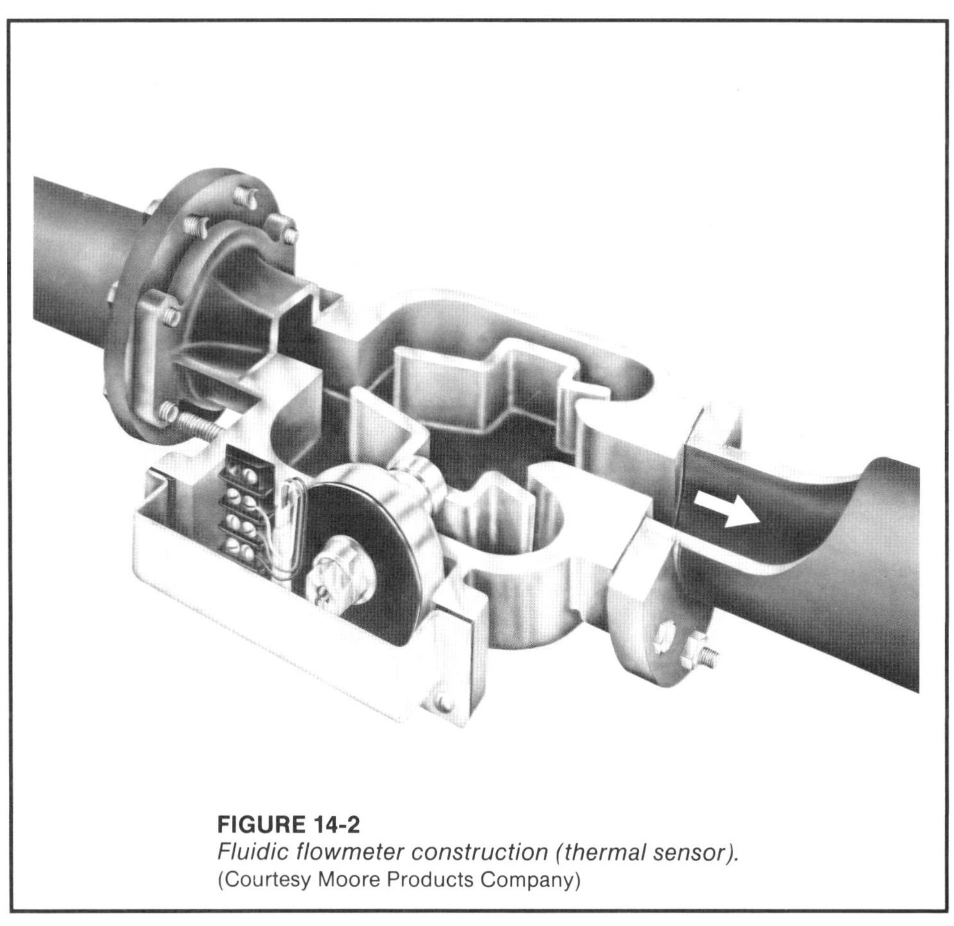

FIGURE 14-2
Fluidic flowmeter construction (thermal sensor).
(Courtesy Moore Products Company)

Sizing Flowmeter sizing for 1 to 4-inch fluidic flowmeters is accomplished by determining whether a given size flowmeter will perform accurately in the desired flow measurement range. The minimum measurable flow can be calculated using Reynolds number constraints, while the maximum velocity constraints of the flowmeter, typically 15 to 25 feet per second, can be used to determine the maximum flow. Flow velocities above 10 to 15 fps are not recommended due to accelerated pipe erosion, excessive pressure drop, and the possibility of cavitation. Minimum and maximum allowable flows are usually tabulated by the manufacturer. However, caution must be exercised when determining the minimum measurable flow, as the data is typical for a liquid with a viscosity of 1 cP and a specific gravity of 1.00.

Installation Fluidic flowmeters are of a wafer design in which the flowmeter is held between two flanges by special length bolts, as shown in Figure 14-3. Some models are available with flanged designs.

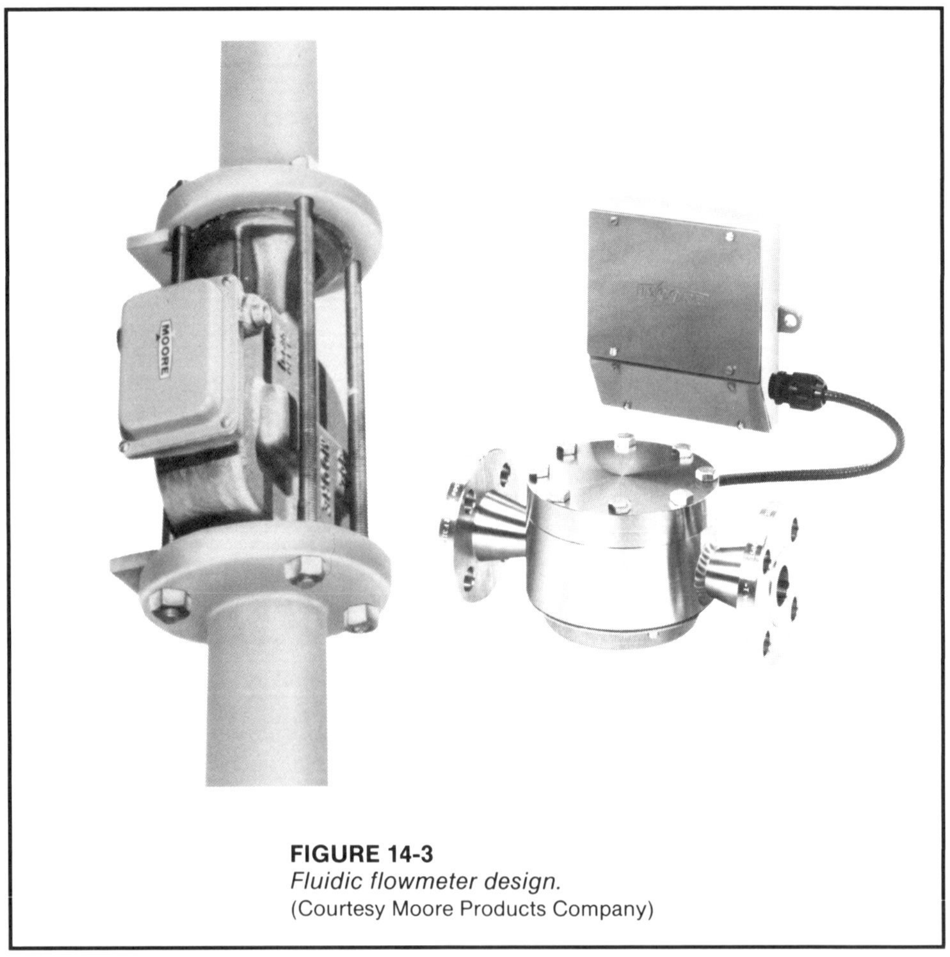

FIGURE 14-3
Fluidic flowmeter design.
(Courtesy Moore Products Company)

Hydraulic Requirements

Fluidic flowmeters, which are sensitive to distortion in the liquid flow profile entering the flowmeter, have upstream and downstream straight run requirements of 9 to $50D/4$ to $5D$, depending on the piping configuration.

Cabling

Fluidic flowmeters are available as 2-wire or 4-wire devices. Subject to some distance constraints, a 4-wire transmitter can be remote-mounted in an area where an external source of power is readily available. In this configuration, only one conduit is required between the sensor in the flowmeter and the transmitter.

Maintenance

Sensor Coating

Routine maintenance may be required if the process liquid has a tendency to coat the thermal sensor. In such a case, a sufficient change in its heat transfer characteristic will cause the sensor to be unable to detect the difference between the effective cooling effects of the liquid flowing in the feedback passage. When this condition exists, the flowmeter can operate sporadically or fail to operate. The deflection sensor design is less susceptible to coating problems.

Electronic Failure

Electronic failures are usually handled by board or sensor replacement.

Spare Parts

Electronic circuit boards and replacement sensors can be stocked for use in case of failure.

Calibration

Calibration of the electronics is performed by adjusting the zero with no flow through the flowmeter and adjusting the span with a pulse generator inputting a frequency signal that simulates the maximum flow.

Vortex Shedding Flowmeters

Advances in and wider acceptance of vortex shedding flowmeters has led to versatile designs that can be utilized in many applications. While some designs are dedicated either to liquid or to gas service, most recent designs are applicable to both. Compromises made in design so that the same hardware can be used for both liquid and gas service are usually not significant unless the application demands full use of all the attributes of the technology. This makes the vortex shedding flowmeter a versatile piece of equipment that is challenging more traditional technologies.

The trend in vortex shedding flowmeter technology is towards flowmeters that can handle liquid, gas, and vapor applications in an attempt to manufacture and apply standard hardware. Vortex shedding flowmeters are applicable to many applications. However, the Reynolds number, velocity constraints, materials of construction, and other limitations of the technology must be thoroughly considered.

Principle of Operation The phenomenon of vortex shedding is described mathematically by the von Karman Effect and is illustrated in Figure 14-4.

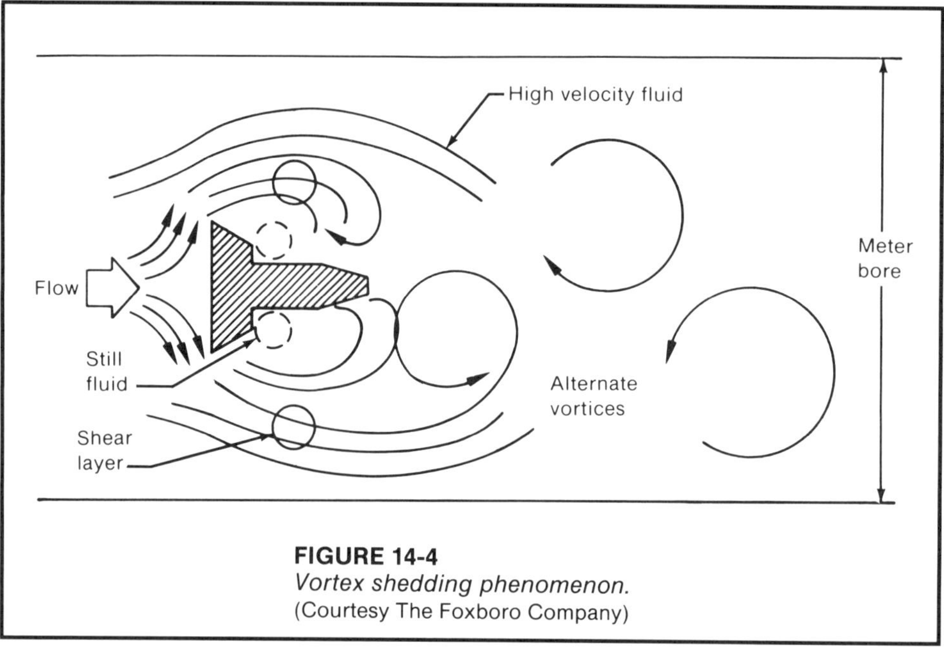

FIGURE 14-4
Vortex shedding phenomenon.
(Courtesy The Foxboro Company)

As the fluid passes a bluff object at low velocity, the flow pattern remains streamlined. As velocity increases, the fluid separates alternately from each side of the bluff body and swirls to form vortices downstream of it. A vortex is an area of swirling motion with high local velocity and hence lower pressure than the surrounding fluid. The frequency of vortex generation is directly proportional to the velocity of the fluid. Examples of von Karman vortex formation are the whistling sound of tree branches or electric cables in the wind, and the waving of a flag in the wind.

The principle of vortex shedding is applied to flowmeters by introducing a bluff body into a pipe and sensing the frequency of vortex generation. This frequency is proportional to the fluid velocity; hence, the flow may be expressed as:

$$Q = A \times v$$

The output of a vortex shedding flowmeter is dependent upon the Strouhal number, which is a dimensionless number. This relationship is represented by:

$$f = \mathrm{St} \times \frac{v}{\text{shedder width}}$$

For practical purposes, the Strouhal number is not a constant and can vary with Reynolds number, such that vortex shedding flowmeters are devices whose operating characteristics are Reynolds number dependent as illustrated in Figure 14-5.

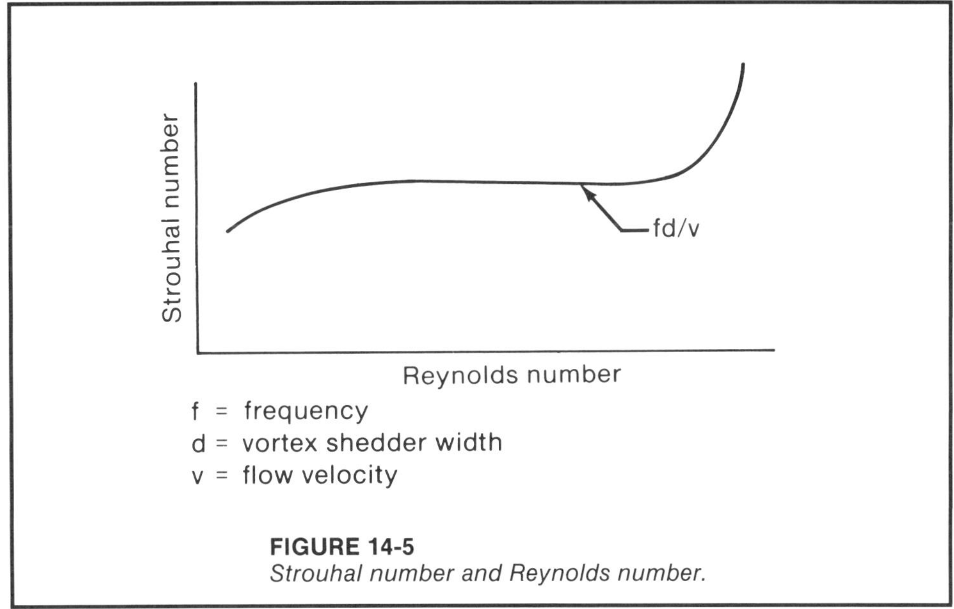

FIGURE 14-5
Strouhal number and Reynolds number.

Linear operation is achieved when the flowmeter is operated in the flat horizontal region of the curve, which virtually dictates the shedder width necessary for linear flowmeter operation and optimal immunity to the effects of shedder wear. As a result, most vortex shedding flowmeters have similar shedder widths. The hydraulic design and performance can and do vary greatly.

While most designs use a single shedder, one design introduces two additional bodies into the flow stream, which are active in vortex formation. These active bodies combine the von Karman vortex shedding phenomenon and the Coanda Effect to form strong and stable vortices by alternately developing a stagnant zone between the shedder and each of the bodies, as shown in Figure 14-6.

Another vortex shedder design exclusively for gas service utilizes a relatively thin wire to generate vortices within the pipe.

Construction The shedder and sensing system are mounted on the vortex shedding flowmeter body.

Shedder

The shedder is the bluff body that is introduced into the fluid flow stream. Shedder design, while typically of similar width independent of manufacture, varies significantly from manufacturer to manufacturer. Manufacturers have optimized trapezoidal, rectangular, triangular, T-shapes, and the like, in an effort to develop a shedder that optimizes that particular manufacturer's design criteria, which include:

- Immunity to upstream pipe hydraulics
- Immunity to pipe vibration

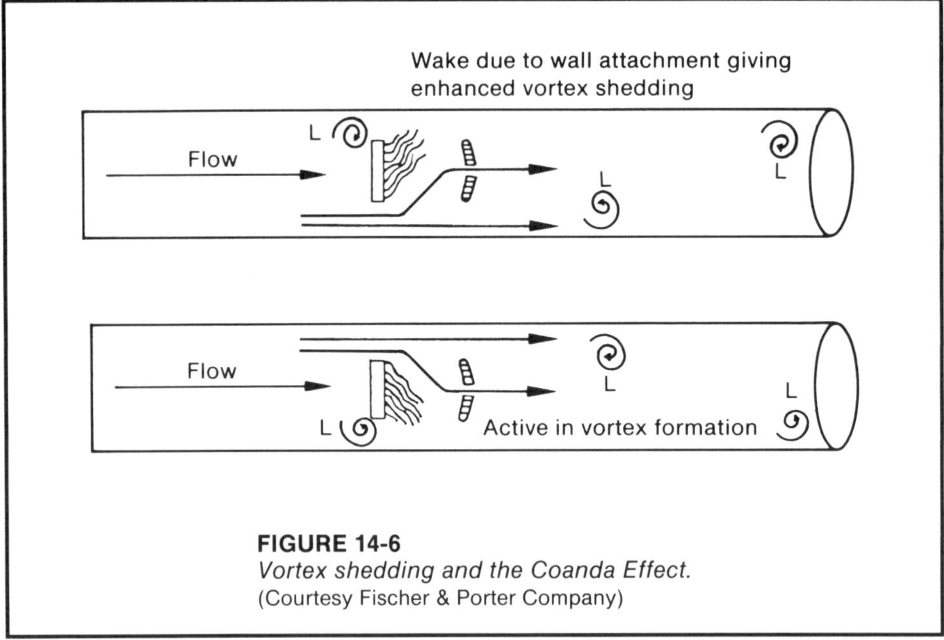

FIGURE 14-6
Vortex shedding and the Coanda Effect.
(Courtesy Fischer & Porter Company)

- Sensitivity of sensing system
- Insensitivity to improper alignment
- Accuracy
- Repeatability
- Linearity

Design tradeoffs are made to achieve the features desired by the manufacturer. Therefore, each design should be examined carefully to determine what tradeoffs were made and how they might affect flowmeter operation in a given application.

Sensing Systems

Vortex sensing systems vary significantly from manufacturer to manufacturer and are usually patented. As the vortices that are formed within the flowmeter body are localized areas of high velocity, they are also localized areas of low pressure and low density relative to the surrounding fluid. The fluid also has momentum through the pipe, some of which can be used to drive the sensing system. These observations are exploited to measure vortex generation, some implementations of which are discussed below.

Dual Body

Vortices are formed by the shedder and are measured downstream by a second body, which contains the sensor. Vortex development between the bodies results in stronger vortices that are easier to measure (see Figure 14-13).

Oscillating Disc

The formation of vortices at the shedder causes alternating pressure of the same frequency to be exerted on the shedder. The pressure is transmitted via passages on each side of the shedder to opposite sides of a disc or diaphragm. As one side of the disc experiences a high pressure while the other side conversely experiences a low pressure, the forces act in the same direction, and the disc will oscillate at the same frequency as the local pressure at the shedder (and hence the same frequency as vortex formation).

Oscillation of the disc is measured by a magnetic sensor located in an assembly near the disc. Recognizing that the passages can plug, that the magnetic sensor can fail, and that the disc is not an integral part of the assembly, the magnetic sensor and disc are removable (see Figure 14-7).

Pivot of Shedder

Formation of vortices at the shedder causes alternating pressure to be exerted on the shedder, resulting in an alternating force that causes a minute twisting of the shedder at the same frequency as the formation of vortices. These twisting motions are measured by a piezoelectric element on the shedder. A removable shedder design is employed when the sensor is bonded to the shedder (see Figure 14-8).

Pressure

The alternating pressure exerted on the shedder by the formation of vortices is measured by pressure sensors that are located in the shedder itself (see Figure 14-9).

The sensor is in the flow stream and can fail in which case sensor removal is required. An isolation manifold is available to enable sensor replacement under flowing conditions.

Temperature

The pressures exerted on the shedder are shunted via a passage from a tap on or near one side of the shedder to another tap on or near the other side of the shedder. As a result, fluid flows alternately back and forth in the passage, and a thermal sensor is used to measure the presence of this flow. In another design, thermal sensors are located in the shedder to measure the change in velocity at the shedder due to vortex formation. See Figure 14-10.

Twist of Torque Tube

The presence of vortices at the shedder and an additional active body in the design shown in Figure 14-11 causes alternating pressures to be exerted on the active body, resulting in an alternating force that causes a minute twisting of the active body assembly, or torque tube, at the same frequency as the formation of vortices. These twisting motions are transmitted to a piezoelectric element located external to the flow stream by a link. As the sensor can be replaced external to the flow stream, the flowmeter is welded, and none of the parts in contact with the fluid are removable.

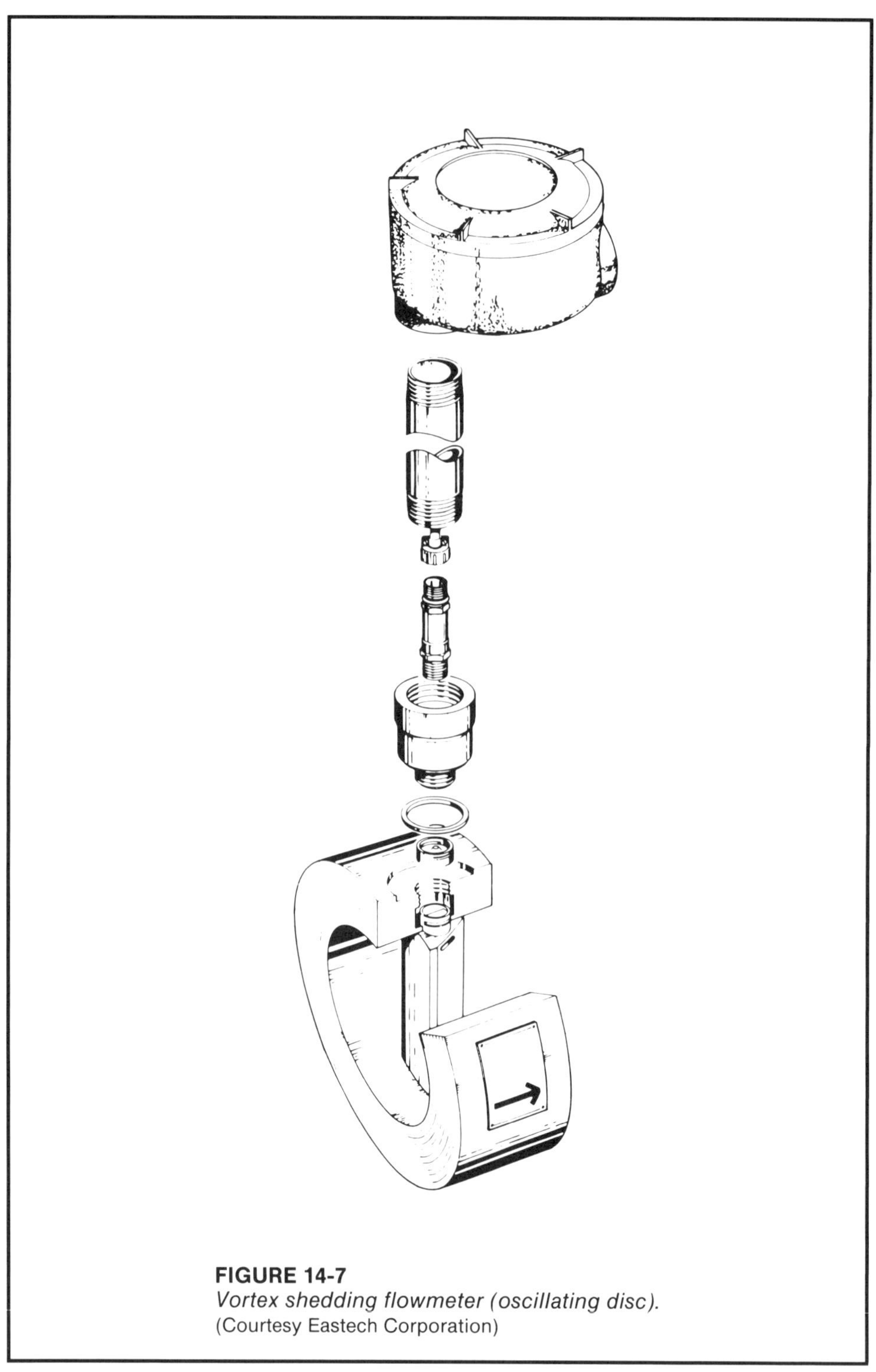

FIGURE 14-7
Vortex shedding flowmeter (oscillating disc).
(Courtesy Eastech Corporation)

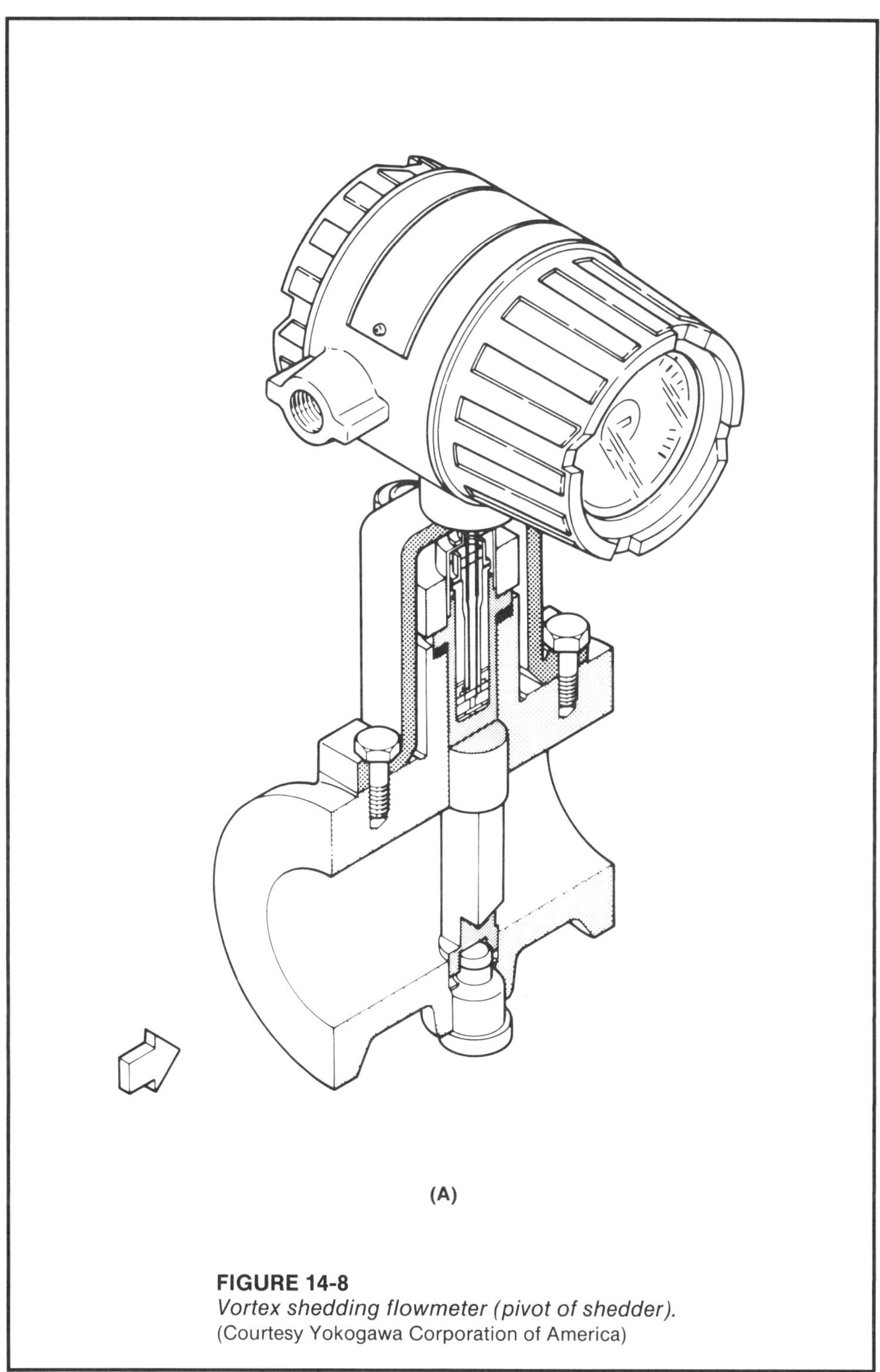

FIGURE 14-8
Vortex shedding flowmeter (pivot of shedder).
(Courtesy Yokogawa Corporation of America)

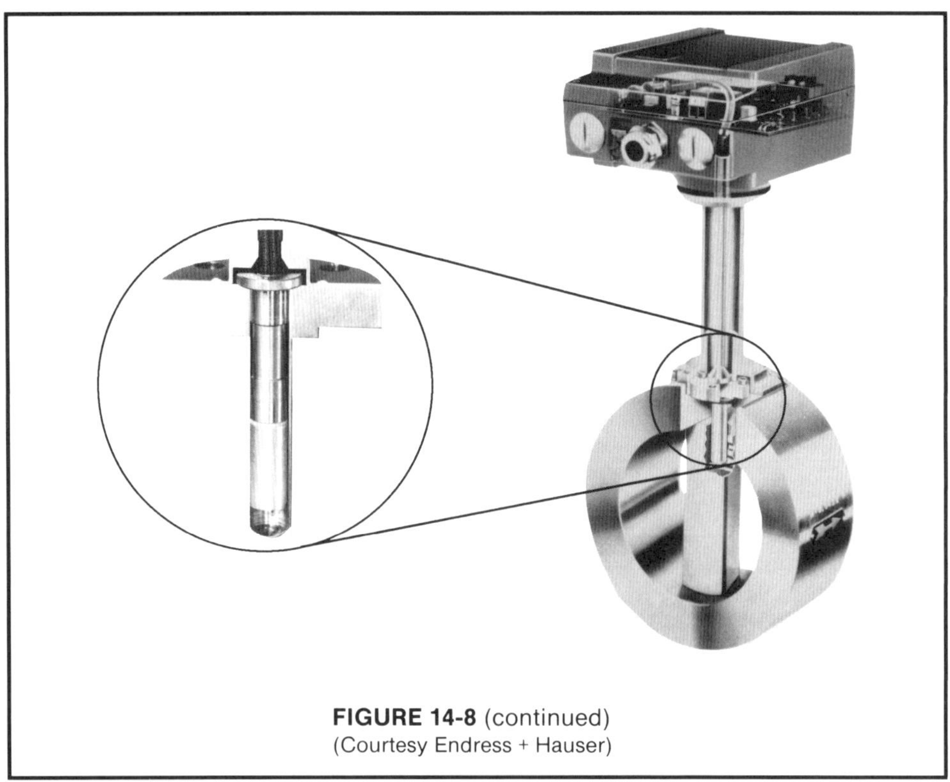

FIGURE 14-8 (continued)
(Courtesy Endress + Hauser)

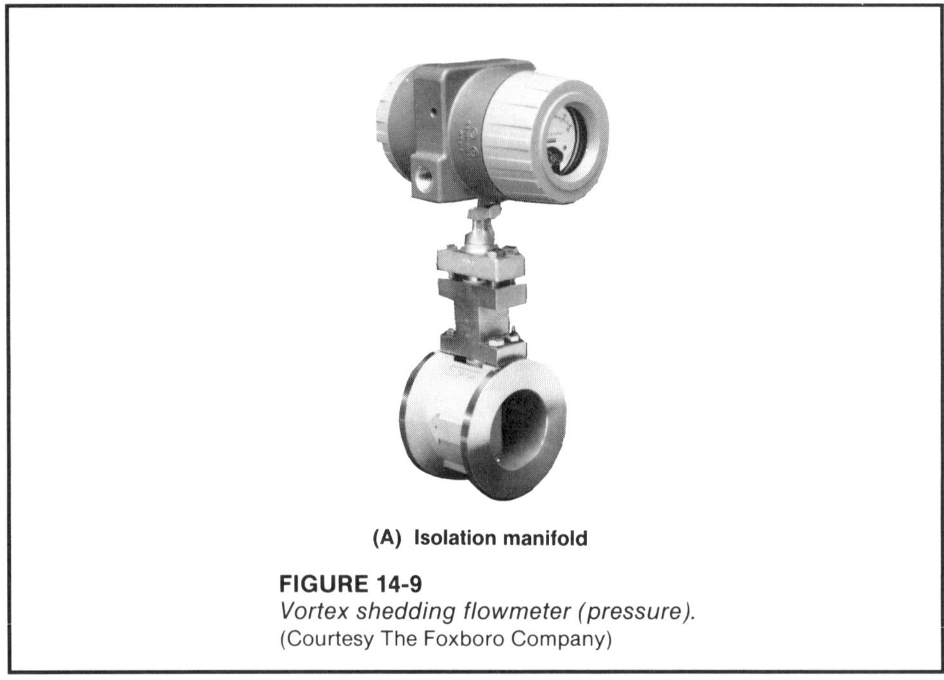

(A) Isolation manifold

FIGURE 14-9
Vortex shedding flowmeter (pressure).
(Courtesy The Foxboro Company)

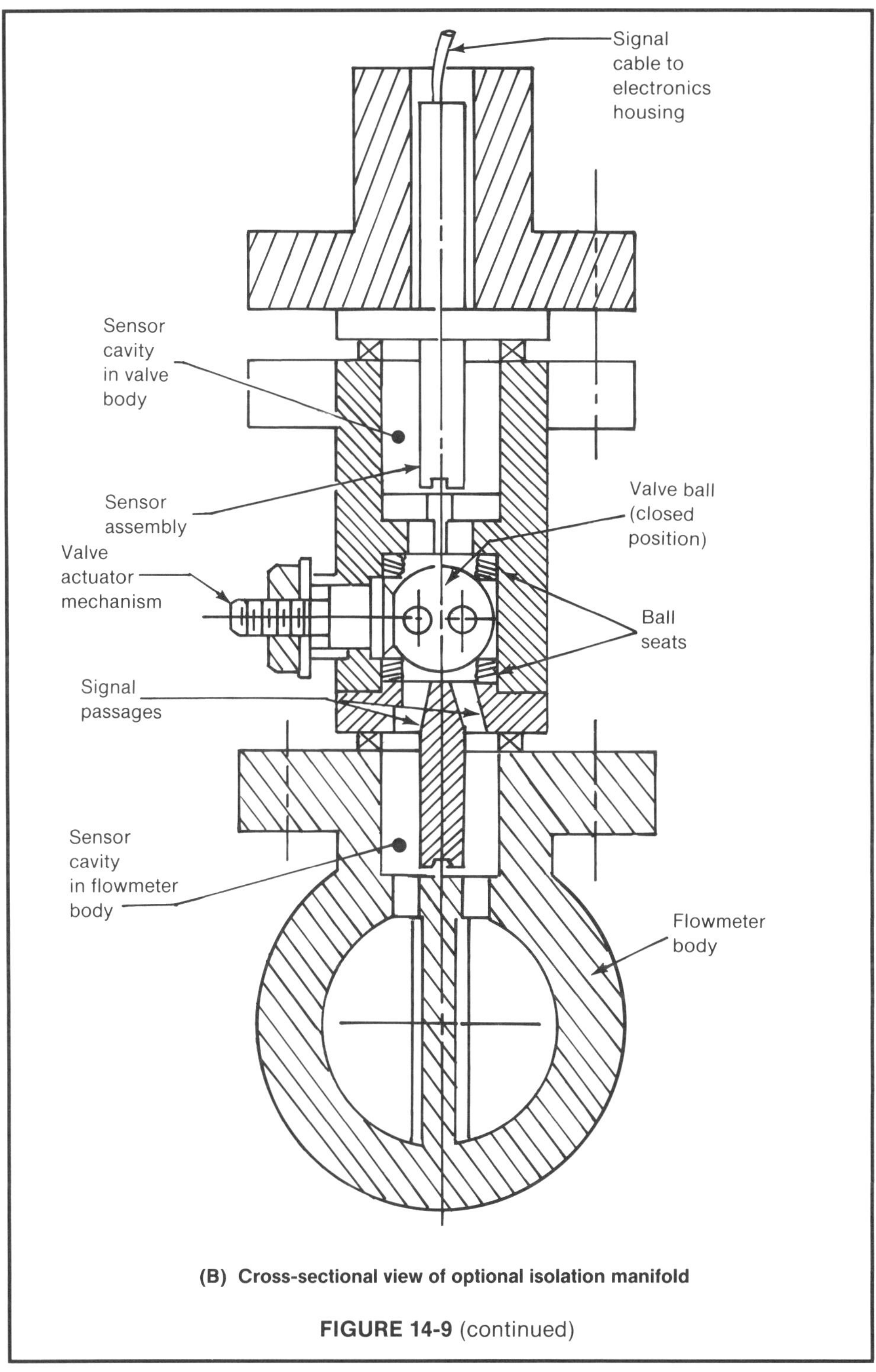

(B) Cross-sectional view of optional isolation manifold

FIGURE 14-9 (continued)

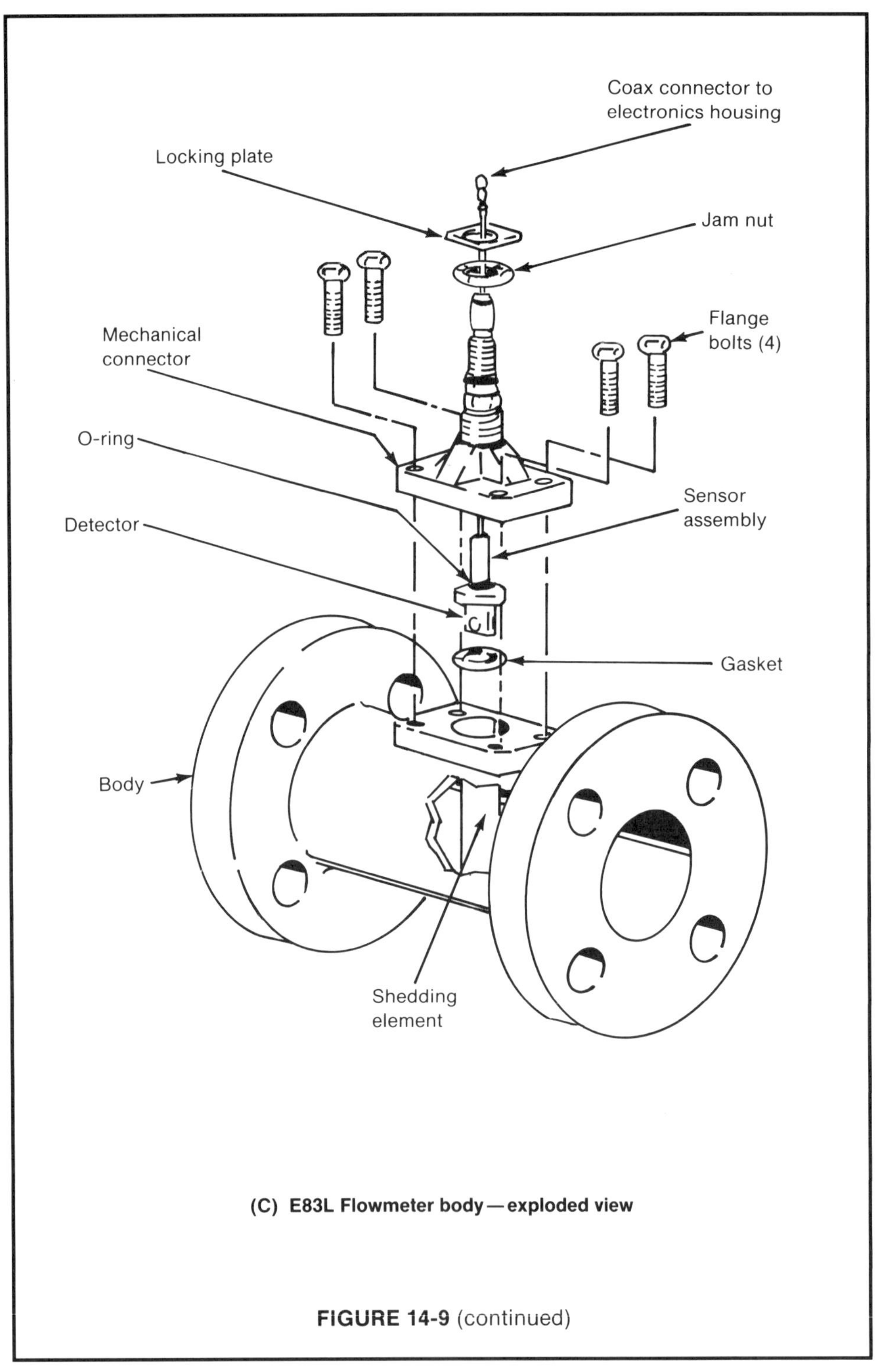

(C) E83L Flowmeter body — exploded view

FIGURE 14-9 (continued)

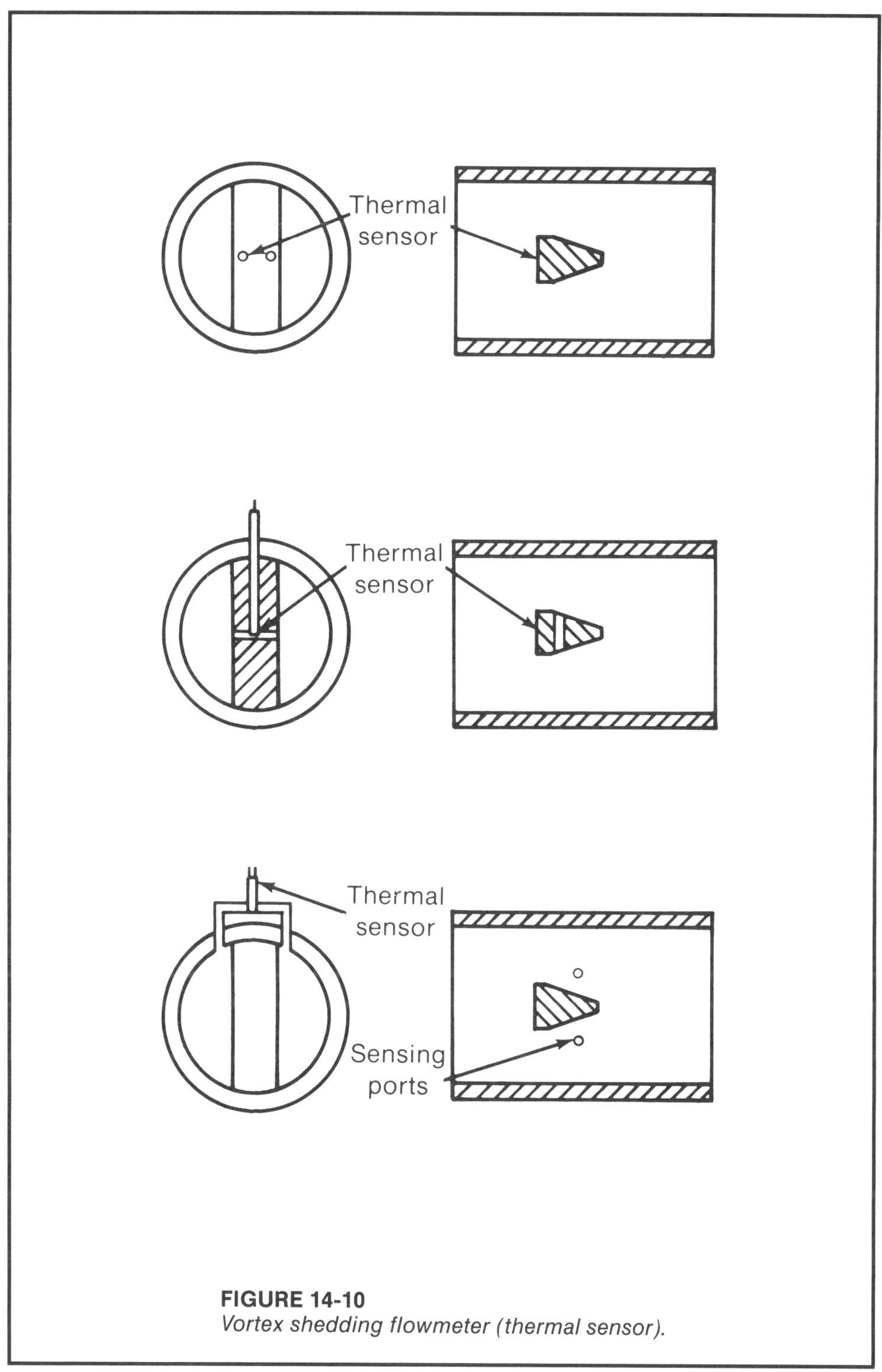

FIGURE 14-10
Vortex shedding flowmeter (thermal sensor).

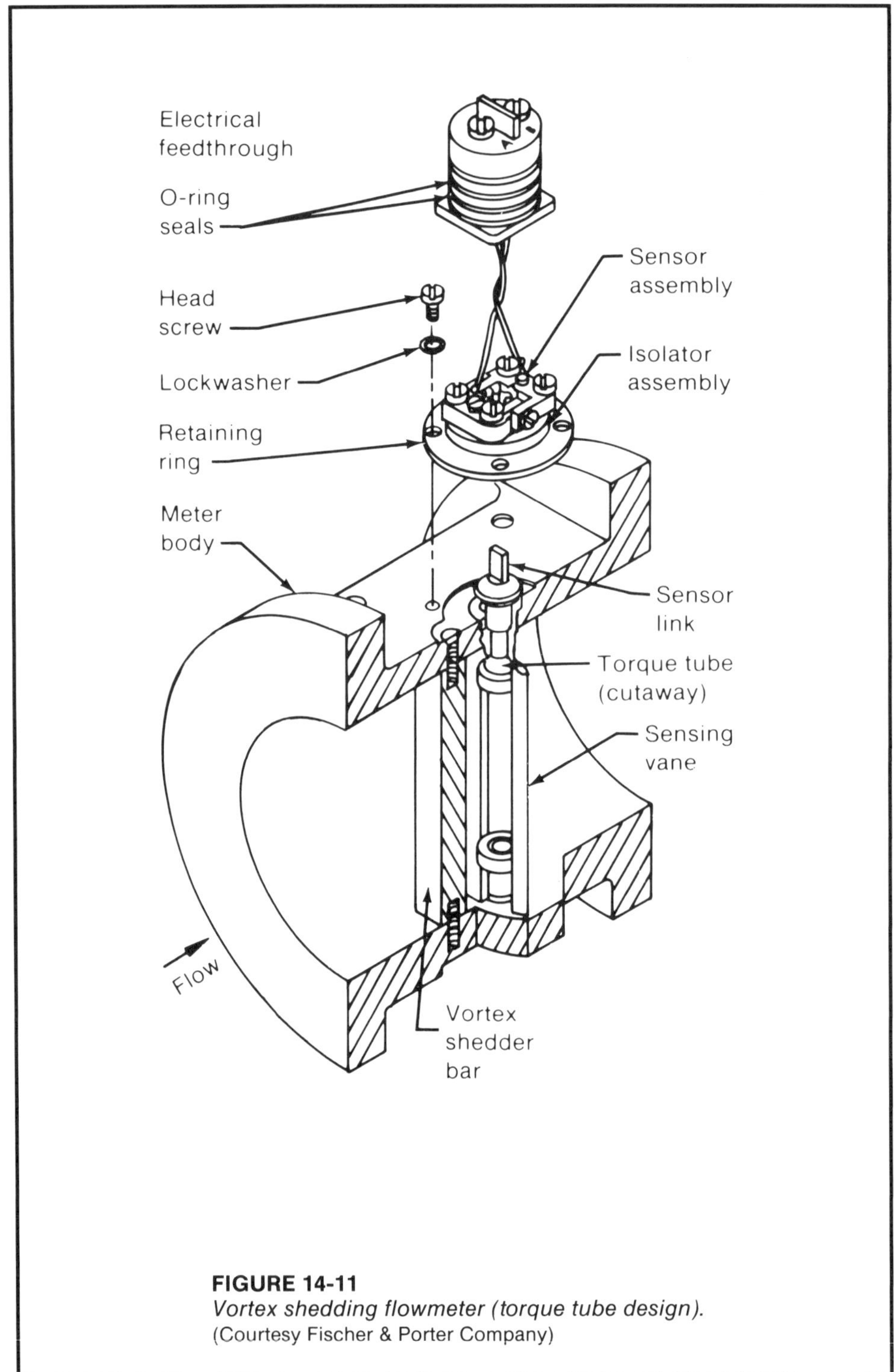

FIGURE 14-11
Vortex shedding flowmeter (torque tube design).
(Courtesy Fischer & Porter Company)

Ultrasonic

Vortices generated by the shedder are allowed to grow and reach maturity downstream of the shedder where the presence of a vortex is sensed using ultrasonic techniques. An ultrasonic beam is transmitted from one side of the shedder and received by a receiver on the other side. The presence of a vortex is sensed by the amplitude modulation of the received signal.

Due to the relative difficulties in coupling ultrasonic energy with gas, the sensors in the gas vortex shedding flowmeter that uses a thin-wire shedder are hermetically sealed assemblies that are mounted to the body of the flowmeter (see Figure 14-12).

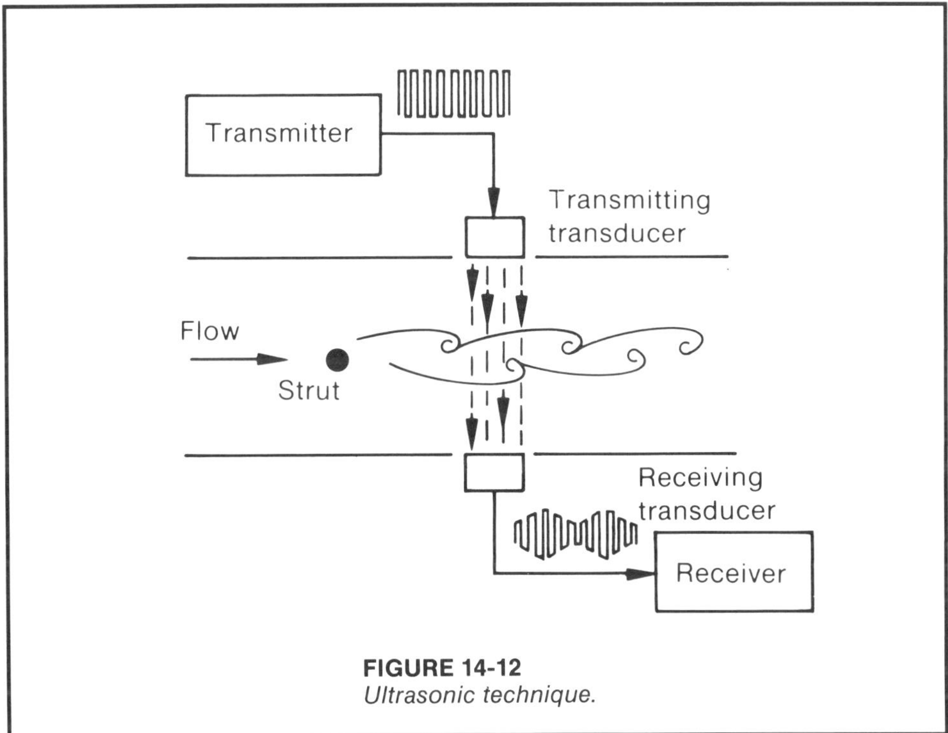

FIGURE 14-12
Ultrasonic technique.

Wetted Parts

Vortex shedding flowmeter wetted parts include the body and the shedder. However, depending on the design of the flowmeter, the sensing assembly and/or sensor may also be wetted. Standard materials of construction of all wetted parts of these flowmeters is stainless steel. One manufacturer does produce a vortex shedder using Hastelloy® construction, which can be very expensive. Another manufacturer offers PVC wetted materials in smaller sizes.

It should be noted that the rate at which the fluid attacks the metal in the flowmeter should be examined closely when considering designs that have relatively thin metal membranes. The flowmeter may be damaged beyond repair if the fluid corrodes the membrane to the point that it leaks.

Operating Constraints Metal vortex shedding flowmeters are available in 1/2 to 12 in. sizes, which effect liquid flow measurements of approximately 3 to 5000 gallons per minute. Gas flows will vary so much with pressure and temperature that a meaningful statement of the measurement range is difficult to express.

Vortex shedding flowmeters are limited by the ratings of the flanges for pressure, while the sensor is usually the factor that limits temperature. Sensor designs will operate in the temperature range of approximately -40 to 400°C, depending on manufacture and design.

Operation of metal vortex shedding flowmeters is dependent upon Reynolds number and the momentum of the fluid in the flowstream. Although Reynolds number constraints vary with size and manufacture, accurate flow measurement is typically achieved at Reynolds numbers greater than 10,000. Most vortex shedders cease to operate below a Reynolds number of approximately 3000 and exhibit nonlinear operation when the flowmeter is operating below the minimum Reynolds number required for accurate measurement.

PVC vortex shedding flowmeters are available in 1/4 to 2 in. sizes and are limited to -25 to 60°C at maximum pressures of 50 to 150 psig, depending on temperature. Accurate measurement can be obtained at Reynolds numbers above approximately 1000 in some sizes. Depending on size, flows of 0.6 to 200 gallons per minute can be measured.

Most sensing systems are passive in nature; that is, the sensing system uses forces or motion generated by the momentum of the fluid through the flowmeter in order to operate the sensing system. As a result, this type of vortex shedding flowmeter has a minimum momentum that the fluid must generate and under which the flowmeter ceases to operate. This is commonly expressed as the minimum velocity constraint of the flowmeter.

For liquid service, the minimum velocity constraint is typically 1 to 2 feet per second and is dependent primarily on the manufacturer and the design. The specific gravity can affect the minimum constraint, but as most liquids have specific gravities between 0.8 to 1.2, the minimum constraint will not vary significantly and is sometimes considered a constant. This is not true in gas service, where the density of different gases can vary over a range of 20:1 or more. Manufacturer's graphs or formulas defining the minimum velocity constraint should be consulted. An example can be seen in Figure 14-14.

Maximum velocity constraints are usually stated to prevent mechanical damage to the flowmeter and to prevent cavitation. Liquid service constraints range from 15 to 20 feet per second, but this is usually degraded when the specific gravity is significantly above 1.0. Maximum velocity constraints are usually not a limiting factor since most liquid applications are designed to operate at considerably lower velocities. In gas service where the density can vary significantly, manufacturer's graphs or formulas defining the maximum velocity constraint should be consulted.

As the operation of vortex shedding flowmeters is dependent upon Reynolds number, the flow measurement (which is volumetric in nature) is not affected by varying specific gravity or viscosity as long as Reynolds number is such that the flowmeter is operating linearly and the minimum velocity constraint is satisfied.

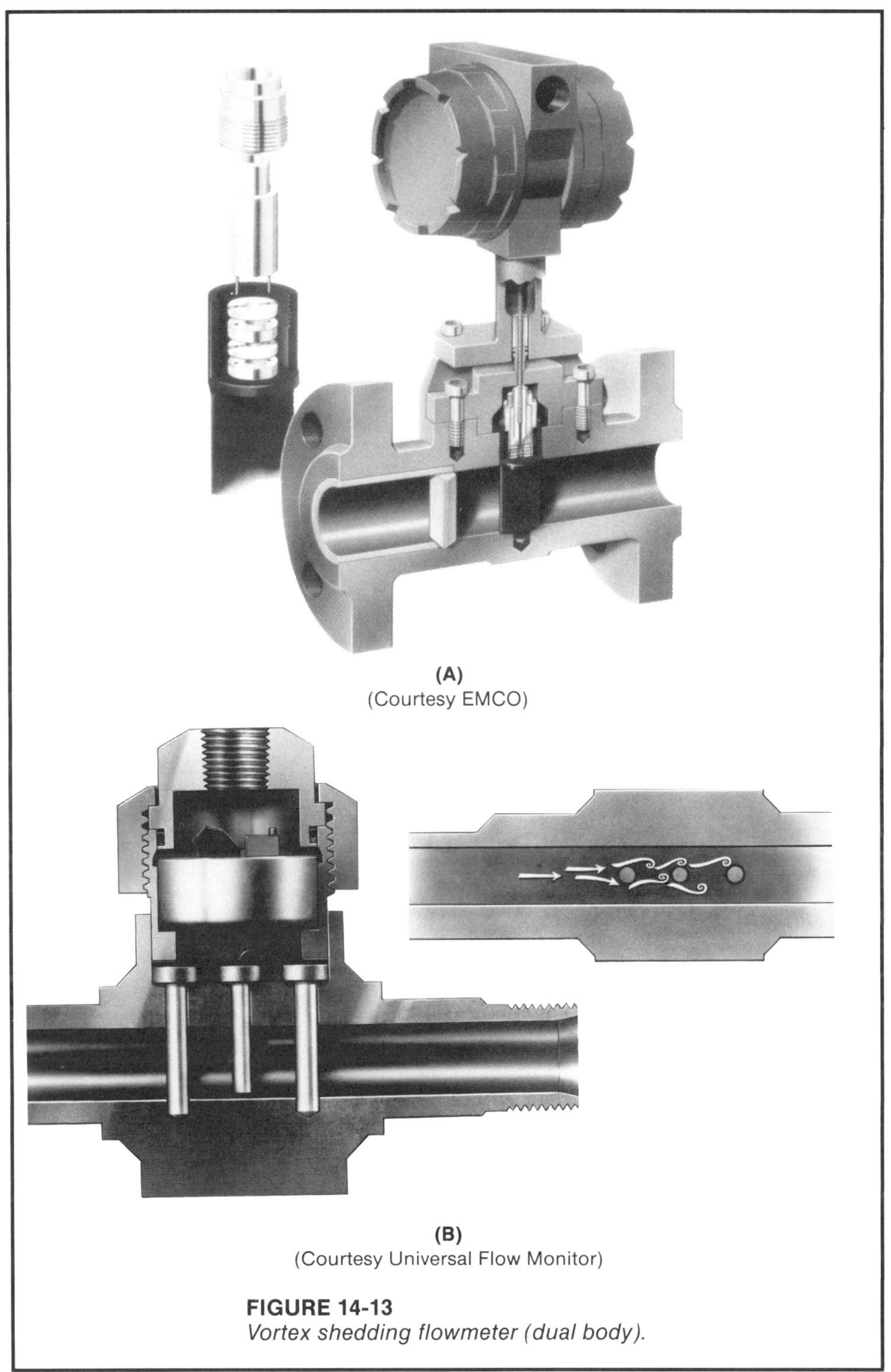

(A)
(Courtesy EMCO)

(B)
(Courtesy Universal Flow Monitor)

FIGURE 14-13
Vortex shedding flowmeter (dual body).

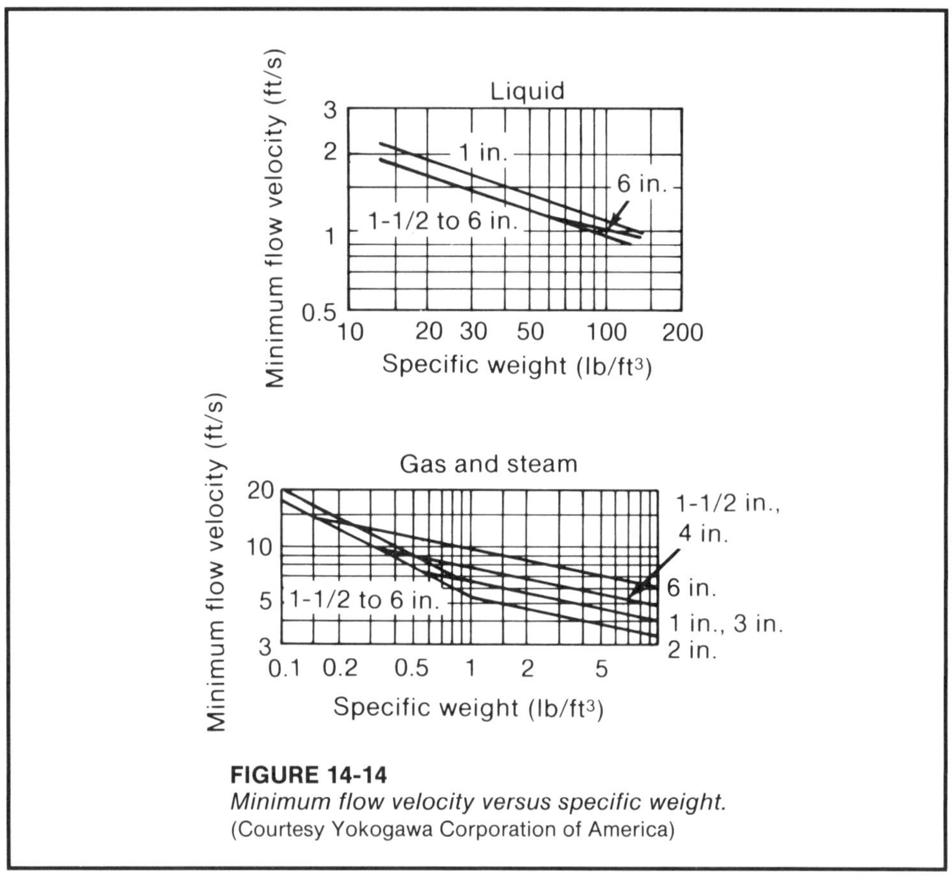

FIGURE 14-14
Minimum flow velocity versus specific weight.
(Courtesy Yokogawa Corporation of America)

This can result in a virtual immunity to specific gravity and viscosity variations in most applications.

Care should be taken in piping design so that the local pressure in the vortices is maintained above the vapor pressure (p_{vapor}) of the liquid to avoid cavitation. One manufacturer suggests that this be done by maintaining the inlet pressure above the minimum inlet pressure, defined by:

$$P_{min}\,(\text{psig}) = 4\,\Delta p + 1.25 \times p_{vapor}\,(\text{psia})$$

With the exception of the design employing a thin wire shedder, the pressure drop across a vortex shedding flowmeter is virtually independent of manufacturer, as the shedder widths and, hence, the obstruction to the flow are virtually the same.

Performance Calibrated vortex shedding flowmeters in liquid service that are operated in the linear region can be accurate to within $\pm 1/2$ to 1 percent rate, while accuracies for gas service are typically between 1.5 to 2 percent rate. PVC vortex shedding flowmeters exhibit accuracies of $\pm 1\%$ of meter capacity. The gas vortex shedding flowmeter that uses a wire shedder is typically accurate to 1 to 2 percent of full scale. Gas flow measurement accuracy is not as good as liquid flow measurement accuracy primarily due to the uncertainties involved with flowmeter operation in the higher Reynolds number ranges (which are difficult or impossible to verify) and the suspected downward sloping characteristic curve at high Reynolds numbers.

The accuracies stated above are based upon water calibrated flowmeters operated in their linear region. Vortex shedding flowmeters can be purchased without calibration with accuracies of approximately 1.5 percent rate for liquids and 2 percent rate for gases, while oil calibrations are available that define the operation of the flowmeter at lower Reynolds numbers in the nonlinear region of operation, allowing the flow signal to be corrected so as to achieve reasonable accuracy.

Vortex shedding flowmeters are affected by the effects of thermal expansion of the flowmeter body, which can affect the K-factor by approximately 0.5 percent rate per 100°C in a stainless steel flowmeter. Vortex shedding flowmeters in services with large temperature excursions can exhibit some loss of accuracy. Modifying the calibration of the transmitter to correct for this effect at the nominal operating temperature will minimize it.

Turndown of vortex shedding flowmeters is usually stated using statements such as "up to 15:1". In practical applications, this turndown is not realized in that

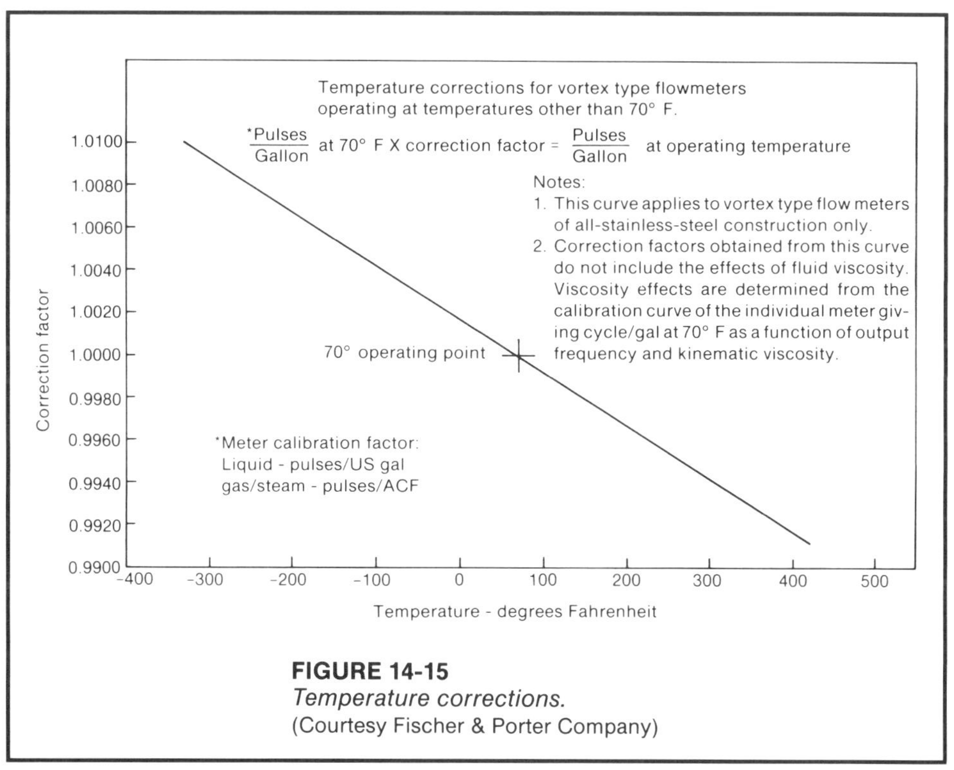

FIGURE 14-15
Temperature corrections.
(Courtesy Fischer & Porter Company)

liquids flowing in pipe are typically designed to operate at velocities of approximately 6 to 8 feet per second. A vortex shedding flowmeter with a minimum velocity constraint of 1.0 foot per second will achieve a turndown of 6 to 8:1, which falls short of the 15:1 that may be implied in manufacturers' literature. Considering that most vortex shedding flowmeters have minimum velocities that are greater than 1.0 foot per second and that Reynolds number constraints have not been considered, the effective turndown can be significantly less than that which the manufacturer implies.

The performance of vortex shedding flowmeters is compromised any time any wetted part that determines the geometry of the flowmeter in the fluid is removed. An example of this is the removal of the shedder bar, which can shift the K-factor of the flowmeter by as much as 1 percent, depending upon the size, due to the inability to relocate the shedder precisely in its original position. The new K-factor of the flowmeter can be determined by recalibrating of the flowmeter in a flow facility.

Applications Vortex shedding flowmeters are applicable to fluids that exceed the minimum velocity and Reynolds number constraints in the desired measurement range. This technology is typically applied to low viscosity liquids that have sufficiently high Reynolds numbers to operate linearly and to pressurized gases that have sufficiently high densities and momentum to operate the flowmeter.

Determination of whether the flow to be measured is constant or variable is important. Often a vortex shedding flowmeter can be applied where a relatively high viscosity but constant flow can be measured with sufficient accuracy, since the Reynolds number constraint is satisfied at that flow and turndown is not necessary.

Sizing Sizing of vortex shedding flowmeters is accomplished by determining that the fluid will supply sufficient momentum for a given size vortex shedder to operate at the minimum desired flow, and that Reynolds number is sufficiently high enough for the flowmeter to operate linearly. Sizing, when performed per literature supplied by the manufacturer, is usually nothing more than velocity and Reynolds number constraints in disguise.

Most vortex shedder manufacturers present sizing data in terms of water, air, or steam flows from which the applicability of a different application is determined. Care should be taken so as to realize that the ranges for liquids above 1 cP and for gases other than air or steam can vary significantly from the ranges of the standard fluids.

The sizing charts in Figure 14-16 are presented to illustrate one approach to sizing vortex shedding flowmeters. Individual manufacturers should be consulted for detailed sizing data and constraints as they may vary significantly from those presented.

For applications other than water, air, and steam, it is usually more convenient to use velocity and Reynolds number constraints for sizing. Constraints should be obtained from the manufacturer. Estimates of flowmeter sizing can be obtained by assuming that the typical minimum velocity constraint is 1.25 feet per second for liquids and that the typical minimum Reynolds number constraint is approximately 10,000, although it may be as high as 40,000. PVC flowmeters can measure flows operating at Reynolds numbers as low as 1000 in some applications. Manufacturer literature should be consulted for minimum velocity constraints for gas, the density of which can vary significantly.

EXAMPLE 14-2

Problem: Size a vortex shedding flowmeter to measure 0 to 50 gpm of a liquid that has a viscosity of 1.0 cP and a specific gravity of 1.02.

Solution: As typical design velocities are 6 to 8 feet per second, a 1-1/2-inch size will be investigated since the liquid velocity is 7.88 feet per second at maximum flow. Assuming that the flowmeter will turn off at 1.25 feet per second, the turndown will be 7.88/1.25, or 6.3:1. Reynolds number at full scale flow is:

$$R_D = \frac{3160 \times 50 \text{ gpm} \times 1.02}{(1.0 \text{ cP} \times 1.610 \text{ in.})} = 100{,}099$$

Assuming that 10,000 is the minimum Reynolds number constraint for linear operation, the turndown based upon Reynolds number is 100,099/10,000, or 10.1:1. Using Reynolds number constraints, the flowmeter would operate over a 100,099/3000 or 33.4:1 range. As the flowmeter must satisfy all constraints simultaneously, a 1-1/2-inch vortex shedding flowmeter will operate with an approximate 6.3:1 turndown, such that the flowmeter will turn off at 100/6.3, or approximately 16 percent of full scale flow due to the minimum velocity constraint.

If the liquid had a viscosity of 3 cP and were measured with a 1-1/2-inch flowmeter, Reynolds number at full scale flow is 100099/3, or 33,366. Turndown due to Reynolds number constraints is 33,366/10,000, or approximately 3.3:1 for accurate measurement, while the flowmeter would operate over a 33,366/3000 or 11.1:1 range if velocity constraints were satisfied. As the performance is determined by both the velocity and Reynolds number constraints, the flowmeter will operate accurately with an approximate turndown of 3.3:1, and it will turn off at approximately 16 percent of full scale flow.

If a larger turndown were required, the next smallest size flowmeter could to be employed provided the maximum flowmeter velocity is not exceeded. This would result in a full scale velocity of 18.6 feet per second and a turndown of 18.6/1.25, or almost 15:1.

EXAMPLE 14-3

Problem: Size a vortex shedding flowmeter to measure 0 to 60 acfm of a gas that has a viscosity of 0.017 cP and a density of 0.5 pound per cubic foot.

Solution: As typical design velocities are 40 to 60 feet per second, a 1-1/2-inch size will be investigated, as the gas velocity is 70.9 feet per second at maximum flow. Reynolds number at full scale flow is:

$$R_D = \frac{379 \times 60 \text{ acfm} \times 0.5 \text{ lb/ft}^3}{(0.017 \text{ cP} \times 1.610 \text{ in.})} = 415{,}418$$

which illustrates that sufficient Reynolds number is usually available in gas service due to low gas viscosity, and that the velocity constraints will be the limiting factor in most gas applications. Using the gas sizing chart in Figure 14-16, the minimum and maximum gas flows for a gas with a density of 0.5 pound per cubic foot is read to be 9 acfm and 200 acfm, respectively, resulting in a turndown of 60/9 or 6.7:1, and the flowmeter will turn off at approximately 15 percent of full scale flow.

If a larger turndown were required, the next smallest size flowmeter could be employed provided the maximum velocity is not exceeded; this would result in a full scale velocity of 167 feet per second in a 1-inch pipe, a minimum flow of 3 acfm, and a turndown of 60/3, or 20:1.

Installation Vortex shedding flowmeters are typically of a wafer design with the wafer sandwiched between two flanges. Flanged designs are also available.

Hydraulic Requirements

Vortex shedding flowmeters, which measure volumetric flow, have varying degrees of sensitivity to upstream and downstream pipe hydraulics depending upon the design of the flowmeter. Upstream and downstream piping requirements range from 10 to 20*D*/5*D* to 25 to 45*D*/8*D*.

Most manufacturers require that piping 4*D* upstream and 2*D* downstream of the flowmeter have a 350 finish free of mill scale, holes, bumps, and the like, while the pipe diameter should not depart from the average by more than 0.33 percent. These requirements exceed ASTM pipe specifications of commonly used pipe, as well as the quality of field welds at the inlet and outlet flanges. The vortex shedder that uses additional active elements does not have these requirements and achieves accuracy claims using commercially available pipe.

Shifts in accuracy can occur when the upstream and downstream piping is not of the same schedule about which a particular size flowmeter was designed, usually schedule 40 or schedule 80. Some manufacturers have *K*-factor correction that can be applied when a different pipe schedule other than that recommended by the manufacturer is used, while some manufacturers have no data on this effect. Again, the vortex shedder that uses additional active elements does not have these requirements and achieves accuracy claims with commercially available schedule 80 or lighter pipe.

Piping

Vortex shedding flowmeters must be properly aligned in the center of the pipe. This is most commonly accomplished by the use of centering spacers that fit over some or all of the bolts that hold the flowmeter in place, thereby centering the

Water flow rates

Nominal size		Minimum and maximum measurable flow rates in U.S. gpm[a]
mm	inch	
25	1	(c) 3.3 (7.3) and 57 (b)
40	1-1/2	6.9 (11.3) and 137
50	2	12 (14.5) and 227
80	3	22 and 440
100	4	39 and 766
150	6	84 and 1670

(a) At standard conditions of 15° C (59° F).
(b) Maximum flow rates are based on 7 m/s (23 ft/s).
(c) The values in parentheses show the minimum linear flow rate (Re = 20,000 or 40,000) when they are higher than the minimum measurable flow rate.

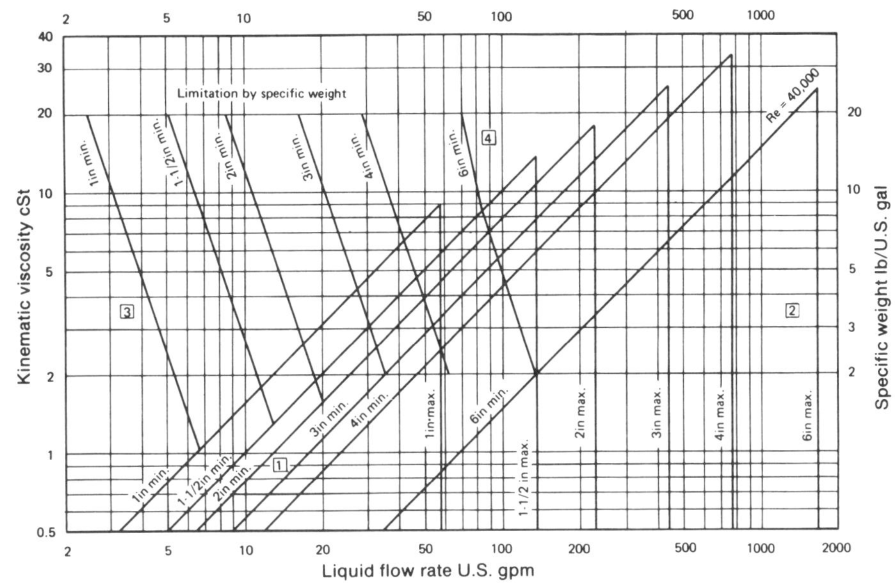

[1] 1 to 4 in: $Q = 6.321 \times D \cdot \nu$ (Re = 20,000)
6 in: $Q = 69.08 \times \nu$ (Re = 40,000)

[2] $Q = 10320 \times S$

[3] 1 in: $Q = 6.668 \times \gamma_f^{-1/3}$
1-1/2 to 6 in: $Q = 5.699 \times D^2 \cdot \gamma_f^{-1/3}$

[4] 6 in: $Q = 128.3 \times \gamma_f^{-1/3}$
(when $\gamma_f > 8.36$ lb/U.S. gal or 62.6 lb/ft)^3

Q: Liquid flow rate (U.S. gpm)
D: Internal diameter (inch)
S: Cross sectional area (ft^2)
γ_f Specific weight of normal operating conditions (lb/U.S. gal)

Note 1. The minimum flow rate is the larger of those defined by curves [1] and [3]

Note 2. μ (cP) = ν (cSt) $\times$ γ (g/cm^3)

(A) Liquid

FIGURE 14-16
Sizing charts.
(Courtesy Yokogawa Corporation of America)

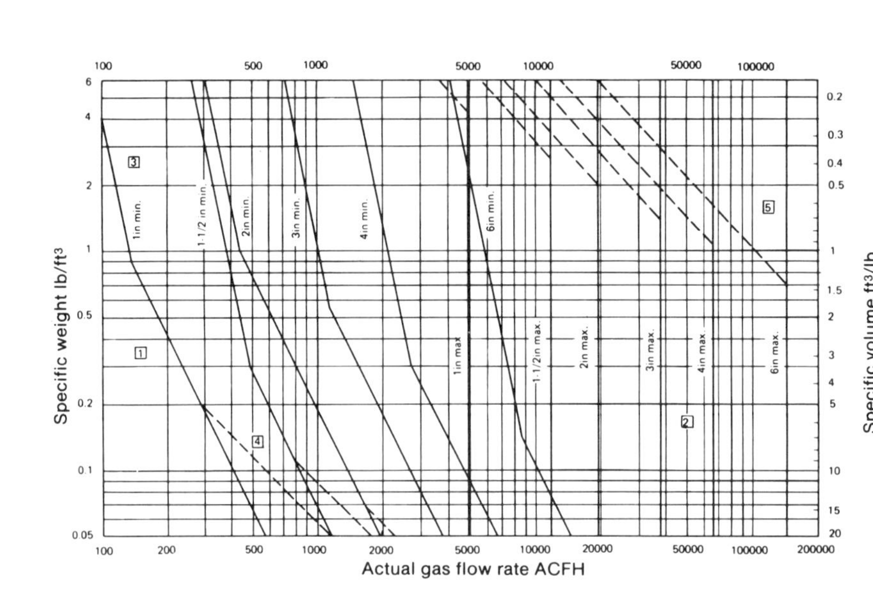

1 — 1 in: $Q_f = \dfrac{128.90}{\sqrt{\gamma_f}}$

1-1/2 to 6 in: $Q_f = \dfrac{107.9 \times D^2}{\sqrt{\gamma_f}}$

2 — $Q_f = 4828 \cdot D^2$

3 —
1 in: $Q_f = 132.7 \cdot \gamma_f^{-1/5}$ (when $\gamma_f > 0.91$ lb/ft³)
1-1/2 in: $Q_f = 376.6 \cdot \gamma_f^{-1/5}$ (when $\gamma_f > 0.30$ lb/ft³)
2 in: $Q_f = 431.7 \cdot \gamma_f^{-1/5}$ (when $\gamma_f > 1.05$ lb/ft³)
3 in: $Q_f = 1013 \cdot \gamma_f^{-1/5}$ (when $\gamma_f > 0.55$ lb/ft³)
4 in: $Q_f = 2102 \cdot \gamma_f^{-1/5}$ (when $\gamma_f > 0.30$ lb/ft³)
6 in: $Q_f = 5828 \cdot \gamma_f^{-1/5}$ (when $\gamma_f > 0.14$ lb/ft³)

4 — 1 to 4 in: $Q_f = \dfrac{3164 \times \mu \cdot D}{\gamma_f}$

6 in: $Q_f = \dfrac{6327 \times \mu \cdot D}{\gamma_f}$

5 — $Q_f = \dfrac{1.107 \times 10^6 \mu \cdot D}{\gamma_f}$

Q_f: Actual flow rate (ACFH)
D: Internal diameter (inch)
γ_f: Specific weight at normal operating conditions (lb/ft³)
μ: Viscosity

Note 1. The dotted lines 4 and 5 are linear flow rate limits (Reynolds Numbers Re = 20,000 and 7,000,000, Re = 40,000 and 7,000,000 for 6-inch meter) for air at 15° C (59° F).

Note 2. γ_f and Q_f corresponding to the scale flow rate are calculated from the following equations.

$$\gamma_f = \gamma_s \times \frac{P + 14.7}{14.7} \times \frac{520}{T + 460}$$

$$Q_f = Q_s \times \frac{14.7}{P + 14.7} \times \frac{T + 460}{520}$$

γ_s: Specific weight at standard conditions (lb/ft³)
P: Normal operating pressure (psig)
T: Normal operating temperature (°F)
Q_s: Flow rate at standard conditions (SCFH)

(B) Gas

FIGURE 14-16 (continued)
Sizing charts.
(Courtesy Yokogawa Corporation of America)

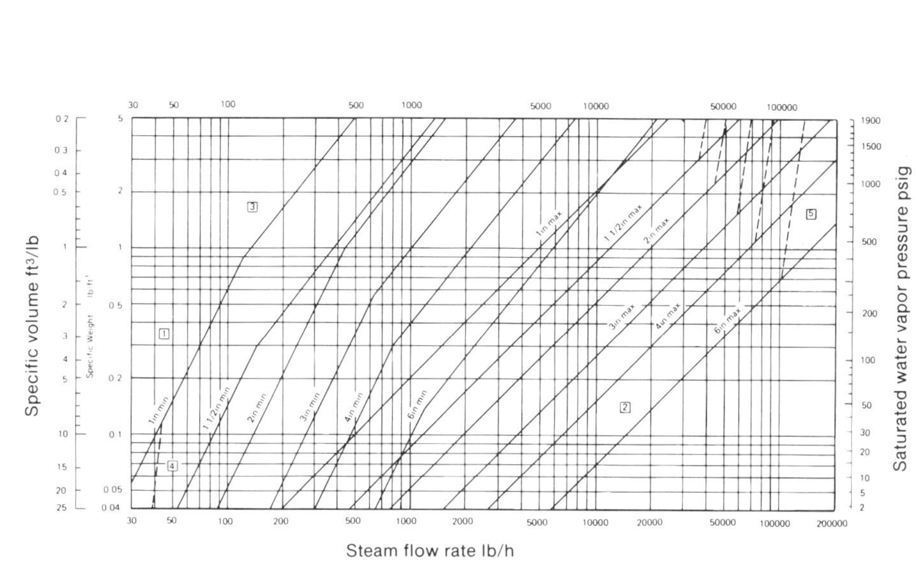

[1] 1 in: $W = 128.9 \sqrt{\gamma_f}$

1-1/2 to 6 in. $W = 107.9 \times D^2 \cdot \sqrt{\gamma_f}$

[2] $W = 4828 \times D^2 \cdot \gamma_f$

[3]

1 in:	$W = 132.7 \times \gamma_f^{4/5}$	(when $\gamma_f > 0.91$ lb/ft³)
1-1/2 in:	$W = 376.6 \times \gamma_f^{4/5}$	(when $\gamma_f > 0.30$ lb/ft³)
2 in:	$W = 431.7 \times \gamma_f^{4/5}$	(when $\gamma_f > 1.05$ lb/ft³)
3 in:	$W = 1013 \times \gamma_f^{4/5}$	(when $\gamma_f > 0.55$ lb/ft³)
4 in:	$W = 2102 \times \gamma_f^{4/5}$	(when $\gamma_f > 0.30$ lb/ft³)
6 in:	$W = 5828 \times \gamma_f^{4/5}$	(when $\gamma_f > 0.14$ lb/ft³)

[4] 1 to 4 in: $W = 3164 \times \mu \cdot D$

6 in: $W = 34550 \times \mu$

[5] $W = 1107 \times 10^3 \times \mu \cdot D$

W: Steam flow rate (lb/h)

D: Internal diameter (inch)

γ_f: Specific weight at normal operating conditions (lb/ft³)

μ: Viscosity (cP)

Note 1. The dotted lines [4] and [5] are linear flow rate limits (Reynolds Numbers Re = 20,000 and 7,000,000, Re = 40,000 and 7,000,000 for 6-inch meter) for saturated water vapor.

Note 2. Steam measurements are influenced by the moisture in the steam.

$$\upsilon = x \cdot \upsilon_g + (1 - x)\upsilon_f$$

where,

υ: Specific volume of steam/water mixture.

υ_g: Specific volume of saturated water vapor in steam.

υ_f: Specific volume of water in steam.

x: Dryness fraction

(1 - x): Wetness fraction

(C) Steam

FIGURE 14-16 (continued)
Sizing charts.
(Courtesy Yokogawa Corporation of America)

flowmeter. The same flowmeter will usually have different centering spacers for each flange size and rating it is capable of mating to, and care must be taken to ensure that the correct centering spacers are ordered and used for installation.

One manufacturer uses an alignment tool to center the vortex shedding flowmeter in the pipe. While use of the tool is not difficult, instructions are necessary to orient the tool to properly center the flowmeter.

Piping Orientation

With the exception of the oscillating disc design, which requires that the flowmeter not be located more than 30° from the vertical in any service, the piping orientation of vortex shedding flowmeters can be in any plane (except downward flow in liquid service). In condensible gas service, it is recommended that the shedder be oriented horizontally so that any condensate in the pipe flows under the shedder and damage to the flowmeter is avoided.

Piping Vibration

Vortex shedder flowmeters are usually susceptible to vibration that can occur parallel to the pipe and normal to the shedder as a result of forces that can be exerted on the sensing system from acceleration that occurs due to vibration. This can cause the sensing system to measure a vortex that does not exist or to delete a vortex that does exist. This effect is most notable at low flow or zero flow conditions since the forces due to vibration effects can be large in relation to forces that are developed by real vortices. These effects usually disappear at higher flows where the force of the vortices overpowers vibration that may be present in the pipe.

Information regarding a flowmeter's immunity to vibration and suggestions to overcome the effects vary with manufacturer. One manufacturer recommends that the vortex shedder be rigidly supported, while another manufacturer suggests positioning the flowmeter so that the vibrations are parallel to the shedder to minimize the effects of vibration. Great care must be taken in this regard since the effects of vibration are not known prior to installation, and changes are often costly and difficult to make after installation.

An enhancement to the vortex shedding flowmeter that uses the pivot of the shedder to sense vortices compensates for vibration by using additional sensors on the shedder that electronically cancel vibration effects. This design, while improved, is still somewhat susceptible to pipe vibration.

The vortex shedder that uses an additional active element to form a torque tube assembly reduces the effects of vibration. It is sensitive to rotational forces that tend to twist the torque tube and not to forces that are related to pipe vibration.

Cabling

Vortex shedding flowmeters are usually specified as 2-wire devices but can be purchased in 3-wire and 4-wire versions. Ultrasonic designs are typically 3-wire or 4-wire devices since power is required to operate the ultrasonic transmitter. Some designs require that the flowmeter body be grounded to the pipe.

While most designs offer either pulse or analog outputs, some manufacturers offer both or field selectability of either. One manufacturer offers an option that allows totalization of flow in the 2-wire loop.

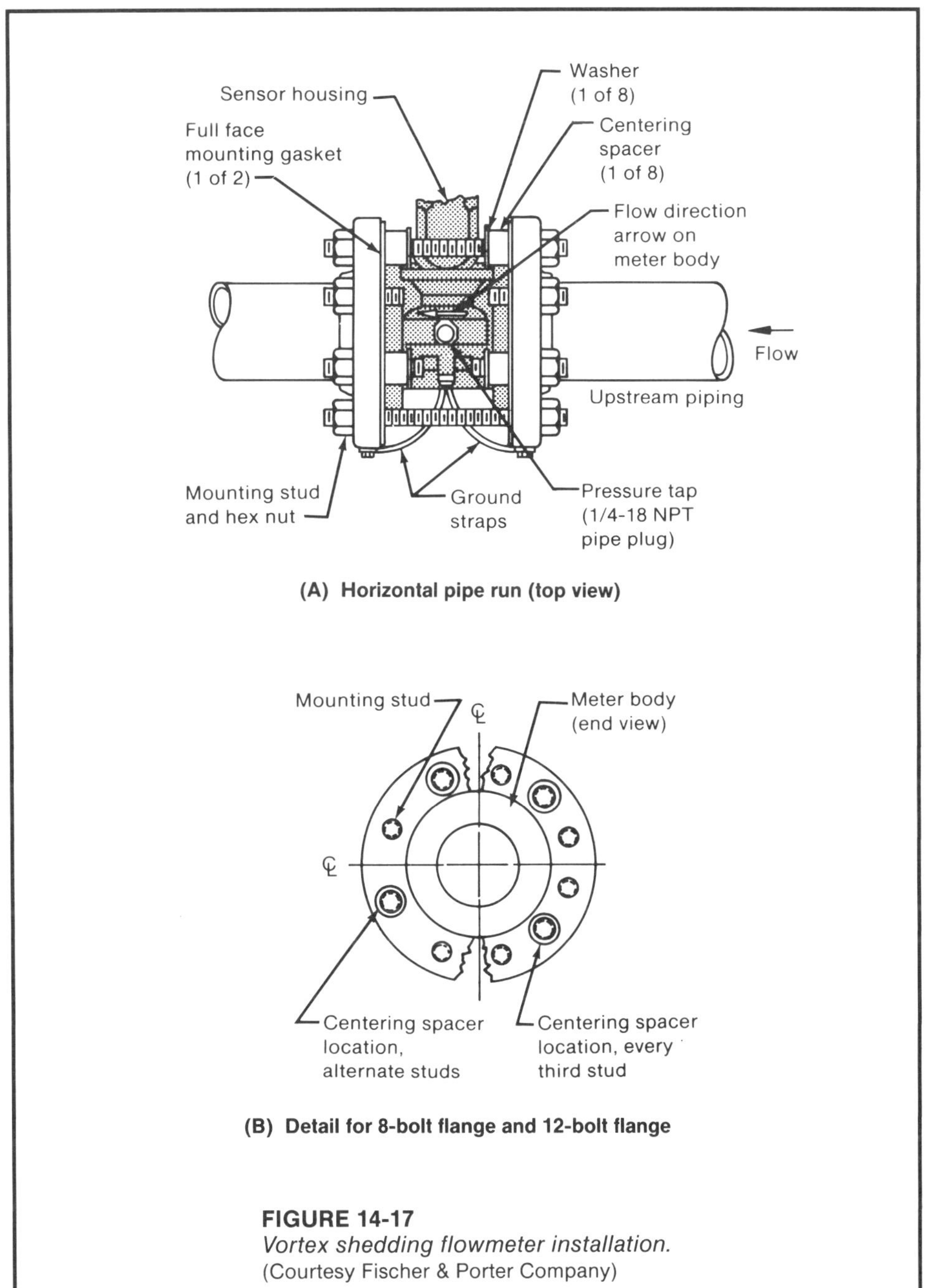

FIGURE 14-17
Vortex shedding flowmeter installation.
(Courtesy Fischer & Porter Company)

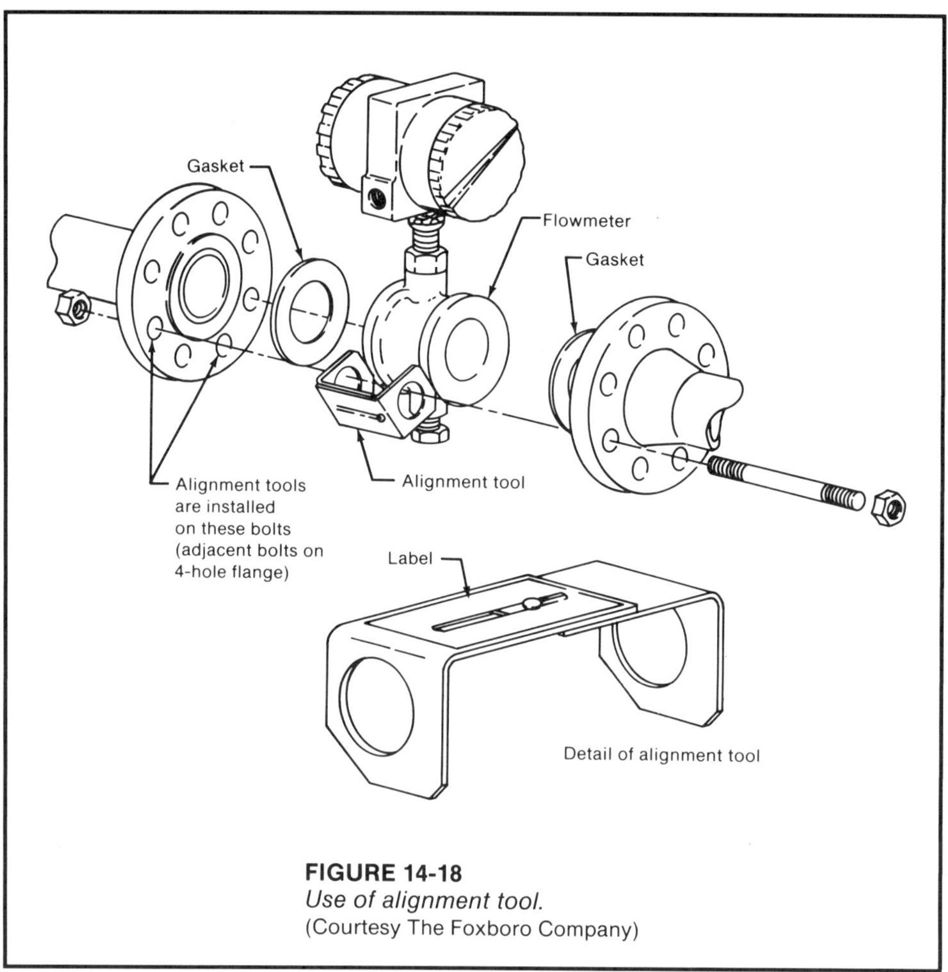

FIGURE 14-18
Use of alignment tool.
(Courtesy The Foxboro Company)

Maintenance No routine maintenance, other than routine calibration checks, is required. Problems such as sensor failure, shedder wear, leaks, and electronic failures can occur.

Shedder Wear

The effects of normal shedder wear on vortex shedding flowmeter performance is usually not a problem. All designs, except that which uses a wire as the shedder, recognize that at certain shedder widths the *K*-factor is almost immune to small variations in shedder width. To minimize the effects of shedder wear, manufacturers use similar shedder widths in their designs to take advantage of this phenomenon. The net result is a flowmeter that is virtually immune to a reasonable amount of wear.

Most vortex shedding flowmeter designs are subject to leaks due to the seals, gaskets, or fitting necessary to remove parts that are positioned in the flowstream, such as the shedder and/or the sensor. The vortex shedding flowmeter that uses the twist of the torque tube has a completely welded design that eliminates this potential problem.

Sensor Failure

Difficulty of sensor replacement is dependent upon flowmeter element design and typically entails opening the flowmeter body. This requires that there be no flow in the pipe and that any pressure present in the pipe be bled off.

The exceptions to this are the vortex shedding flowmeters that use the torque tube design, which is of all welded construction, and the vortex shedding flowmeter with an isolated manifold, which allows the sensor to be replaced with the pipe in service. This is not only convenient from a maintenance standpoint but may also eliminate bypass valves and piping that would be necessary if the flow could not be shut down for sensor replacement.

Electronic Failure

Electronic failures can occur and are usually remedied by board replacement. Many vortex shedding flowmeters have built-in circuits that can be used to verify the operation of approximately 90 percent of the components in the transmitter. This is usually accomplished by moving a link or jumper with the flowmeter in service so as to inject a known frequency reference signal into the electronics. Electronic failures can often be quickly located and the board replaced if the output of the transmitter does not agree with the calculated output corresponding to the reference frequency input.

Spare Parts

Spare parts requirements vary with flowmeter design. In general, when the sensor or shedder is wetted and is removable for replacement, spare parts for each size flowmeter should be stocked. Transmitters can usually be used for all sizes and services, which minimizes spare parts requirements for circuit boards.

Calibration

Vortex shedding flowmeters do not require zero adjustment. The span adjustment is typically performed with field-changeable links or thumbwheel switches. When an analog output is used, both zero and span of the analog circuit should be performed.

As the calibration of the primary flowmeter is determined at manufacture, there are no adjustments or checks of primary flowmeter performance. The transmitter can be calibrated in a number of ways, dependent upon transmitter design.

One method of calibration is to inject frequency signals specified by the manufacturer into the transmitter and make the appropriate adjustments. This allows verification of the thumbwheel adjustments as well as fine adjustment of the zero and span of the analog circuit. When a reference frequency source is available internal to the transmitter, it is often more expedient to perform a rough calibration using the reference source and a zero input to verify the thumbwheel settings as well as the zero and span of the analog output. Minor adjustments to the zero and span can be made using the reference frequency. However, the reference source is subject to larger uncertainty than the test equipment that would be used to perform a bench calibration.

It should be noted that vortex shedding flowmeters are sensitive to expansion of the flowmeter body due to temperature. It should be verified that calibration data includes temperature correction about the nominal operating temperature, when the nominal operating temperature is significantly different from the temperature at which the *K*-factor is referenced.

EXERCISES

14.1 What are the typical materials of construction of vortex shedding flowmeters? Fluidic flowmeters?

14.2 What is the advantage of a fluidic flowmeter over a vortex shedding flowmeter regarding Reynolds number?

14.3 Why do oscillatory flowmeters have accuracy statements that are usually expressed as a percent of rate?

14.4 Why are oscillatory flowmeters not generally applicable to liquids with viscosities of greater than a few centipoise?

14.5 Why should a vortex shedder in steam service be positioned horizontally?

14.6 Size a vortex shedding flowmeter and determine the turndown when measuring a flow of 0 to 600 gpm of a liquid with a specific gravity of 1.18 and a viscosity of 0.83 cP.

14.7 Size a vortex shedding flowmeter and determine the turndown when measuring a flow of 0 to 500 acfm of a gas with a density of 0.23 pound per cubic foot and a viscosity of 0.014 cP.

14.8 Size a vortex shedding flowmeter and determine the turndown when measuring a flow of 0 to 300 pounds per minute of 650-pound saturated steam that has a density of 1.45 lbs/ft^3 at a temperature of 497°F.

15

Positive Displacement Flowmeters

Introduction Positive displacement flowmeters continue to be applied in many flowmeter applications that other flowmeter technologies can handle as well as in applications where there are no viable alternatives to effectively measure flow.

Since positive displacement flowmeters have many attributes in common, a generic design is considered in the following paragraphs. The remainder of this section will present descriptions of specific positive displacement flowmeter technologies.

Principle of Operation In principle, positive displacement flowmeters repeatedly entrap a known quantity of fluid as it passes through the flowmeter. When the number of times the fluid is entrapped is known, the quantity of fluid that has passed through the flowmeter is also known.

In practice, this type of flowmeter senses the entrapped fluid by generating pulses, each of which represents a fraction of the known quantity entrapped. When a flow signal rather than a totalized signal is desired, pulse frequency is converted to an analog signal. There are no Reynolds number constraints.

Sizing The interaction of viscosity and pressure drop across the flowmeter should be understood before one attempts to size a positive displacement flowmeter. As the viscosity increases, the pressure drop across the flowmeter increases, often dramatically. Slippage through the flowmeter decreases with increasing viscosity, allowing more accurate measurement of lower flows. As a result, with increasing viscosity the maximum capacity of the flowmeter is reduced when the pressure drop across the flowmeter is excessive. However, the minimum measurable flow is also reduced due to decreasing slippage. The pressure drop across the flowmeter usually constrains the maximum operating flow of the flowmeter in high viscosity service.

Installation Mounted in the pipe, typically with flanged or screwed connections, positive displacement meters have no upstream or downstream piping requirements.

Due to the nature of the operation, any gas that may be present in the liquid will cause the flowmeter to read the gas volume as if it were liquid. This problem can often be resolved by installing an appropriately sized air eliminator upstream of the flowmeter.

These flowmeters may become plugged or damaged by dirt that may find its way into the high tolerance workings. A strainer upstream of the flowmeter will reduce this problem, but in high viscosity service the pressure drop across the strainer can be significant.

Various types of receiver electronics are available. The pulse output of the flowmeter primary is often transmitted directly to the receiver instrument without local amplification. This results in a cabling configuration where 2 wires carry a pulse signal to a central location, thereby avoiding the added expense of cabling the 4-wire system that could result if the transmitter were located at the flowmeter primary.

Maintenance In addition to the possibility of electronic failures, positive displacement flowmeters are subject to deterioration due to wear, corrosion, exposure to a dirty liquid, and abrasion, as would any flowmeter that relies upon high tolerance moving parts to maintain performance. Pluggage can occur if the flowmeter is exposed to a dirty liquid. The flowmeter must then be disassembled and thoroughly cleaned. Line cleaning prior to commissioning a new system is recommended.

Most positive displacement flowmeters should not be exposed to steam, which is often used to clean pipes, as this can result in damage. When steam must be introduced to the flowmeter, internal parts must often be removed prior to the introduction of steam in order to avoid damage.

Wear

Corrosion, abrasion, and exposure to a dirty liquid can cause premature wear. Failure due to corrosion or abrasion usually results in excessive slippage, while exposure to a dirty liquid may cause the flowmeter to bind up and cease to operate. The flowmeter components can usually be replaced by disassembling the flowmeter, removing the old parts, and installing the new parts. Gasket sets are often required to reassemble the flowmeter.

Part replacement may be necessary due to failure of the magnet or to metal embedded in the part, which can be the effect of a cracked part that performs satisfactorily but allows liquid to enter and corrode the metal embedded within.

Bearing Wear

Overspinning the flowmeter as well as incompatibility with the liquid being measured are likely causes for premature bearing and seal failure. Bearings can usually be replaced by disassembling the flowmeter, pressing out the old bearings, and pressing in the new bearings, or sending the flowmeter back to the manufacturer for refurbishing. Gasket sets are often required to reassemble the flowmeter.

Leaks

These flowmeters are subject to leakage due to the gaskets and seals that may be required for the body and sensor assembly.

Sensor Failure

Sensor failure can occur. Replacement is usually performed under flow conditions, external to the pipe.

Electronic Failure

Electronic failures can occur and are usually remedied by board replacement.

Spare Parts

Spare parts requirements vary with the design of the flowmeter, but replacement rotor, bearings, sensor, and electronics are typically required. Most of the mechanical parts vary with meter size, thereby increasing the spare parts inventory.

Calibration

The meter constant of the flowmeter primary is fixed by design and cannot be calibrated as such. The K-factor establishes the relationship between the frequency output of the flowmeter, the volumetric flow, and the output of the converter. A frequency signal that corresponds to the output of the flowmeter primary at a known flow is injected into the converter so as to verify operation of the converter and set zero and span.

Helical Gear Positive Displacement Flowmeter

Helical gear flowmeters are typically used on extremely viscous liquid service where it is often difficult to apply other flowmeters because of Reynolds number constraints. This design is somewhat tolerant of dirt, as there are few passages that are easily plugged, but is susceptible to overspeed and bearing damage.

Principle of Operation Two radially-pitched helical gears are used to continually entrap liquid as it passes through the flowmeter, causing the rotors to rotate in the longitudinal plane. Flow through the flowmeter is proportional to the rotational speed of the gears.

As the sealing surfaces are in the longitudinal plane, the forces required to operate the flowmeter are relatively small and the pressure drop through the flowmeter is relatively low. Use on high viscosity liquids is possible without establishing undue constraints on the piping system.

Construction The body of the flowmeter is the assembly in which the rotors are mounted and on which the sensing system is housed. Gaskets and/or O-rings are used to seal the flowmeter body assembly where wetted parts are removable for access during manufacture and maintenance.

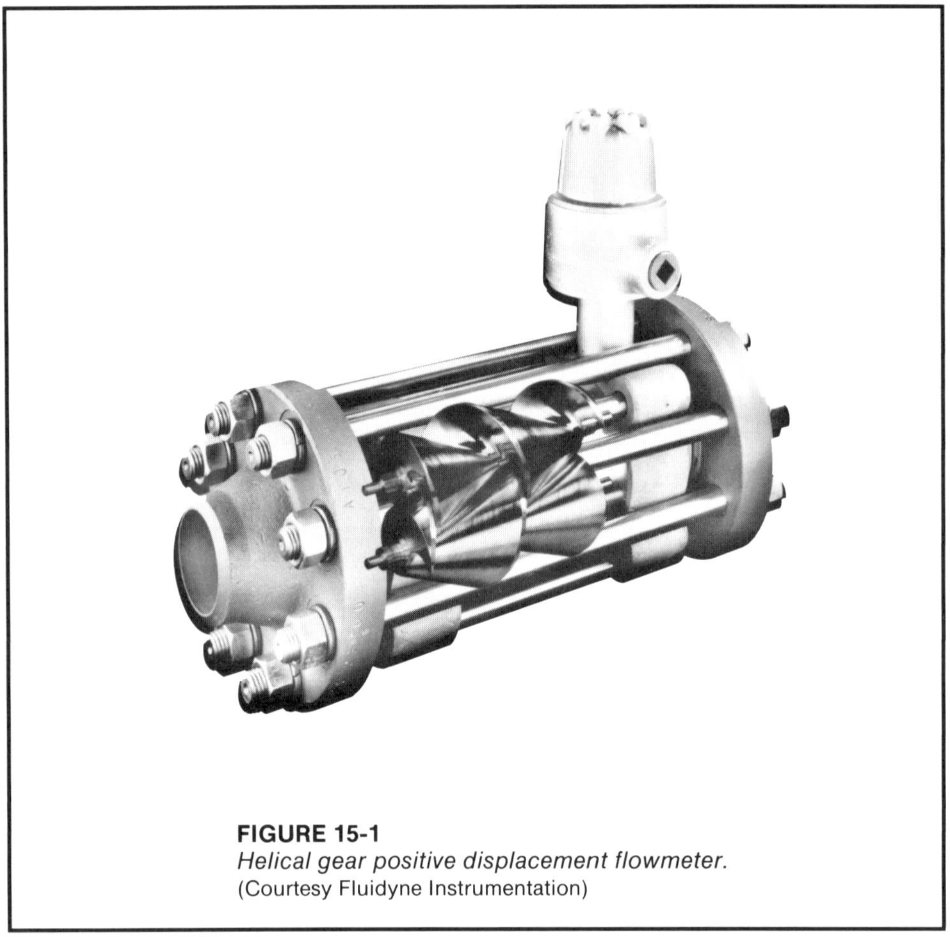

FIGURE 15-1
Helical gear positive displacement flowmeter.
(Courtesy Fluidyne Instrumentation)

Rotor

The rotors rotate on their shafts due to the forces exerted by the flow of liquid through the flowmeter. As the rotors must mesh and form a seal with each other as well as with the flowmeter body, these parts are manufactured to tight tolerances that must be maintained over the life of the flowmeter in order to maintain performance and to reduce slippage or blow-by, i.e., flow that gets past these seals and is not measured.

Bearings

The rotors require bearings, which are typically pressed into the body.

Sensing System

Magnetic and optical sensing systems are prevalent in helical gear flowmeter designs. The magnetic sensing system illustrated in Figure 15-2 employs a magnetic gear, the teeth of which are sensed by a magnetic pickup and amplified.

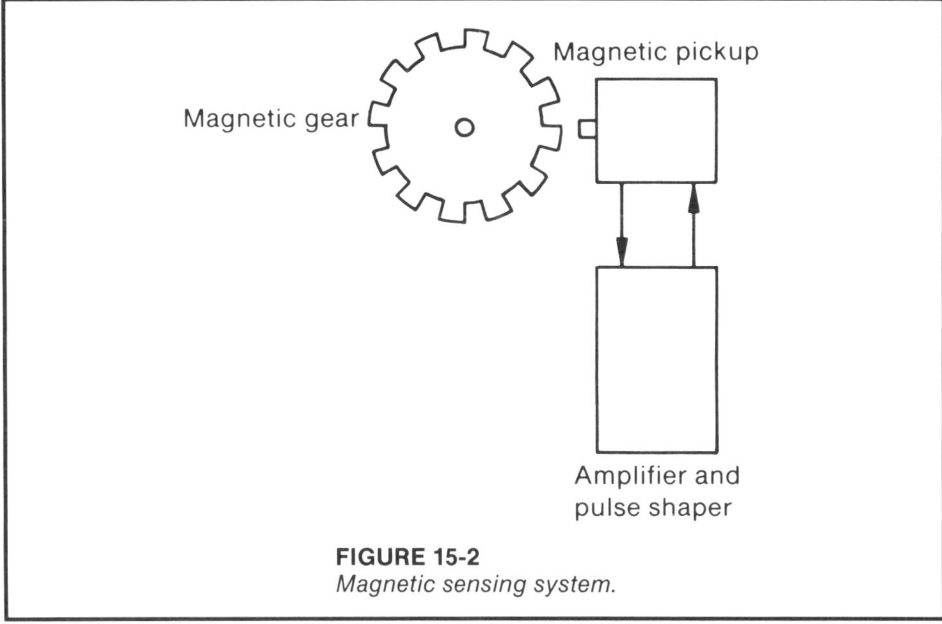

FIGURE 15-2
Magnetic sensing system.

The optical sensing system utilizes a magnetically driven optically encoded disc, the rotation of which is sensed by an optical pickup in order to sense a pulse each time a portion of a revolution occurs, as illustrated in Figure 15-3.

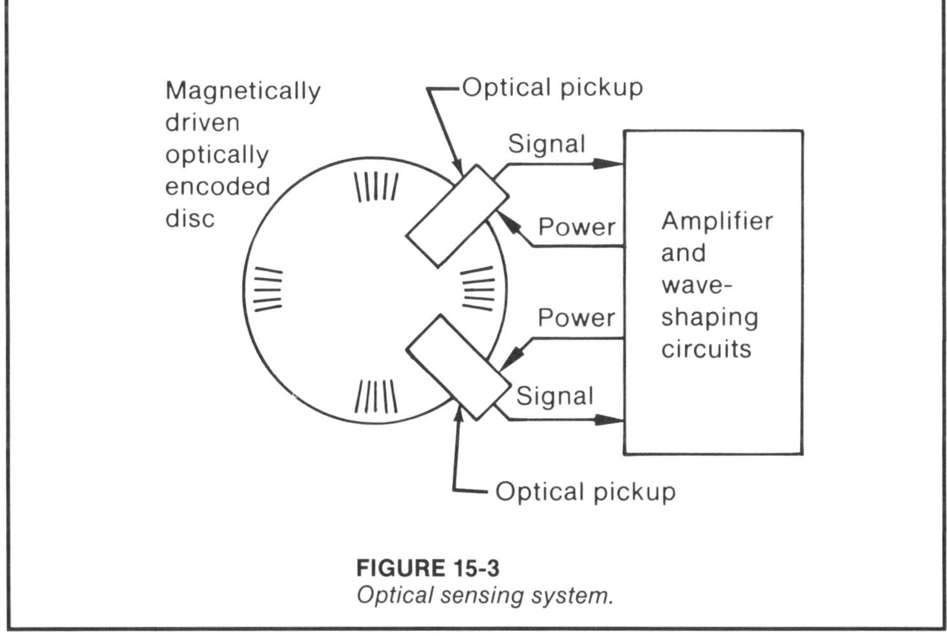

FIGURE 15-3
Optical sensing system.

Wetted Parts

Wetted parts of helical gear flowmeters include the body, O-rings, rotors, and bearings. Bodies are available in stainless steel 4 in. and under, and carbon steel 6 in. and over. Rotors are typically constructed of stainless steel or aluminum in the smaller sizes, and either carbon steel or aluminum in the larger sizes. Standard bearings are typically made of stainless steel, the grade of which may be unsuitable for some common applications such as water, aqueous solutions, bases, or salts. Helical gear flowmeters are often limited by the bearing materials of construction.

As with all flowmeters, the compatibility of the materials of construction of each component should be investigated, as the effects of wear and corrosion on the performance of the flowmeter are significant when slippage becomes excessive, the bearings fail, or the seals leak.

Operating Constraints Available in 1-1/2 to 10 in. sizes, helical gear flowmeters are usually pressure- and temperature-limited by the flange ratings and the temperature ratings of the sensor, which can be up to approximately 300°C in certain applications. Flow can range from 5 to 4000 gpm with viscosities to several hundred thousand centipoise.

Pressure drop across these flowmeters should typically be kept below approximately 30 psid so as not to cause excessive bearing wear and premature bearing failure. Exact pressure drop limitations for each flowmeter are available from the manufacturer.

Performance Volumetric flows can be measured with an accuracy that ranges from approximately ±0.2 to 0.4 percent rate, depending on the application and the flowmeter design. Nonviscous flows are generally measured less accurately than viscous flows due to errors caused by increased slippage through the flowmeter at low viscosities.

The accuracy statements above represent ideal operating conditions. Changes in viscosity can cause shifts in the accuracy.

Figure 15-4 illustrates how accuracy is affected by viscosity changes. Note that the graph shows flowmeter accuracy as a function of the maximum rated flow of the flowmeter and not of the desired flow range. As viscous liquids can exhibit relatively large variations in viscosity over a relatively small temperature range, inaccuracies caused by viscosity changes may be larger than the stated accuracy of the flowmeter.

Low liquid lubricity can adversely affect bearing and rotor tolerances, which must be maintained in order to maintain accuracy. Turndown can be as high as 100:1 in certain applications, although lower turndowns are typical of actual applications.

Applications Helical gear flowmeters are generally applicable to non-abrasive lubricious liquids with viscosities from approximately 3 cP to 300,000 cP. Slippage can pose a problem in low viscosity applications, especially if there is any wear of machined parts, so most applications are on high viscosity liquids. The relatively

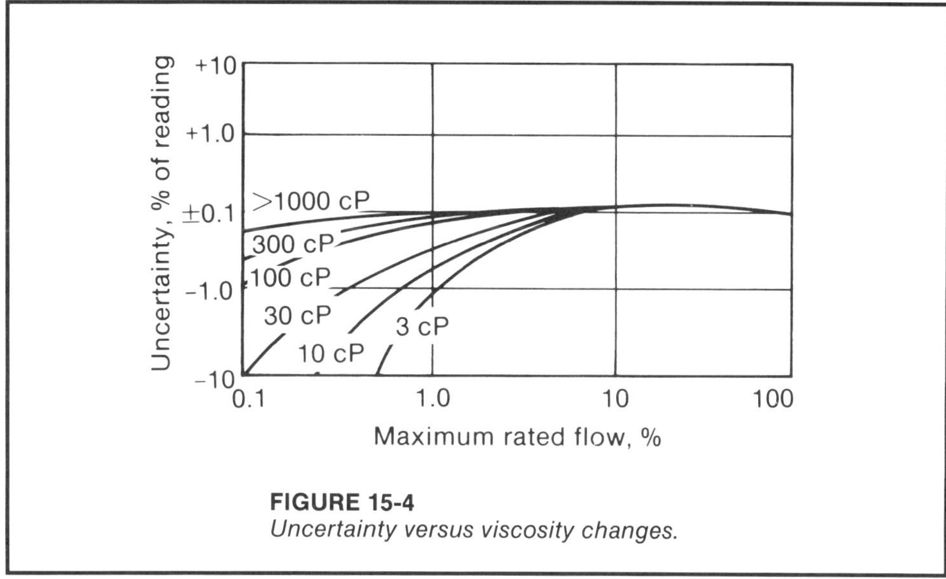

FIGURE 15-4
Uncertainty versus viscosity changes.

low pressure drop introduced into the piping system makes this flowmeter design attractive for high viscosity applications.

Sizing Figure 15-5 illustrates the relationship between flowmeter capacity, maximum pressure drop across the flowmeter, and viscosity. This graph can be used in sizing helical gear flowmeters.

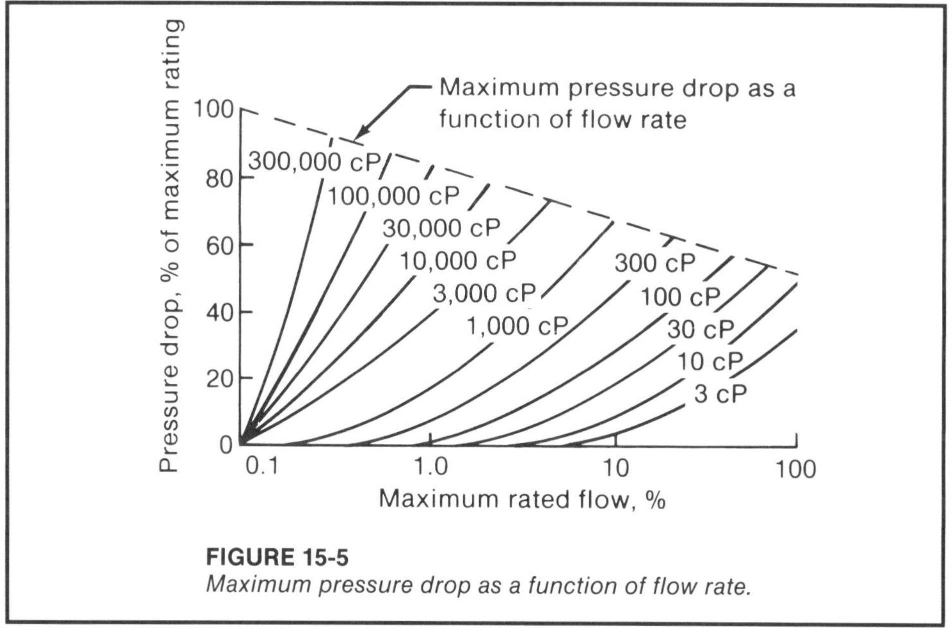

FIGURE 15-5
Maximum pressure drop as a function of flow rate.

The relatively horizontal accuracy curves show the ability of the flowmeter to accurately measure highly viscous liquids at low flow. The nonlinearities of the low viscosity curves typify the slippage that can occur in this service. The pressure drop curves show how the pressure drop across the flowmeter constrains the maximum operating flow of the flowmeter in high viscosity service.

EXAMPLE 15-1

Problem: Size a helical gear flowmeter for a maximum flow of 100 gpm of a liquid with a viscosity of 1000 cP, given the following flowmeter capacity data:

Size	Maximum Flow
1-1/2 in.	50 gpm
2-1/2 in.	150 gpm
4 in.	450 gpm
6 in.	1350 gpm
10 in.	4000 gpm

Solution: A 2-1/2-inch flowmeter would appear to be applicable. However, examination of the typical performance graph shows that the maximum differential pressure limits the maximum flow to approximately 10 percent of the flowmeter capacity, or 15 gpm. A 4-inch flowmeter is similarly limited to 45 gpm, while a 6-in. flowmeter is limited to 135 gpm and would be applicable. The 6-in. flowmeter has a minimum flow that is approximately 0.1 percent of flowmeter capacity, or 1.35 gpm, resulting in a turndown of 100 gpm/1.35 gpm, or approximately 74:1.

Nutating Disc Positive Displacement Flowmeter

Nutating disc flowmeters are most often applied to water service to effect economic flow measurement where accuracy is not of great importance. This design is somewhat tolerant of dirt, as there are few passages that are easily plugged.

Principle of Operation A nutating disc flowmeter utilizes a cylindrical measurement chamber in which a disc is allowed to wobble, or nutate, as flow passes through the flowmeter, causing the spindle to rotate. The motion of the spindle is transmitted to a magnet assembly that is used to drive a following magnet external to the flowstream. This rotation can be used to drive a register or a transmitter. Operation of the nutating disc flowmeter is illustrated in Figures 15-6 and 15-7.

As the flowmeter entraps a fixed quantity of liquid each time the spindle is rotated, the rate of flow is proportional to the rotational velocity of the spindle.

Construction The body is the housing in which the nutating disc is mounted and on which the sensing system is housed. O-rings and/or gaskets are used to seal the flowmeter body assembly where wetted parts are removable for access during manufacture and maintenance.

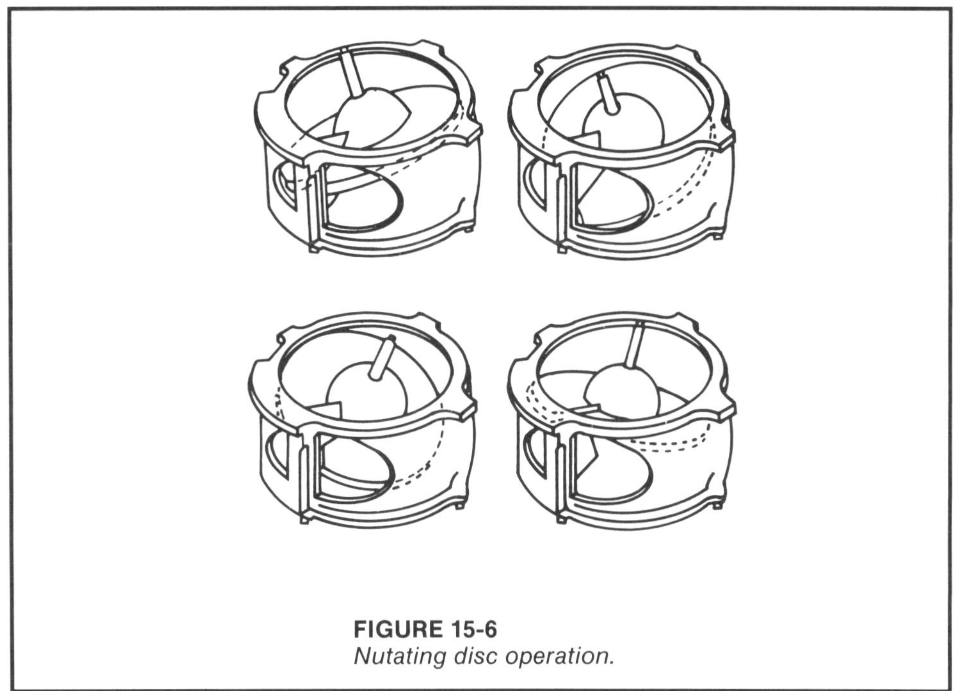

FIGURE 15-6
Nutating disc operation.

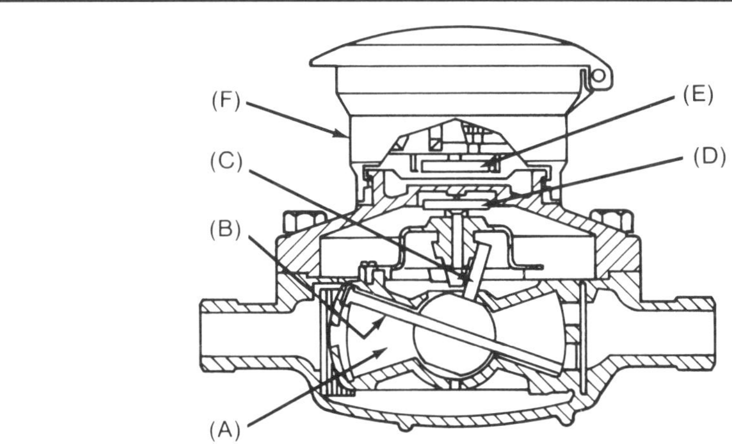

Liquid flowing through the meter chamber (A) causes a disc (B) to nutate or wobble. This motion, in turn, results in the rotation of a spindle (C) and drive of magnet (D). Rotation is transmitted through the wall of the meter to a second magnet (E) which operates the meter register (F) or a pulse transmitter.

FIGURE 15-7
Nutating disc operation.
(Courtesy Badger Meter, Inc.)

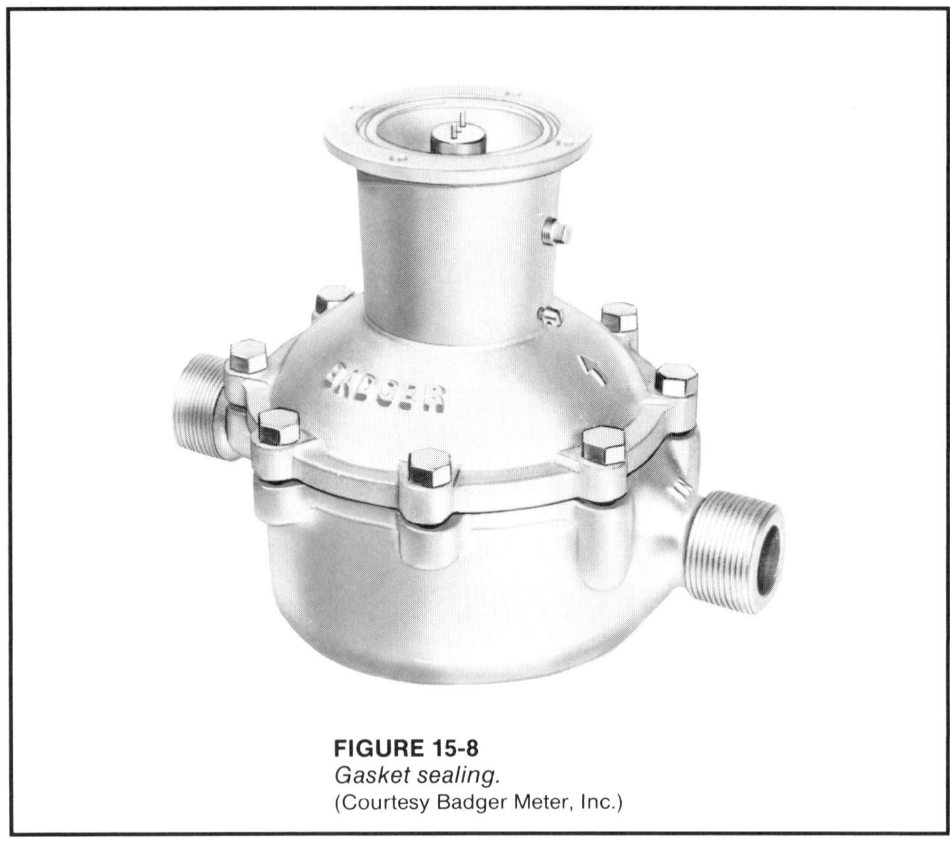

FIGURE 15-8
Gasket sealing.
(Courtesy Badger Meter, Inc.)

Disc

The disc rotates on its shaft due to the forces exerted by the upstream pressure and the flow of liquid through the flowmeter. Because the disc must form a seal with a partition in the measuring chamber as well as with the flowmeter body, these parts are manufactured to tight tolerances that must be maintained over the life of the flowmeter in order to maintain performance and reduce slippage or blow-by.

The spindle protrudes from one side of the disc so as to transmit the wobbling motion of the disc to the sensing system.

Sensing System

Magnetic sensing systems are prevalent in nutating disc flowmeter designs. The magnetic sensing system employs driving magnets that rotate at the same rate at which the disc wobbles. These magnets rotate a following magnet external to the flowstream, which can drive a local indicator, a totalizer, or a transmitter.

Wetted Parts

Wetted parts of nutating disc flowmeters include the body, O-rings, disc, spindle, and magnet assembly. Nutating disc flowmeter bodies and metal parts must be nonmagnetic and are generally available in bronze. The disc may be constructed of such materials as bronze, rubber, or aluminum. O-rings are constructed of materials such as Viton®, BUNA-N® and neoprene. Compatibility of construction materials should be investigated.

Operating Constraints Nutating disc flowmeters, which are available in 5/8-in. to 2-in. sizes, are generally pressure-limited to 150 psig. The temperature limit of these flowmeters is approximately 120°C, while flow ranges are typically 1 to 160 gpm in intermittent service and 1 to 120 gpm in continuous service. The flowmeter can be damaged by excessive operation in the intermittent flow region, and flow ranges can be reduced for some liquids, depending upon temperature.

Pressure drop across nutating disc flowmeters is typically kept below 15 psid so as not to cause damage to the flowmeter.

Performance Nutating disc flowmeters can measure volumetric flows with an accuracy of approximately ±2 percent rate, depending on the application. Nonviscous flows are generally measured less accurately than viscous flows due to errors caused by increased slippage through the flowmeter at low viscosities.

The accuracy statement above represents ideal operating conditions. Viscosity changes can cause shifts in the accuracy of the flowmeter due to varying amounts of slippage at different viscosities. As viscous liquids can exhibit relatively large variations in viscosity over a relatively small temperature range, the inaccuracies caused by viscosity changes may be on the order of the stated accuracy of the flowmeter.

The maximum flow through nutating disc flowmeters is usually a function of usage. These flowmeters generally have different flow ratings for continuous and intermittent service. The intermittent service rating is the maximum flow that can be maintained for short time durations.

Turndown in an intermittent liquid application is typically 5 to 20:1, while in a continuous application the maximum turndown can be as low as 5 to 10:1 when the desired flow range coincides with the flow range of the flowmeter. The larger turndowns stated above represent rubber disc construction, which has better sealing characteristics than metal disc construction and therefore enables accurate measurement over a wider range.

Applications Nutating disc flowmeters are generally applicable to clean nonabrasive liquids. Slippage can pose a problem in low viscosity applications, but because it can measure low viscosities reasonably well and economically, this design has many applications in water service.

Sizing Typical capacity information is given in Table 15-1 for a nutating disc flowmeter.

EXAMPLE 15-2

Problem: Size a nutating disc flowmeter to measure a full scale flow of 0 to 20 gpm of a cold liquid.

TABLE 15-1
Typical Nutating Disc Capacities

	Operating Pressure—150 psi					
	Cool liquids to 120° F		*Hot water to 250° F*		*Chemical and oils to 250° F*	
	Hard rubber disc		*Syn. rubber disc*		*Bronze or aluminum disc*	
Flowmeter size	*Normal flow range, gpm*	*Max. cont. flow, gpm*	*Normal flow range, gpm*	*Max. cont. flow, gpm*	*Normal flow range, gpm*	*Max. cont. flow, gpm*
5/8 in.	1-20	10	2-20	10	2-10	10
5/8 in.	1-20	10	2-20	10	2-10	10
3/4 in.	2-30	12	3-30	12	3-15	12
1 in.	3-50	36	5-50	36	5-25	25
1 in.	3-50	36	5-50	36	5-25	25
1-1/2 in.	5-100	83	8-100	83	10-50	50
2 in.	8-160	120	12-160	120	16-80	80

Solution: 20 gpm is greater than the maximum continuous flow of a 3/4-inch flowmeter; therefore, a 1-inch flowmeter should be considered. The desired flow range is within the normal flow range of all disc types; the turndown for a hard rubber disc is 20/3, or 6.7:1, while the turndown for a metal or synthetic rubber disc is 20/5, or 4:1.

Oscillating Piston Positive Displacement Flowmeter

Oscillating piston flowmeters are typically used on viscous liquid service where turndown is not of great importance. This positive displacement flowmeter design is somewhat tolerant of dirt, as there are few easily plugged passages, but large or abrasive solids can be compressed between the piston and the flowmeter body, thereby distorting the piston, altering the seal it must make to achieve flowmeter accuracy.

Principle of Operation A cylindrical measurement chamber with a partition plate separating the inlet from the outlet port is used. The piston is also cylindrical and has numerous holes in its supports so that liquid is free to flow on both sides of the piston as well as on both sides of a slot for the partition plate. The piston is guided within the measuring chamber by rotation around a control roller. The motion of the piston is transmitted to a magnet assembly that is used to drive a follower magnet external to the flowstream. It can also be used to drive a register or a transmitter. The motion of the piston is oscillatory: the center of the piston moves around the control roller and the slot in the piston can only operate in one plane. Operation of the oscillating piston flowmeter is presented in Figure 15-9.

As the flowmeter entraps a fixed quantity of liquid each time the meter is rotated, the rate of flow is proportional to the rotational velocity of the piston.

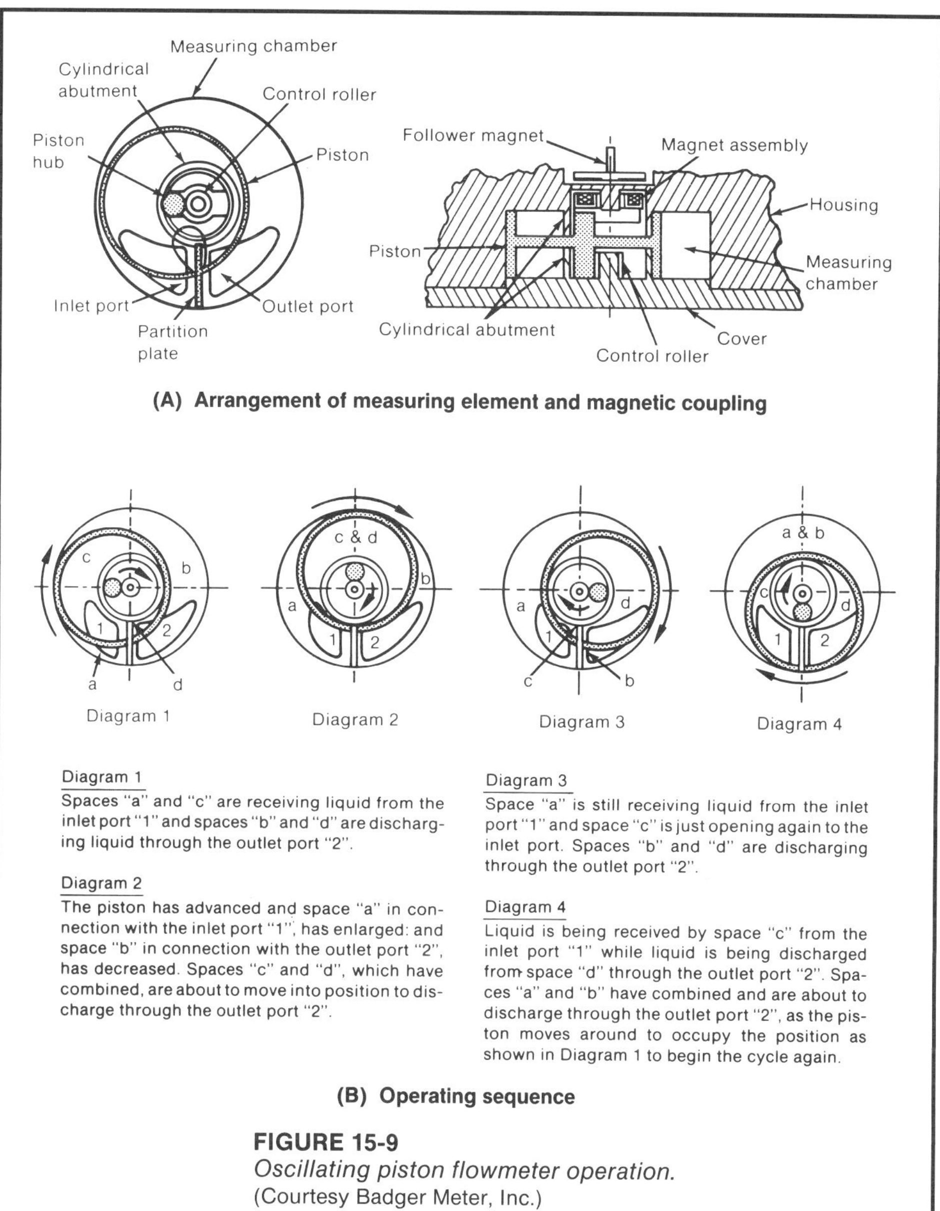

(A) Arrangement of measuring element and magnetic coupling

(B) Operating sequence

FIGURE 15-9
Oscillating piston flowmeter operation.
(Courtesy Badger Meter, Inc.)

Construction The body is the housing in which the piston is mounted and on which the sensing system is housed. O-rings and/or gaskets are used to seal the flowmeter body assembly where wetted parts are removable for access during manufacture and maintenance. The body is usually designed so that disassembly can be accomplished quickly without removing the flowmeter from the pipe.

Piston

The piston rotates on its shaft due to the forces exerted by the upstream pressure and the flow of liquid through the flowmeter. As the piston must form a seal with the partition as well as with the flowmeter body, these parts are manufactured to tight tolerances that must be maintained over the life of the flowmeter in order to maintain performance and reduce slippage or blow-by. Materials of construction are limited to those that exhibit the property of low expansion due to temperature change. Slotted pistons are available to handle liquids that are lumpy or contain a small amount of solids.

Control Roller

The control roller is the guide in which the center of the piston, called the piston hub, rotates.

Sensing System

Magnetic sensing systems are prevalent in oscillating piston flowmeter designs. The magnetic sensing system employs driving magnets in the magnet assembly to rotate a following magnet external to the flowstream, which can drive a local indicator, a totalizer, or a transmitter.

Wetted Parts

Wetted parts include the body, O-rings, piston, control roller, and magnet assembly. Oscillating piston flowmeter bodies and metal parts must be non-magnetic and are generally available in stainless steel, aluminum, bronze, and Alloy 20. The piston may be constructed of such materials as polypropylene, Kynar®, Ryton®, and carbon. O-rings are constructed of materials such as Vito®, BUNA-N®, neoprene, and Teflon®. If the above materials of construction are not compatible with the service, plastic-lined oscillating piston flowmeters can be considered. Materials of construction of each component must be compatible with the fluid being measured.

Operating Constraints Oscillating piston flowmeters, which are available in 1/2-in. to 3-in. sizes, are generally pressure-limited to between 100 to 150 psig, depending upon the materials of construction of the body although ratings up to 5000 psig are available. Due to the expansion characteristics of the piston, the range of temperatures over which the flowmeter will operate accurately may be limited to approximately 60°C to approximately 120°C. Flows from 1 to 100 gpm in intermittent service and 1 to 65 gpm in continuous service can be measured. The flowmeter can be damaged by operation in the intermittent flow region for more than a specific time period during a 24-hour period.

Pressure drop should be kept below 35 psid in most designs.

Performance Oscillating piston flowmeters can measure volumetric flows with an accuracy of up to approximately ±0.5 percent rate, depending on the application. Nonviscous flows are generally measured less accurately due to errors caused by increased slippage through the flowmeter at low viscosities.

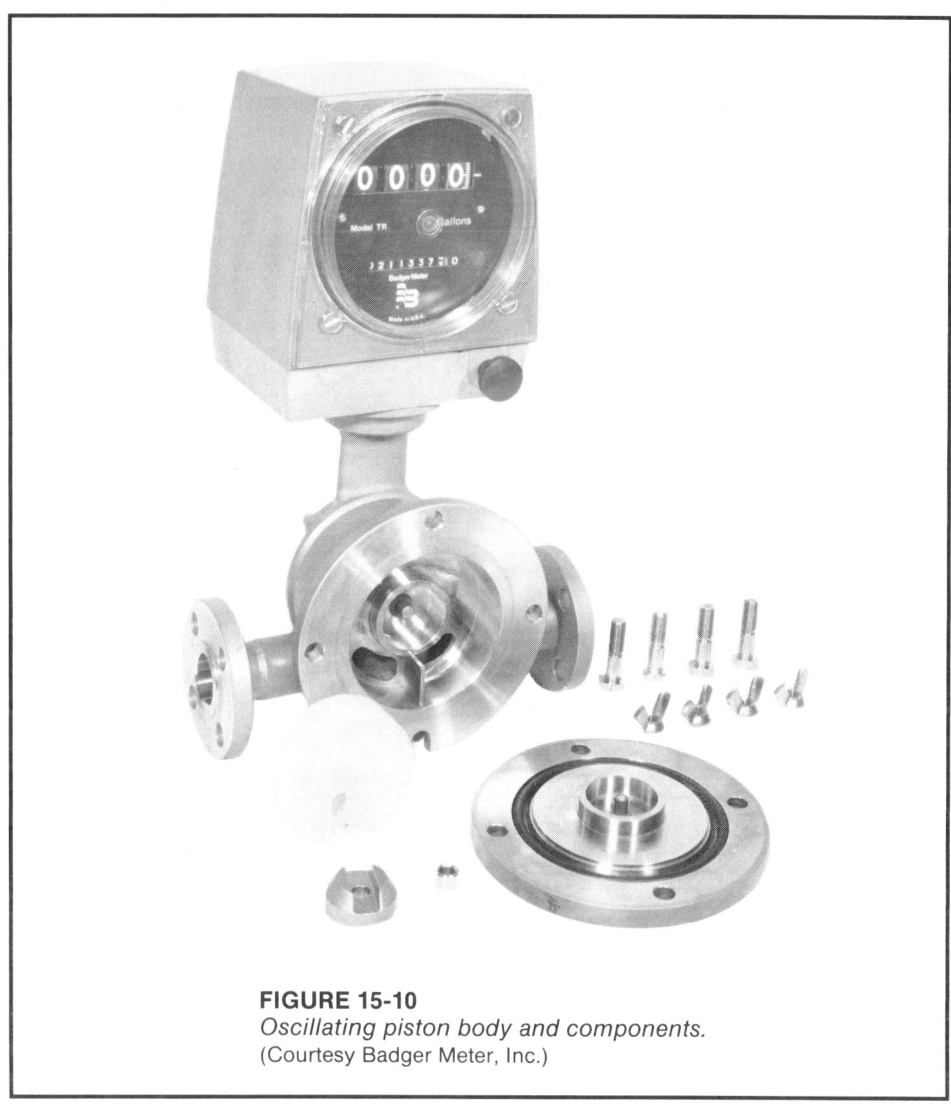

FIGURE 15-10
Oscillating piston body and components.
(Courtesy Badger Meter, Inc.)

The accuracy statement represents ideal operating conditions. Viscosity changes can cause shifts in the accuracy of the flowmeter due to varying amounts of slippage at different viscosities. As viscous liquids can exhibit relatively large variations in viscosity over a relatively small temperature range, the inaccuracies caused by viscosity changes may be larger than the stated accuracy of the flowmeter.

The maximum flow through oscillating piston flowmeters is usually a function of usage. These flowmeters generally have different flow ratings for continuous and intermittent service where the intermittent service rating is the maximum flow that can be maintained for short durations, often not exceeding several minutes per day.

Turndown in an intermittent liquid application is typically 5 to 6:1. In a continuous application, the maximum turndown can be as low as 3 to 4:1, when the desired flow range coincides with the flow range of the flowmeter.

Applications This design is generally applicable to clean non-abrasive liquids with viscosities from approximately 0.2 cP to 10,000 cP. Slotted pistons are available to handle liquids that are lumpy or contain a small amount of solids. Slippage can pose a problem in low viscosity applications, especially if there is any wear of machined parts; therefore, most applications are on higher viscosity liquids.

Sizing Figure 15-11B depicts the typical measurement accuracy and pressure drop across an oscillating piston flowmeter for a low viscosity liquid over its operating flow range. Figure 15-11C shows the pressure drop across a 1-in. flowmeter for liquids of varying viscosities throughout the flow range. The pressure drop across the flowmeter constrains the maximum operating flow of the flowmeter in high viscosity service.

EXAMPLE 15-3

Problem: Size an oscillating piston flowmeter for a maximum flow of 30 gpm of a lubricating liquid with a viscosity of 1.0 cP.

Solution: A 1-inch flowmeter is not sufficiently large, as the maximum continuous flow rating is 20 gpm. The 2-inch flowmeter could be applied, but the turndown would be 30 gpm/20 gpm, or 1.5:1.

EXAMPLE 15-4

Problem: Size an oscillating piston flowmeter for a maximum flow of 20 gpm of a lubricating liquid with a viscosity of 500 cP.

Solution: A 1-inch flowmeter has a maximum continuous flow rating of 20 gpm, which is satisfactory. The pressure drop across the flowmeter is 20 psi, which is also satisfactory. It will have a turndown of 20 gpm/2.5 gpm, or 8:1.

Meter size	Normal flow range, ± 0.5%	Short duration max. flow	Continuous max. flow	Oper. press., psi	Max. oper. temp.
1/2 in.	1-6 gpm	6 gpm	4 gpm	150	250° F
1 in.	5-30 gpm	30 gpm	20 gpm	100-150	250° F
2 in.	20-100 gpm	100 gpm	65 gpm	100-150	250° F

(A) Measurement accuracy and pressure drop

FIGURE 15-11
Typical oscillating piston flowmeter performance.

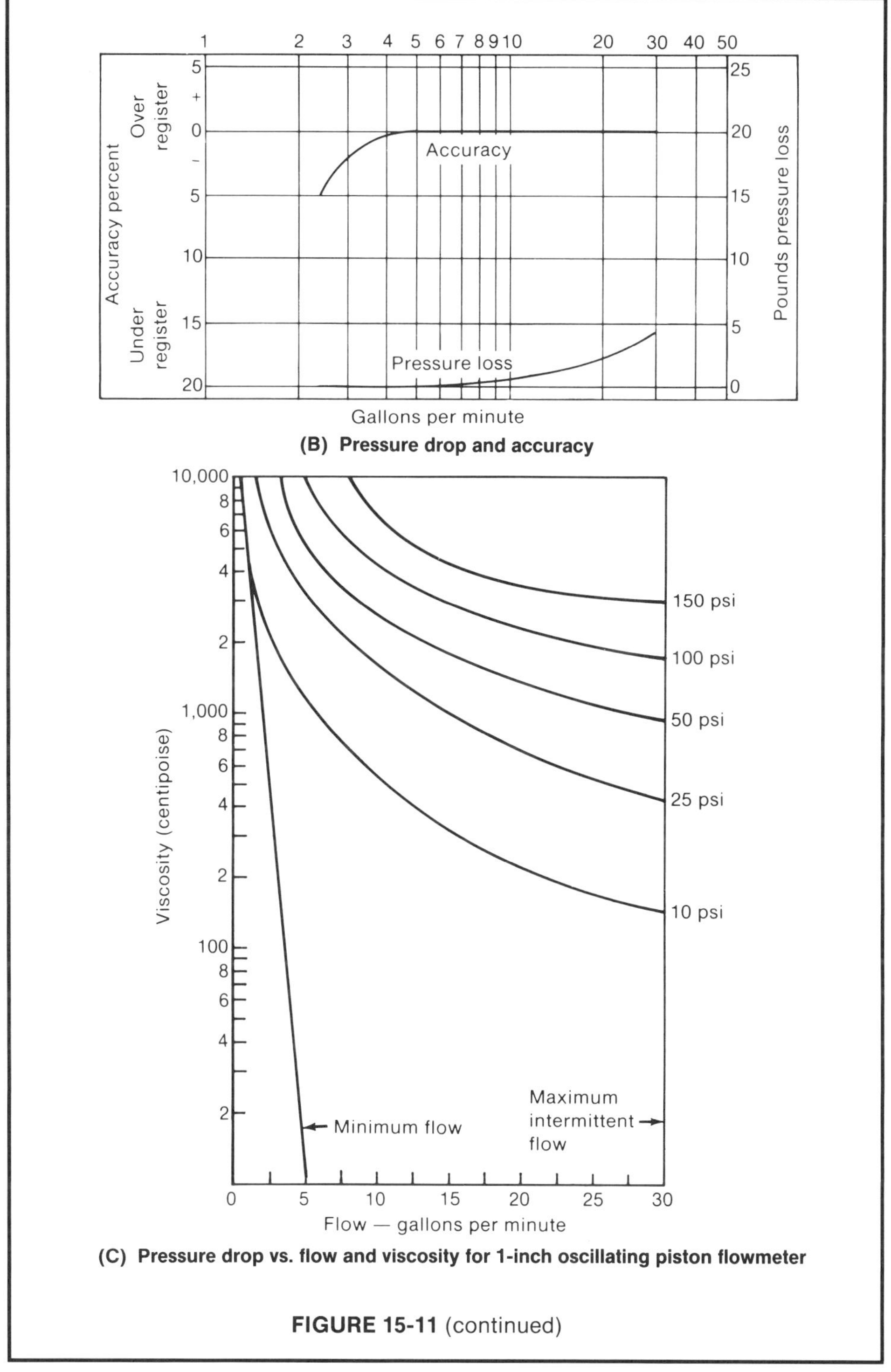

(B) Pressure drop and accuracy

(C) Pressure drop vs. flow and viscosity for 1-inch oscillating piston flowmeter

FIGURE 15-11 (continued)

Oval Gear Positive Displacement Flowmeter Oval gear flowmeters are typically used on viscous liquid service where it is often difficult to apply other flowmeters due to Reynolds number constraints. This design is somewhat tolerant of dirt as there are few passages that are easily plugged; however, dirt can damage the precision manufactured gear teeth.

Principle of Operation The differential pressure across the flowmeter causes forces to be exerted on a pair of oval gears, often called rotors, causing them to rotate. See Figure 15-12.

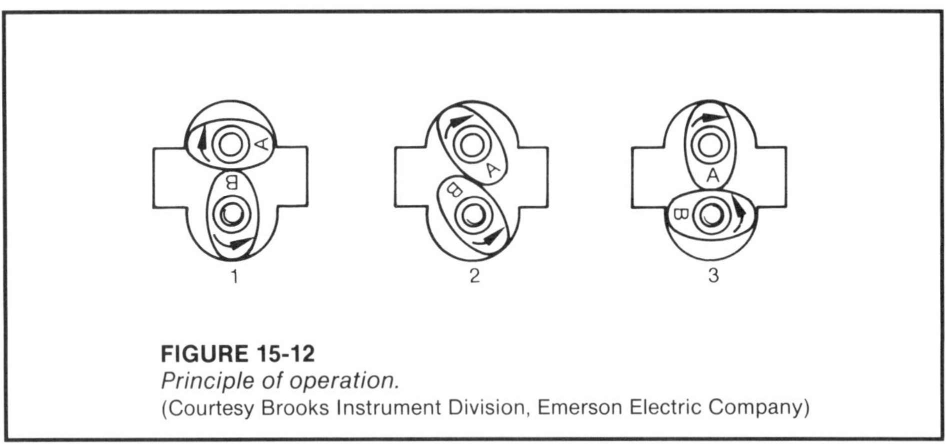

FIGURE 15-12
Principle of operation.
(Courtesy Brooks Instrument Division, Emerson Electric Company)

In position 1, uniform forces are exerted on each broad face of rotor B. Even though these forces are different, rotor B is hydraulically balanced and will not rotate. Rotor A has a uniform force exerted on its upper face, which has entrapped a known quantity of liquid between the rotor and the flowmeter body, but a uniform force is not exerted on the lower face. As the upstream pressure is greater than the downstream pressure, the force exerted on the upstream end of the lower face of rotor A is greater than on the downstream end. This tends to make rotor A rotate in a clockwise direction and rotor B in a counterclockwise direction to position 2.

In position 2, liquid enters the space between rotor B and the flowmeter body, as liquid that was entrapped between rotor A and the body simultaneously leaves the area of entrapment. The larger upstream forces oppose the smaller downstream forces at the ends of rotor A and rotor B, which tend to make rotor A and rotor B continue to rotate in clockwise and counterclockwise directions, respectively, to position 3.

In position 3, a known quantity of liquid has become entrapped between rotor B and the flowmeter body. This operation is then repeated with each revolution of the rotors representing the passage of four times the quantity of liquid that fills the space between the rotor and the flowmeter body. Therefore, the flow is proportional to the rotational speed of the gears.

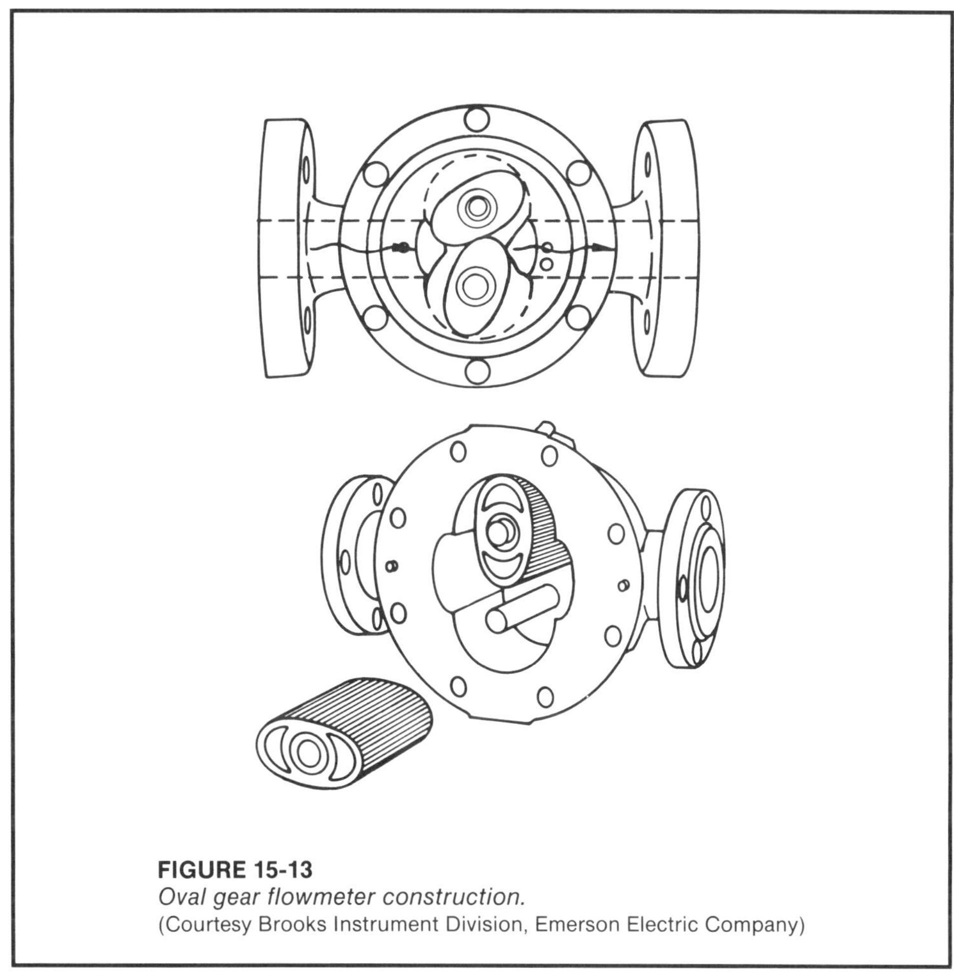

FIGURE 15-13
Oval gear flowmeter construction.
(Courtesy Brooks Instrument Division, Emerson Electric Company)

Construction The body is the housing in which the rotors are mounted and on which the sensing system is housed. O-rings and/or gaskets are used to seal the flowmeter body assembly where wetted parts are removable for access during manufacture and maintenance.

Rotor

The rotors rotate on their shafts due to the forces exerted by the upstream pressure and the flow of liquid through the flowmeter. As the rotors must mesh and form a seal with each other as well as with the flowmeter body, these parts are manufactured to tight tolerances that must be maintained over the life of the flowmeter.

Rotors are available with different cuts and wider tolerances for higher viscosity service due to reduced slippage that occurs at higher operating viscosities. This also effectively keeps the pressure drop across the flowmeter reasonable in most applications.

Bearings

The rotors require bearings, which are typically pressed into the body, on which to rotate. Standard bearings are typically hard carbon. For high temperature applications, special bearings may have to be specified. Specially constructed bearings can be specified when required, typically due to the incompatibility of the liquid with the standard bearings.

Sensing System

Magnetic and electrical sensing systems are prevalent in this design. The magnetic sensing system shown in Figure 15-14 employs driving magnets embedded in the rotor to rotate a following magnet external to the flowstream, which can drive a local indicator, a totalizer, or a transmitter.

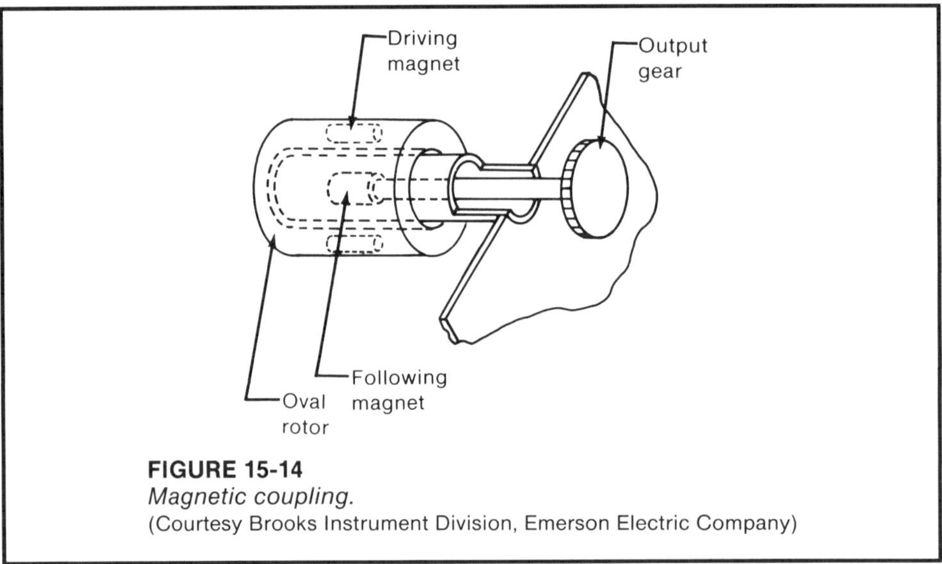

FIGURE 15-14
Magnetic coupling.
(Courtesy Brooks Instrument Division, Emerson Electric Company)

Electrical sensing systems utilize a magnetic pickup external to the flowstream to sense the presence of metal that is embedded in one rotor as it passes by the pickup. The presence of the piece of metal near the pickup causes the electrical properties of the pickup to change, resulting in the generation of an output pulse.

Wetted Parts

Wetted parts of oval gear flowmeters include the body, O-rings, rotors, and bearings. Oval gear flowmeter bodies are generally available in steel, stainless steel, or Alloy 20. The O-rings are available in such materials as Viton®, BUNA-N®, and Teflon®. Although the rotors and shafts are typically constructed of stainless steel or Alloy 20 to be compatible with the body, some designs construct the rotors from resins such as glass-filled phenol. Carbon bearings are usually standard, but special constructions are available.

Compatibility of the materials of construction of each component should be investigated.

Operating Constraints Oval gear flowmeters, which are available in 1/8-in. to 3-in. sizes, are generally pressure- and temperature-limited by the flange ratings and the temperature ratings of the bearings and O-ring materials, which can be up to approximately 300°C in certain applications. Flow can range from approximately 0.02 to 265 gpm in intermittent service and 0.02 to 175 gpm in continuous service.

Pressure drop is typically kept below 15 psid so as not to cause excessive bearing wear and premature bearing failure.

Performance This flowmeter can measure volumetric flows with an accuracy that ranges from approximately ±0.25 to 1 percent rate, depending on the application and flowmeter design. Nonviscous flows are generally measured less accurately due to errors caused by increased slippage through the flowmeter at low viscosities.

The accuracy statements above represent ideal operating conditions. Changes in viscosity can cause shifts in flowmeter performance.

Curves A and B in Figure 15-15 show the shifts in accuracy that can occur due to viscosity variations when the flowmeter is calibrated at 1 cP. Curves C and D show the effects of viscosity variations on a flowmeter calibrated at 10 cP. The maximum shift that can occur when the viscosity is varied from 1 cP to 100 cP is approximately +1.2 percent and +0.5 percent, if the flowmeter were to be calibrated at 1 cP and 10 cP, respectively. As viscous liquids can exhibit relatively large variations in viscosity over a relatively small temperature range, the inaccuracies caused by viscosity changes may be larger than the stated accuracy of the flowmeter.

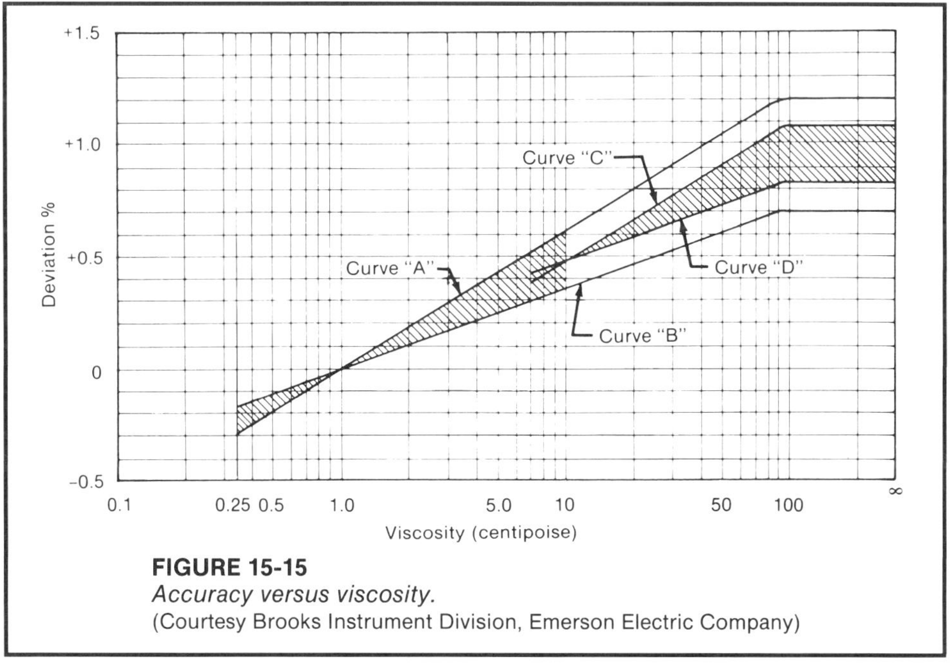

FIGURE 15-15
Accuracy versus viscosity.
(Courtesy Brooks Instrument Division, Emerson Electric Company)

The lubricity of the liquid can adversely affect the turndown of oval gear flowmeters. A liquid that does not lubricate well will cause the maximum rotor speed to be derated to minimize bearing wear, thereby decreasing the maximum operating flow of the flowmeter. Such a liquid makes it difficult for the rotors to operate at low flows, thereby increasing the minimum operating flow of the flowmeter. The net effect can be a significant reduction in the range over which the flowmeter will measure accurately.

The maximum flow through oval gear flowmeters is usually a function of usage. These flowmeters generally have different flow ratings for continuous and intermittent service and an absolute flow limit, all of which are dependent upon the lubricity and viscosity of the liquid. The intermittent service rating is the maximum flow that can be maintained 8 hours per day, while the absolute limit is the point at which the flowmeter will be damaged.

Turndown in an intermittent lubricious liquid application can be as high as 20:1. The same flowmeter in an intermittent non-lubricious liquid application may have a maximum turndown of 4:1. In continuous operation, the turndown is further reduced to approximately 3:1.

Applications Oval gear flowmeters are generally applicable to non-abrasive lubricious liquids with viscosities from approximately 0.2 cP to 500,000 cP. Slippage can pose a problem in low viscosity applications, especially if there is any wear of machine parts. Therefore, most applications are on higher viscosity liquids.

Sizing Oval gear flowmeters are typically sized using capacity data for different services, an example of which is presented in Table 15-2. Note that lubricating liquids such as petroleum-based liquids are generally rated for larger maximum flows than non-lubricating liquids such as hot water.

Figure 15-16 shows the relationship between pressure drop across the flowmeter at maximum flow and viscosity for specific rotors. The factor by which the maximum flow is reduced is shown in Figure 15-17 for two rotor designs.

Figure 15-18 illustrates the effects of viscosity on the accuracy of the flowmeter throughout the flow range, as well as the pressure drop across the flowmeter for various liquids of different viscosities throughout the flow range.

The horizontal accuracy curves show the ability of the flowmeter to accurately measure highly viscous liquids at low flow. The nonlinearities of the low viscosity curves typify the slippage that can occur in this service. The pressure drop curves show how the pressure drop across the flowmeter constrains the maximum operating flow of the flowmeter in high viscosity service.

EXAMPLE 15-5

Problem: Size an oval gear flowmeter for a maximum flow of 70 gpm of an oil that has a viscosity of 2000 cP.

Solution: From the capacity table, a 1-1/2-inch flowmeter, which has a continuous flow range of approximately 4 to 71 gpm, appears to be applicable. Due to viscosity effects, the maximum flow is reduced by a factor of

TABLE 15-2
Over Gear Flowmeter Capacity Data

		Water		*Petroleum products*				
	Liquid viscosity	*140° F*	*140–230° F*	*Up to 0.2 cP*	*0.2–2.0 cP*		*2–200 cP*	
Bore (inches)	*temp.*	*cold water*	*hot water*	*LPG*	*Gasoline*	*Kerosene*	*Light oil*	*Heavy oil*
3/8	Cont.	.016–.177			.016–.117	.008–.117		
	Inter.	.016–.175			.016–.175	.008–.175		
3/8	Cont.	.088–.59			.088–.59	.043–.59		
	Inter.	.088–.88			.088–.88	.043–.88		
1/2	Cont.	0.22–0.97		0.35–1.14	0.26–1.14	0.22–1.14	0.10–1.06	0.04–1.06
	Inter.	0.22–1.41		0.35–1.59	0.26–1.59	0.22–1.59	0.20–1.59	0.04–1.59
1/2	Cont.	0.66–2.64	0.88–2.20	1.32–3.08	0.88–3.08	0.66–3.08	0.31–4.40	0.18–4.40
	Inter.	0.66–2.86	0.88–2.64	1.32–4.40	0.88–4.40	0.66–4.40	0.31–5.28	0.18–5.28
1/2	Cont.	1.32–4.40	1.76–3.52	1.76–5.28	1.76–5.28	1.32–5.28	0.66–7.05	0.35–7.05
	Inter.	1.32–6.60	1.76–4.40	1.76–7.93	1.76–7.93	1.32–7.93	0.66–8.81	0.35–8.81
1	Cont.	2.64–8.80	3.52–7.04	3.52–10.56	3.52–10.56	2.64–10.56	1.32–14.10	0.70–14.10
	Inter.	2.64–13.20	3.52–8.80	3.52–15.86	3.52–15.86	2.64–15.86	1.32–17.62	0.70–17.62
1	Cont.	4.40–22.0	5.28–17.6	7.93–24.2	5.28–24.2	4.40–24.2	1.76–35.2	1.14–35.2
	Inter.	4.40–30.8	5.28–22.0	7.93–37.4	5.28–37.4	4.40–37.4	1.76–44.0	1.14–44.0
	Limit	39.6	26.4	44.0	44.0	44.0	44.0	44.0
1–1/2	Cont.	8.81–44.0	11.0–35.2	15.4–48.4	11.0–48.4	8.81–48.4	3.96–10.5	2.64–70.5
	Inter.	8.81–61.6	11.0–44.0	15.4–70.5	11.0–70.5	8.81–70.5	3.96–88.1	2.64–88.1
	Limit	79.3	52.8	88.1	88.1	88.1	88.1	88.1
2	Cont.	17.6–88.1	22.0–66.0	35.2–96.9	22.0–96.9	17.6–96.9	8.81–141.0	5.28–141.0
	Inter.	17.6–132.0	22.0–88.1	35.2–154.0	22.0–154.0	17.6–154.0	8.81–176.0	5.28–176.0
	Limit	154.0	110.0	176.0	176.0	176.0	176.0	176.0
3	Cont.	35.0–176.0	44.0–141.0	66.0–198.0	44.0–198.0	35.0–198.0	17.6–286.0	11.0–286.0
	Inter.	35.0–264.0	44.0–176.0	66.0–308.0	44.0–308.0	35.0–308.0	17.6–352.0	11.0–352.0
	Limit	308.0	220.0	352.0	352.0	352.0	352.0	352.0

(Courtesy Brooks Instrument Division, Emerson Electric Company)

approximately 0.73 to approximately 52 gpm, so that a 2-inch flowmeter with a capacity of 0.73 × 141, or 103 gpm would be required. A special cut rotor would be required, as the viscosity is greater than 500 cP.

Due to the relatively high viscosity, the minimum measurable flow will also be reduced. Assuming that the minimum flow is reduced by the same factor by which the maximum flow was reduced, the minimum flow is 0.73 × 8.8, or 6.4 gpm, and the turndown is 70/6.4, or 11:1.

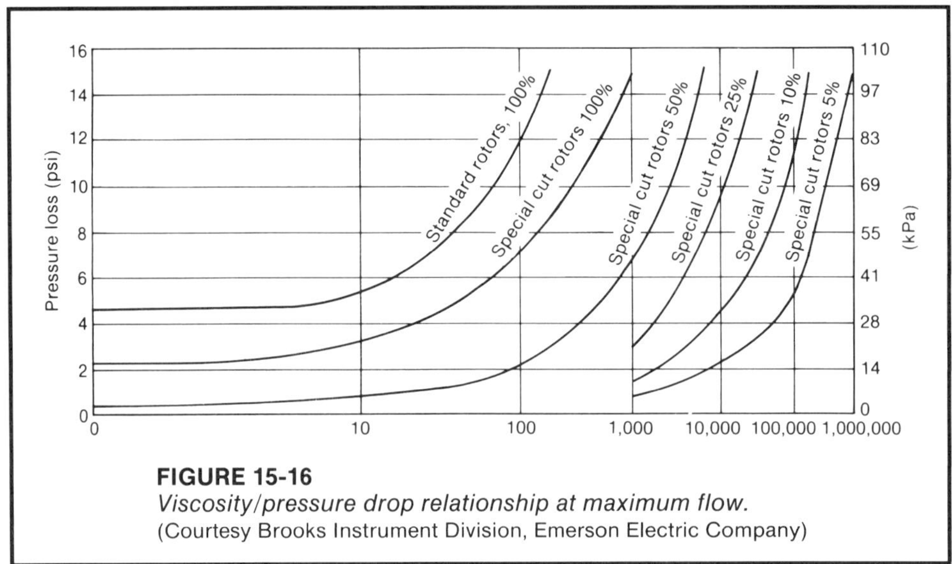

FIGURE 15-16
Viscosity/pressure drop relationship at maximum flow.
(Courtesy Brooks Instrument Division, Emerson Electric Company)

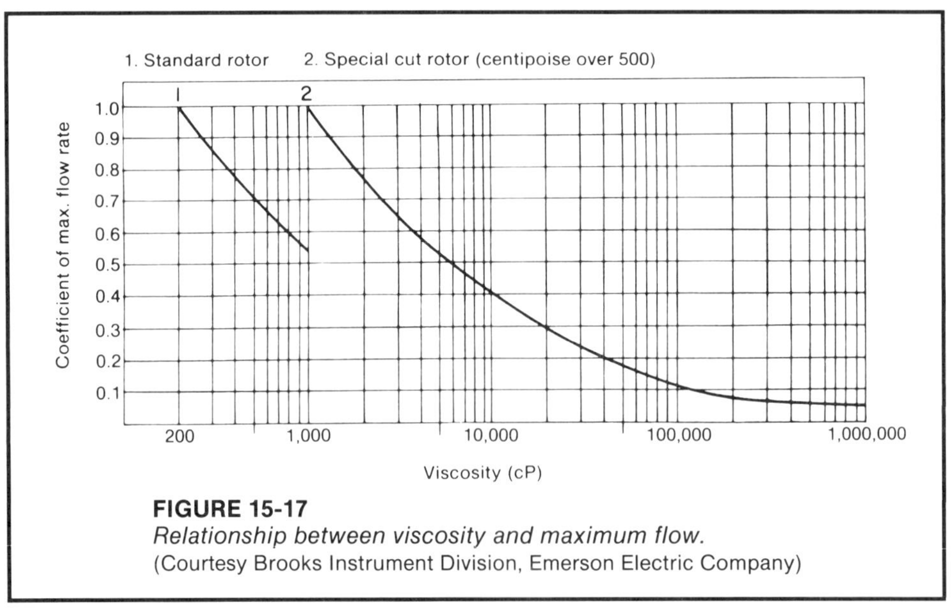

FIGURE 15-17
Relationship between viscosity and maximum flow.
(Courtesy Brooks Instrument Division, Emerson Electric Company)

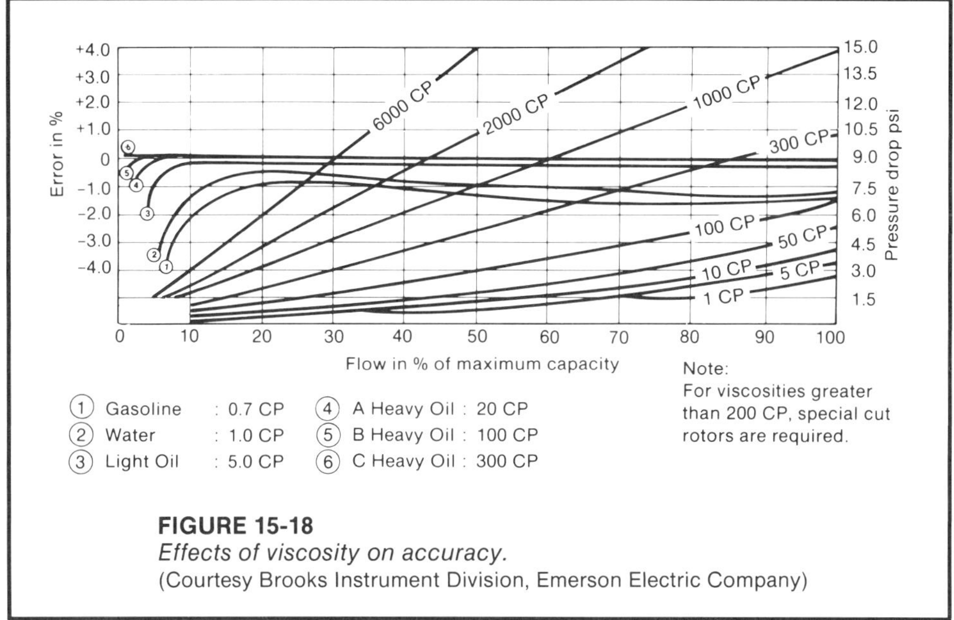

FIGURE 15-18
Effects of viscosity on accuracy.
(Courtesy Brooks Instrument Division, Emerson Electric Company)

Piston Positive Displacement Flowmeter Piston flowmeters are typically used in lower flow viscous liquid service where it is often difficult to apply other flowmeters due to Reynolds number constraints. This positive displacement flowmeter design is not tolerant of dirt in the liquid, as there are small passages that can be easily plugged.

Principle of Operation Piston positive displacement flowmeters utilize four pistons to entrap liquid as it passes through the flowmeter, causing a crankshaft to rotate, as shown in Figure 15-19. The flow through the flowmeter is proportional to the rotational speed of the crankshaft.

Construction The body of the flowmeter is the assembly in which the pistons are mounted and on which the sensing system is housed. O-rings and/or gaskets are used to seal the flowmeter body assembly where wetted parts are removable for access during manufacture and maintenance.

Piston

The pistons rotate on the crankshaft due to the forces exerted by the flow of liquid through the flowmeter. As the pistons form a seal with the cylinder walls as well as seal passages, these parts are manufactured to tight tolerances that must be maintained over the life of the flowmeter.

Bearings

The pistons require a bearing on the crankshaft so that the assembly is free to rotate.

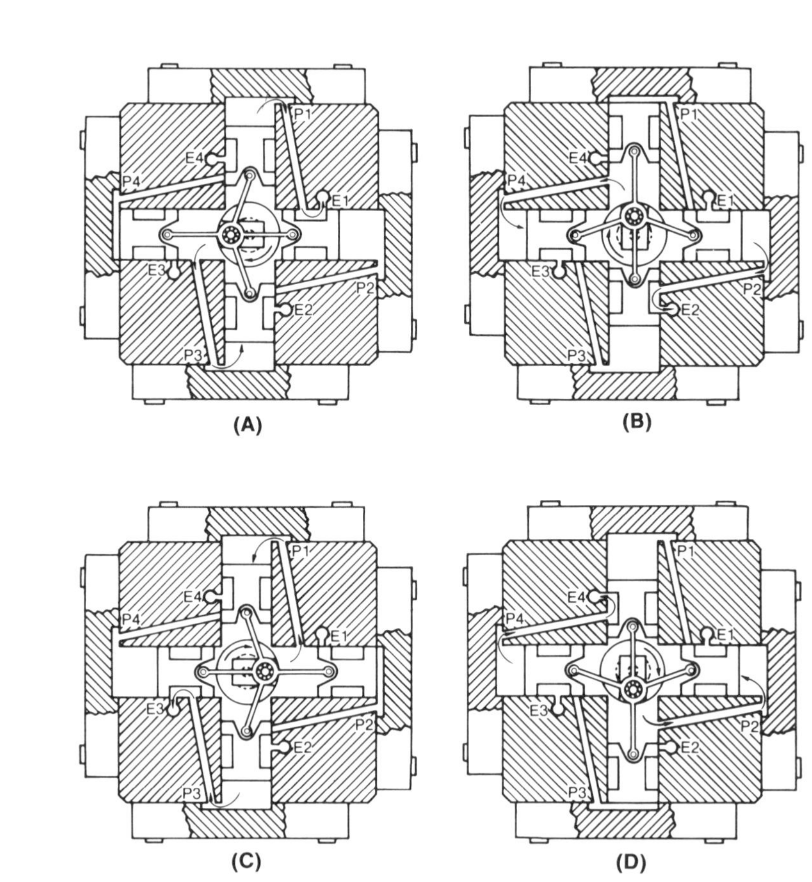

Fluid is admitted to the area surrounding the crankshaft through a port in the inlet block of the transducer assembly. In (A), the position of the crankshaft and pistons is such that port P3 is open to the fluid inlet acting on the face of piston 3. Port P1 is open to exhaust port E1. Thus piston 1 displaces the fluid within its particular chamber through a collector ring connected to the exhaust ports to the outlet of the meter. Movement of piston P3 drives the crankshaft in a clockwise direction.

As the crankshaft reaches position 2, port P4 is now open to inlet pressure driving piston 4 to the right, the crankshaft further clockwise, and piston 1 exhausts fluid from its chamber to the outlet of the flowmeter. Continuing this process in (C), piston 3 is now ported to exhaust connection 3 and inlet fluid ported through P1 drives piston 1 downward.

Finally, in (D), 270° of the 360° possible rotation has been accomplished and piston 2 is driven to the left as a result of application of pressure through port P2 with the exhaust drive from the chamber of piston 4. Thus it can be seen that each piston acts as a 3-way valve to port the counterclockwise adjacent piston.

FIGURE 15-19
Piston flowmeter operation.
(Courtesy Fluidyne Instrumentation)

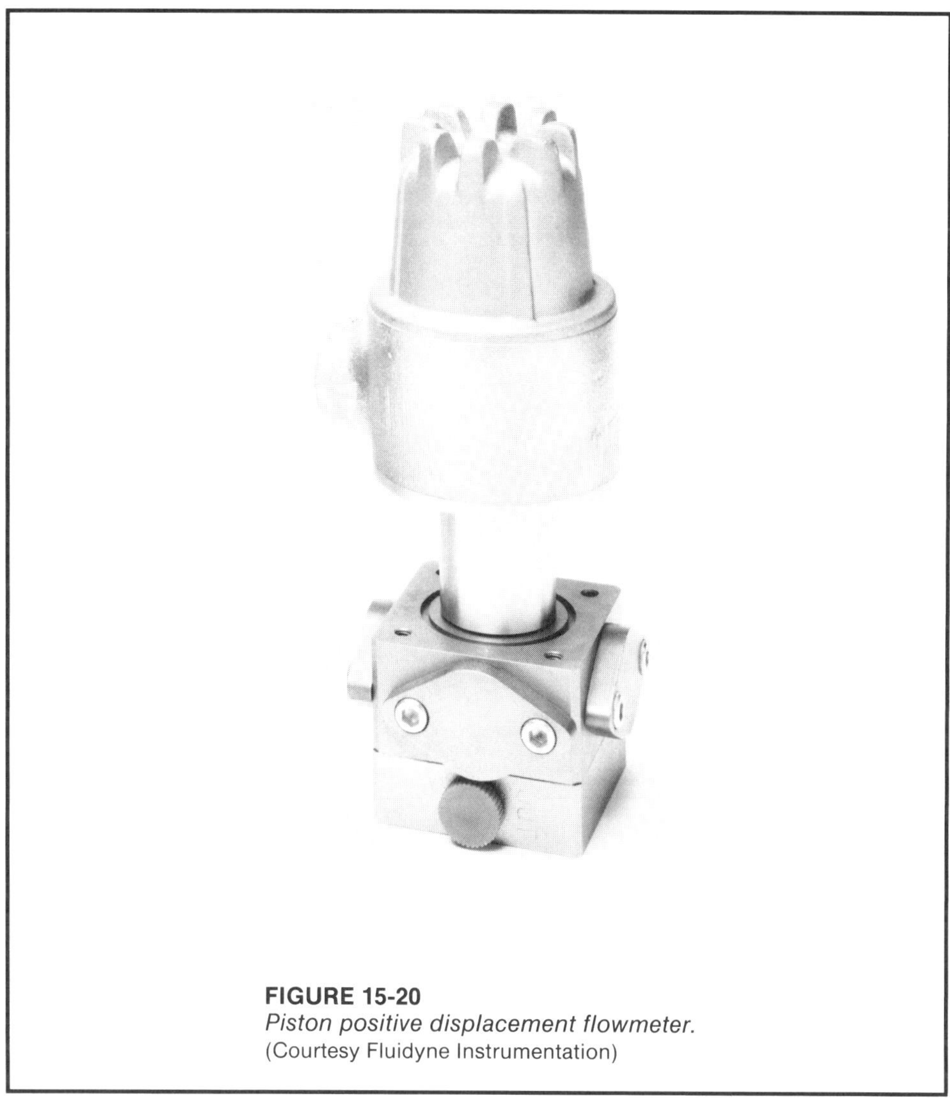

FIGURE 15-20
Piston positive displacement flowmeter.
(Courtesy Fluidyne Instrumentation)

Sensing System

Tachometer and optical sensing systems are used in piston flowmeter designs. The tachometer sensing system employs a driving magnet that rotates in a tachometer assembly, thereby generating an analog signal, which is proportional to the rotational velocity of the crankshaft, as shown in Figure 15-21.

The optical sensing system utilizes a magnetically driven optically encoded disc, the rotation of which is sensed by an optical pickup so as to sense a pulse each time a portion of a revolution occurs, as shown in Figure 15-22.

Bidirectional sensing systems that sense the direction of flow and subtract the reverse flow from the forward flow pulses are also available. Sensing is done optically as shown in Figure 15-23.

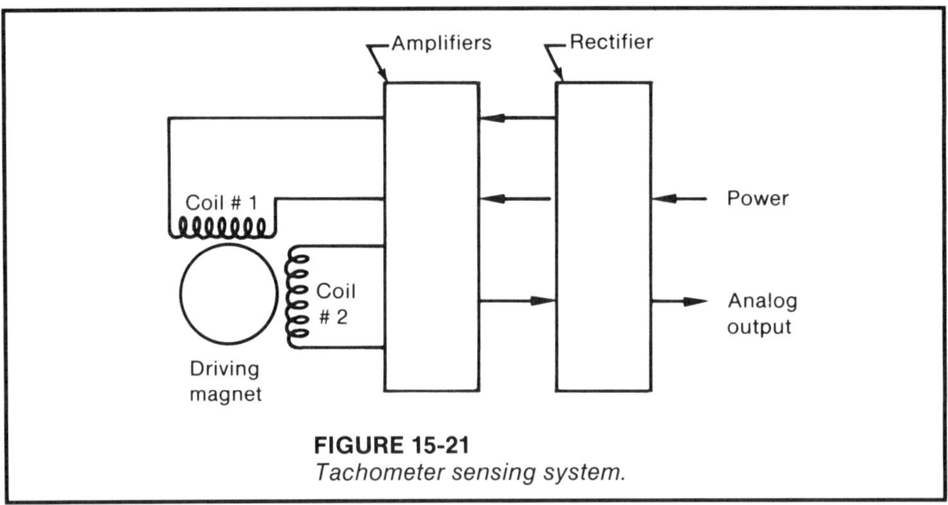

FIGURE 15-21
Tachometer sensing system.

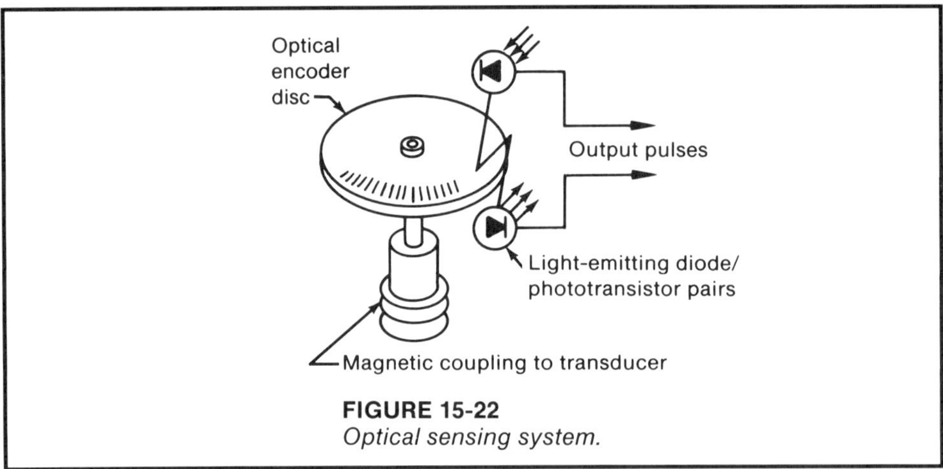

FIGURE 15-22
Optical sensing system.

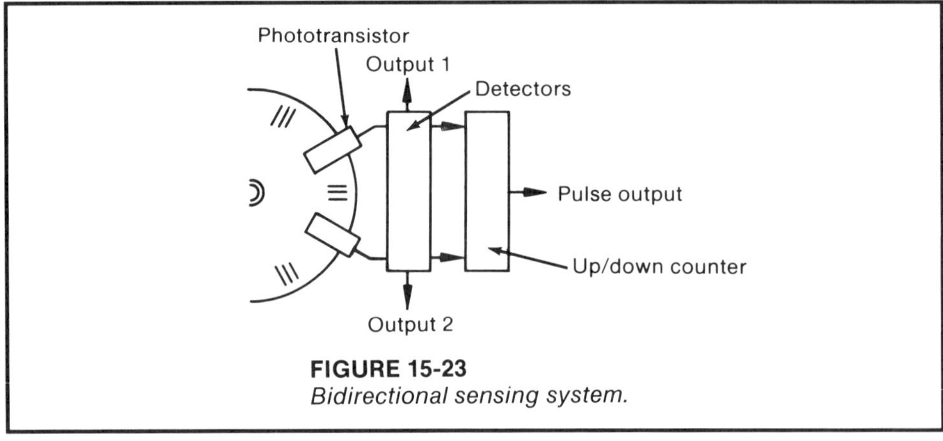

FIGURE 15-23
Bidirectional sensing system.

Wetted Parts

Wetted parts of piston flowmeters include the body, pistons, O-rings, crankshaft, and bearings. Bodies and crankshafts are available in stainless steel. Standard bearings are typically made of stainless steel, the grade of which may be unsuitable for some common applications such as water, aqueous solutions, bases, or salts. While carbon bearings are available to handle additional applications, piston flowmeters are typically limited by the bearing materials of construction.

Careful investigation should be made of the compatibility of the materials of construction of each component.

Operating Constraints Piston flowmeters, which are available in 1/8-in. to 3/4-in. sizes, are generally pressure-limited to approximately 1000 to 3000 psig and temperature limited to the ratings of the sensor, which can be up to approximately 200°C in certain applications. Flow can range from 0.26 to 25 gpm.

Pressure drop across piston flowmeters is typically kept below approximately 30 psid in order to avoid excessive bearing wear and premature bearing failure.

Performance Volumetric flows can be measured with an accuracy of approximately ±0.5 to 1 percent rate, depending on the application. Nonviscous flows are generally measured less accurately than viscous flows due to errors caused by increased slippage through the flowmeter at low viscosities.

The accuracy statements above represent ideal operating conditions. Changes in viscosity can cause shifts in the accuracy of the flowmeter.

Figure 15-24 illustrates how accuracy is affected by viscosity changes. Note that the graph shows the accuracy as a function of the maximum rated flow of the flowmeter and not of the desired flow range. As viscous liquids can exhibit relatively large variations in viscosity over a relatively small temperature range, the inaccuracies caused by viscosity changes may be larger than the stated accuracy of the flowmeter.

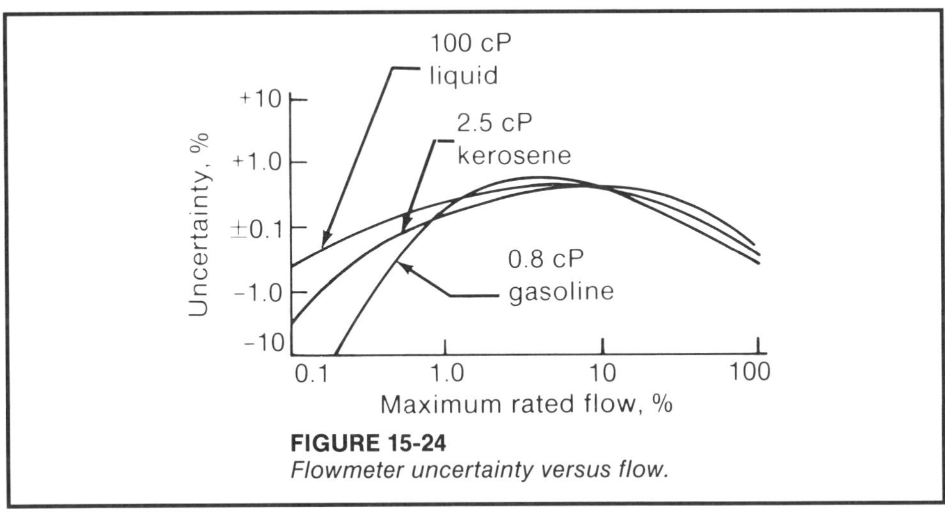

FIGURE 15-24
Flowmeter uncertainty versus flow.

Low liquid lubricity can adversely affect piston tolerances, which must be maintained in order to maintain accuracy. Turndown can be as high as 100:1 in certain applications, although lower turndowns are typical of actual applications.

Applications Piston flowmeters are generally applicable to clean non-abrasive lubricious liquids with viscosities from approximately 0.5 cP to 10,000 cP. Slippage can pose a problem in low viscosity applications, especially if there is any wear of machined parts, and most applications are on medium viscosity liquids.

Sizing Figure 15-25 illustrates the relationship between flowmeter capacity, maximum pressure drop across the flowmeter, and viscosity. Such curves can be used to size piston flowmeters.

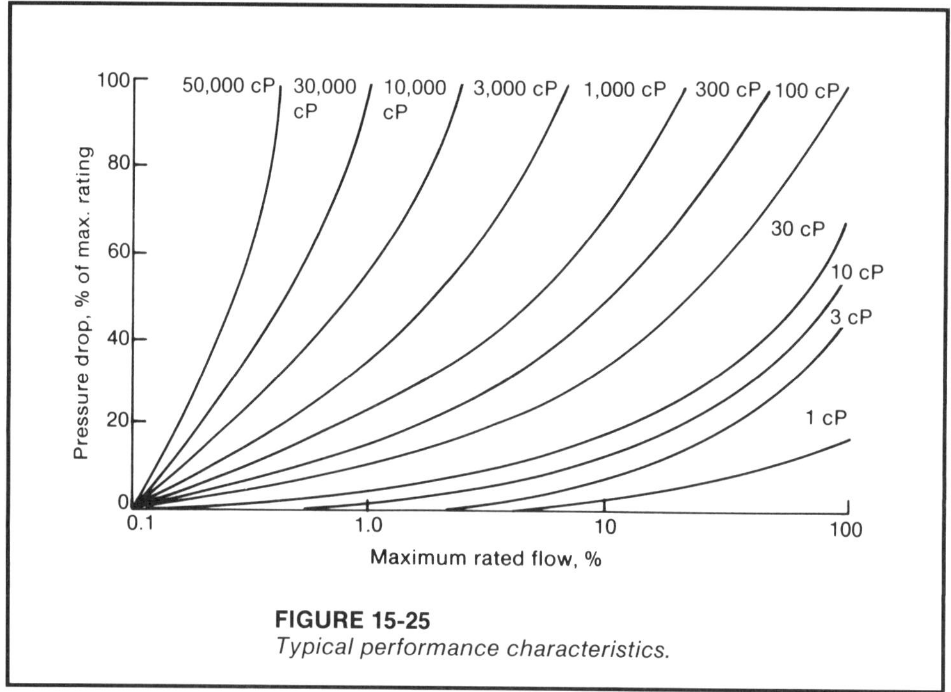

FIGURE 15-25
Typical performance characteristics.

Piston flowmeters generally have different flow ratings for continuous service and for intermittent service, as well as an absolute flow limit above which the flowmeter will be damaged. Operating under intermittent flow conditions can result in significant reduction in the life of the flowmeter. Therefore the continuous flow ratings should be used in sizing piston flowmeters.

The relatively horizontal accuracy curves show the ability of the flowmeter to accurately measure highly viscous liquids at low flows. The nonlinearities of the low viscosity curves typify the slippage that can occur in this service. The pressure drop curves show how the pressure drop across the flowmeter constrains the maximum operating flow of the flowmeter in high viscosity service.

EXAMPLE 15-6

Problem: Size a piston flowmeter for a maximum flow of 1 gpm of a liquid with a viscosity of 3000 cP, given the following flowmeter capacity data:

Size	Maximum Flow
1/8 in.	0.32 gpm
3/8 in.	1.58 gpm
1/2 in.	7.93 gpm
3/4 in.	26.42 gpm

Solution: A 3/8-inch flowmeter would appear to be applicable, but examination of the typical performance graph shows that the maximum differential pressure limits the maximum flow to approximately 8 percent of the flowmeter capacity, or 0.13 gpm. A 1/2-inch flowmeter is similarly limited to 0.63 gpm. A 3/4-in. flowmeter is limited to 2.11 gpm and would be applicable. The 3/4-in. flowmeter has a minimum flow that is approximately 0.1 percent of flowmeter capacity, or 0.026 gpm, resulting in a turndown of 1 gpm/0.026 gpm, or 38:1.

Rotary Positive Displacement Flowmeter

Rotary flowmeters are typically used on liquid service where accuracy is of importance but turndown is not. This positive displacement flowmeter design is somewhat tolerant of dirt since there are few passages that are easily plugged. However, large or abrasive solids can be compressed between the internal parts and the flowmeter body, thereby distorting the internal parts and the seal that must be made to achieve flowmeter accuracy.

Principle of Operation

Rotary positive displacement flowmeters can be implemented in a number of ways. Each converts the entrapment of liquid into a rotational velocity proportional to the flow through the flowmeter, as shown in Figure 15-26.

As the flowmeter entraps a fixed quantity of liquid each time the meter is rotated, the flow is proportional to the rotational velocity of one of the rotating parts.

Construction

The body is the housing in which the rotating parts are mounted and on which the sensing system is housed. O-rings and/or gaskets are used to seal the flowmeter body assembly where wetted parts are removable for access during manufacture and maintenance.

Internal Parts

The internal parts of the flowmeter rotate on shafts due to the forces exerted by the flow of liquid through the flowmeter. As the internal parts must form a seal with each other as well as with the flowmeter body, these parts are manufactured to tight tolerances that must be maintained over the life of the flowmeter. Where more than one part rotates, the parts are typically geared together.

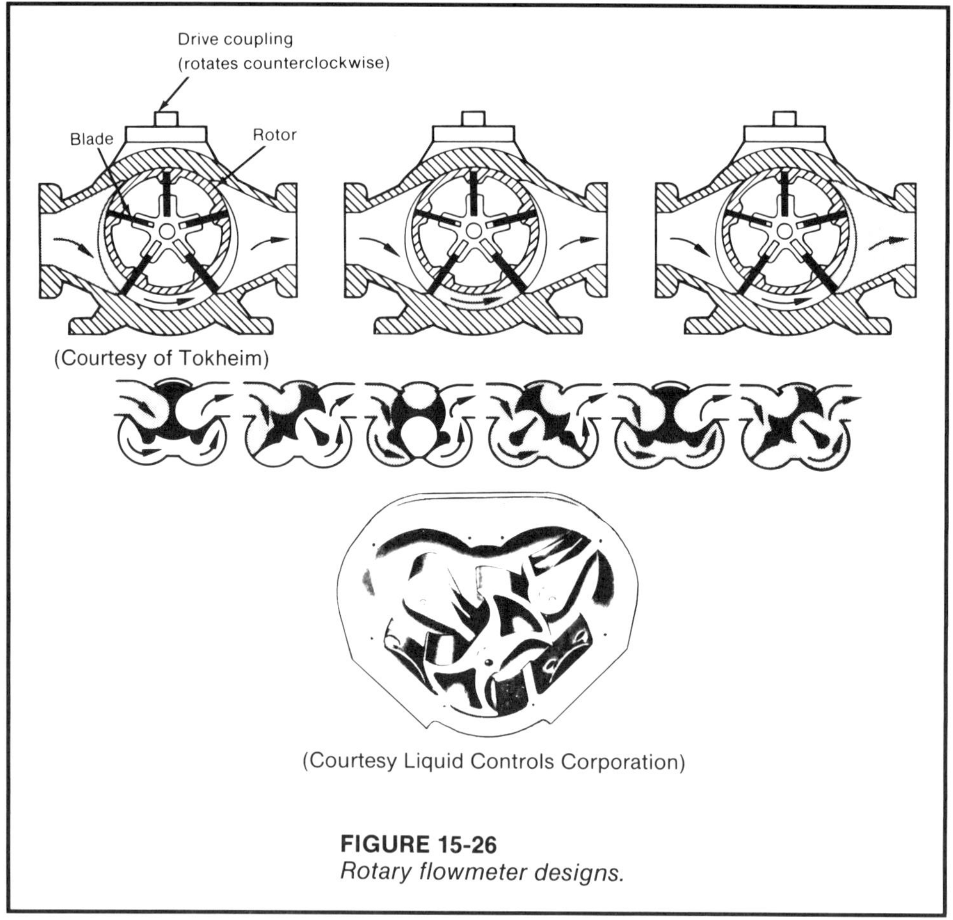

(Courtesy of Tokheim)

(Courtesy Liquid Controls Corporation)

FIGURE 15-26
Rotary flowmeter designs.

Sensing System

Rotation of the internal parts of the flowmeter is mechanically coupled, either by gears or by the shaft of one of the rotating parts, external to the flowmeter. The rotation of the shaft is sensed by a transmitter that typically outputs a pulse for each revolution of the shaft external rotating part.

Wetted Parts

Wetted parts include the body, O-rings, and other internal parts. Rotary flowmeter bodies and metal parts are generally available in carbon steel and stainless steel. The internal parts may be constructed of the same metal as the body or of other materials such as Teflon® or Ryton®. O-rings are constructed of materials such as Viton®, BUNA-N®, neoprene, and Teflon®.

It should be noted that the compatibility of the materials of construction with each component should be investigated, as the effects of wear and corrosion on the performance of the flowmeter are significant when slippage becomes excessive, parts fail, or seals leak.

Operating Constraints Rotary flowmeters, which are available in 1-in. to 6-in. sizes, are usually pressure-limited to 150 psig, although higher ratings are available. Flow measurement ranges are from 5 to 1000 gpm, while the maximum operating temperature is approximately 230°C.

The liquid to be measured lubricates the internal rotating parts of the flowmeter; therefore, the life of the flowmeter can be diminished by measuring liquids that are not lubricious.

Performance This design can measure volumetric flows with an accuracy that ranges from approximately ±0.1 to 0.2 percent rate, depending on the application. Nonviscous flows are generally measured less accurately than viscous flows due to errors caused by increased slippage through the flowmeter at low viscosities.

The accuracy statement above represents ideal operating conditions. Changes in viscosity can cause shifts in the accuracy of the flowmeter due to varying amounts of slippage at different viscosities. As viscous liquids can exhibit relatively large variations in viscosity over a relatively small temperature range, the inaccuracies caused by viscosity changes may be larger than the stated accuracy of the flowmeter. It should be noted that the performance of some rotary flowmeters can be within the above stated accuracy specifications on liquids less than 1 cP.

The maximum turndown of this type of flowmeter is typically in the range of 5 to 10:1 when the desired range coincides with the flowmeter capacity. However, practical applications usually have less than the above stated turndowns.

Applications Rotary flowmeters are generally applicable to clean non-abrasive liquids with viscosities from approximately less than 1 cP to 25,000 cP. Slippage can pose a problem in low viscosity applications, especially if there is any wear of machined parts. Therefore, many applications are on higher viscosity liquids, but good accuracy can be achieved in low viscosity applications.

Sizing Figure 15-27 illustrates the accuracy across a typical rotary flowmeter for liquids throughout the flow range.

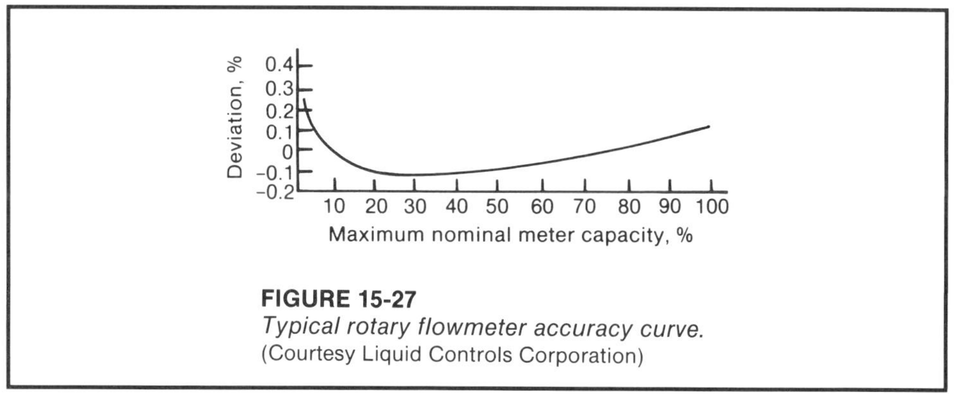

FIGURE 15-27
Typical rotary flowmeter accuracy curve.
(Courtesy Liquid Controls Corporation)

The pressure drop across the flowmeter may constrain the maximum operating flow of the flowmeter in high viscosity service. A graph of the pressure drop across a typical rotary flowmeter and a graph used to obtain a conversion factor for viscosities above 1 cP are shown in Figure 15-28.

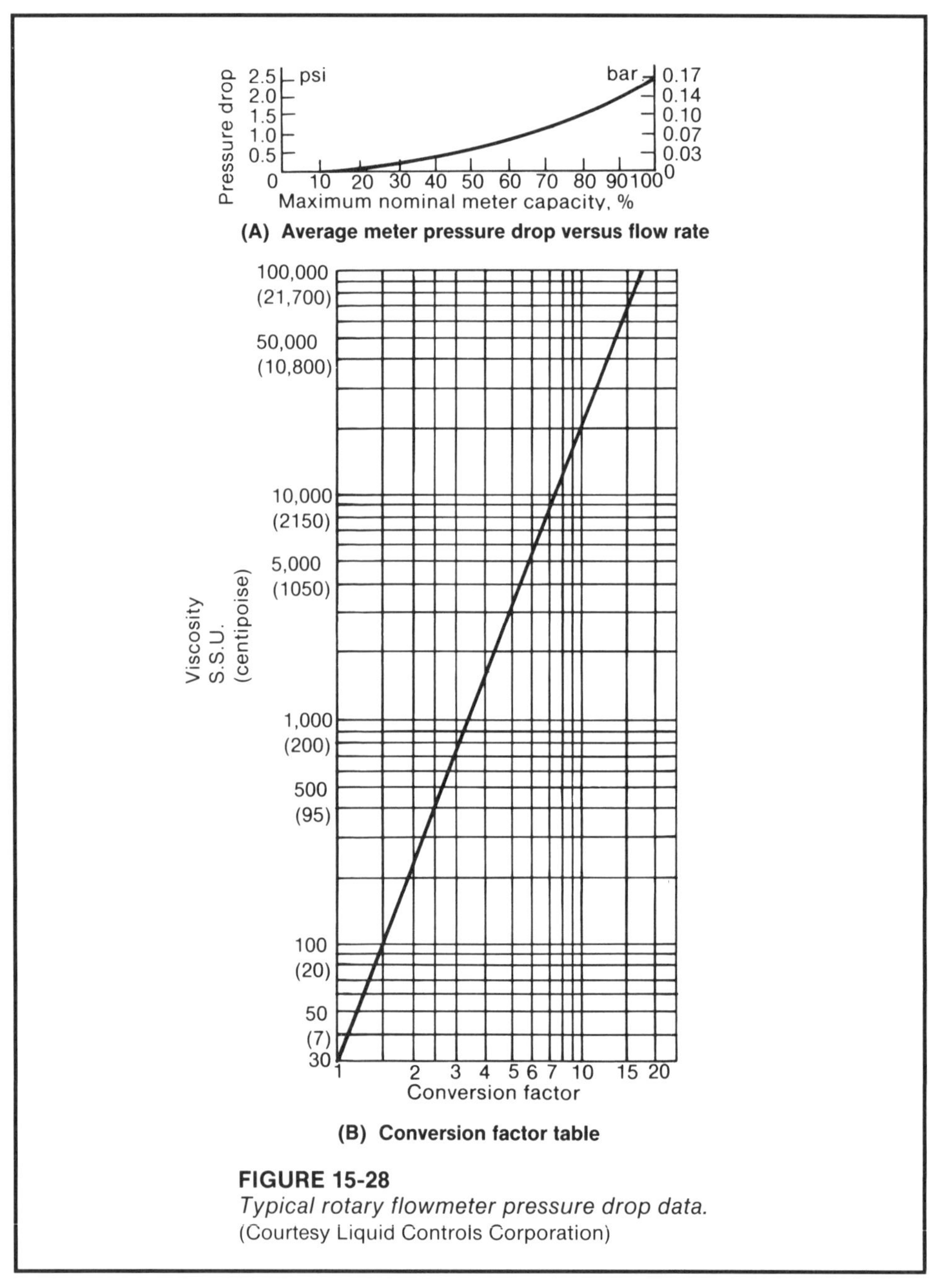

(A) Average meter pressure drop versus flow rate

(B) Conversion factor table

FIGURE 15-28
Typical rotary flowmeter pressure drop data.
(Courtesy Liquid Controls Corporation)

EXAMPLE 15-7

Problem: Size a rotary flowmeter for a maximum flow of 100 gpm of a liquid with a viscosity of 4000 cP when the maximum allowable pressure drop across the flowmeter is 15 psi, given the following flowmeter capacity data:

Size	Maximum Flow
3/4 in.	30 gpm
1-1/2 in.	50 gpm
2 in.	100 gpm
3 in.	200 gpm
4 in.	350 gpm

Solution: The conversion factor for calculating the pressure drop across the flowmeter is approximately 9: the pressure drop is 9 times that shown in Figure 15-28. A 2-inch flowmeter would appear to be applicable; however, it is operated at 100 percent of flowmeter capacity and the pressure drop is 9×2.2 psi or 19.8 psi. A 3-inch flowmeter would operate at 50 percent of its capacity and would result in a pressure drop of 9×0.5 psi or 4.5 psi, which is acceptable.

It should be noted that rotary flowmeters of different type of manufacture may have different sizing procedures.

EXERCISES

15.1 Size a helical gear flowmeter for a maximum flow of 300 gpm of a liquid with a viscosity of 500 cP.

15.2 Size a nutating disc flowmeter to measure a full scale flow of 0 to 50 gpm of hot water.

15.3 Size an oscillating piston flowmeter for a maximum flow of 15 gpm of a lubricating liquid with a viscosity of 200 cP.

15.4 Size an oval gear flowmeter for a maximum flow of 10 gpm of an oil that has a viscosity of 200 cP.

15.5 Size a piston flowmeter for a maximum flow of 5 gpm of a liquid with a viscosity of 300 cP.

15.6 Size a rotary flowmeter for a maximum flow of 30 gpm of a liquid with a viscosity of 2000 cP when the maximum allowable pressure drop across the flowmeter is 5 psi.

16

Target Flowmeters

Introduction Target flowmeters represent a viable economic alternative to the measurement of liquid and gas flow streams, especially in large pipe sizes. While the principle of operation of target flowmeters remains essentially the same regardless of manufacture, performance claims will vary considerably.

Target flowmeter technology is usually considered a means by which to achieve only a rough flow measurement, but this degree of accuracy is often all that is necessary. Target flowmeters can be used in many applications, but Reynolds number, velocity, materials of construction constraints, and drift constraints can be encountered.

Principle of Operation Target flowmeters operate on the principle of the measurement of the force exerted on a body, called the target, suspended in the flowstream. Acceleration of the fluid around the target results in a reduced pressure at the rear of the target. The force exerted on the target is the difference between the upstream and downstream pressures, integrated over the area of the target.

In the turbulent flow regime, the force is represented as:

$$F = \text{constant} \times \rho \times A_{\text{target}} \times v^2$$

The flowmeter output, which is a linear representation of the force exerted on the target, is proportional to the square of the velocity and, hence, the square of the flow. In the laminar regime, the force can be represented by

$$F = \text{constant} \times \mu_{\text{cP}} \times v \times A_{\text{target}}/G$$

The flow is therefore directly proportional to the force exerted on the target and directly proportional to the viscosity of the fluid. As small temperature changes can cause large fluctuations in viscosity, operation is generally not linear in the nonturbulent flow regime.

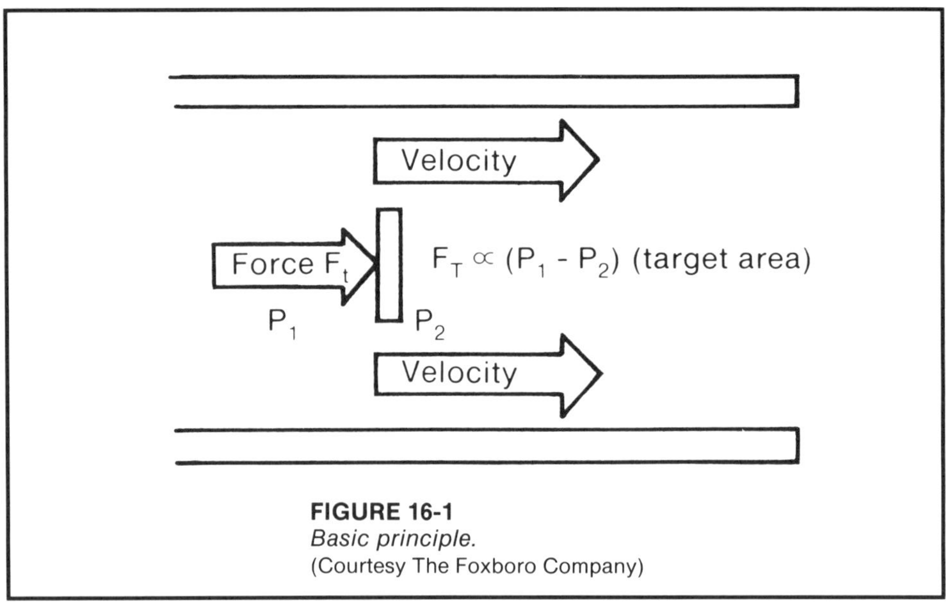

FIGURE 16-1
Basic principle.
(Courtesy The Foxboro Company)

EXAMPLE 16-1

Problem: Determine the percentage change in flow measurement if the specific gravity increases by 2 percent when flow is in the turbulent flow regime.

Solution: Solving for the velocity,

$$v = [F/(\text{constant} \times A_{\text{target}} \times \rho)]^{1/2}$$

In a squared output flowmeter, a 1 percent change in specific gravity is estimated to affect the flow measurement by -1/2 percent. Therefore, the 2 percent increase in specific gravity will cause an estimated 1 percent decrease in the flow measurement.

Construction The body of the flowmeter is the housing, which is typically of stainless steel construction and is inserted into the pipe as an in-line wafer or with screwed or flanged connections. Target flowmeters for larger size pipes, considered insertion flowmeters, are usually screwed into a flange or a coupling in the pipe and typically have no body.

Target The target, typically of stainless steel construction, is the drag body that is inserted in the flowstream upon which the force is exerted. It is typically a round disc positioned with its front face at right angles to the direction of flow. The beta ratio of the target flowmeter is illustrated in Figure 16-3.

FIGURE 16-2
Target flowmeter construction.
(Courtesy Hersey)

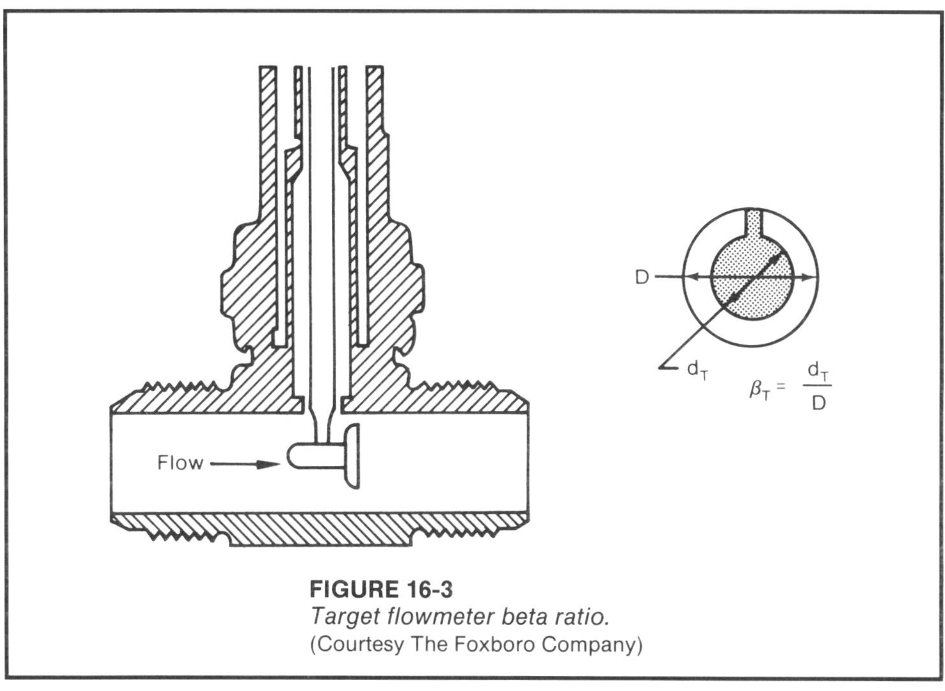

FIGURE 16-3
Target flowmeter beta ratio.
(Courtesy The Foxboro Company)

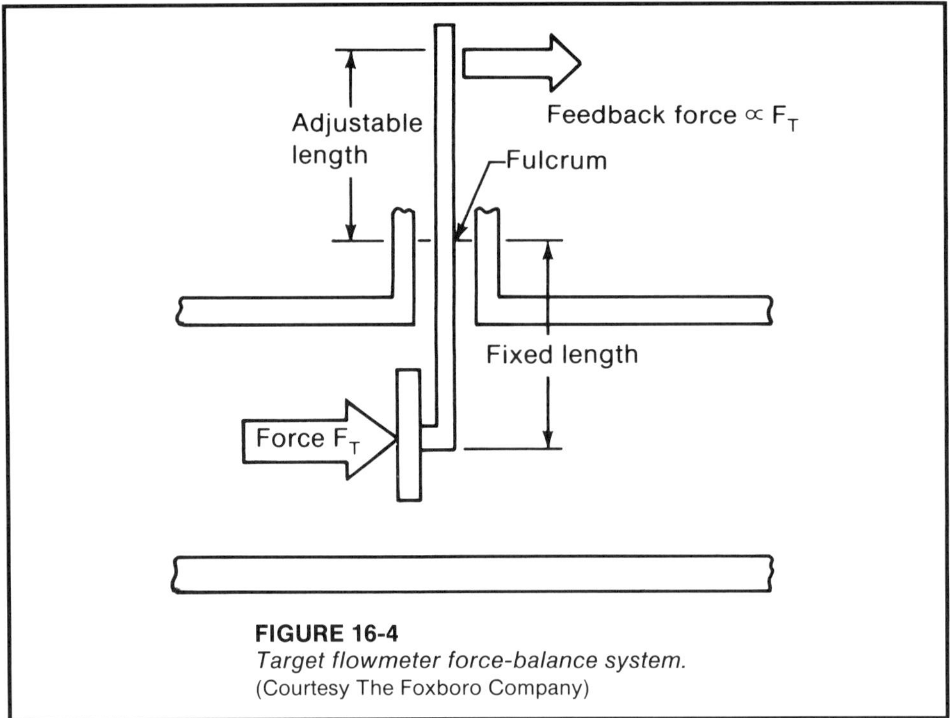

FIGURE 16-4
Target flowmeter force-balance system.
(Courtesy The Foxboro Company)

Seals Most designs are sealed either mechanically or hermetically. Seals are required in some designs to prevent leakage of the flowmeter. Such leakage can occur where the arm that links the target in the flowstream to the sensing system located outside the pipe passes through the pipe wall.

Sensing Systems Sensing the force that is created by the differential pressure developed across the target is typically accomplished using force balance or strain gage techniques. In the force-balance system, the target and its connecting rod are constructed in a lever configuration.

The displacement of the arm external to the flowstream is proportional to the force exerted on the target. A balancing device, whose output is directly proportional to the torque necessary to position the arm in the balanced position, is employed to generate the flowmeter output.

Strain gage techniques translate the force exerted on the target to the arm external to the flowstream where strain gages are located to sense the magnitude of the forces. The output of the flowmeter is derived from the output of the strain gages.

Wetted Parts Wetted parts include the body, the target, and any seals or gaskets that may be required. Metal parts are typically of stainless steel construction, although other materials such as Hastelloy® and Inconel® are available from some manufacturers.

The compatibility of the materials of construction of each component should be investigated, as the effects of wear, coating, and corrosion on the performance of the flowmeter are significant.

Operating Constraints Target flowmeters, which are available in virtually any pipe size above approximately 3/8 inch, are generally pressure-limited by the flange rating, but some designs are rated for as high as 10,000 psig. The temperature limit is set by the flange rating and the sensor design to 150°C. However, certain designs can handle up to 400°C applications. Applicable measurable flows range from 0.1 gpm and up, as the target flowmeter can be inserted in large diameter pipes.

The long-term accuracy can be shifted by applications in which the pipe wall can become caked or the target can be eroded away, significantly changing the beta ratio. Vibration can cause the flowmeter to generate false flow signals, while specific gravity shifts will affect the output of the flowmeter by approximately -1/2 percent rate for each 1 percent shift in specific gravity.

Performance Accuracy specifications vary substantially with manufacturer and are typically between ±1 and ±6 percent full scale over a 3.5 to 10:1 turndown. Accuracies of ±1/2 to 1 percent FS can be achieved with a water calibration of the flowmeter. Linearity of target flowmeters is approximately 0.1 percent FS, while the repeatability is approximately ±0.25 percent FS. The flowmeter span can be influenced by ambient temperature up to ±0.75 percent per 55°C in some designs. Static pressure can affect the zero by up to ±0.1 percent per 100 psi.

As the equation for the force generated by the target is different for the laminar and the turbulent flow regimes, the target flowmeter is generally applied to flows in the turbulent regime above a Reynolds number of approximately 4000, such that the force is proportional to the square of the flow and not affected by viscosity.

EXAMPLE 16-2

Problem: Calculate the turndown of a 2-inch target flowmeter based on Reynolds number for a full scale flow of 100 gpm of a liquid with a specific gravity of 1.03 and a viscosity of 4.0 cP.

Solution: At full scale, Reynolds number is

$$R_D = (3160 \times 100 \times 1.03) / (4.0 \times 2.067) = 39{,}366$$

Assuming that a target flowmeter will operate accurately above a Reynolds number of 4000, the turndown is 39,366/4000, or 9.8:1. This turndown, which is based upon pipe hydraulics, may not be achieved as the flowmeter turndown is also limited by the turndown specifications of the flowmeter and is typically 3.5:1.

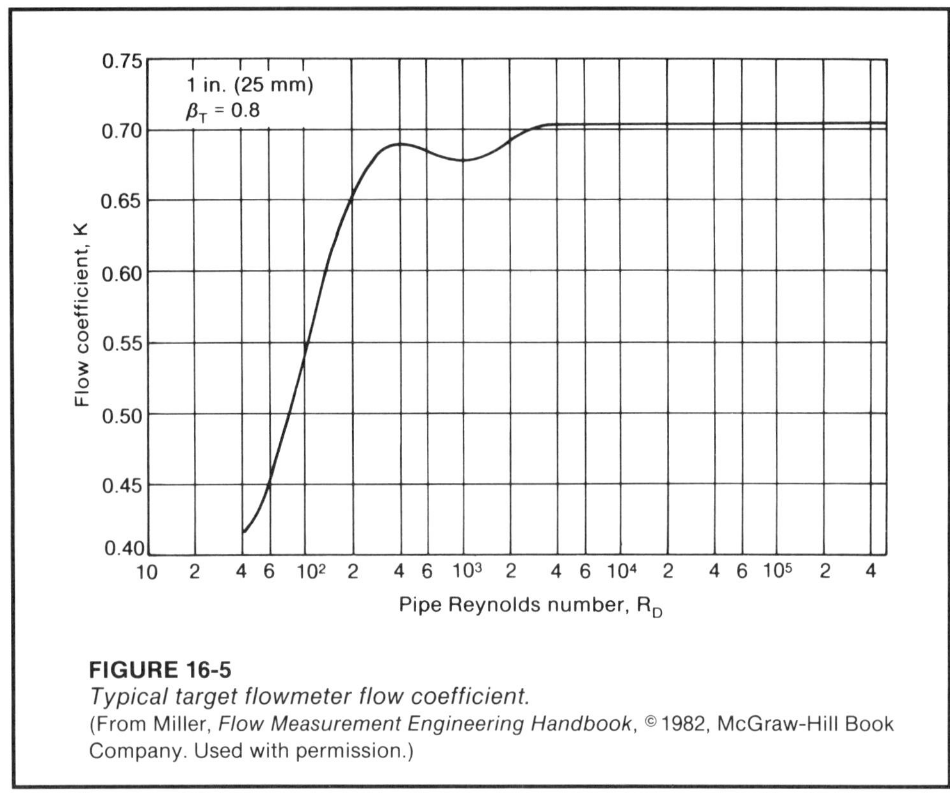

FIGURE 16-5
Typical target flowmeter flow coefficient.
(From Miller, *Flow Measurement Engineering Handbook*, © 1982, McGraw-Hill Book Company. Used with permission.)

Applications Target flowmeters are applied to fluids that have sufficient momentum to exert enough force on the target for the sensing system to operate.

Applications in which the pipe wall can become caked or the target can be eroded away, significantly changing the beta ratio, should be avoided. The effects of a change in beta ratio can shift the long-term accuracy of the flowmeter and make it difficult, if not impossible, to calibrate accurately.

Sizing Methods for sizing target flowmeters vary with manufacturer, but the basic concept is to select the appropriate size based on water and air flow test data. Calculations are used to relate the performance of the flowmeter to the properties of the fluid being measured. If one or more of the measurement requirements are not met by the selected size flowmeter, such as minimum and maximum flows, the process is repeated.

The Reynolds number for all measurable flows should be kept above a minimum of approximately 4000 to maintain the accuracy of the flowmeter. Care should be taken in the measurement of liquids with viscosities of greater than 1 cP.

Table 16-1 is an example of target flowmeter sizing data for air and water service.

TABLE 16-1
Flowmeter Size, Target-to-Bore (Beta) Ratio, Input Range Limits, and Target Force Limits

Flowmeter size, nominal		Target-to-bore, beta ratio	Flow ranges at referenced conditions; lower range value is zero, upper range value selected between — Based on water[b]			Base on air		Target force limits select calibration upper range value between[e]	
mm	*in.*	*d/D*[a]	m^3/h	*L/s*	*U.S. gpm*	m^3/h[c]	*scfh*[d]	*N*	*lb*
15	1/2	0.800	1.6 and 3.4	0.44 and 0.95	7 and 15	44 and 95	420 and 900	6.67 and 68.06	1.5 and 15.3
25	1	0.800	2.7 and 8.4	0.76 and 2.33	12 and 37	75 and 235	700 and 2210	5.78 and 58.27	1.3 and 13.1
50	2	0.658 0.807	12 and 35 5.3 and 19	3.33 and 9.77 1.51 and 5.35	53 and 155 24 and 85	333 and 980 150 and 537	3130 and 9215 1395 and 5040	8.45 and 85.40	1.9 and 19.2
80	3	0.523 0.750	27 and 87 11.1 and 36	7.57 and 24.3 3.09 and 9.77	120 and 385 49 and 155	755 and 2430 310 and 984	7085 and 22,810 2900 and 9240	7.12 and 72.06	1.6 and 16.2
100	4	0.414 0.665	46 and 152 19.4 and 61	13.3 and 42.3 5.42 and 17.0	210 and 672 86 and 270	1329 and 4250 550 and 1720	12,480 and 39,890 5085 and 16,120	6.23 and 60.94	1.4 and 13.7

[a] Beta ratio (β) is target diameter (d) divided by bore diameter (D).

[b] Water at base and flowing temperature 15.6° C (60° F), and specific gravity (relative density) = 1.00.

[c] Air at base and flowing conditions of 0° C, 1 bar (100 kPa), and specific gravity = 1.00.

[d] Air at base conditions of 59° F, 14.7 psia, and specific gravity = 1.00; and flowing conditions of 59° F, 1 psia, and 89 = 1.00.

[e] Lower range values are zero; upper range values are listed. These upper range values of target force limits do not necessarily correspond exactly to the listed upper range value flow rates, which are practical limits.

(Courtesy The Foxboro Company)

To estimate the correct flows in Table 16-1 for other liquids and gases, the following relationships are used:

$$Q_{liquid} = Q_{water} \times SG^{1/2}/SG_{base}$$

$$Q_{gas} = Q_{air} \times [(T_{base} \times P) / (T \times SG)]$$

Viscosity, Reynolds number, and gas compressibility factors are taken into account in precise calculations performed by the manufacturer.

EXAMPLE 16-3

Problem: Calculate the full scale limits of a 1-inch target flowmeter for a liquid that has specific gravity of 1.5 at base and 1.46 at flowing conditions, and a viscosity of 1.0 cP.

Solution: The equivalent water flow limits for this flowmeter are 12 and 37 gpm. Therefore,

$$Q_{min} = 12 \times \frac{1.46^{1/2}}{1.50} = 9.7 \text{ gpm}$$

$$Q_{max} = 37 \times \frac{1.46^{1/2}}{150} = 29.8 \text{ gpm}$$

Reynolds number at 9.7 gpm can be calculated as:

$$R_D = 3160 \times 9.7 \times 1.46 / (1.0 \times 1.049) = 42662$$

which is sufficient to operate the flowmeter in the turbulent flow regime.

EXAMPLE 16-4

Problem: Calculate the full scale limits of a 2-inch target flowmeter with a beta ratio of 0.807 for a liquid that has a specific gravity of 1.1 at a pressure of 50 psig and a temperature of 150°F.

Solution: The equivalent air flow limits for this flowmeter are 1395 and 5040 scfh. Therefore,

$$Q_{min} = 1395 \times \frac{[519 \times (50 + 14.7)]}{[(150 + 460) \times 1.1]^{1/2}} = 9868 \text{ scfh}$$

$$Q_{max} = 5040 \times \frac{[519 \times (50 + 14.7)]}{[(150 + 460) \times 1.1]^{1/2}} = 35{,}654 \text{ scfh}$$

Installation Typically of wafer, flange, or screwed design, target flowmeters are inserted directly into the pipe.

Hydraulic Requirements There are upstream and downstream piping requirements due to the sensitivity of the flowmeter to the velocity profile and Reynolds number dependency of the design. The requirements vary between the lengths required for an orifice plate with a beta ratio of between 0.6 and 0.8, and 9 to $50D/4$ to $8D$, depending upon size and manufacture.

Piping The target of the flowmeter should be properly aligned in the pipe in order to perform accurately. This is not normally a problem, as the body usually aligns the flowmeter within the allowable tolerances.

Piping Vibration Since this device measures forces that are generated in one plane, vibration in that plane and other perturbations can also be measured. Care should be taken to locate the flowmeter in the pipe such that pipe vibration, pulsation, and turbulence are minimized. Some designs electronically filter the signal to aid in minimizing the effects of these factors.

Cabling Target flowmeters are available with 2-wire and 4-wire designs. Some 4-wire designs require that only two wires be run between the flowmeter primary and remote-mounted electronics.

Maintenance

No routine maintenance other than routine calibration checks is required, but problems such as sensor failure, drift, wear, coating, leaks, and electronic failures can occur.

Target Wear Target wear can cause a gradual shift in the calibration of the flowmeter. The calibration can sometimes be modified to compensate for some of the wear if the amount of wear is known and the geometry of the flowmeter has not been greatly altered. If this cannot be done, the accuracy will be reduced.

Piping Coating Pipe coating can affect the accuracy of the flowmeter by effectively changing the beta ratio. Any coating on the pipe that may have accumulated within the upstream and downstream piping requirements of the flowmeter should be removed.

Leaks Target flowmeters with seals and gaskets are subject to leaks, which are usually repaired by removing the flowmeter from service and replacing the leaking part.

Drift Drift can occur due to process deviations, and the support for the target can permanently change its characteristics over time due to the continual forces acting upon it. Any sudden forces that are exerted on the target, such as a surge when a pump is turned on, can cause a shift in the calibration of the flowmeter. Drift can be detected by performing periodic calibration checks.

Electronic Failure Electronic failures can occur and are usually remedied by board replacement.

Spare Parts Spare parts are usually limited to electronic parts, the target, and seals. As a result, spare parts are usually kept to a minimum, but different targets may be required for each flowmeter size.

Calibration Calibration is performed by removing the flowmeter from the line and simulating the force that would be exerted on the target by the fluid at a given flow. This is achieved by placing known weights on the upstream face of the target and making the appropriate electronic adjustments per procedures provided by the manufacturer.

EXERCISES

16.1 What are the Reynolds number constraints of a target flowmeter?

16.2 Is a target flowmeter affected by viscosity?

16.3 Size a target flowmeter for 0 to 100 gpm of a liquid that has a viscosity of 1.5 cP and a specific gravity of 1.10 at base and 1.04 at flowing conditions.

16.4 Size a target flowmeter for 0 to 10,000 scfh of a gas that has a specific gravity of 0.97 and viscosity of 0.015 cP at an operating pressure of 100 psi and a temperature of 70°F.

17
Thermal Flowmeters

Introduction Thermal flowmeters use thermal properties of the fluid (usually clean gases) to measure flow by measuring velocity or mass flow, depending upon flowmeter design. As some thermal flowmeters measure mass flow by inference from the thermal behavior and properties of the fluid instead of by measuring mass directly, these flowmeters are examined as thermal flowmeters with outputs that represent mass flow, rather than as mass flowmeters that use thermal properties to effect a mass flow measurement.

These devices can be applied to a number of flow measurement applications that are difficult using other technologies. For example, flowmeter output is dependent upon the thermal as opposed to the physical properties of the fluid, which allows applications to fluids that are not dense enough to be sensed by technologies that use the mechanical properties of the fluid.

Principles of Operation

Hot Wire Anemometer Thermal flowmeters based upon the principle of hot wire anemometers have probes inserted into the flowstream, which are usually ruggedized for industrial operation. These probes are usually part of an electronic bridge circuit.

In one configuration, one of two probes is controlled in such a way that it is heated at a fixed temperature above the second probe, which measures the temperature of the fluid. As flow increases, heat is removed from the heated probe by the fluid, and more current is needed to maintain it at the correct temperature. The current reflects the energy input required to compensate for the heat loss from the probe to the fluid and is indicative, after linearization, of the mass flow through the flowmeter.

This principle works while the thermal conductivity (that is, the ability of the heat to be transferred or conducted from the probe to the fluid) and the heat capacity of the fluid (the quantity of heat that a given mass requires to raise its temperature a specified amount) are assumed to be constant.

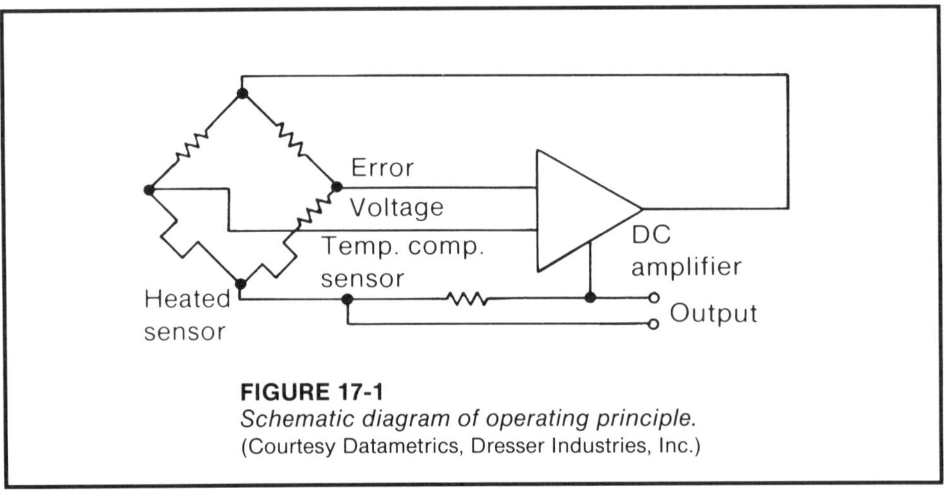

FIGURE 17-1
Schematic diagram of operating principle.
(Courtesy Datametrics, Dresser Industries, Inc.)

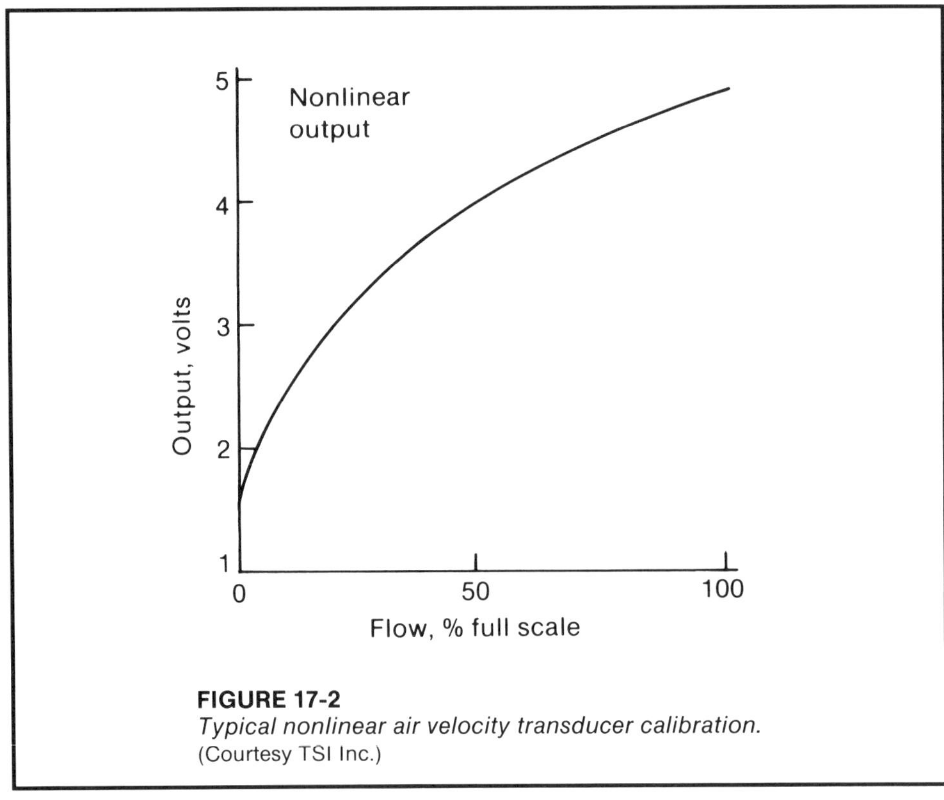

FIGURE 17-2
Typical nonlinear air velocity transducer calibration.
(Courtesy TSI Inc.)

Thermal Profile This technology utilizes a capillary tube that is uniformly heated by a transformer. At zero flow conditions, the temperature profile is symmetrical about the midpoint and two thermocouples produce equal outputs, as shown in Figure 17-3.

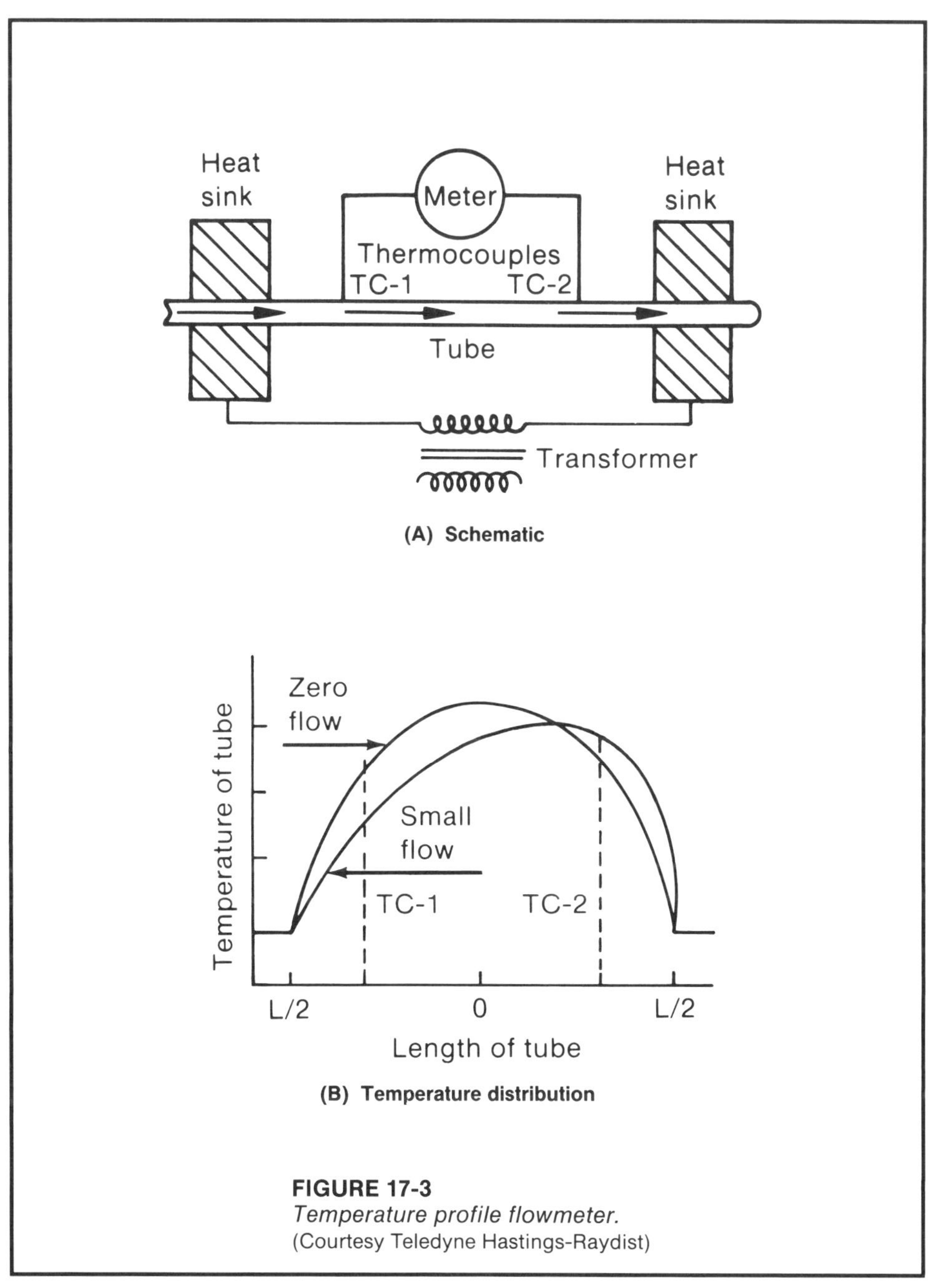

FIGURE 17-3
Temperature profile flowmeter.
(Courtesy Teledyne Hastings-Raydist)

Under flowing conditions, heat is transferred to the gas and then back again to form an asymmetrical temperature profile. With a constant power input, the difference in the temperatures at the sensing points is a linear function of the mass flow of the gas and the heat capacity of the gas. Since the heat capacity is virtually constant over wide ranges of pressure and temperature, the flowmeter can be calibrated directly in mass flow units as illustrated in the typical calibration curve in Figure 17-4.

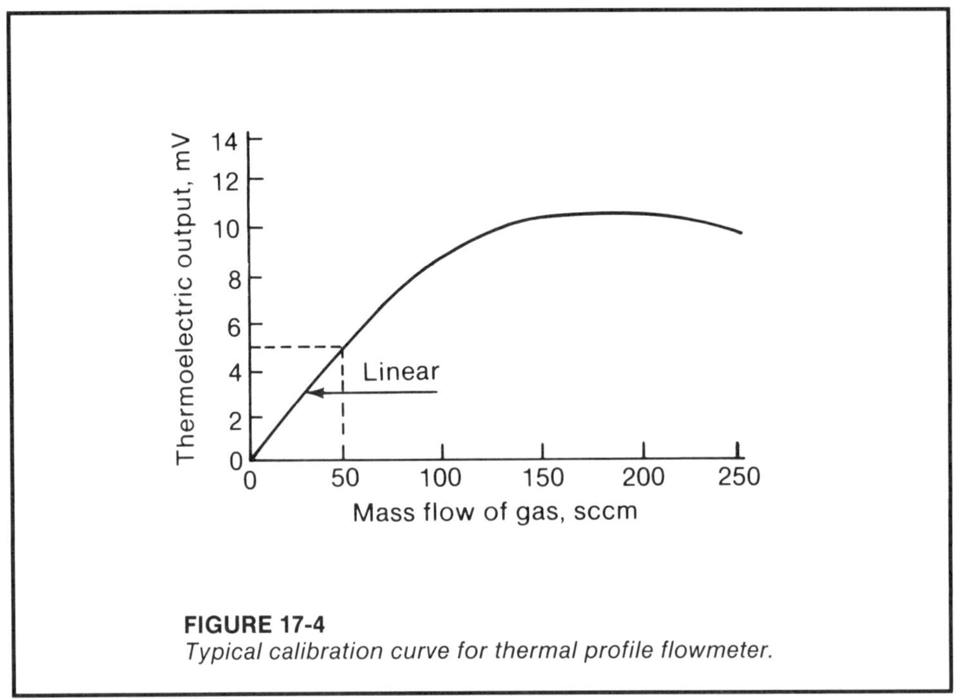

FIGURE 17-4
Typical calibration curve for thermal profile flowmeter.

Construction The body, where the thermal sensing system is housed, is typically of stainless steel construction although other materials such as brass, Monel®, and aluminum are used by some manufacturers.

Some bodies are flow primaries into which thermal sensors, usually point sensors, are placed to measure the flow.

Probes The probes contain or hold the thermal elements, depending upon the design of the flowmeter, and are typically of stainless steel construction. In the configuration where the thermal sensor and/or heater is located inside the probe, the thermal conductivity between the fluid and the probe can vary if film or deposits form on the probe, thereby affecting the flow measurement. As a result, some manufacturers recommend routine cleaning of the probe.

Probes with replaceable tips are point-sensitive and can be replaced when contaminated or, in some designs, can be electrically cleaned by burning off any film or deposit that may have accumulated (see Figure 17-6).

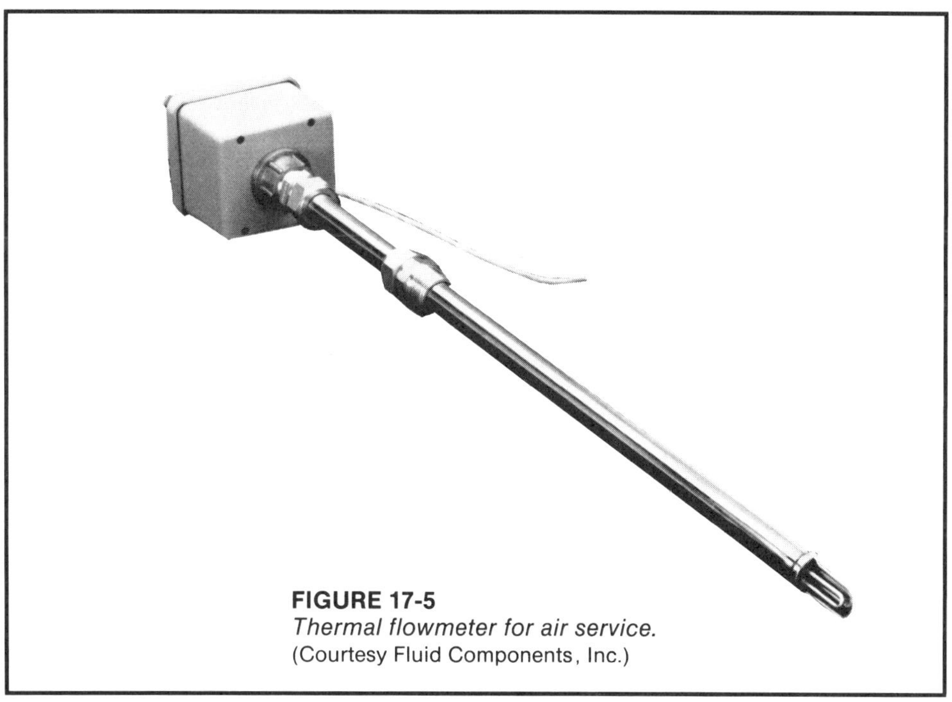

FIGURE 17-5
Thermal flowmeter for air service.
(Courtesy Fluid Components, Inc.)

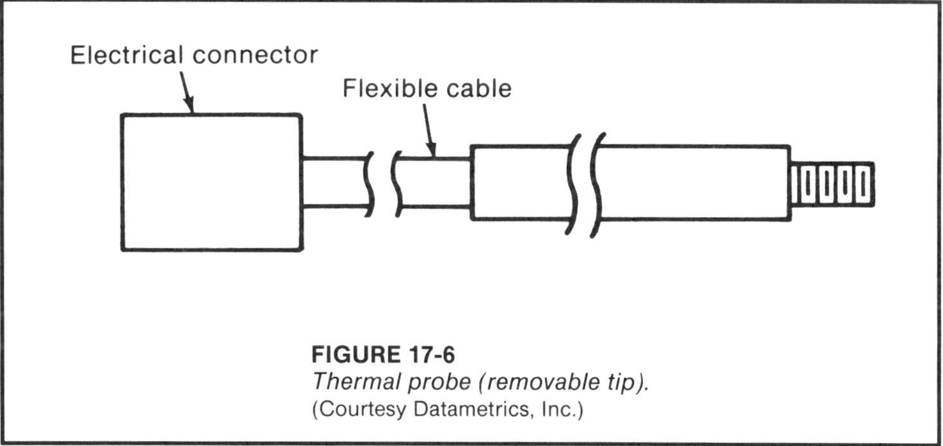

FIGURE 17-6
Thermal probe (removable tip).
(Courtesy Datametrics, Inc.)

Capillary Tubes Capillary tubes are used in thermal profile flowmeters as a means to effect heat transfer from a transformer to the fluid and to the thermal sensors. Capillary tubes are commonly constructed of constantan.

Wetted Parts The wetted parts generally include the body and the probes, which are typically constructed of stainless steel, with other materials of construction available in some designs. Capillary tubes are typically constructed of constantan to maintain sufficient thermal conductivity. It should be noted that

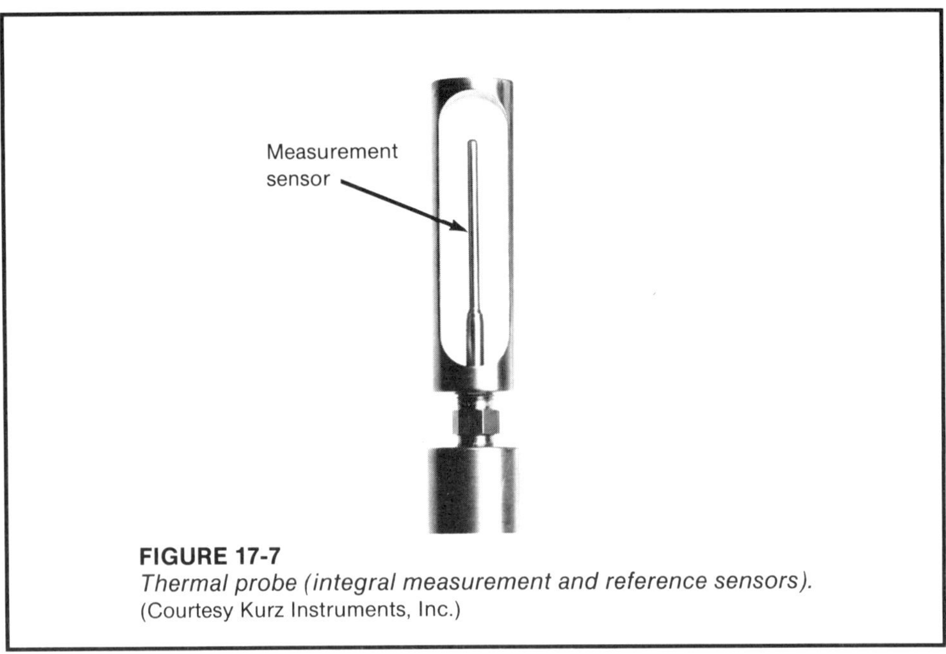

FIGURE 17-7
Thermal probe (integral measurement and reference sensors).
(Courtesy Kurz Instruments, Inc.)

the probe material should be compatible with the fluid that is being measured in order to minimize contamination of the probe. Such contamination may affect the thermal conductivity between the probe and the fluid before catastrophic failure and cause the flowmeter to drift. This situation can usually be remedied by routine sensor cleaning. Often the expense of a more exotic sensor can be offset by reduced maintenance requirements.

Operating Constraints Thermal flowmeter designs are available that can measure flow at temperatures as high as 450°C, although most have a maximum temperature rating between 100 to 150°C. Pressure ratings are normally limited by the pressure rating of the flange or connection.

Some designs require that the temperature and/or pressure be held within tolerances, some as tight as 15 psi ±3 psi, which would be considered restrictive for most industrial flow measurement applications. Measurement shifts of as much as 0.3 percent rate per degree centigrade are possible in some designs. There are no Reynolds number constraints as velocity or mass is sensed directly, and data are available from the manufacturer as to the minimum and maximum flows that can be accurately measured by each sensor.

Performance Hot wire anemometer thermal flowmeters for gas service have accuracy statements that are typically the higher of ±2 percent rate plus ±0.2 percent FS, and 0.5 percent of meter capacity. Repeatability statements are typically the higher of ±1 percent rate and 0.25 percent of meter capacity. Thermal profile and thermal dispersion flowmeters have accuracies of ±2 percent FS

($\pm$1 percent FS when pressure and temperature fluctuations are held to $\pm$20 percent and $\pm$2 to 5 percent FS in gas service, respectively).

Applications Applications of thermal flowmeters are limited to those fluids that have known heat capacities. This technology is usually applied to clean, pure gases or clean mixtures of pure gases of known composition; therefore, the heat capacity is known and constant during flowmeter operation. Probe coating, which can affect the thermal conductivity between the probe and the fluid, should be held to a minimum.

Typical applications include air, nitrogen, oxygen, ammonia, Freon®, helium, hydrogen, carbon monoxide, methane, nitric oxide, carbon dioxide, ethane, nitrous oxide, argon, and the like.

Liquid applications are less common because liquid is generally dirtier than gas. The heat transfer rate can change when bubbles collect on the sensing surface, and in some designs problems can arise due to the presence of a current-carrying wire in a conducting or a corrosive medium.

Sizing Sizing is accomplished by converting the desired flow range to its equivalent flow range of air or water (using a conversion factor supplied by the manufacturer) and selecting the flowmeter with the proper capacity. The equivalent flow range for fluid mixtures should be calculated on a weighted average basis before selecting the proper flowmeter size from a manufacturer's literature. The gas conversion factors for a typical thermal profile flowmeter are given in Table 17-1.

EXAMPLE 17-1

Problem: Given that thermal profile flowmeters are available in full scale air flow ranges of 5, 10, 25, 50, 500, 200, and 500 scfm, determine which size flowmeter can be used to measure a maximum flow of 60 scfm of argon. What is the full scale flow of the flowmeter?

Solution: From Table 17-1, the conversion factor for argon is 1.43. Therefore, a 50 scfm air flowmeter could be used in this application, as the full scale flow range is 50×1.43, or 71.5 scfm of argon.

The above sizing information will vary with manufacture as other thermal flowmeters may use different sizing methods and conversion factors.

It should be noted that as the principle of operation is thermal and not momentum related, the conversion factor represents differences in heat capacity. Even light gases such as hydrogen can be measured using this technique because its heat capacity is sufficiently high.

Installation Thermal flowmeters are available with threaded, screwed, and flanged connections.

TABLE 17-1
Gas Conversion Factors

Gas	*Symbol*	*Conversion factor*	*(1) Density, gr/liter*	*(2) Specific gravity*
Acetylene	C_2H_2	0.67	1.09	0.90
Air		*1.00	1.20	1.00
Ammonia	NH_3	0.77	0.71	0.59
Argon	A	*1.43	1.66	1.38
Arsine	AsH_3	0.76	3.25	2.70
Bromine	Br_2	0.88	5.98	4.96
Butane	C_4H_{10}	0.30	2.51	2.08
Butene 1	C_4H_8	0.34	2.40	1.99
Carbon dioxide	CO_2	0.73	1.84	1.53
Carbon monoxide	CO	*1.00	1.17	0.97
Chlorine	Cl_2	0.85	2.98	2.47
Chlorine trifluoride	ClF_3	0.45	3.78	3.14
Cyclopropane	C_3H_6	0.52	1.75	1.45
Diborane	B_2H_6	0.50	1.15	0.95
Ethane	C_2H_6	0.56	1.26	1.05
Ethene (ethylene)	C_2H_4	0.69	1.17	0.97
Ethylene oxide	C_2H_4O	0.60	1.79	1.49
Fluorine	F_2	0.93	1.58	1.31
Freon 11	CCl_3F	0.36	5.93	4.92
Freon 12	CCl_2F	*0.36	5.13	4.26
Freon 13	$CClF_3$	0.42	4.59	3.81
Freon 14	CF_4	0.48	3.65	3.04
Freon 22	$CHClF_2$	*0.43	3.65	3.03
Freon 114	$CClF_2$	*0.22	6.99	5.80
Helium	He	*1.43	0.17	0.14
Hydrogen	H_2	*1.03	0.08	0.07
Hydrogen chloride	HCl	1.01	1.48	1.23
Hydrogen fluoride	HF	1.00	1.53	1.27
Hydrogen sulfide	H_2S	0.85	1.43	1.19
Isobutane	C_4H_{10}	0.31	2.48	2.06
Krypton	Kr	1.39	3.49	2.90
Methane	CH_4	*0.69	0.68	0.56
Neon	Ne	1.38	0.84	0.70
Nitric oxide	NO	1.00	1.24	1.03
Nitrogen	N_2	*1.02	1.17	0.97
Nitrous oxide	N_2O	0.75	1.85	1.54
Oxygen	O_2	*0.97	1.33	1.10
Pentaborane	B_5H_9	0.15	2.83	2.35
n-Pentane	C_5H_{12}	0.22	3.18	2.64
Phosphine	PH_3	0.79	1.53	1.27
Propane	C_3H_8	*0.32	1.89	1.57
Silane	SiH_4	0.68	1.33	1.10
Sulfur dioxide	SO_2	0.70	1.72	2.26
Sulfur hexafluoride	SF_6	0.28	6.43	5.34
Tungsten hexafluoride	WF_6	0.23	8.22	6.82
Uranium hexafluoride	UF_6	0.23	14.65	12.16
Water vapor	H_2O	0.80	0.76	0.63
Xenon	Xe	1.37	5.54	4.60

*Empirical data; other theoretical. (Courtesy Teledyne Hastings-Raydist)

(1) Density in grams/liter at 20° C and 1 atmosphere.
(2) Specific gravity (air = 1.00).

Hydraulic Requirements Hot wire anemometer flowmeters typically require 8 to $10D/3D$ to set up the proper velocity profile in the pipe, while thermal profile flowmeters require $4D/2D$. A filter upstream of the thermal profile flowmeter is recommended to minimize the possibility of capillary tube pluggage.

Cabling Most thermal flowmeters limit the cable length between the sensor and transmitter to 15 to 100 feet, depending on design, effectively locating the transmitter near the sensor in an industrial environment. As a result, thermal flowmeters are usually cabled as 4-wire transmitters.

Maintenance

Thermal flowmeters usually require some amount of routine maintenance to clean thermally conductive surfaces. Sensor failure, sensor wear, leaks, and electronic failures can also occur.

Routine Maintenance As fluid flows over the probes or other thermally conductive surfaces of a thermal flowmeter, contamination can occur. Many manufacturers recommend that these surfaces by cleaned on a routine basis to maintain flowmeter performance.

Sensor Failure Some thermal flowmeters are designed with replaceable tips that allow for relatively simple sensor replacement. Other flowmeter designs have integral sensors that often cannot be replaced. This results in replacement of the entire sensor assembly when the sensor fails.

Sensor Wear As thermal flowmeters are sensitive to the thermal conductivity between the probes and the fluid, any change in the area of these probes will affect the measurement. For this and other reasons, thermal flowmeters are generally not applied to abrasive fluid service where sensor wear may be a factor in flowmeter performance.

Leaks Flowmeter leakage can occur, typically at the point where penetrations are made for probes. Leaks can be handled using standard piping practices.

Electronic Failure Electronic failures can occur and are usually handled by board replacement.

Spare Parts Spare parts requirements vary with design and include replaceable probes, when applicable, and often the entire flowmeter assembly, as some designs have few replaceable parts.

Calibration In general, thermal flowmeters are calibrated for the specified range by the manufacturer and cannot be adjusted after the flowmeter leaves the factory.

EXERCISES

17.1 Can the full scale flow range of a thermal flowmeter be adjusted in the field?

17.2 Given that thermal profile flowmeters are available for full scale air flows of 10, 50, 100, and 500 sccm, which size flowmeter should be used to measure a maximum flow of 90 sccm of propane? What is the full scale flow of the flowmeter?

18

Turbine Flowmeters

Introduction Turbine flowmeters have been widely accepted as a proven technology that is applicable for measuring flow with high accuracy and repeatability, even though moving parts are inherent in this design and any physical alteration or damage to the flowmeter results in a loss of accuracy.

The accuracy of turbine flowmeters can be superior to other technologies in the turbulent flow regime. As a result, the trend towards flowmeters that have no moving parts appears not to have displaced the turbine flowmeter where high accuracy is desired.

Axial Turbine Flowmeters

Principle of Operation

Turbine Flowmeter

Turbine flowmeters consist of a rotating device, called a rotor, that is positioned in the flowstream in such a manner that the rotational velocity of the rotor is proportional to the fluid velocity and hence the flow through the flowmeter, as illustrated in Figure 18-1.

Dual Rotor Turbine Flowmeter

Dual rotor turbine flowmeters are available that compensate errors that are inherent in the single rotor design, as illustrated in Figure 18-2.

Construction The body is where the rotor assembly and sensing system are mounted. Vanes that are used to aid in characterizing the flow at the flowmeter inlet are often welded into the body.

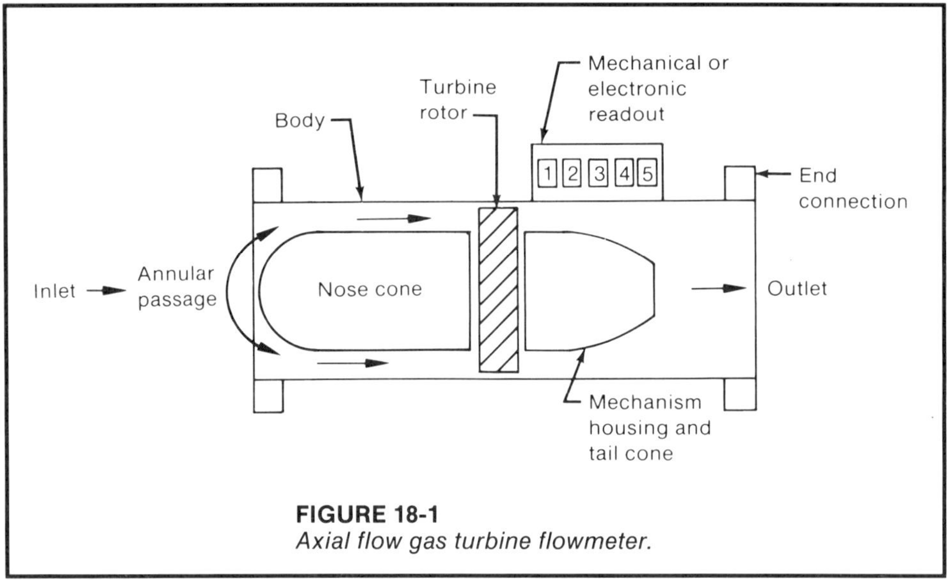

FIGURE 18-1
Axial flow gas turbine flowmeter.

Rotor

The rotor is the part of the turbine flowmeter that rotates at a velocity that is proportional to the fluid flow. Rotor designs vary with manufacturer, as do materials of construction and the type of bearings on which they rotate.

Rotors are generally designed to be as light as possible so that the momentum of the fluid is large in relation to the mass of the rotor. As a result of low rotor mass, low flow ranges can be measured more accurately and changes in fluid velocity can be detected more rapidly. This results in increased sensitivity of the flowmeter to fluctuations in flow.

Rotor Bearings

Rotor bearings are the parts of the flowmeter on which the rotor rotates, the design of which varies with manufacturer. The spinning of the rotor in many process fluids can cause the bearings to wear and eventually fail. As a result, good bearing design and proper application of the flowmeter are necessary to achieve good bearing life. Turbine flowmeter bearings are usually self-lubricating, but lubricated bearing designs are available.

Care should be taken not to expose the flowmeter to an incompatible fluid nor to overspin the rotor for any reason, as the bearing can burn up and fail in a matter of seconds. Blowing out a pipe with nitrogen or steam can pose this sort of problem. The flowmeter should not be subjected to any sudden surges of flow as the whole rotor assembly can be damaged by forces that are applied suddenly.

Sensing Systems

Mechanical or non-contact magnetic and RF-proximity techniques are typically used to transform the rotation of the rotor into usable signals suitable for transmission.

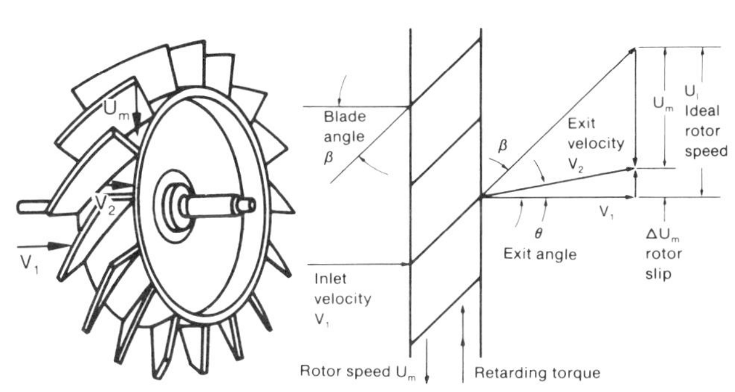

$$\text{Meter accuracy} \propto \left(\frac{U_m}{U_I}\right) = \left(\frac{U_I - \Delta U_m}{U_I}\right) = 1 - \left(\frac{\Delta U_m}{U_I}\right) \text{ Fractional rotor slip}$$

$$= 1 - \left(\frac{\operatorname{Tan}\theta}{\operatorname{Tan}\beta}\right) \text{ Dependent on } \theta$$

(A) Standard turbine meter

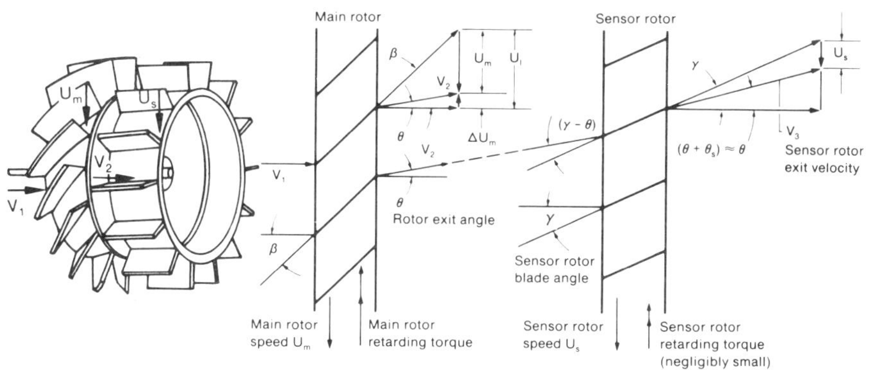

$$\text{Define auto-adjust turbo-meter accuracy} \propto \frac{U_m - U_s}{U_I} = \left(\frac{U_m}{U_I}\right) - \left(\frac{U_s}{U_I}\right)$$

$$= \left(1 - \frac{\operatorname{Tan}\theta}{\operatorname{Tan}\beta}\right) - \left(\frac{\operatorname{Tan}\gamma}{\operatorname{Tan}\beta} - \frac{\operatorname{Tan}\theta}{\operatorname{Tan}\beta}\right)$$

$$= \left(1 - \frac{\operatorname{Tan}\gamma}{\operatorname{Tan}\beta}\right) = \text{Constant}$$

(B) Dual rotor turbo-meter

FIGURE 18-2
Standard and dual rotor flowmeters.
(Courtesy Rockwell International)

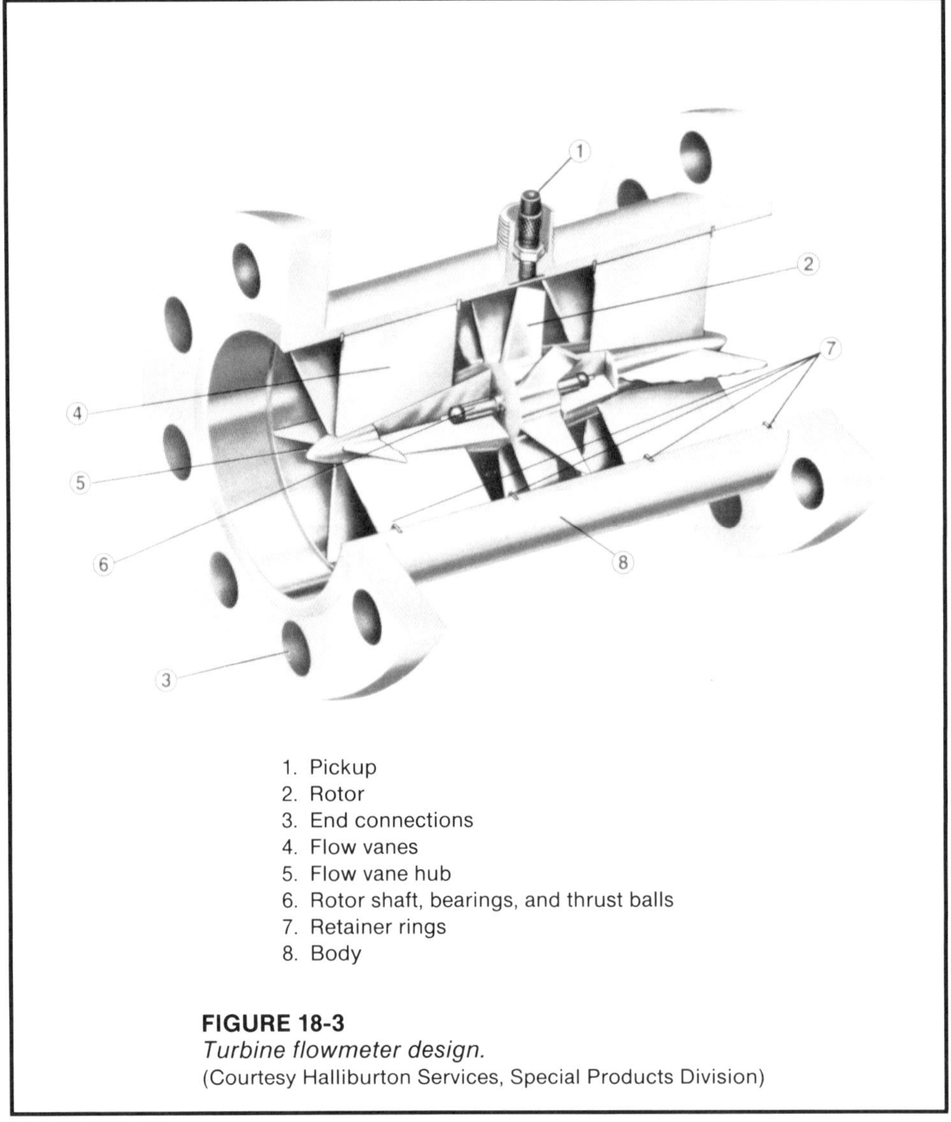

1. Pickup
2. Rotor
3. End connections
4. Flow vanes
5. Flow vane hub
6. Rotor shaft, bearings, and thrust balls
7. Retainer rings
8. Body

FIGURE 18-3
Turbine flowmeter design.
(Courtesy Halliburton Services, Special Products Division)

In mechanical systems, the rotor is geared to a shaft that protrudes from the flowmeter body. The shaft can be coupled to a transmitter, pulser, local indicator, and/or local totalizer.

Magnetic systems utilize a permanent magnet in the pickup, which results in voltage variations on a pickup coil. These variations are indicative of the passage of the rotor blades and are used to determine the velocity with which the rotor is spinning.

RF techniques use a pickup that generates a high frequency signal that is amplitude-modulated by the passage of the turbine rotor blades. This signal can then be connected to a transmitter.

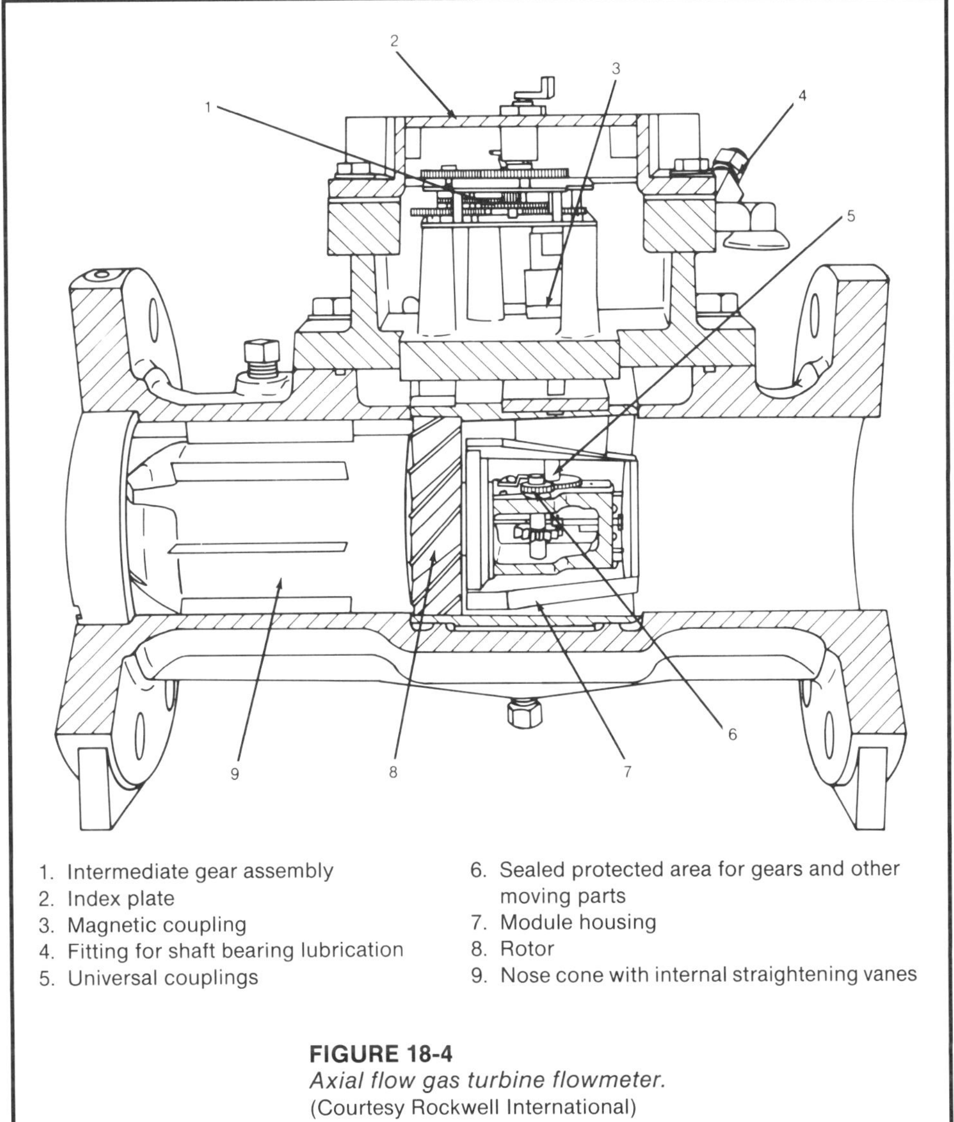

1. Intermediate gear assembly
2. Index plate
3. Magnetic coupling
4. Fitting for shaft bearing lubrication
5. Universal couplings
6. Sealed protected area for gears and other moving parts
7. Module housing
8. Rotor
9. Nose cone with internal straightening vanes

FIGURE 18-4
Axial flow gas turbine flowmeter.
(Courtesy Rockwell International)

Wetted Parts

Wetted parts include the body, the rotor, and the rotor bearings. Flowmeter bodies are commonly available in aluminum, steel, and stainless steel, but other materials such as Hastelloy®, Monel® titanium, and PVC are also available. Standard rotors are typically stainless steel, although rotors can be made compatible with body materials of construction when more exotic materials are specified. Rotor bearing types and materials of construction include stainless steel ball bearings, tungsten carbide sleeve bearings, and graphite sleeve bearings. Rotor bearings of proprietary design are offered by some manufacturers.

Operating Constraints Turbine flowmeters are available in 1/2-in. to 24-in. sizes, which effect a liquid flow measurement of 0.06 to 50,000 gallons per minute. Pressure is typically limited by the flange rating, while temperatures as high as 450°C can be handled. It should be noted that designs that are marketed in industries where elevated temperatures are not encountered can have maximum operating temperatures of as low as 75°C.

Operation of the turbine flowmeter is dependent upon Reynolds number and, to some degree, the momentum of the fluid in the flowstream, which must be sufficient to operate the rotor. Reynolds number constraints will vary with design. Most turbine flowmeters, however, operate linearly in the turbulent regime above a minimum Reynolds number that varies with manufacture between 4000 and 20,000. Turbine flowmeters achieve a 10 to 100:1 turndown depending upon flowmeter and sensor design and upon fluid viscosity.

Some turbine flowmeters are designed to operate repeatably, not linearly, from the turbulent regime down into the transition and laminar flow regimes. Due to the repeatable operation of the flowmeter primary, with proper calibration techniques and sufficient information concerning the fluid being measured, the system can be linearized electronically to produce accurate measurements at Reynolds numbers as low as 400. As a result, the maximum turndown can be as high as 500:1 in some designs.

Care must be taken not to operate the flowmeter at flows greater than those recommended by the manufacturer, as overspinning the rotor can destroy the rotor bearings in seconds. Turbine flowmeters should be specified with care, and a bypass should be considered where there is a possibility of cleaning the pipe with steam or other gas since this might overspin and damage the flowmeter. Sudden surges of liquid flow, such as are produced when a pump is started or a valve is opened, when the flowmeter has no liquid in it should be avoided. The sudden forces that are exerted on an empty flowmeter can be so large as to damage the flowmeter.

Some designs require that the fluid continually lubricate the bearings to provide longer bearing life. Rotor failure can occur within minutes if a nonlubricated turbine flowmeter is improperly applied. Lubricated designs are available and are used in natural gas and other services. Some designs are usable in saturated steam service.

Performance Accuracies in liquid service can exceed ±0.25 percent rate within limited flow ranges. The expected accuracy of a turbine flowmeter primary is typically 1 percent rate over a nominal 10:1 range, although not all designs achieve this goal. Measurement accuracy in gas service is typically lower than that of liquid service due to the additional uncertainties characteristic of gas flow measurement. Repeatability is typically ±0.05 percent rate in liquid service and ±0.1 percent rate in gas service.

Pressure drop across the flowmeter will vary with flow and service, but maximum pressure drops can vary from 0.2 to 85 psi in gas service to 1 to 20 psi in liquid service, depending upon flowmeter design and operating conditions.

Applications Turbine flowmeters are generally applicable on lubricating fluids operating at Reynolds numbers in excess of 4000 to 20,000, depending on manufacture. Not all will operate in gas service where close attention to bearing design is necessary due to the higher rotor velocities encountered.

Determination of the flow range over which the flowmeter is to operate is important, as the *K*-factor can be selected to minimize errors within this flow range.

Sizing Turbine flowmeters, which measure the actual volume of fluid that passes through the flowmeter, are usually sized by using tables supplied by the manufacturer. These tables list the minimum and maximum flows and pressure drop of specific liquids and gases such as water, air, and natural gas at nominal operating conditions for each size flowmeter.

Liquid sizing can be readily performed from capacity data supplied by the manufacturer. It should be noted that the pressure drop across the flowmeter will vary with the specific gravity of the liquid.

EXAMPLE 18-1

Problem: Size a liquid turbine flowmeter for 0 to 1000 gpm of a liquid that has a specific gravity of 1.18, given the following sizing data:

Size	Minimum Flow	Maximum Flow
1 in.	5 gpm	50 gpm
2 in.	22 gpm	225 gpm
3 in.	65 gpm	650 gpm
4 in.	125 gpm	1250 gpm
6 in.	300 gpm	3000 gpm
8 in.	850 gpm	8500 gpm

Solution: A 4-inch turbine flowmeter would be applicable and would measure from 125 to 1000 gpm.

In gas applications, operation is limited by minimum and maximum flows that are functions of the rotor drag and gas density and maximum rotor velocity, respectively. As turbine flowmeters measure actual gas velocity and volume, the gas laws may be applied to convert the maximum desired flow to an equivalent volumetric flow at the nominal operating conditions specified by the manufacturer.

$$Q_{\text{max}} = Q_{\text{max@ref}} \times \frac{P}{P_{\text{ref}}} \times \frac{T_{\text{ref}}}{T} \times \frac{Z_{\text{ref}}}{Z}$$

The minimum flow varies as the square root of the gas law factor, as follows:

$$Q_{\text{min}} = Q_{\text{min@ref}} \times \left[\frac{P}{P_{\text{ref}}} \times \frac{T_{\text{ref}}}{T} \times \frac{Z_{\text{ref}}}{Z} \times \frac{SG_{\text{ref}}}{SG}\right]^{1/2}$$

EXAMPLE 18-2

Problem: Given the following turbine flowmeter capacity data for natural gas service with a base specific gravity of 0.60 at reference conditions of 0.25 psi and 60°F, when standard conditions are 14.73 psia and 60°F.

Size	Minimum Flow	Maximum Flow
4 in.	20 scfm	300 scfm
6 in.	30 scfm	500 scfm
8 in.	50 scfm	1000 scfm
10 in.	95 scfm	2300 scfm

Calculate the minimum and maximum flows through a 6-inch turbine flowmeter in natural gas service with a base specific gravity of 0.67 that is operated at 500 psi and 90°F when the ratio of the Z-factor at base conditions to the Z-factor at operating conditions is 1.0625.

Solution: The maximum flow is limited by the maximum turbine velocity and hence by the maximum actual volumetric flow, above which the flowmeter may be damaged. Using the equation above,

$$Q_{max} = 500 \text{ scfm} \times \frac{(500 \text{ psi} + 14.73 \text{ psi})}{(14.73 \text{ psi} + 0.25 \text{ psi})} \times \frac{(460°\text{F} + 60°\text{F})}{(460°\text{F} + 90°\text{F})} \times (1.0625)$$

$$= 17259 \text{ scfm}$$

The minimum flow is a function of gas density and rotor drag, which partially offset each other, so that

$$Q_{min} = 30 \text{ scfm} \times \left[\frac{(500 \text{ psi} + 14.73 \text{ psi})}{14.73 \text{ psi} + 0.25 \text{ psi})} \times \frac{(460°\text{F} + 60°\text{F})}{(460°\text{F} + 90°\text{F})} \times \frac{(0.60)}{(0.67)}\right] \times (1.0625)^{1/2}$$

$$= 167 \text{ scfm}$$

The original turndown was 500 scfm/30 scfm, or 16.7:1, which is increased to 17259 scfm/167 scfm, or 103:1 as a result of the increased forces of the denser high pressure gas overcoming the drag forces of the rotor at a lower flow.

EXAMPLE 18-3

Problem: Size a turbine flowmeter to measure a gas flow of 0 to 2000 scfm of a gas with a Z-factor of 0.985 at base conditions and a base specific gravity of 1.30, which is operated at 100 psi and 120°F with an operating Z-factor of 0.968.

Solution:

$$Q_{max} = 2000 \text{ scfm} \times \left[\frac{(100 \text{ psi} + 14.73 \text{ psi})}{(14.73 \text{ psi} + 0.25 \text{ psi})} \times \frac{(460°\text{F} + 60°\text{F})}{(460°\text{F} + 120°\text{F})} \times \frac{(0.985)}{(0.968)}\right]^{-1}$$

$$= 286 \text{ acfm}$$

A 4-inch flowmeter that has a maximum flow of 300 scfm at reference conditions would be applicable. The minimum reference flow of 20 scfm at actual operating conditions can be calculated as

$$Q_{min} = 20 \text{ acfm} \times \left[\frac{(100 \text{ psi} + 14.73 \text{ psi})}{(14.73 \text{ psi} + 0.25 \text{ psi})} \times \frac{(460°\text{F} + 60°\text{F})}{(460°\text{F} + 120°\text{F})} \times \frac{(0.985)}{(0.968)} \times \frac{(0.60)}{(1.30)}\right]^{-1/2}$$

$$= 35.9 \text{ scfm}$$

The flowmeter would perform accurately from 36 to 2000 scfm.

It should be noted that most manufacturers do not specify Reynolds number requirements in their literature. For liquid applications over 1 cP and gas applications with viscosities greater than that of the gas in the data presented by the manufacturer, the user should investigate Reynolds number constraints to avoid nonlinear operation of the flowmeter.

Installation Turbine flowmeters may be of screwed, flanged or wafer design, depending upon the size and manufacture of the flowmeter.

Hydraulic Requirements

These meters are sensitive to the velocity profile of the fluid as it enters and leaves the flowmeter, as well as to any swirl that may be present at the flowmeter inlet. Straight run upstream and downstream of the flowmeter is recommended. Standard requirements are $10D/5D$, with straightening vanes to stabilize the velocity profile that enters the flowmeter and reduce swirl.

Optional short coupled installations of a minimum of $4D/-$ with straightening vanes can be used, but will degrade flowmeter accuracy.

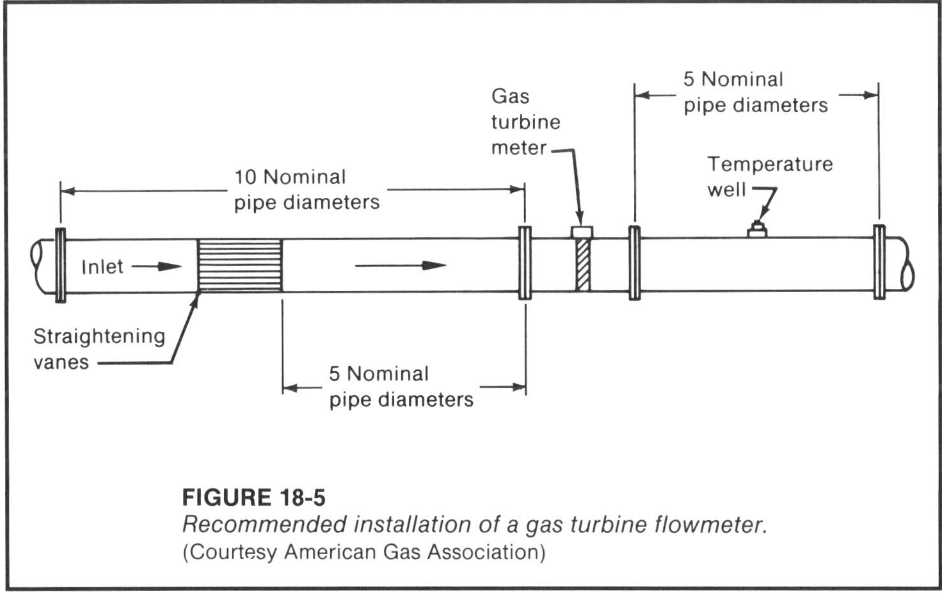

FIGURE 18-5
Recommended installation of a gas turbine flowmeter.
(Courtesy American Gas Association)

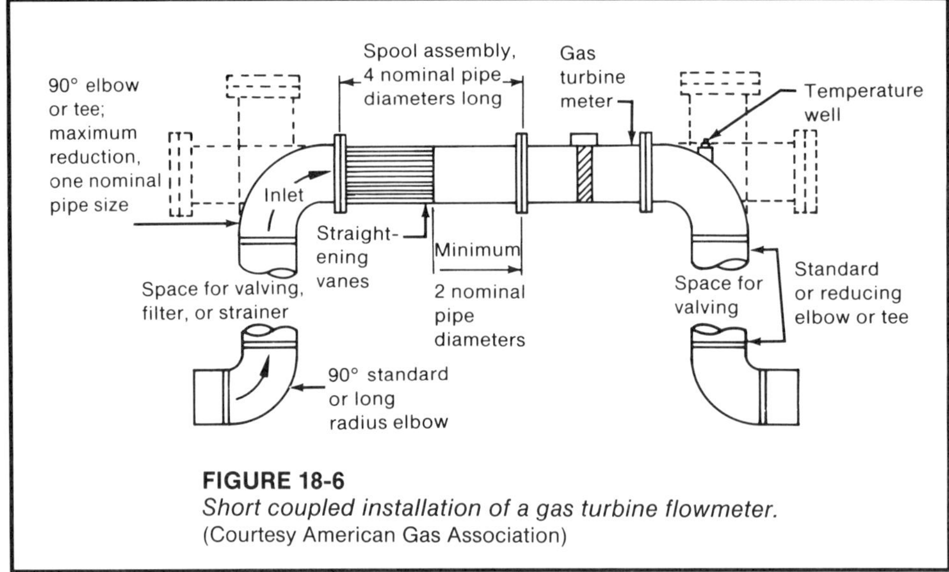

FIGURE 18-6
Short coupled installation of a gas turbine flowmeter.
(Courtesy American Gas Association)

Piping

The piping arrangement should be such that an empty flowmeter is not subject to sudden surges of liquid flow, such as the starting of a pump or the opening of a valve when the flowmeter is empty. Such a sudden large force can damage the flowmeter. An example of where this can occur is a self-draining pipe.

The pipe upstream and downstream of the flowmeter should have the same inside diameter as that of the flowmeter flange.

Cabling

Many turbine flowmeters are available as 4-wire transmitters. However, most manufacturers can mount the electronics remote from the flowmeter primary, thus eliminating the requirement for power at the flowmeter primary. This results in cabling requirements similar to those of a 2-wire transmitter, with the exception that use of a special cable may be required.

Maintenance Typically, no routine maintenance is required, but problems such as leaks, sensor failure, rotor bearing wear, rotor wear, and electronic failure can occur. Some designs require that the bearings be periodically lubricated.

Leaks

Leakage can occur at any connection that is machined into the body, such as grease fittings. Standard practices for sealing pipe leaks can be used on these connections.

Sensor Failure

Diagnosis of sensor failure, as opposed to rotor, bearing, or electronic failure, is important in order to avoid unnecessary work. Sensor failure should be suspected when flow is known to exist in the pipe, but zero flow is indicated at the transmitter output.

Transmitter operation may be verified by providing a pulsed, variable frequency signal at the flow primary and observing the output of the transmitter. If the transmitter functions electrically, then the fault probably lies in the rotor, the bearing, or the sensing element. Generally, the flowmeter must be removed in such a case for checkout in a shop environment.

Bearing wear will tend to cause the rotor to drag at low flows where the momentum of the fluid is not sufficient to overcome the frictional forces of the worn bearing. This results in measurement errors that may not be immediately apparent in the normal operation of the flowmeter. Excessive wear can cause the rotor to eventually stop rotating and fail completely. While most turbine flowmeter failures are catastrophic in nature (that is, the rotor ceases to operate), this is not always the case.

Rotor bearing wear can be caused by factors other than those attributed to wear due to normal use. Excessive wear can occur in applications where the fluid is relatively non-lubricating. It can also be caused by practices such as overspinning the rotor by blowing air through a liquid turbine flowmeter to flush out the lines or by starting up the pump with unfilled lines.

Rotor Wear

In an industrial environment, rotor wear is a difficult problem to detect, but it is known to occur under certain conditions. Abrasive fluids or fluids with solids can erode as well as deform the rotor. It can be corroded by corrosive fluids and damaged by debris. In general, any change that occurs in the geometry of the flowmeter after calibration adds additional uncertainty to the measurement, and the rotor should be examined for such occurrences whenever it is disassembled.

Electronic Failure

Electronic failures can occur and are usually remedied by board replacement.

Spare Parts

Spare parts vary with flowmeter design, but the rotor, the rotor bearings, and the sensor should be stocked for each size flowmeter to minimize downtime when a failure occurs. Transmitter parts, which are typically the same for all flowmeter sizes of given manufacture, should also be stocked.

Calibration

The flowmeter primary is factory calibrated, and turbine transmitter calibration is performed by adjustment to properly interpret the frequency output of the primary. The span adjustment is made by simulating the frequency that the primary would generate at maximum flow and adjusting the transmitter output for full scale. The zero adjustment is made by simulating a zero frequency input, which corresponds to zero flow conditions, and zeroing the transmitter.

Other Turbine Flowmeter Designs

Other designs are available that can be used to measure flow, most of which do not perform as well as standard turbine flowmeters, and some are specified as nonlinear devices.

Paddle Wheel Paddle wheel flowmeters are available in an integral configuration and an insertion configuration that mounts flush with the pipe wall to effect flow measurements larger than 0.05 gallons per minute. This technology has typical linearity and repeatability of ± 1 percent of full scale and ± 0.5 percent of full scale, respectively.

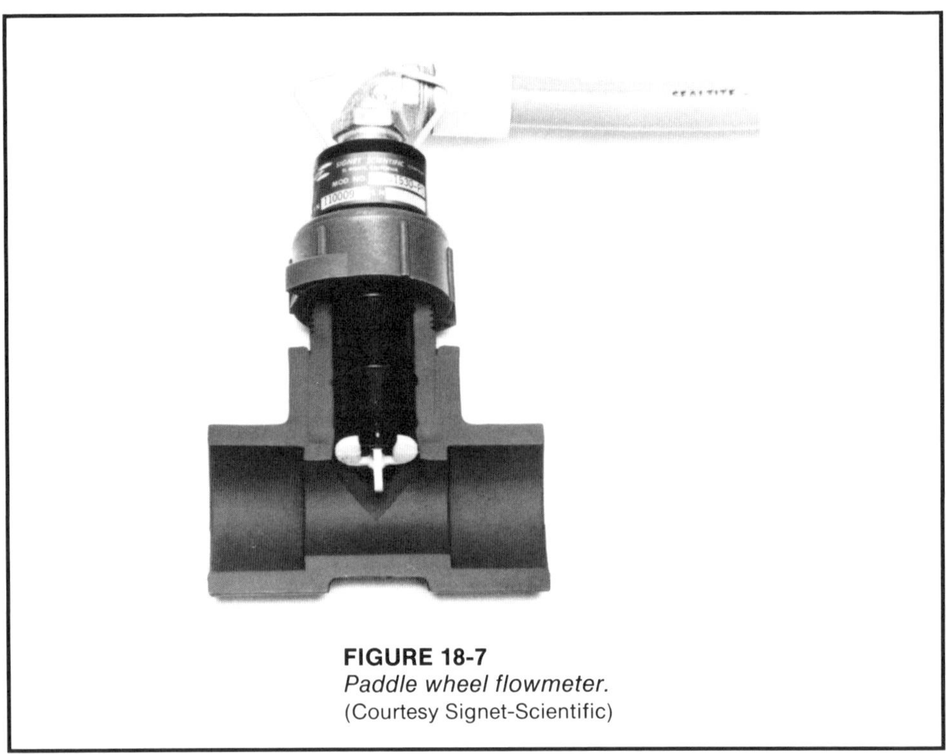

FIGURE 18-7
Paddle wheel flowmeter.
(Courtesy Signet-Scientific)

Propeller Propeller flowmeters are constructed so that the rotor, commonly called the propeller, is suspended in the flowstream and coupled external to the flowmeter as shown in Figure 18-8. This technology can be mechanically coupled to a shaft located external to the flowmeter and used to operate a local indicator, totalizer, or transmitter. Accuracy of the flowmeter is typically ± 2 percent rate.

Tangential Turbine The tangential turbine flowmeter can be used for liquid flow measurement from approximately 0.001 to 5 gallons per minute and in clean gas service. The flowmeter uses an orifice internal to the flowmeter to tangentially shoot liquid at a rotor, as illustrated in Figure 18-9. The output of this device is nonlinear, but repeatability is typically ± 0.1 percent rate for liquid service and ± 0.2 percent rate for gas service.

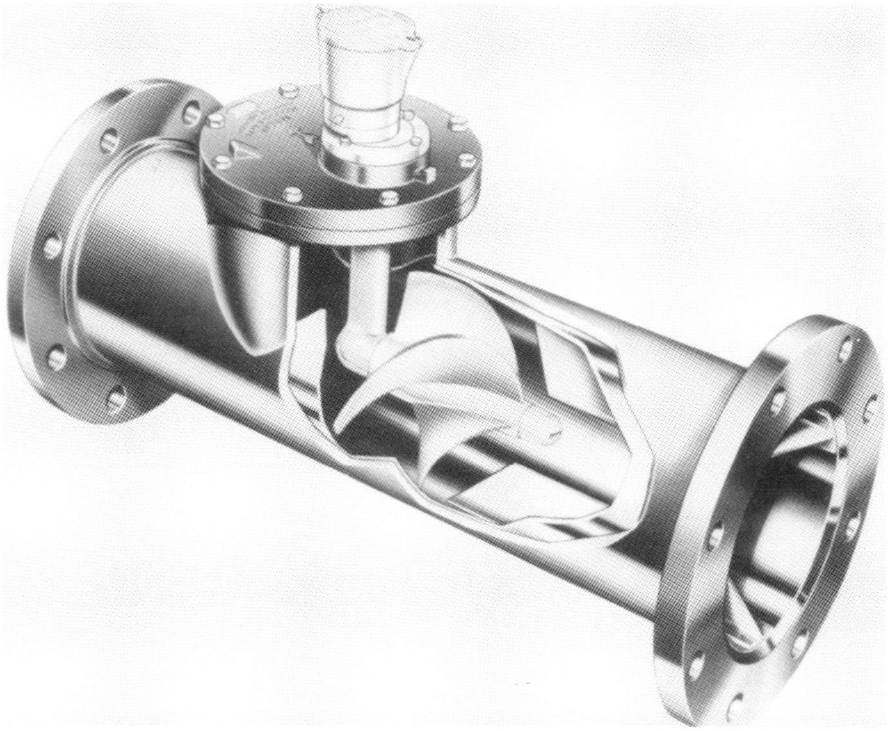

FIGURE 18-8
Propeller flowmeter construction.
(Courtesy Badger Meter, Inc.)

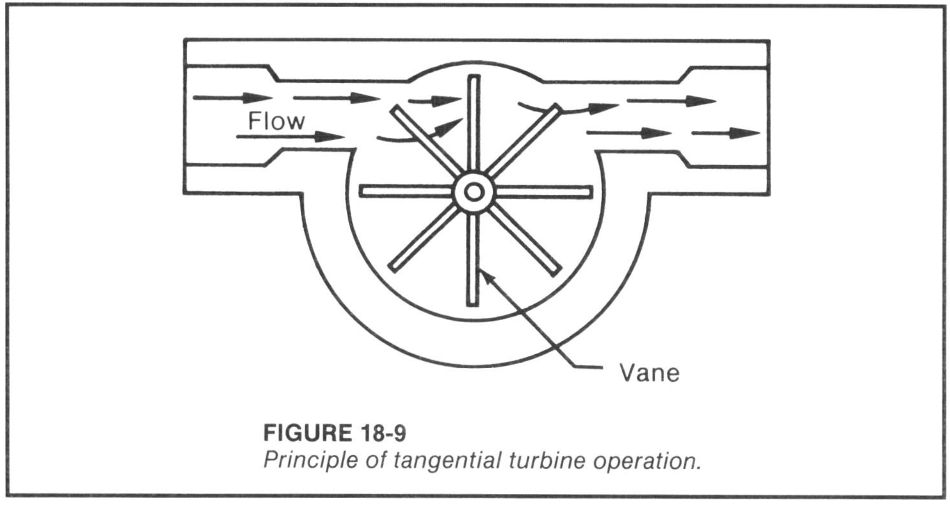

FIGURE 18-9
Principle of tangential turbine operation.

EXERCISES

18.1 Size a turbine flowmeter to measure a gas flow of 0 to 4000 scfm of an ideal gas and a base specific gravity of 1.05, operated at 30 psi and 70°F.

18.2 Size a turbine flowmeter to measure a gas flow of 0 to 400 scfm of an ideal gas and a base specific gravity of 1.19, operated at 5 psi and 50°F.

18.3 Size a turbine flowmeter to measure a gas flow of 0 to 30,000 scfm of natural gas with a base specific gravity of 0.61 that is operated at 1500 psi and 60°F, given that the ratio of *Z*-factors between base and operating conditions is 1.2157.

18.4 What are the expected accuracy and repeatability of a turbine flowmeter?

19

Ultrasonic Flowmeters

Introduction A relative newcomer to the field of flow measurement, ultrasonics shows considerable promise as a viable flowmeter technology for liquid applications and some gas applications. Some designs allow measurements to be made external to the pipe and utilize no wetted parts, while other designs require that the sensor be in contact with the flowstream. As a result, in some designs the sensor is clamped onto the flowstream pipe, while in other designs a section of pipe is supplied by the manufacturer with the sensors already mounted for insertion into the flowstream.

While ultrasonic flowmeters may function reasonably well in the flow laboratory, only mixed success has been achieved thus far in industrial flow applications. The user should be aware of manufacturer claims and is well advised to experiment, if practical, with a clamp-on style to determine if a permanent application, perhaps with wetted transducers and requiring piping modifications, is feasible.

Principle of Operation Ultrasonic flowmeters use acoustic waves or vibrations to detect the flow traveling through a pipe. Ultrasonic energy is typically coupled to the fluid in the pipe using transducers that may be wetted or non-wetted, depending upon the design of the flowmeter. Time of flight and Doppler measurement techniques are available.

Doppler The Doppler effect can be illustrated by the change in frequency that occurs when a vehicle approaches a bystander with its horn on. As the vehicle approaches, the horn is perceived by the bystander to be higher pitched since the velocity of the vehicle causes the sound waves to be more closely spaced than if the vehicle were standing still. Likewise, the horn is perceived to be lower pitched as the vehicle moves away from the bystander; the sound waves tend to become farther apart, resulting in a lower frequency. The Doppler shift is proportional to the relative velocity along the path between the source and the observer.

Doppler ultrasonic flowmeters utilize the Doppler effect to detect and measure flow in a pipe. A transducer transmits continuous or pulsed (modulated) acoustic energy into the flowstream to a receiver. Under no flow conditions, the frequency received is identical to the frequency at the transmitter; however, when there is flow, the frequency reflected from particles or bubbles in the fluid is altered linearly with the amount of flow through the pipe due to the Doppler effect. The net result is a frequency shift between the transmitter and the receiver that is linearly proportional to flow. The two signals are then "beat" together to generate a frequency signal at the difference between the transmitted and received frequencies, which is then converted to an analog signal proportional to flow.

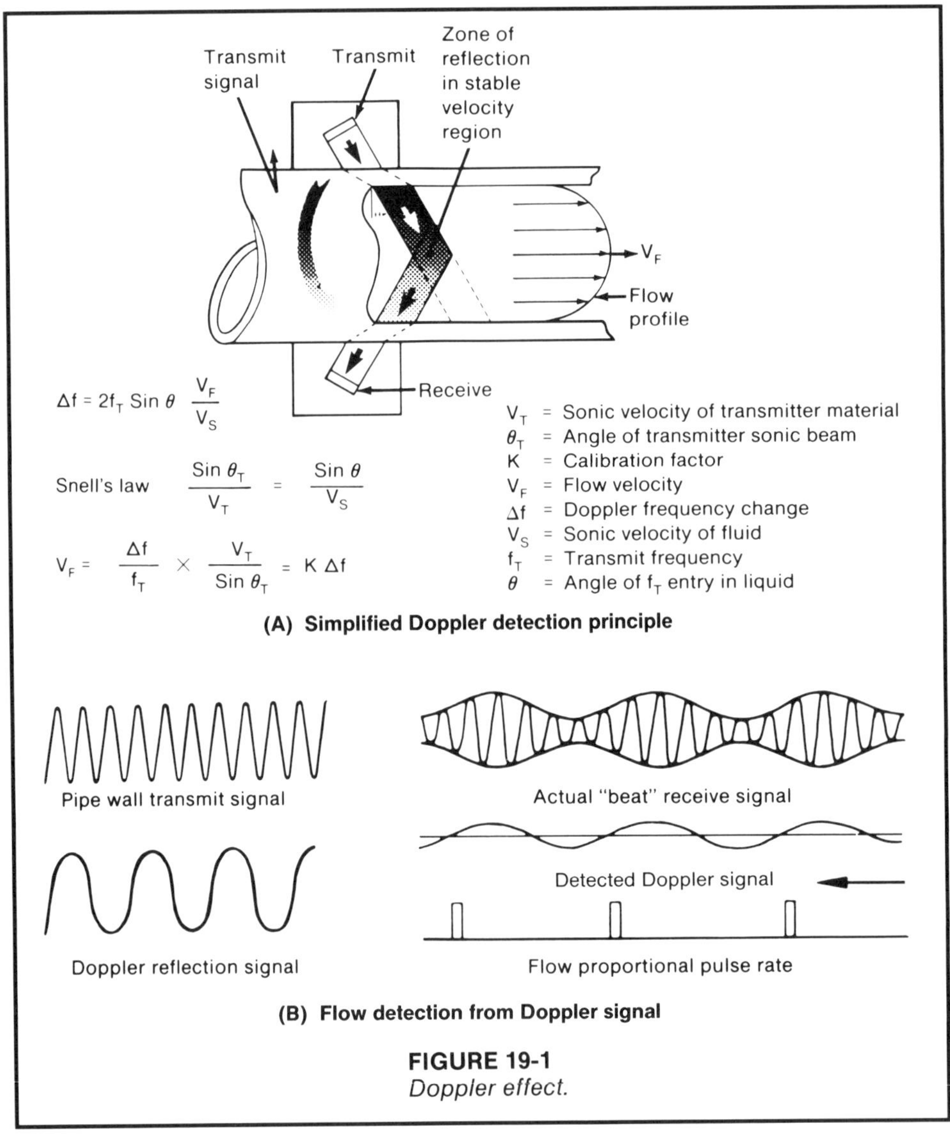

FIGURE 19-1
Doppler effect.

Most designs have two transducers, one each for transmitting and receiving, while some designs utilize a common transducer to achieve both functions.

Time of Flight Time of flight ultrasonic flowmeters measure the difference in travel time between pulses transmitted along and against the fluid flow and beamed at an angle in the pipe. One transducer is located upstream of the other and the times of transit of the ultrasonic beam in the upstream and downstream directions are measured over the same path and used to calculate the flow through the pipe, as illustrated in Figure 19-2.

Clamp-on transducers that utilize the time of flight principle are usually capable of retransmitting sooner and operating faster as the sonic echo is away from the receiver and is not caught up in an "echo chamber" as is the case with inserted transducers that face each other. However, inserted transducers typically make better sonic contact with the fluid. Any variations in sonic velocity due to fluid property changes will affect the performance of the flowmeter.

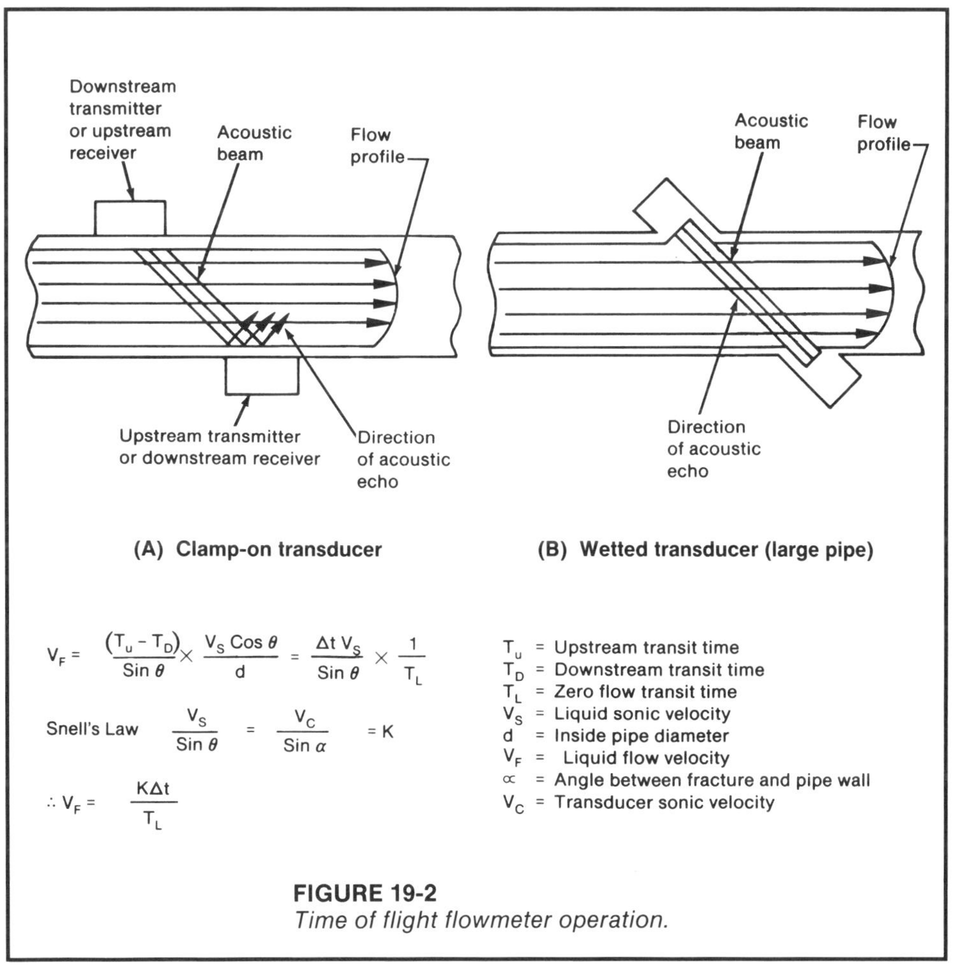

FIGURE 19-2
Time of flight flowmeter operation.

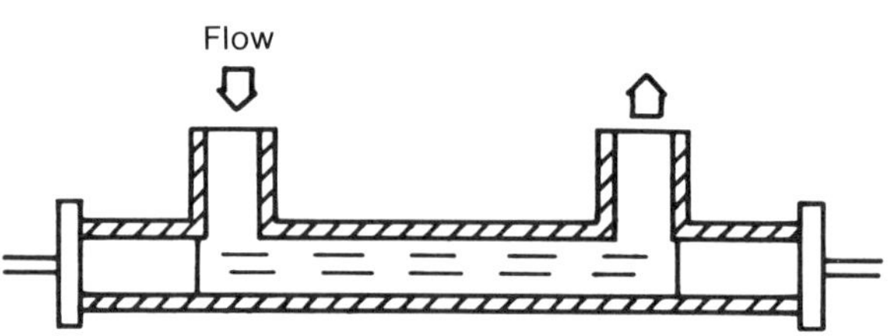

(C) Wetted transducer (small pipe)

Mathematical Proof*

Let L = Distance between transducers
θ = Angle at which transducers are mounted
V = Average flow velocity in the pipe
C = Speed of sound in the fluid
ΔF = Frequency difference

The component measured by the flowmeter in the downstream direction (with the flow) is:

$$\text{Downstream} = \frac{(C + V\cos\theta)}{L}$$

The upstream component (against the flow) is:

$$\text{Upstream} = \frac{(C - V\cos\theta)}{L}$$

Substracting the upstream from the downstream measurement:

$$\Delta F = \frac{(C + V\cos\theta)}{L} - \frac{(C - V\cos\theta)}{L}$$

$$= \frac{C}{L} + \frac{V\cos\theta}{L} - \frac{C}{L} + \frac{V\cos\theta}{L}$$

$$= \frac{2V\cos\theta}{C}$$

*C (speed of sound) disappears in the final equation. This is because the distance (L) and relative angle (θ) between the two transducers remain constant, making the frequency difference (ΔF) proportional to the average flow velocity in the pipe.

Figure 19-2
(continued)

Clamp-on transducers can be designed to generate shear or axial beams in the pipe wall (see Figure 19-3). Each type has its advantages and limitations; for example, shear mode ultrasonic energy is transmitted into the fluid so that the beam signal at the receiver shifts in time and position as the flow varies. As a result, if the sonic properties of the liquid vary significantly, the beam could conceivably miss the receiver and not be sensed. Axial beam injection avoids this potential problem by transmitting the ultrasonic energy axially along the pipe. As a result, the placement of the receiver is not critical, and the flowmeter is not very sensitive to changes in liquid sonic velocity.

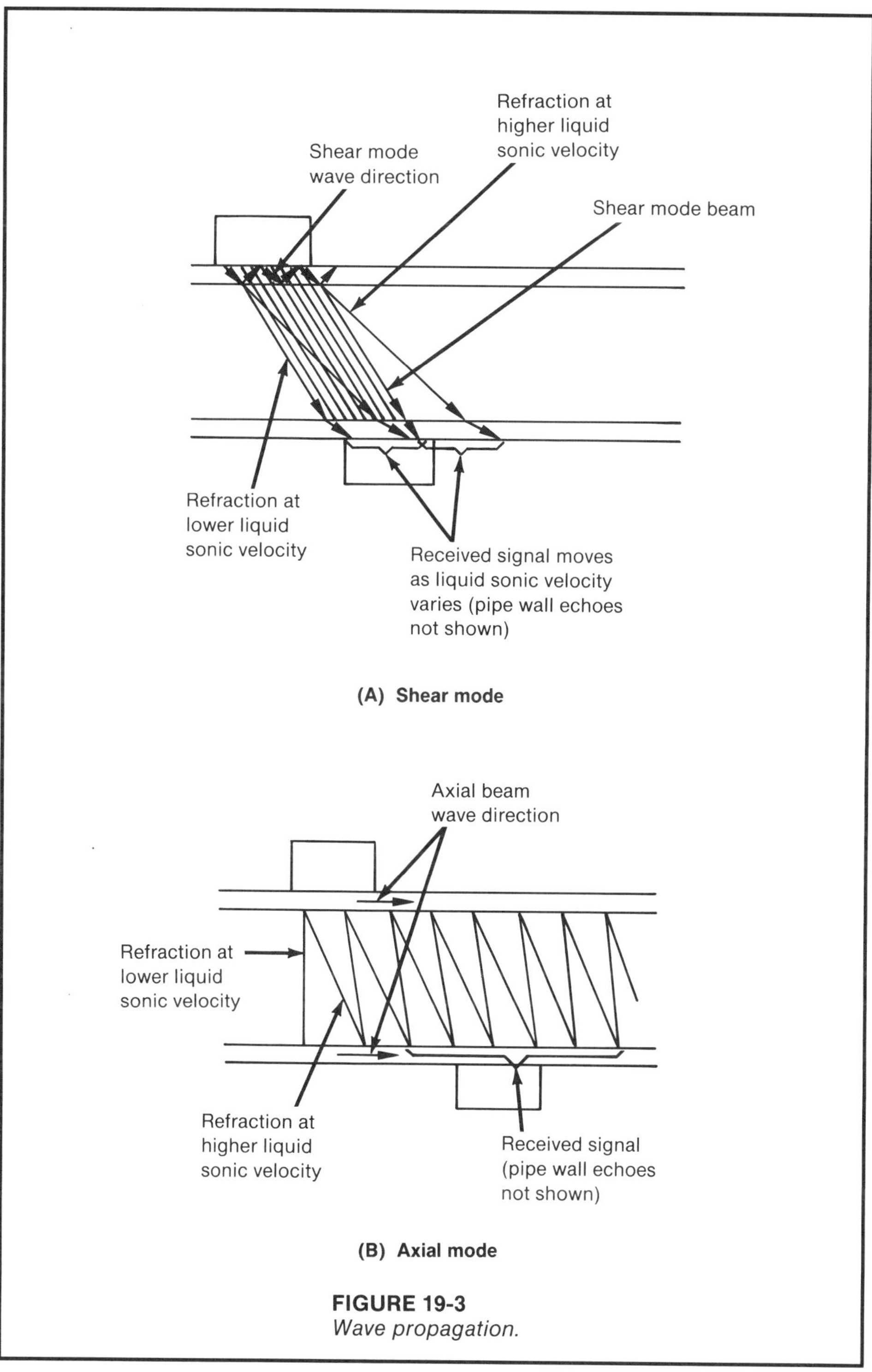

FIGURE 19-3
Wave propagation.

Differential Frequency Differential frequency ultrasonic flowmeters are time of flight devices in which the transducer is positioned so that the ultrasonic energy is beamed at an angle in the pipe. One transducer is located upstream of the other, and the frequencies of the ultrasonic beam in the upstream and downstream directions are detected and used to calculate the flow through the pipe, as shown in Figure 19-2(c).

The frequency shift is linearly proportional to the velocity of the fluid and independent of the velocity of sound in the fluid.

Construction

Construction of ultrasonic flowmeters can be classified by the mounting of the transducers as either clamp-on or wetted. Clamp-on transducers offer convenience and, in some cases, rather good accuracy. Wetted transducers are usually required for more accurate liquid measurement, especially when multiple ultrasonic paths are needed. Time of flight measurement of gas flow virtually always requires wetted transducers.

Clamp-on Transducer Clamp-on transducers are attached to the pipe externally, typically with a pipe clamp on a small pipe. As there are no wetted parts, fluid compatibility is not a consideration. Clamp-on designs typically employ one or two transducers, depending upon manufacturer (see Figure 19-4).

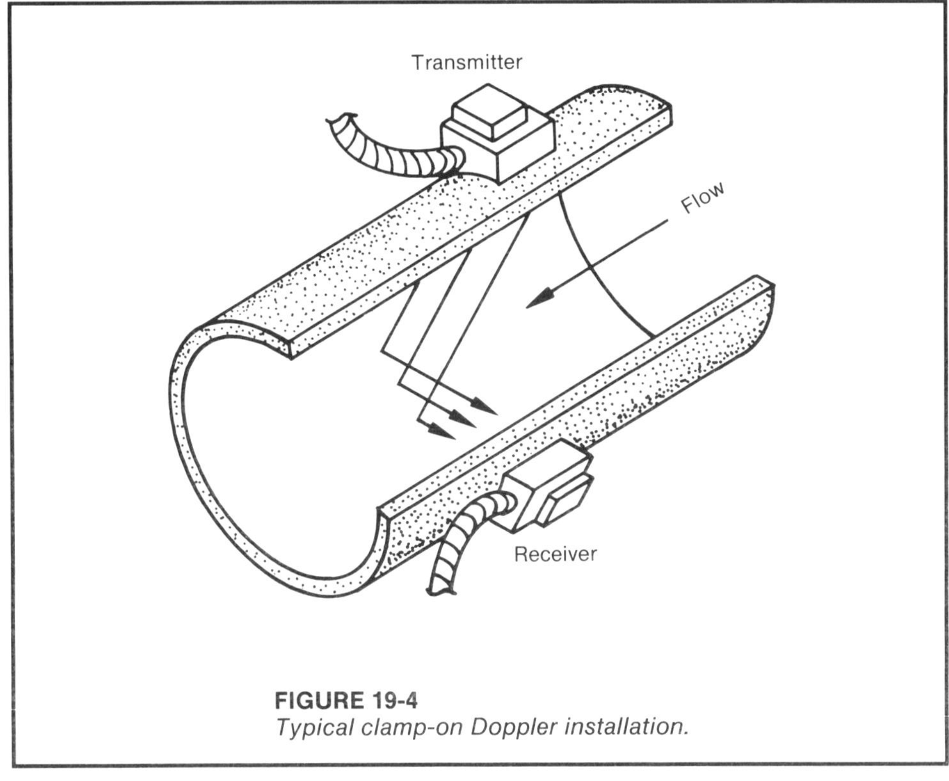

FIGURE 19-4
Typical clamp-on Doppler installation.

Wetted Transducer The flowmeter body houses the transducers in their proper orientation and permits direct contact with the fluid, usually resulting in a superior signal-to-noise ratio. Typically, two or more transducers are required for this design. As the body and the transducers are wetted, attention must be paid to the materials of construction, which are typically stainless steel in the smaller sizes and carbon steel in the larger sizes. Other materials are also available. Some designs allow removal of transducers from the body while the liquid is flowing through the pipe, while others require that the flow be interrupted and the pipe be drained for transducer removal.

Wetted Parts

Clamp-on transducer designs have no wetted parts, other than those exposed to the surrounding atmosphere, while the wetted transducer design requires that the transducer, any required seals, and the flowmeter body be wetted. Metal parts are typically of stainless steel, although other materials of construction are available.

Operating Constraints

Ultrasonic flowmeters are available for measurement in sonically conductive pipes greater than 1/8 inch in size. They can measure flows greater than approximately 0.1 gpm at temperatures of up to approximately 260°C. Pressure for wetted transducers is limited by the flowmeter flange rating and sensor design. While most ultrasonic flowmeters measure liquid flow only, time of flight designs that measure gas flows are available.

As the ultrasonic energy passes through only part of the liquid being measured, Reynolds number affects the performance of the flowmeter. Some Doppler and differential time flowmeters require minimum Reynolds numbers of 4000 and 10,000, respectively, in order to perform within their stated specifications. The differential frequency design can operate in the laminar flow regime at Reynolds numbers of less than 2000, and in the turbulent flow regime at Reynolds numbers of over 4000. Most flowmeters are nonlinear in the transition regime of Reynolds numbers of 2000 to 4000.

Most Doppler flowmeters require that some entrained gas or particles be present in the fluid to reflect ultrasonic energy to indicate the velocity of the flowstream to the receiving transducer. Maximum allowable entrained gas or solids varies with design of the flowmeter and is usually not specified. Should the percentage be above the maximum, the ultrasonic energy will not sufficiently penetrate the flowstream, resulting in a loss of accuracy.

The velocity of the fluid must be above the threshold velocity of the flowmeter, which is typically between 0.1 foot per second for the time of flight and differential frequency designs, and 0.5 foot per second for the Doppler design.

Clamp-on ultrasonic flowmeters typically require that the thickness of the pipe wall be small in relation to the distance that the ultrasonic energy passes through the fluid. As a rule of thumb, the ratio of the pipe diameter to the wall thickness should be greater than 10:1.

Performance

Ultrasonic flowmeter accuracy, repeatability, and linearity are typically in the ranges of ±0.5 to 10 percent FS, 0.1 percent FS and 0.15 to 0.5

percent FS, respectively. It should be noted that manufacturers often state flowmeter performance in terms of percent without stating whether this is percentage of rate, full scale, or meter capacity. As some flowmeters are specified as a percent of meter capacity, the manufacturer should be consulted when there is any doubt as to which specification is intended. A flowmeter that is specified as a percent of meter capacity will exhibit significant errors at velocities encounted in typical applications, as meter capacity typically represents a velocity of 40 feet per second. It should also be noted that some specifications may reflect operation of the flowmeter under simulated conditions as opposed to operating conditions, which does not accurately define the expected performance of the flowmeter.

There is little independent flow test data for ultrasonic flowmeters to confirm or deny manufacturers' accuracy claims. Nevertheless, the differential frequency and time of flight technologies generally achieve better performance than flowmeters using Doppler technology.

The time of flight technologies transmit signals that usually travel through the entire flowstream between transducers on opposite sides of the pipe, while the Doppler technology relies on reflections of ultrasonic energy from particles or entrapped gas in the flowstream. Doppler technology sometimes has the added uncertainty of the depth of penetration of the ultrasonic energy; the velocity profile, fluid properties, or fluid composition change can result in errors of greater than 30 percent under process conditions. In other words, there is uncertainty as to whether a flowmeter using Doppler technology is measuring the average velocity in the pipe or some other velocity, since the depth to which the ultrasonic energy penetrates the flowstream is not well defined, especially as the amount of particles or entrapped gas varies. Slurries are particularly susceptible to large shifts in accuracy; the particles can cause the slurry to be opaque to ultrasonic energy, causing lack of penetration into the flowstream and hence a considerable loss of accuracy or loss of signal.

Applications Doppler flowmeters can be applied to fluids that have some amount of entrained gas or particles to reflect ultrasonic energy. Differential frequency and time of flight technologies can measure flows of clean liquids as well as liquids that contain up to approximately 30 percent solids, depending upon manufacture. Clamp-on sensor designs require that the pipe be sonically conductive, as the ultrasonic energy must be efficiently transmitted to and received from the liquid being measured.

Ultrasonic flowmeters can be applied to pipes of all sizes. Since the flowmeter element is virtually the same above certain sizes, this technology has economic advantages over other flowmeter technologies in applications in large pipe.

Sizing In general, ultrasonic flowmeters are the same size as the pipe size to take advantage of the obstructionless design of the flowmeter, unless the flow is such that the Reynolds number and velocity constraints are not satisfied. In such a case the flowmeter size may be altered as necessary. Compensation for pipe size is usually performed electronically in the transmitter, so field modification of the transmitter to another size pipe is usually possible. Wetted transducers, typically applied to differential frequency, time of flight, and some Doppler designs, require that the flow primary be changed when the pipe size is changed. Clamp-on sensors may require replacement if a different amount of ultrasonic power is required to penetrate a different size pipe.

EXAMPLE 19-1

Problem: Size a Doppler flowmeter for a 100 gpm full scale flow of a liquid with a specific gravity of 1.0 and a viscosity of 1.0 cP

Solution: Assuming that Doppler flowmeters operate in a velocity range of 0.5 to 40 feet per second and typical liquid velocities are 6 to 8 feet per second, a 2-inch flowmeter could be applied and would operate at a velocity of 9.56 feet per second at full scale. Reynolds number can be calculated as follows:

$$R_D = (3160 \times 100 \text{ gpm} \times 1.0) / (1.0 \text{ cP} \times 2.067 \text{ in.}) = 152{,}879$$

which is sufficiently high to ensure that the flow operates in the turbulent flow regime for all applicable flows.

EXAMPLE 19-2

Problem: Size a Doppler flowmeter for a 60 gpm full scale flow of a liquid with a specific gravity of 1.2 and a viscosity of 40 cP.

Solution: Typical liquid design velocity is 6 to 8 feet per second, so a 2-inch flowmeter could be applied and would operate at 5.74 feet per second at full scale flow. Reynolds number is calculated by:

$$R_D = (3160 \times 60 \text{ gpm} \times 1.2) / (40 \text{ cP} \times 2.067 \text{ in.}) = 2752$$

and is found to be in the transition flow regime, which is unsatisfactory.

If the size of the flowmeter were decreased, the velocity as well as Reynolds number will increase. The increase in Reynolds number will not be sufficient to ensure that part of the desired flow measurement range will not be in the transition flow regime, so another alternative should be pursued.

Increasing the size of the flowmeter to 3 inches reduces Reynolds number as well as the velocity, so that the flowmeter can be operated totally in the laminar flow regime with a maximum velocity of 2.60 feet per second. As the differential frequency technology can be applied to laminar flow and can measure velocities as low as 0.1 foot per second, a 3-inch flowmeter would be applicable:

$$R_D = (3160 \times 60 \text{ gpm} \times 1.2) / (40 \text{ cP} \times 3.068 \text{ in.}) = 1854$$

Installation

Proper installation of ultrasonic flowmeters is important to proper operation. As the performance of most ultrasonic flowmeters cannot be verified in typical applications involving large pipe sizes, manufacturer recommendations should be followed as closely as possible to achieve the best performance possible.

Hydraulic Requirements Ultrasonic flowmeters, which are sensitive to the velocity profile entering the flowmeter, require 10 to 30 D/5 to 10 D upstream and downstream straight run, depending upon manufacture and technology. In general, increasing the straight run of the flowmeter will decrease the possibility of shifts in measurement due to an improperly developed velocity profile at the inlet of the flowmeter.

Piping Orientation As gas or solids collecting at or flowing on a transducer can affect the transmission of ultrasonic energy into the flow, thereby affecting the accuracy of the measurement, ultrasonic transducers should be orientated in a manner to eliminate this possibility. This can be accomplished by locating the transducers in the horizontal plane.

Piping Vibration The frequencies at which ultrasonic flowmeters operate are usually selected to be outside the realm of frequencies at which pipes will vibrate. Nevertheless, it may be possible for the receiving transducer to respond to shock or high intensity vibration. As many transducers are temperature- and moisture-sensitive, care should be exercised to avoid attributing all "unidentifiable responses" to vibration.

Sensor Mounting Clamp-on transducers typically require that a coupling material be applied to the pipe and/or transducer before installation so as to provide satisfactory acoustic contact.

Cabling Most ultrasonic flowmeters are 4-wire devices that have maximum distance limitations between the transmitter and the transducers. Special cable is usually required between the transmitter and the transducers to minimize attenuation of signals.

Maintenance

Ultrasonic flowmeters require no routine maintenance other than routine calibration checks. Problems such as transducer failure, lack of sufficient contact between the transducer and the pipe wall, and electronic failures can occur.

Transducer Failure Difficulty of transducer replacement is dependent upon transducer design. Replacement may require interrupting flow and opening the pipe, such as in the case of wetted transducers that have no valving arrangement with which to isolate the transducer from the pipe. Clamp-on transducers, which are mounted externally to the pipe, can be replaced without interrupting flow.

Loss of Contact between Transducer and Pipe Wall Materials used to improve acoustic coupling between a clamp-on transducer and the pipe can become ineffective over a period of time due to dehydration or material loss. The lack of proper coupling reduces the ultrasonic energy by reflection, often to the point of causing the flowmeter to cease to operate. This condition can be corrected by removing the transducers and replacing the conducting material per manufacturer specifications.

Electronic Failure Electronic failures can occur and are usually remedied by board replacement. It should be noted that process data and calibration information may need to be entered into a replacement board.

Spare Parts Spare parts inventory varies with flowmeter design, but the transducer and any associated mounting hardware such as gaskets should be stocked. Identical transducers are usually used for many pipe sizes, while the transmitter for a particular design is typically identical for all pipe sizes, both of which minimize spare parts requirements.

Calibration Calibration of ultrasonic flowmeters is performed by electronically simulating the signals that would be present under flow conditions and making the necessary adjustments to the transmitter. A better calibration could be obtained if the flowmeter were calibrated at the manufacturer's flow facility.

EXERCISES

19.1 Size an ultrasonic flowmeter for a flow of 0 to 1400 gpm of a liquid with a specific gravity of 0.98 and a viscosity of 3.3 cP. Can Doppler technology be applied? Why or why not?

19.2 Size an ultrasonic flowmeter for a flow of 0 to 600 gpm of a liquid with a specific gravity of 1.13 and a viscosity of 150 cP. Can Doppler technology be applied? Why or why not?

19.3 Why must Doppler technology be applied to fluids with particles or bubbles?

20

Variable Area Flowmeters

Introduction In the past, variable area flowmeters have been one of the mainstays in flowmeter technology since they provide economical local readouts and control of gases and nonviscous liquids in pipes up to approximately 3 inches in size. While variable area flowmeters have been displaced to some degree by other technologies, the technology has maintained its place in many applications due to its design simplicity and its ability to be tailored to each application by judicious selection of components that comprise the flowmeter.

Rotameters, having once been used in a great percentage of flowmeter applications, have been superseded in many areas where other technologies have eliminated restrictions in flowmeter mounting and the requirement for moving parts. Rotameters still retain many applications where reasonable performance at an economical cost is desired.

Principle of Operation Rotameters operate on the principle of generating a condition of dynamic balance within the flowmeter in which a float is positioned in accordance with the flow through the flowmeter. The float remains in dynamic balance when the sum of the forces acting on the float are zero. Therefore, when the weight of the float less the weight of the fluid that it displaces is equal to the upward force on the float due to fluid velocity, the float is in dynamic balance, as illustrated in Figure 20-1.

With an increase in flow, the float will tend to rise in the metering tube, since the upward fluid force and the buoyant effect of the float exceeds the downward force of gravity. As the float rises, the annular area between the float and the tapered metering tube increases until the upward and downward forces are equalized in dynamic balance. The level of the float in the metering tube is indicative of the flow through the flowmeter.

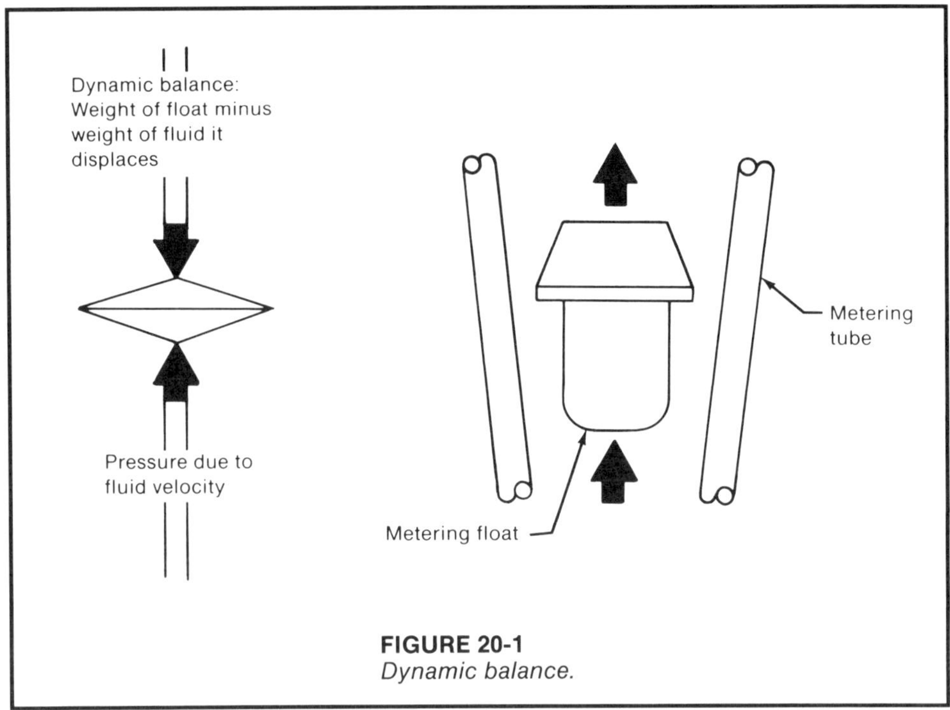

FIGURE 20-1
Dynamic balance.

Construction Rotameters can be classified as being of standard or of armored construction. The standard material of construction for the metering tube is typically borosilicate glass, while armored construction utilizes more rugged materials, typically stainless steel although other metals may be available.

Metering Tube Standard borosilicate glass metering tubes can be read directly as the float is visible in the tube. However, the glass is subject to breakage since glass has many natural enemies in an industrial environment such as tools, falling objects, and elbows. Due to the possibility of breakage as well as the possibility of bursting from overpressure, a plastic protection tube that covers the metering tube is often specified. This can avoid injuries to personnel if the glass fails.

In certain applications such as hazardous or dangerous fluids, high temperatures, high pressures, and flows that exhibit high shock levels, glass metering tubes are not desirable. These applications can be safely measured with rotameters of armored construction and compatible materials such as stainless steel.

Metering tubes are available in various shapes, which determine the type of scale the rotameter will have, and in various cross sections, which aids in guiding the float, as shown in Figure 20-4.

In smaller sizes, numerous metering tubes fit the same end connections, resulting in the ability to change the range of the flowmeter by changing the metering tube.

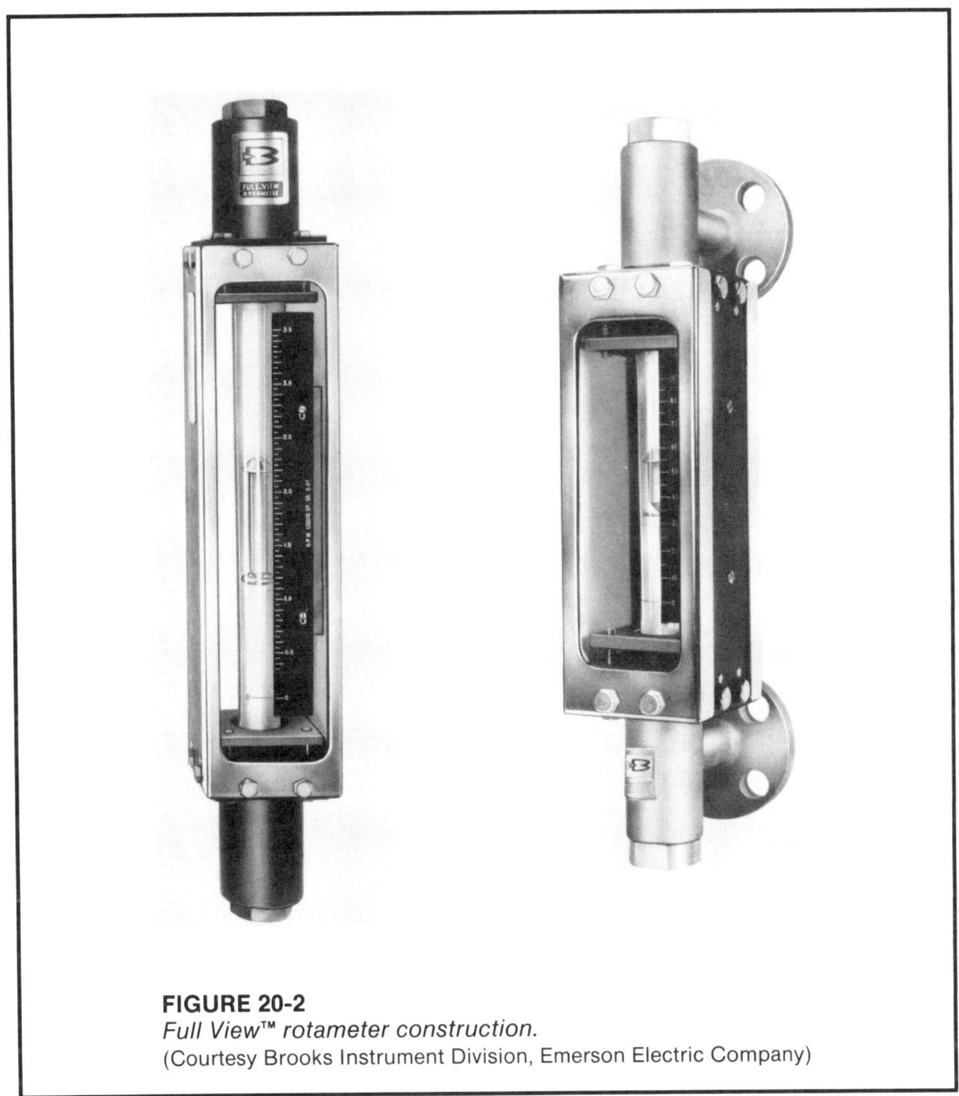

FIGURE 20-2
Full View™ rotameter construction.
(Courtesy Brooks Instrument Division, Emerson Electric Company)

Floats Various float designs are available that can be used to tailor a rotameter to a particular application, as illustrated in Figure 20-5. The ball float is used in low flow rotameters, also called purge meters. The streamlined float is most economical in the larger sizes, as its shape provides high flow capacity in a given metering tube size. Floats with varying degrees of viscosity compensation are available. However, as the annular area is reduced in size, the size of the flowmeter is generally larger than that of a rotameter with a streamlined float that can be used to measure the same flow.

Floats are available with different materials of construction, each having a different density. The range of a rotameter can be changed by changing the

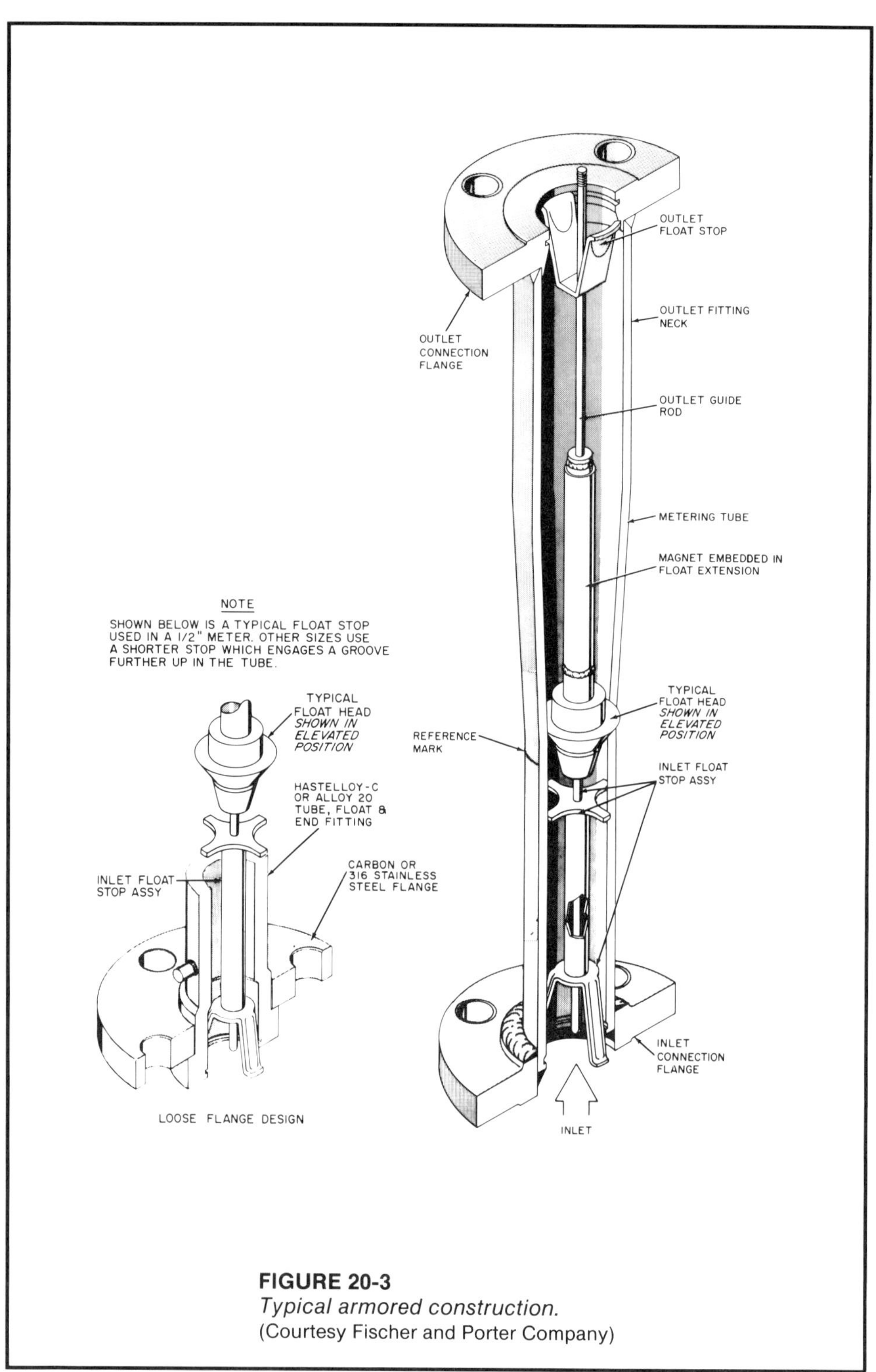

FIGURE 20-3
Typical armored construction.
(Courtesy Fischer and Porter Company)

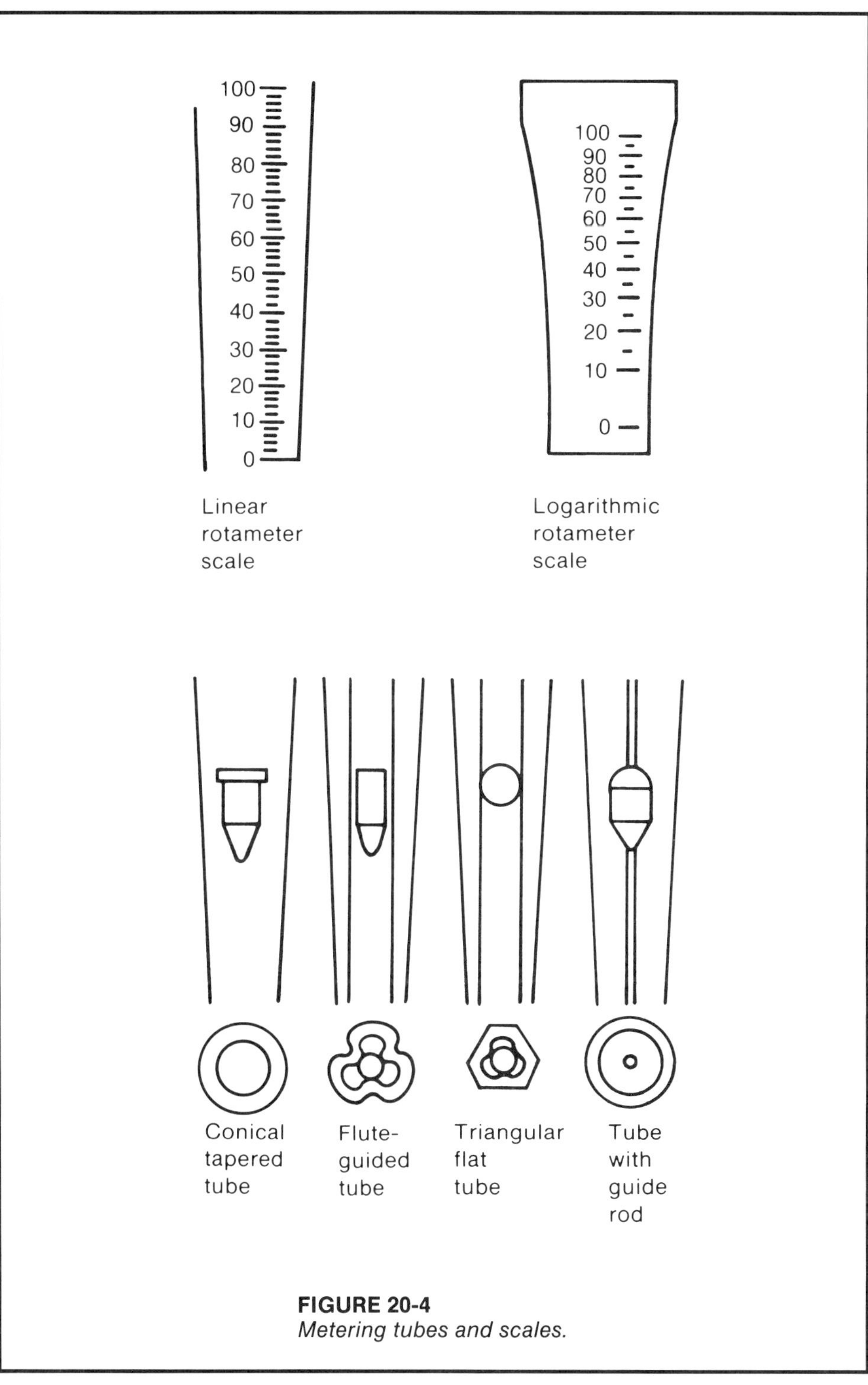

FIGURE 20-4
Metering tubes and scales.

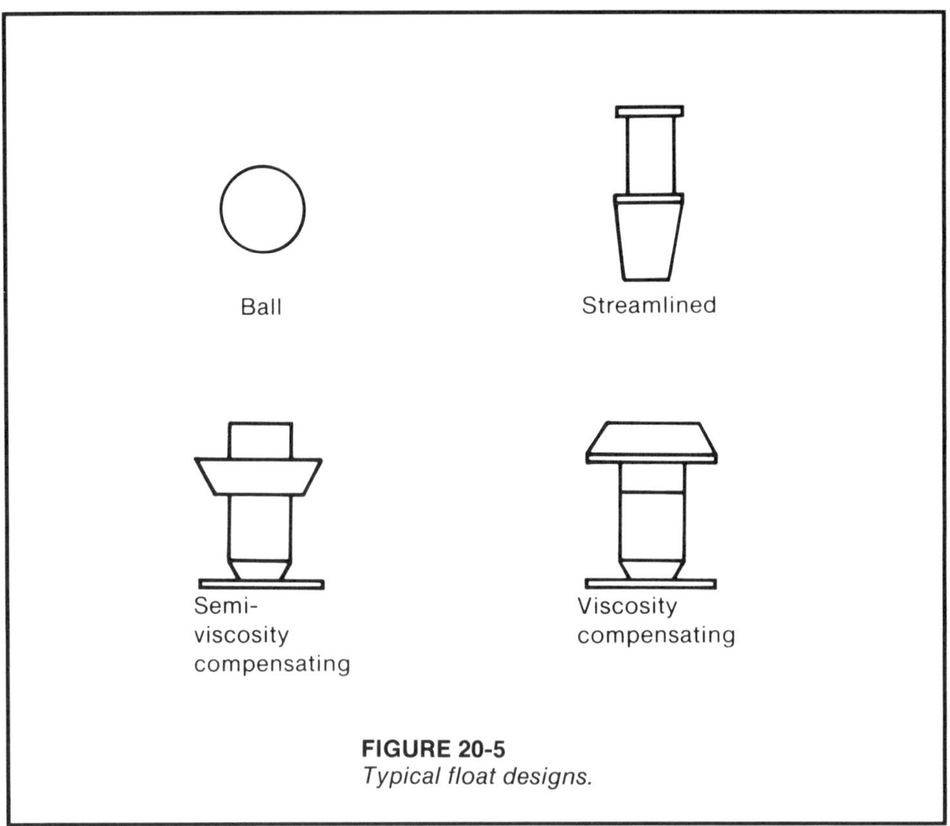

FIGURE 20-5
Typical float designs.

materials of construction of the float, provided that the materials are compatible with the fluid. Therefore, rotameters are flexible in the sense that the range can be changed, although the rotameter must be removed from the pipe and disassembled in order to do so.

Scales Scale lengths range from 1-1/2 in. to 24 in. depending on application. Small scales are used for economic low flow applications, while scales up to 10 in. are typically used for industrial applications. Larger scales are generally used to achieve higher accuracy and resolution, usually in laboratory applications.

End Fittings Mechanical strength is provided to most rotameters by mounting the metering tube in a metal case that also includes the inlet and output connections. Depending upon design, the user may be able to specify side, bottom, or rear inlet and outlet connections. The end fitting assembly at each end of the flowmeter includes the inlet or outlet connection as well as a sealing or packing arrangement between the metering tube and the case. Many of these configurations utilize O-rings for positive sealing. Seal adjustments may be made without disassembling the flowmeter when packing glands, and the like, are accessible external to the case.

Sensing Systems Most metal tube rotameters magnetically translate motion of the float in the fluid into motion external to the flowmeter. Usually movement of the float with a magnet causes a following magnet external to the flowmeter to move or rotate. The position or orientation of the following magnet is sensed and transmitted (see Figure 20-6).

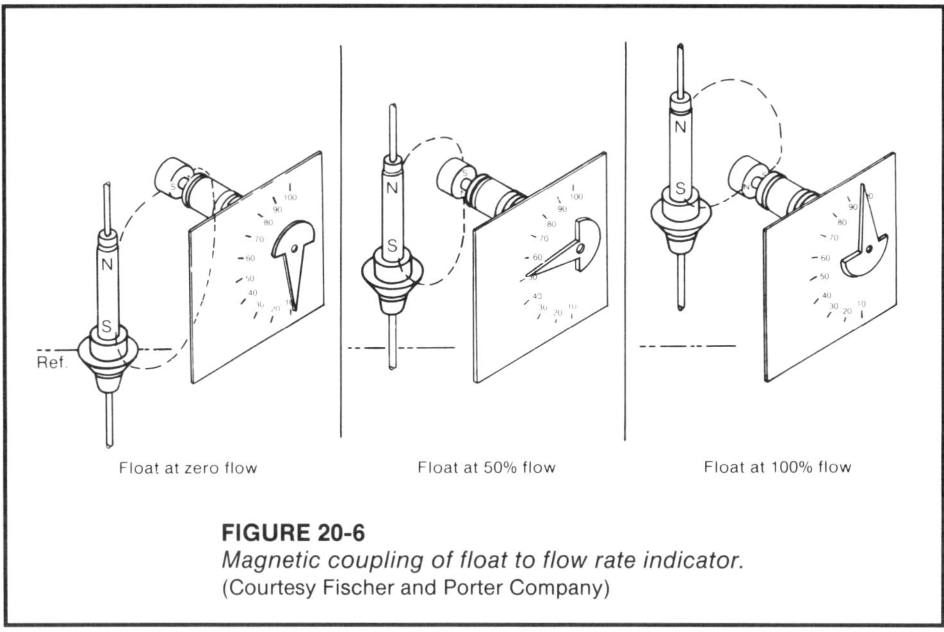

FIGURE 20-6
Magnetic coupling of float to flow rate indicator.
(Courtesy Fischer and Porter Company)

Wetted Parts Wetted parts include the metering tube, the float assembly, and the end connections. The metering tube is generally available in borosilicate glass or stainless steel, although other metals are available. The float assembly is available in a number of materials such as stainless steel, tantalum, Hastelloy®, Monel®, glass, Teflon® and Alloy 20, and is selected to yield the correct flow range and to be compatible with the fluid. End connections are available in materials such as steel, stainless steel, and other materials. Packing materials include neoprene, Teflon®, and Viton®.

Operating Constraints

Rotameters, which are available in sizes up to 3 in., can be used to make liquid flow measurements of approximately 0.05 to 200 gpm on liquids with a viscosity of less than approximately 30 cP. With the proper float design, the minimum Reynolds number is between 250 and 950 for flows under 90 gpm, and 4400 for flows above 90 gpm. Gas flows have no Reynolds number constraints but must be of sufficient density to operate the flowmeter.

Armored rotameters are pressure-limited by their flange rating, and some designs measure fluids with temperatures of up to 320°C. Borosilicate glass rotameters of standard design are pressure-limited by the strength of the glass, typically to 50 psi.

Performance Wet calibrated industrial rotameters can be accurate to within ±1 to 2 percent FS, while some flowmeters can be calibrated to ±1/2 percent FS. The accuracy of purge meters is typically ±5 percent FS, which is usually more than sufficient for many applications. Changes in the density of the fluid cause a change in the upward force of the fluid on the float, resulting in a measurement shift that is predictable when the percentage density change is known. A 1 percent change in density results in a -1/2 percent change in the flow measurement.

Rotameters are typically accurate from 10 to 100 percent of the calibrated range, which effects a 10:1 turndown. The calibrated range cannot be changed unless another type of float or metering tube is installed in the flowmeter, in which case additional uncertainties are added into the measurement since the flowmeter is not wet calibrated with the new parts. As a result, the range of a rotameter can be modified, but the flowmeter must be disassembled to effect the change and that will result in a loss of accuracy. Loss of accuracy will also occur when the flowmeter is operated at less than the minimum Reynolds number constraint.

Applications These devices are applicable to liquids that exceed the applicable Reynolds number constraints and have a viscosity of less than approximately 30 cP. The liquid need not be clean, but any deposits that are allowed to form on the metering tube or float will affect measurement and visibility through a glass metering tube.

Due to the ample selection of tubes and floats, rotameters can handle virtually all gas flow applications where the gas is operated at sufficient density and flow to raise the float.

Sizing Sizing consists of the selection of the proper metering tube and float for a given application and is performed by converting the fluid flow to an equivalent flow of water or air. A rotameter is selected by comparing the calculated equivalent flows to capacity information for air or water flow and viscosity immunity data supplied by the rotameter manufacturer.

The following equations can be used to convert the desired flow of the fluid to equivalent flows in gpm of water and scfm of air for a rotameter with a 316 stainless steel float, which is consistent with the data presented by most manufacturers.

$$Q_{\text{water}} = Q_{\text{gpm}} \times [(7.04 \times \text{SG})/(\text{SG}_{\text{float}} - \text{SG})]^{1/2}$$

$$Q_{\text{air}} = Q_{\text{acfm}} \times [(8.04\ \text{SG}_{\text{ref}} \times T \times P_{\text{ref}})/(\text{SG}_{\text{float}} \times T_{\text{ref}} \times P)]^{1/2}$$

$$= 3.65\ Q_{\text{lb/min}} \times \rho_{\text{lb/ft}^3}{}^{-1/2}$$

The specific gravity of other float materials is listed as follows:

Aluminum	2.80
Durimet®	8.02
Hastelloy B®	9.24
Hastelloy C®	8.94
Monel®	8.84
Nickel	8.91
316 stainless steel	8.04
Tantalum	16.60
Teflon®	2.20
Titanium	4.50

In liquid service, the maximum viscosity should be calculated to determine the viscosity immunity of the rotameter under operating conditions. Caution should be used in gas applications since pressure and temperature fluctuations will result in flow measurement error. A gas pressure regulator with a local pressure indicator is often used upstream of a rotameter to control the inlet pressure.

EXAMPLE 20-1

Problem: Calculate the equivalent water full scale flow range of a rotameter with a stainless steel float for a full scale flow of 100 gpm of a liquid with an operating specific gravity of 0.90.

Solution:

$$Q_{water} = 100 \text{ gpm} \times [(7.04 \times 0.90)/(8.04 - 0.90)]^{1/2}$$
$$= 94.2 \text{ gpm}$$

Assuming that a rotameter with an equivalent full scale range of 110 gpm of water is selected, the full scale flow of the liquid would be 100 × 110/94.2, or 116.8 gpm.

EXAMPLE 20-2

Problem: Calculate the equivalent air full scale flow range of a rotameter with a stainless steel float for a full scale flow of 1000 scfm of a gas with a specific gravity of 1.04 at reference conditions of 70°F and 14.7 psia, when the gas is operated at 50 psi and 110°F.

Solution:

$$Q_{air} = 1000 \text{ scfm} [(8.04 \times 1.04 \times [460 + 110] \times 14.7)/8.04 \times (460 + 70) \times (14.7 + 50)]^{1/2}$$
$$= 4256 \text{ scfm}$$

Assuming that a rotameter with an equivalent full scale range of 5000 scfm of air is selected, the full scale flow of the liquid would be 5000 × 1000/4256, or 1175 scfm.

When flowing conditions change from the original design conditions, the effects can be calculated by taking the ratio of the above equations.

EXAMPLE 20-3

Problem: Determine the effects on flow measurement of a rotameter that is designed to operate on a 500 scfm gas flow at 50 psi, 60°F, and 1.06 specific gravity, if the gas operates at 45 psi, 90°F, and 1.02 specific gravity.

Solution: Taking the ratio of the above equation and cancelling terms yields the correction factor

$$[(SG_{nom} \times T_{nom} \times P) / (SG \times T \times P_{nom})]^{1/2}$$

Substituting the above values yields a correction factor of 0.920; therefore, the full scale flow becomes 500 × 0.920, or 460 scfm, which corresponds to an 8 percent measurement error.

Installation

Hydraulic Requirements Rotameters have no hydraulic constraints and therefore require no upstream or downstream straight run in addition to that required to physically mount the flowmeter. It should be noted that rotameters that are guided with a rod may require some amount of straight run downstream to allow for physical movement of the rod during operation.

Piping Orientation As rotameter operation employs gravitational forces, the flowmeter must be mounted in the vertical position with the inlet flow at the bottom.

Piping Vibration Due to the dynamic equilibrium condition that exists within the flowmeter, rotameters are susceptible to vibration along their axis that may be present in pipes, as well as vibration from other sources such as tools, elbows, and the like. They are not particularly susceptible to transverse vibration. Pipe vibration can be minimized by adding pipe supports where necessary. Rotameters located in high traffic areas where they can be inadvertently bumped are susceptible to damage.

Cabling Most rotameter transmitters are 2-wire devices.

Maintenance

Rotameters require no routine maintenance, but material can build up on the metering tube and float, the flowmeter can plug, the glass metering tubes can fail, and electronic failure can occur.

Material Buildup Some fluids leave a residue that can build up on the metering tube and float. The result of buildup is that the annular opening becomes smaller while the weight of the float becomes larger, causing the flowmeter to measure incorrectly. If the buildup causes the float to stick or bind in its guide, the flowmeter will cease to operate. Flowmeters with glass metering tubes may become difficult or impossible to read locally.

Rotameters should not be applied to fluids where buildup can be a problem. Even relatively clean fluids will exhibit some amount of buildup in time, which can be removed by disassembling and cleaning the flowmeter.

Flowmeter Pluggage Under certain conditions sufficient dirt may be introduced into the flowmeter to plug it and cause it to cease functioning. Examples include low purge supply pressure, which allows dirty process fluid instead of clean purge fluid into the flowmeter, and the introduction of dirt into the purge fluid for whatever reason.

One way to resolve the problem of a totally plugged flowmeter is to remove and clean it. This may not be necessary in all cases. Often, decreasing the pressure at the outlet of the flowmeter (such as open to atmosphere) while increasing the pressure at the inlet of the flowmeter using clean fluid may be sufficient to break up the pluggage and free the movement of the flowmeter without disassembly. In other cases, physical persuasion such as tapping an armored flowmeter can be attempted before removing and disassembling the flowmeter.

Metering Tube Failure Glass metering tubes can fail for a number of reasons, the most likely being overpressure or mechanical damage. Metering tube replacement is effected by removing and disassembling the flowmeter, removing the damaged metering tube, inserting the new metering tube, and reassembling the flowmeter. Armored designs have all-metal construction, which virtually eliminates the possibility of metering tube failure.

Electronic Failure Electronic failure can occur and is usually remedied by board replacement.

Spare Parts Spare parts include glass metering tubes, which are subject to damage. In general, many different metering tubes could be required as each size flowmeter can accept various metering tubes. Floats may also be stocked as spare parts.

Calibration The transmitter can be calibrated for zero and span by manipulating the float to the zero and full scale positions and making the necessary zero and span adjustments, respectively.

EXERCISES

20.1 Under what conditions can a rotameter with a glass metering tube be a personnel safety hazard?

20.2 Calculate the equivalent water full scale flow range of a rotameter with a Teflon® float for a full scale flow of 60 gpm of a liquid with an operating specific gravity of 1.21.

20.3 Calculate the equivalent air full scale flow range of a rotameter with a Monel® float for a full scale flow of 150 acfm of a gas with a specific gravity of 0.97 at reference conditions of 70°F and 14.7 psia, when the gas is operated at 125 psi and 150°F.

20.4 Calculate the equivalent air full scale flow range of a rotameter designed for 700 pounds per hour of saturated steam operating at 150 psi and a density of 0.364 pounds per cubic foot.

21 Insertion Flowmeters

Introduction An insertion flowmeter can generally be described as one that utilizes technology in which the flow through a pipe is inferred from one or more strategically located transducers. Most insertion flowmeters measure the velocity at either the critical position, which is representative of the average velocity in the pipe, or the centerline position, which is representative of the velocity at the center of the pipe. Once the average velocity is measured or inferred from the centerline velocity, the flow in the pipe can be inferred and calculated mathematically. Flow computers are often justified when applying insertion flowmeters due to the large fluid volumes that are measured. Flow computers can minimize errors due to hydraulic considerations that are inherent in insertion flowmeter technology.

There are many applications for insertion flowmeters, but careful attention must be paid to hydraulic and piping design to minimize piping effects. These flowmeters are typically applied in large pipes where they result in a negligible fluid pressure drop and are more economical than full-bore flowmeters. When variations in fluid properties are to be compensated for, additional measurements such as density, pressure, or temperature may be necessary.

Principle of Operation The velocity of an ideal fluid is constant throughout the cross section of the pipe. The velocity profile can be thought of as piston-like in nature and is represented as a straight line, as illustrated in Figure 21-1.

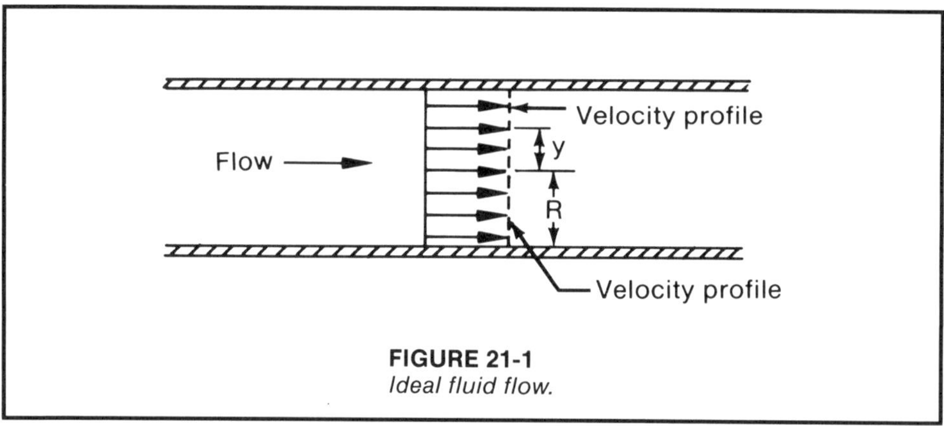

FIGURE 21-1
Ideal fluid flow.

Therefore, sampling of the velocity at any point in the flowstream represents the average velocity of the fluid in the pipe since the local and average velocities are the same. The total flow in the pipe can then be calculated using

$$Q = A \times v$$

EXAMPLE 21-1

Problem: Determine the flow of an ideal liquid through a 2-inch pipe when the liquid velocity is 6.8 feet per second.

Solution: As the velocity of an ideal liquid is constant throughout the pipe,

$$Q = A \times v = 1/4\ \pi\ (2.067\text{ in.}/12\text{ in./ft})^2 \times 6.8\text{ ft/sec} \times 7.48\text{ gal/ft}^3 \times 60\text{ sec/min} = 71.16\text{ gpm}$$

In real applications, the viscous forces of the fluid tend to create drag within the fluid and to develop frictional forces that tend to slow the fluid down at the boundary between the fluid and the pipe wall. The net result is a velocity profile that is parabolic in nature, in which the fluid velocity is lower at the pipe wall than at the center of the pipe, as illustrated in Figure 21-2.

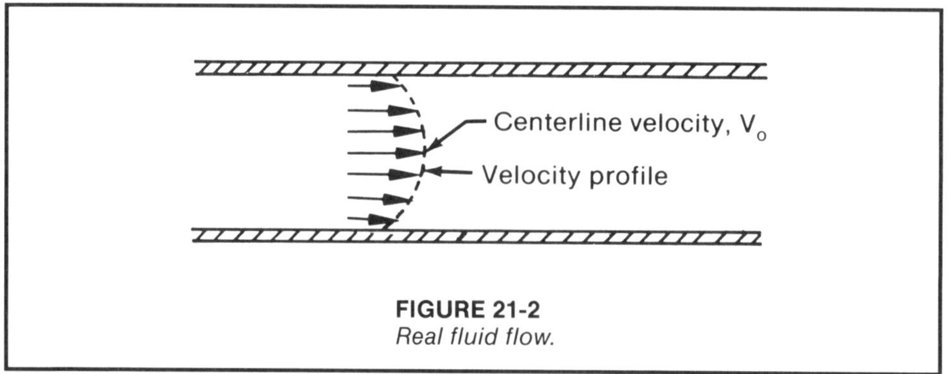

FIGURE 21-2
Real fluid flow.

The curvature of the velocity profile is a function of the viscous and momentum forces acting on the fluid in the pipe and, hence, of Reynolds number. The local velocity at a radial position in the pipe can be expressed mathematically in terms of the velocity at the center of the pipe as:

$$v_y = v_o (1 - y/R)^{1/n}$$

where n is a function of Reynolds number (selected values of which are summarized in the following table).

$$n = 3.299 + 0.3257 \ln R_D \quad \text{for } R_D \text{ under } 400{,}000$$

$$n = 5.5365 + 5.498 \times 10^{-6} (\ln R_D)^5 \quad \text{for } R_D \text{ over } 400{,}000$$

Reynolds Number	n
4×10^3	6.00
1×10^4	6.30
4×10^4	6.75
1×10^5	7.05
4×10^5	7.50
1×10^6	8.30
4×10^6	10.00

Figure 21-3 illustrates how shifts in the velocity profile occur due to Reynolds number changes. The average velocity can be found by using the following equation, which shows that the average velocity is proportional to the velocity at the center of the pipe and a coefficient that is a function of Reynolds number.

$$v_{ave} = v_o \times \frac{2n^2}{(2n + 1) \times (n + 1)}$$

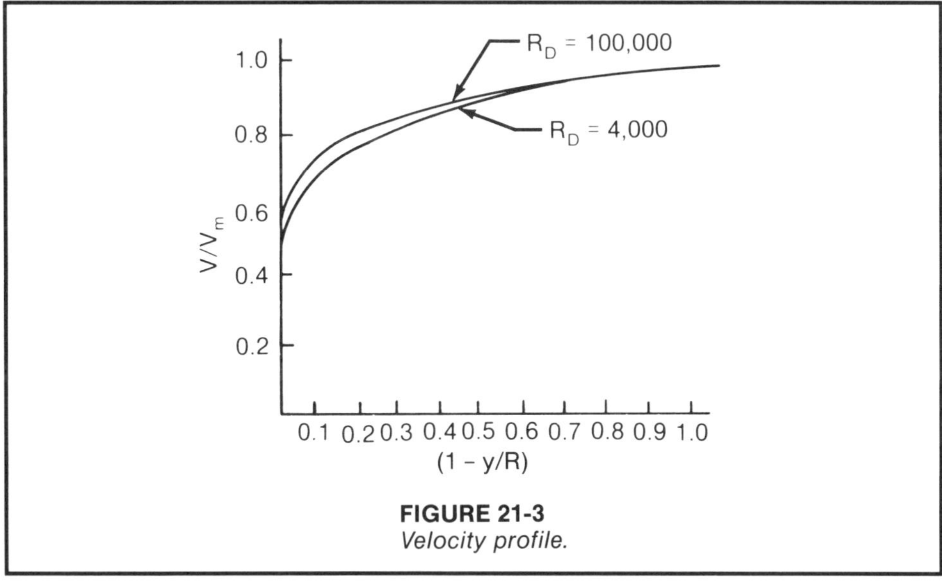

FIGURE 21-3
Velocity profile.

EXAMPLE 21-2

Problem: Calculate the velocity of a liquid 0.5 inch from the pipe wall of a 2-inch pipe operating at a Reynolds number of 40,000 and a flow of 71.16 gpm.

Solution: From the previous example, the average fluid velocity is 6.8 feet per second. Solving the above equation for the centerline velocity and substituting 6.75 for n at a Reynolds number of 40,000

$$v_o = v_{ave} \times \frac{(2n + 1) \times (n + 1)}{2n^2} = 8.4 \text{ ft/sec}$$

$$y / R = (1.033 \text{ in.} - 0.5 \text{ in.}) / 1.033 = 0.516$$

$$v_y = v_o (1 - y/R)^{1/n} = 8.4 \text{ ft/sec} (1 - 0.516)^{1/6.75} = 7.54 \text{ ft/sec}$$

Ideally, a velocity transducer should be radially located at the critical position in order to measure the average velocity directly and have minimum sensitivity to changes in velocity profile that will occur with changing Reynolds number due to changing flow and fluid properties.

Critically Positioned Applications In the case of large diameter pipes, the size of the transducer is small in relation to the overall inside diameter of the pipe. As a result, it is practical to locate the transducer close to the critical position. To determine the location of the critical position, the average velocity is set equal to the expression that describes the velocity profile as follows:

$$v_{ave} = v_o \times \frac{2n^2}{(2n + 1) \times (n + 1)} = v_o \times (1 - y/R)^{1/n}$$

Solving for the critical positions yields

$$(y/R)_{critical} = 1 - \left[\frac{2n^2}{(2n + 1) \times (n + 1)}\right]^n$$

The relationship between critical position and Reynolds number is tabulated below.

Reynolds Number	Critical Position (y/R)
4×10^3	0.755
1×10^4	0.755
4×10^4	0.757
1×10^5	0.757
4×10^5	0.759
1×10^6	0.762
4×10^6	0.763

The effects of a Reynolds number varying from 4,000 to 4,000,000 changes the velocity profile sufficiently to shift the critical position by approximately 1 percent of the radius of the pipe. The critical position is almost independent of the flow, and the amount of error due to the fixed positioning of the transducer operating with different Reynolds numbers is relatively small. Therefore, accurately installed critically positioned transducers are relatively insensitive to changes in flow and Reynolds number.

EXAMPLE 21-3

Problem: Determine the critical position of a flowmeter in a 16-inch pipe with a 1/2-inch wall when the fluid operates at a Reynolds number of 4,000,000.

Solution: Using the above equation and substitution 10.00 for n yields:

$$(y/R)_{\text{critical}} = 1 - \left[\frac{2n^2}{(2n+1)\times(n+1)}\right]^n = 0.763$$

The critical position is calculated as (8.0 − 0.5) (1 − 0.763), or 1.78 inches from the wall of the pipe.

Centerline Positioned Applications Accurate positioning of the transducer in a small pipe is usually difficult if not impossible to perform, as the transducer is large in relation to the inside diameter of the pipe. To minimize Reynolds number effects due to flow and changes in fluid properties, the transducer can be located at the center of the pipe where the centerline velocity is measured. The average velocity can be calculated using the relation

$$v_{\text{ave}} = v_o \times \frac{2n^2}{(2n+1)\times(n+1)}$$

$$= v_o \times F_p$$

$$\text{where } F_p = \frac{2n^2}{(2n+1)\times(n+1)}$$

Reynolds Number	Profile Factor (F_p)
4×10^3	0.791
1×10^4	0.800
4×10^4	0.811
1×10^5	0.818
4×10^5	0.827
1×10^6	0.841
4×10^6	0.866

It should be noted that when fluid properties are constant, the centerline position profile factor, tabulated above as a function of Reynolds number, can vary by more than 2 percent over a 10:1 Reynolds number range and, hence, a 10:1 flow range. Therefore, the linearity of the output of insertion flowmeters is influenced by pipe hydraulics and is dependent upon the linearity of the profile factor. Hydraulic factors alone can result in nonlinearity of over 2 percent when Reynolds number is well defined and higher when Reynolds number is not well defined.

EXAMPLE 21-4

Problem: Flow of a fluid that operates at a Reynolds number of 400,000 is measured in the centerline pipe position. Calculate the relationship between the centerline and average velocities.

Solution: At a Reynolds number of 400,000, $n = 7.50$ and the profile factor is:

$$F_p = \frac{2n^2}{(2n + 1) \times (n + 1)} = 0.8270$$

and

$$v_{\text{ave}} = 0.8270\, v_o$$

At 10 percent of the above flow, which corresponds to a Reynolds number of 40,000, n becomes 6.75, and the profile factor can be calculated to be 0.8109, which illustrates that the theoretical linearity limit is 2 percent due to hydraulic considerations that are independent of flowmeter errors.

Flow Computers Microprocessor-based flow computers can be used to reduce nonlinearities caused by pipe hydraulics associated with critical and centerline insertion flowmeter technology when enough process data are available to sufficiently describe flow.

The basic concept behind the application of flow computers to insertion flowmeter technology is to develop enough information from process measurements or calculations to enable accurate calculation of Reynolds number. Once the Reynolds number has been calculated, equations such as presented above can be used to compensate for the hydraulic effects on a real-time basis.

To determine Reynolds number, the flow, density, and viscosity of the fluid at operating conditions should be known with reasonable accuracy. The flow can be approximated by the flowmeter measurement. The density and viscosity may be assumed to be constant, measured directly, or inferred from other secondary parameters such as pressure and temperature on a real-time basis, utilizing algorithms developed from experimental or theoretical data.

Available Technologies

Numerous technologies are available to measure the velocity at a strategically located point in the flowstream, including differential pressure, magnetic, oscillatory, target, thermal, and turbine.

Care should be taken when examining the accuracy of these flowmeters. Accuracy statements are expressed in terms of the *accuracy of the velocity sensor* and tend to disregard or downplay any hydraulic nonlinearities or piping effects that may be present. Manufacturers can also make statements of *point accuracies*, which reflect the accuracy of the flowmeter under one set of well defined operating conditions. This type of accuracy statement, which is invariably better than the accuracy of the flowmeter over its operating range, can tend to be misleading, and may not be indicative of the real performance of the flowmeter.

Differential Pressure Pitot tubes, which are applicable to liquids and gases, generate a differential pressure across the upstream and downstream pressure-sensing ports, which have a squared relation to the flow at the sensing point.

The principle of operation performance of an Annubar® is essentially the same as that of a Pitot tube except that the upstream pressure is averaged, accuracy statements are better, and Reynolds number constraints are less restrictive.

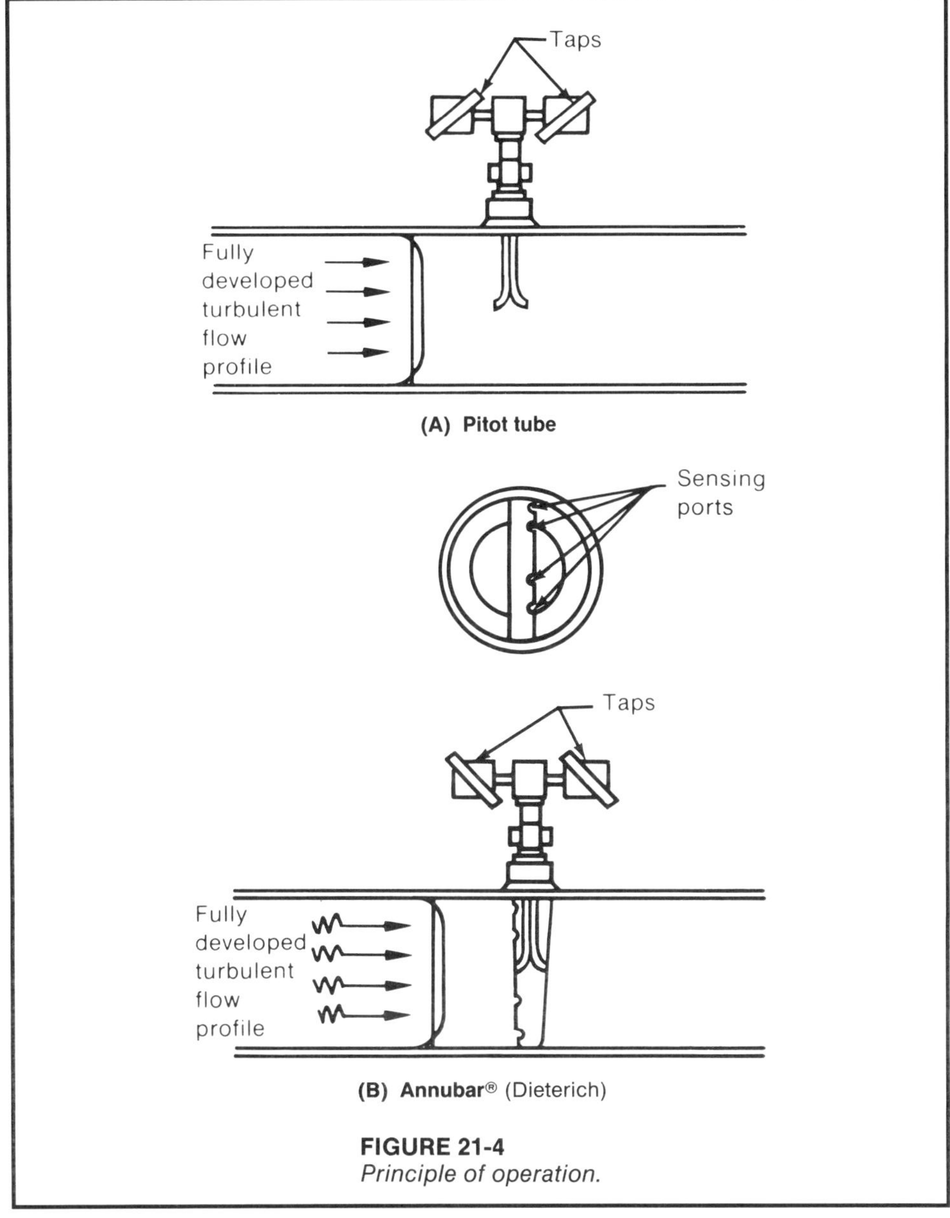

FIGURE 21-4
Principle of operation.

Fluidic Fluidic techniques are used to effect an insertion flowmeter that generates a differential pressure that is linearly dependent upon the flow of gas through the pipe. A jet of inert gas is blown into the flowstream normal to the direction of flow and is sensed by two nozzles located opposite, as illustrated in Figure 21-5.

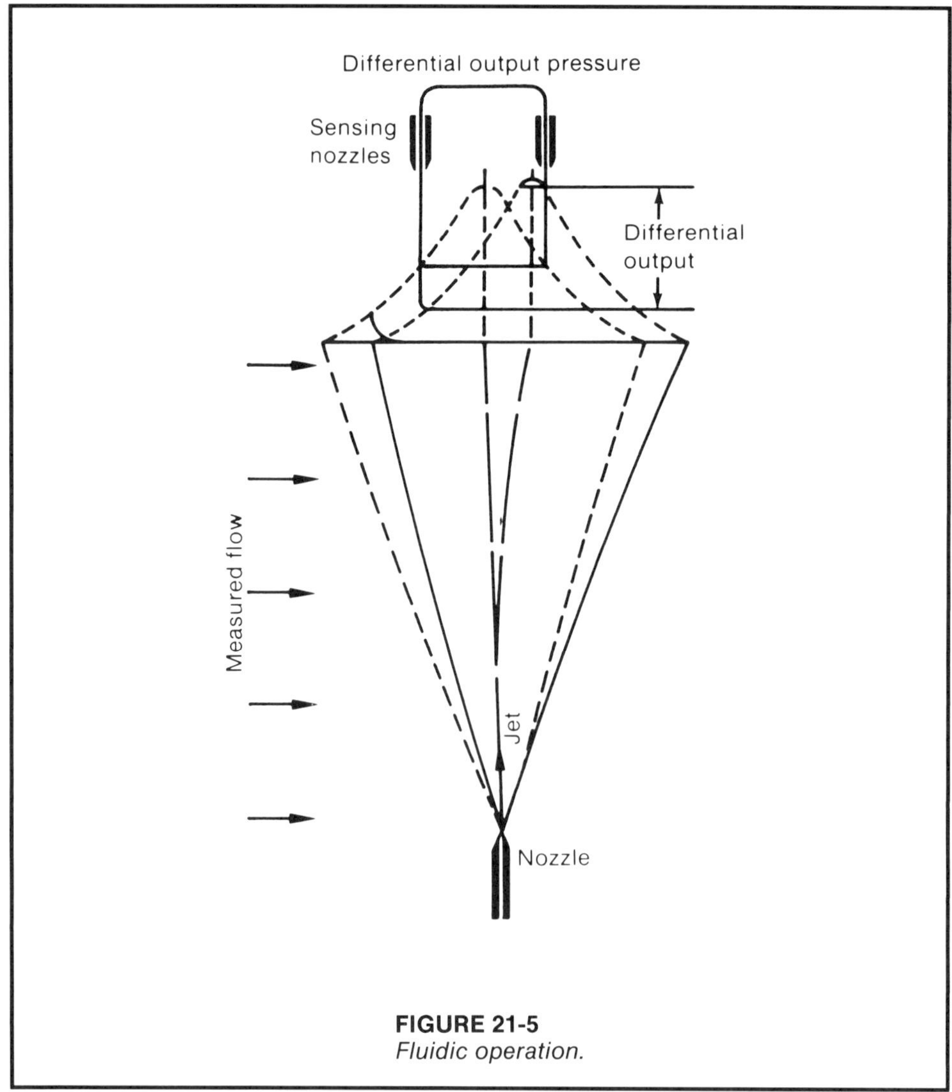

FIGURE 21-5
Fluidic operation.

At no flow conditions, the differential pressure is zero since the jet of inert gas impinges equally on both sensing nozzles. As flow increases, the jet of inert gas is diverted, and the jet of inert gas impinges on the sensing nozzles unequally. This results in the generation of a differential pressure that is linearly proportional to flow. The accuracy that can be achieved by this technology is approximately ±2 percent FS.

Magnetic Magnetic flow probes and insertion magnetic flowmeters can be used as insertion flowmeters for liquid service. They use the same technology as full-bore magnetic flowmeters, except that the electrodes are typically mounted on a probe, as illustrated in Figure 21-6.

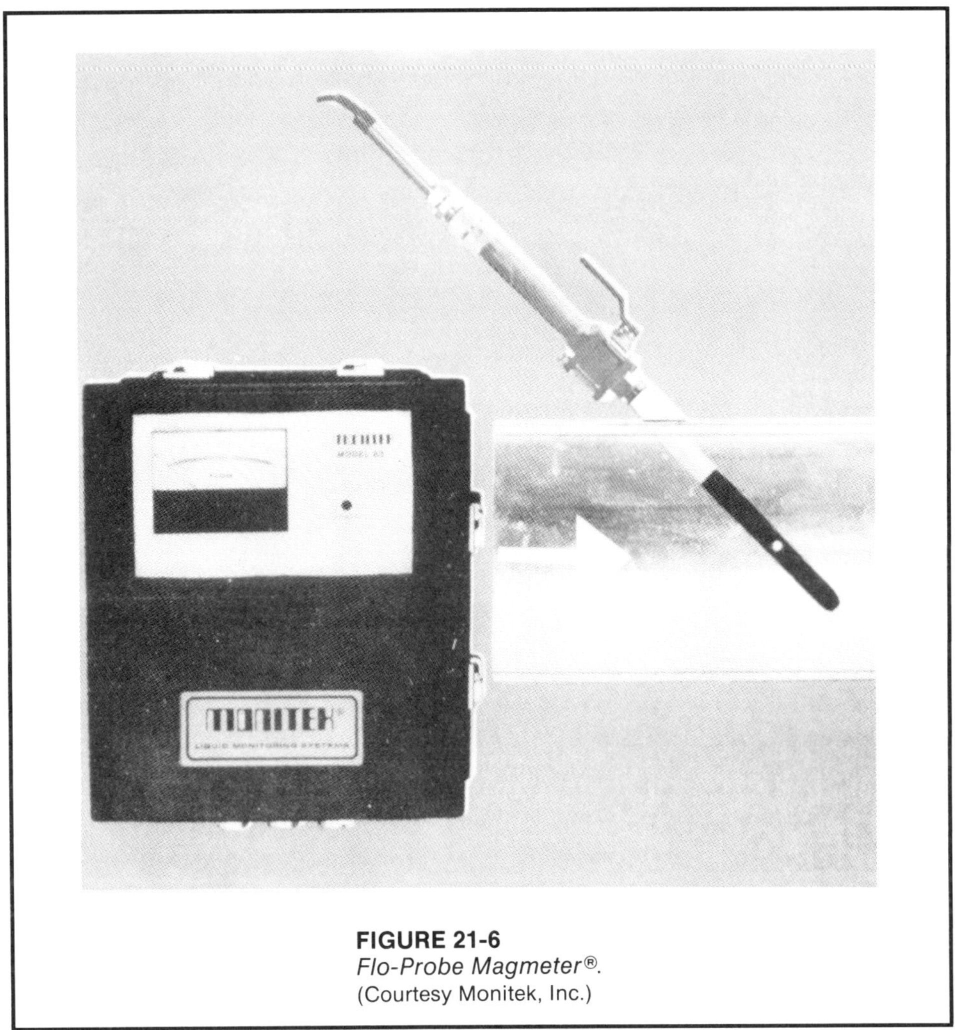

FIGURE 21-6
Flo-Probe Magmeter®.
(Courtesy Monitek, Inc.)

Accuracy, linearity, and repeatability of these flowmeters are typically stated to be ±1 to 2 percent FS, ±0.5 to 1 percent FS, and ±0.2 to 0.5 percent FS, respectively.

Oscillatory Vortex shedding oscillatory flowmeters, which use ultrasonics to sense vortices formed by an insertion-type thin wire shedder, can be applied as an insertion flowmeter in gas service. The principle of operation is the same as that of a full-bore ultrasonic vortex shedder but using a thin wire shedder. Accuracy, linearity, and repeatability of these flowmeters is typically ±2 percent rate, ±0.5 percent rate, and ±0.1 percent rate, respectively over a 10:1 flow range (see Figure 21-7).

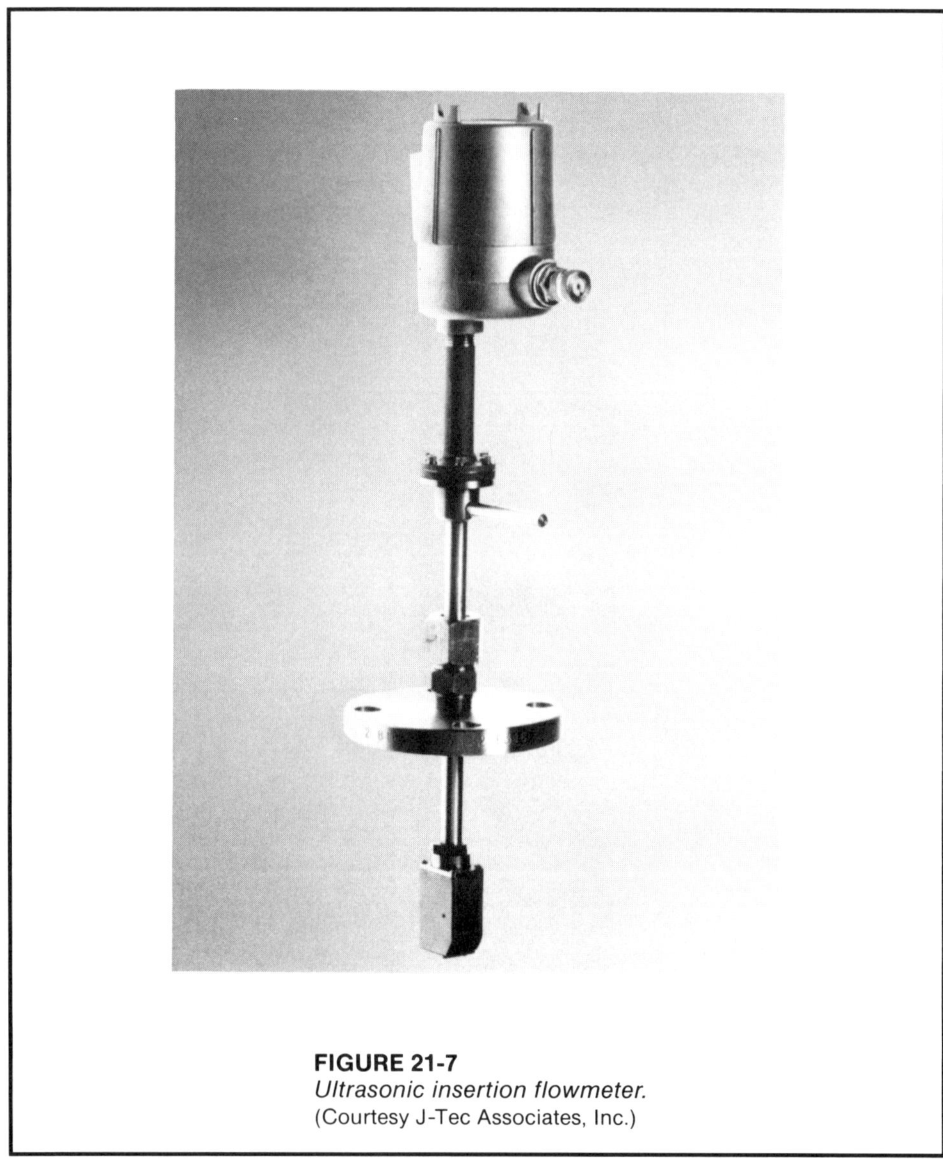

FIGURE 21-7
Ultrasonic insertion flowmeter.
(Courtesy J-Tec Associates, Inc.)

Target Target flowmeters are available for liquid and gas service. They insert into the side of the pipe as illustrated in Figure 21-8.

The principle of operation and performance of target insertion flowmeters are the same as those of full-bore target flowmeters.

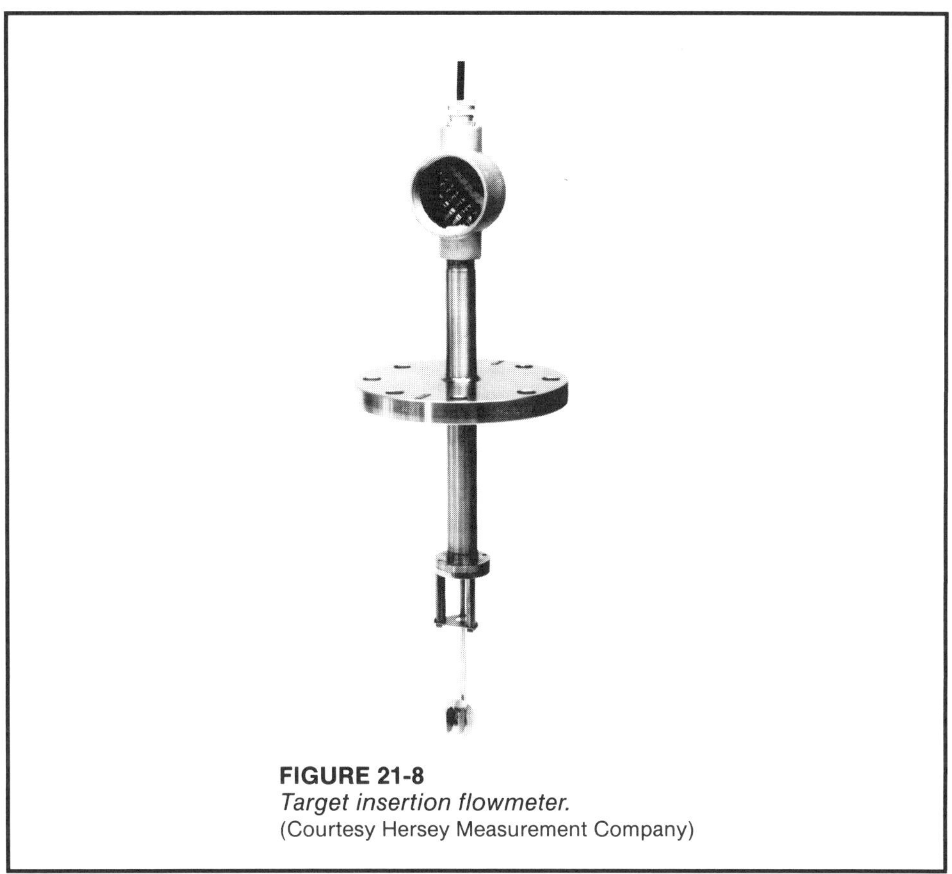

FIGURE 21-8
Target insertion flowmeter.
(Courtesy Hersey Measurement Company)

Thermal Thermal insertion flowmeters use thermal velocity-sensing techniques to measure fluid flow. The most common application of this technology is the measurement of air flow. However, other applications are measured with success (see Figure 21-9). Multipoint thermal insertion flowmeters can be used to measure flows in large pipes where some non-uniformity of the velocity profile exists (see Figure 21-10).

Turbine Insertion turbine flowmeters are available for liquid and gas service and utilize the same principle of operation and have approximately the same performance as full-bore turbine flowmeters. Temperature constraints of the insertion turbines are as high as 400°C for some designs, (see Figure 21-11).

Other Technologies Flowmeters that utilize other technologies, such as the Annubar,® thermal flowmeters with probes, and ultrasonic flowmeters, are not generally thought to be insertion flowmeters but are in fact insertion flowmeters in disguise. These sense the flow of part of the fluid in the pipe, from which the flow in the pipe is inferred. Most thermal flowmeters sense flow at the probe only, while most ultrasonic flowmeters sense flow in line with the transducers.

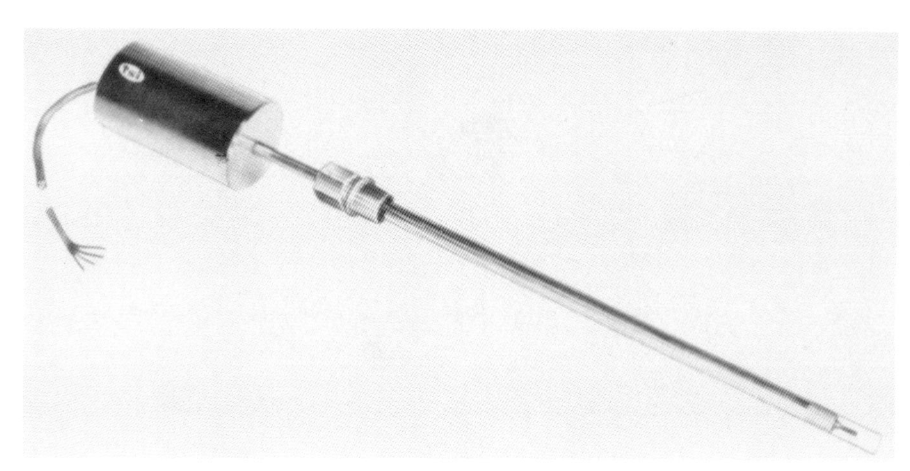

(A)
(Courtesty TSI Incorporated)

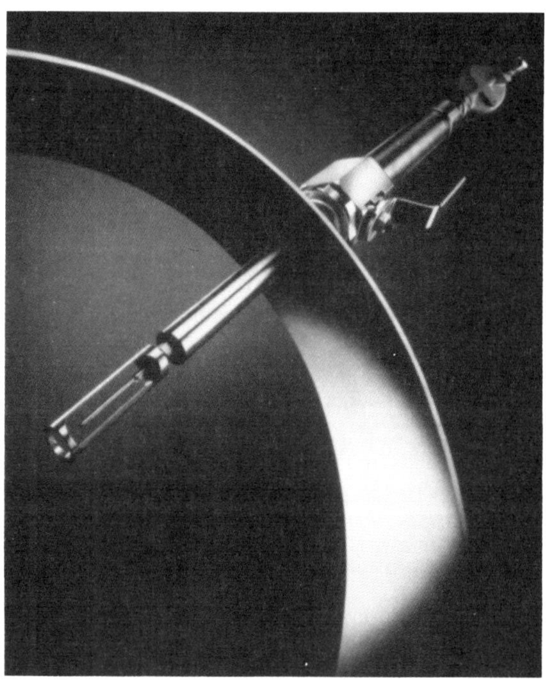

(B)
(Courtesy Kurz Instruments, Inc.)

FIGURE 21-9
Thermal insertion flowmeter.

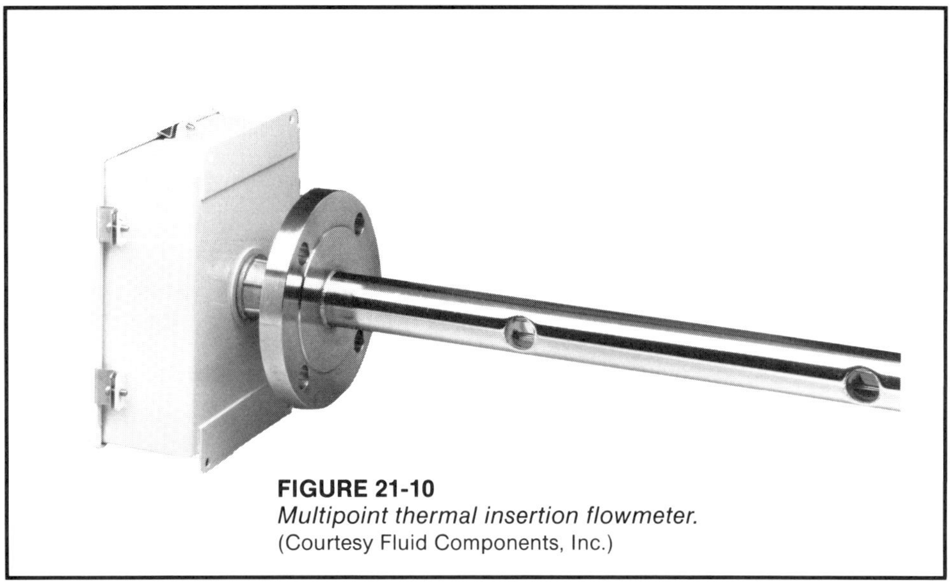

FIGURE 21-10
Multipoint thermal insertion flowmeter.
(Courtesy Fluid Components, Inc.)

Operating Constraints Insertion flowmeters generally require well developed turbulent flow typified by Reynolds numbers greater than approximately 4000 in order to adequately predict the flow in the pipe from strategically located sensing points. Assuming that Reynolds number constraints are satisfied, nonlinearities of over 2 percent can exist when the transducer is located at the centerline position and is operated over a 10:1 range. Accurately installed critically positioned transducers usually result in considerably less error. This error can, however, become significant at higher Reynolds numbers as the velocity profile flattens out and it becomes more difficult to accurately locate the transducer at the critical position.

The requirements of the applicable technology, such as velocity, density, and conductivity constraints, must also be satisfied in order for the flowmeter to function.

Performance The basic accuracy of insertion flowmeters is limited by the hydraulic changes that occur with changing flow and varying operating conditions, which often can be compensated for by utilizing a flow computer, as well as the performance of the technology.

Applications Insertion flowmeters are usually applied to fluids in pipes that are 2 in. and larger, operate with Reynolds numbers of greater than 4000, and satisfy the constraints of the technology employed to effect the flow measurement. Insertion flowmeters are applicable to liquids and gases, but the transducer technology determines whether a given design is applicable to liquids, gases, or both.

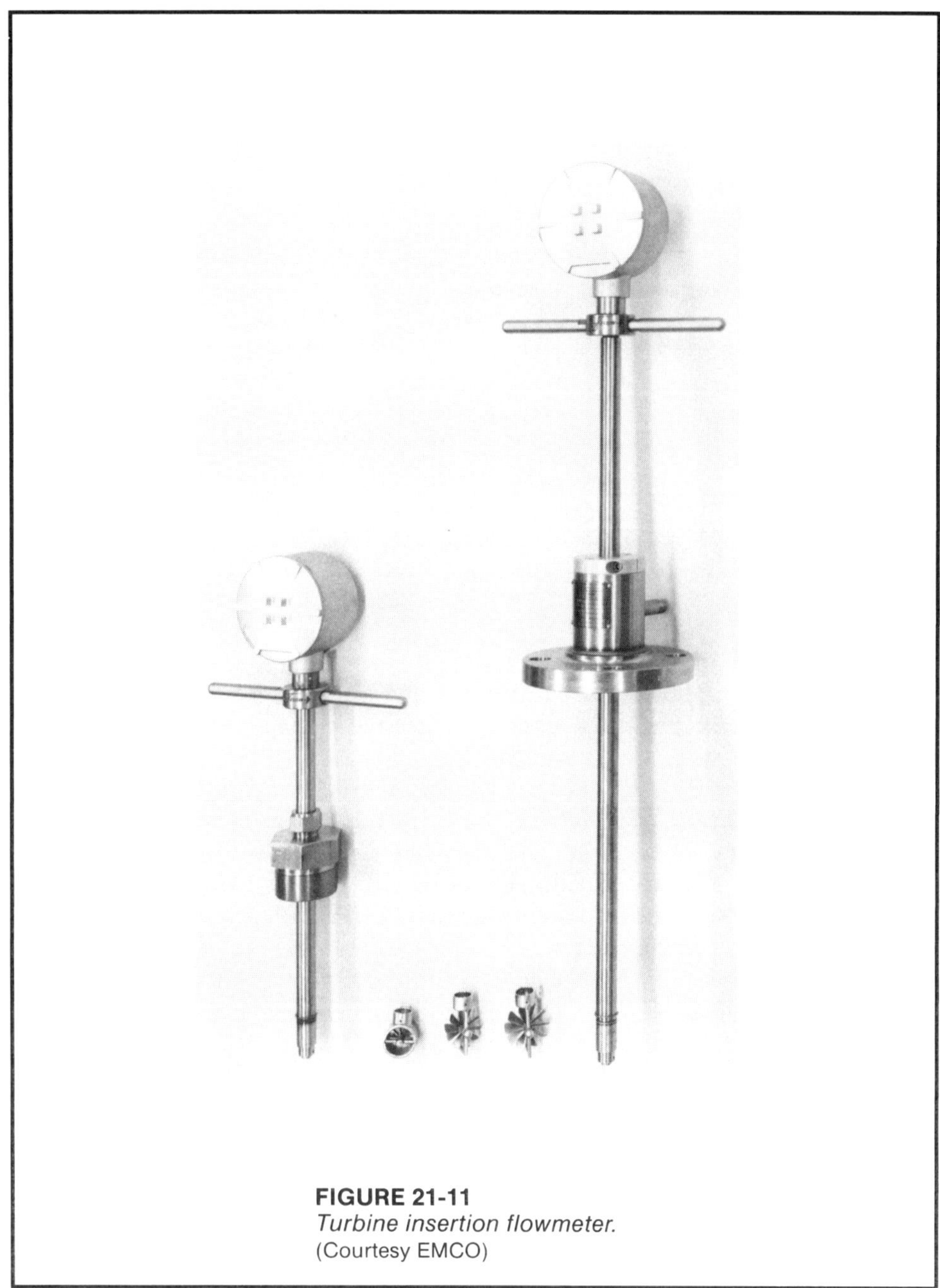

FIGURE 21-11
Turbine insertion flowmeter.
(Courtesy EMCO)

Sizing Sizing of insertion flowmeters is dependent upon the operating conditions of the fluid, which affect selection of the actual transducer for the application. The insertion length will vary depending upon pipe size, Reynolds number, and whether the probe senses at the centerline or the critical position.

Installation

Hydraulic Requirements Straight run requirements are typically stated to be 15 D/5 D, but this should be considered marginal since the flowmeter is highly profile dependent and any jetting or perturbations in the velocity profile at the position of measurement will affect the measurement. Careful attention should be paid to the upstream hydraulics and the nature and location of mechanical pieces of equipment. This is necessary to (1) eliminate all potential sources of turbulence and (2) set up a hydraulically stable and predictable velocity profile with sufficient pipe Reynolds number.

Piping Orientation Insertion flowmeters should be installed either from the side or from the top of the pipe to avoid the problems of condensate or dirt collecting on the flowmeter.

Transducer Positioning Errors in the position of the transducer will cause an error in the flow measurement as the velocity sensed by the transducer would not be indicative of the velocity at the critical or centerline positions. The effects of this are illustrated in Figure 21-12.

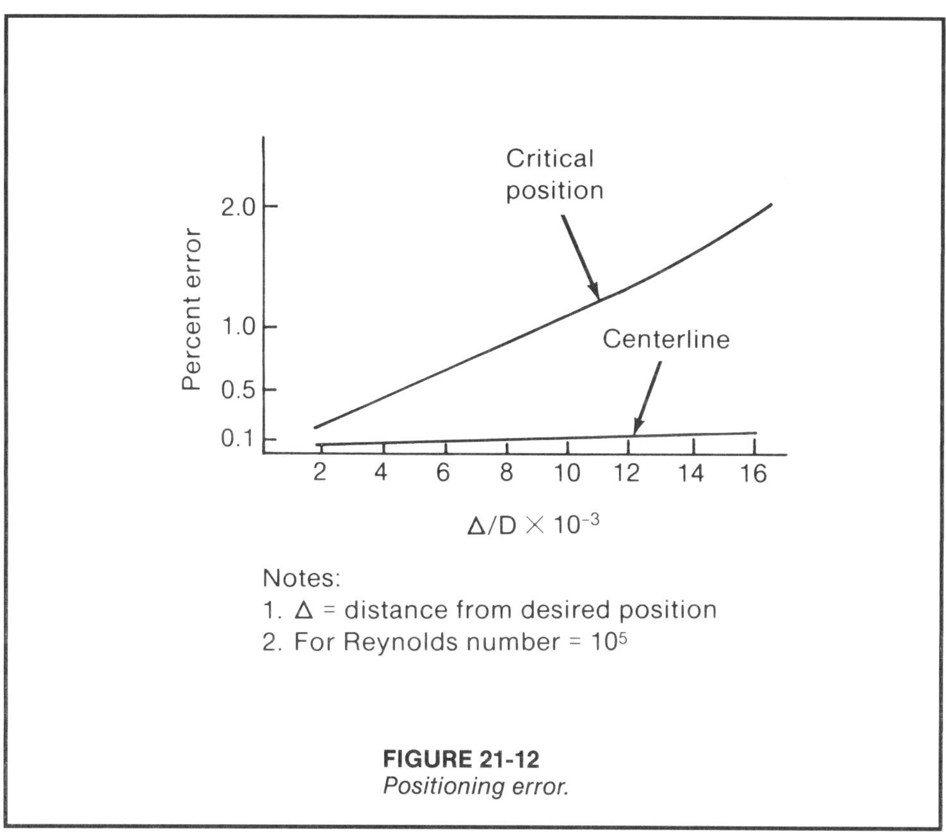

FIGURE 21-12
Positioning error.

Errors in the positioning of a critically positioned insertion flowmeter produce significantly larger measurement errors than those associated with a centerline positioned insertion flowmeter. This emphasizes the necessity to carefully locate critically positioned transducers.

Pipe Area The cross-sectional area of the pipe where the measurement is being made is a potential source of error. Tolerances of pipe manufactured to ASTM standards are tight enough so that the pipe can be fitted together and welded and provide nominal dimensions, weights, and the like for industrial use. They are not tight enough, however, for accurate flow measurement purposes. For example, some pipe specifications state that the minimum pipe wall thickness must be greater than 0.875 times the nominal pipe wall thickness. This is in addition to the ovality specification, which states that variations in the outside diameter of the pipe can be as high as 1.5 percent of the outside diameter. These result in manufacturing tolerances that introduce additional uncertainty in the flow measurement, as the concentricity and inside diameter are not well defined.

EXAMPLE 21-5

Problem: Calculate the error introduced by a perfectly round 16-inch pipe with a nominal 1/2-inch wall thickness that exactly meets the minimum manufacturing wall thickness requirements.

Solution: The effect of reduced wall thickness can be determined by calculating the ratio of the areas of the thinner walled pipe to a pipe of nominal wall thickness, which would be used for flowmeter calculations. As constant terms cancel, the ratio of the areas is equal to the square of the ratio of the inside diameters:

$$(15 \text{ in.} + [2 \times 1/2 \text{ in.} (1\text{–}0.875)]/15.00 \text{ in.})^2 = 1.017$$

The effect of the above tolerance is a 1.7 percent error. As there are no specifications limiting pipe wall thickness, pipe manufactured with thicker pipe walls may result in larger measurement errors.

Cabling Insertion flowmeters can be 2-wire, 3-wire, or 4-wire devices, dependent upon the technology employed.

Maintenance

Maintenance requirements and procedures for insertion flowmeters are similar to those of the technology employed. Retractable insertion flowmeter designs are available: the flowmeter element may be installed in an operating pipe and removed from the pipe without affecting the flow (see Figure 21-13).

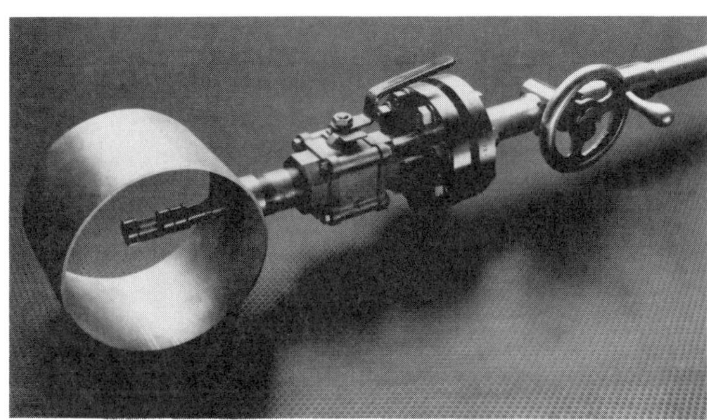

FIGURE 21-13
Retractable insertion flowmeter.
(Courtesy Fluid Components, Inc.)

EXERCISES

21.1 Calculate the flow of an ideal liquid flowing in a 6-inch pipe at 4.5 feet per second.

21.2 Calculate the velocity at a point 2 inches from the wall of a 6-inch pipe in which 1000 acfm of gas is flowing at a Reynolds number of 3,000,000.

21.3 Calculate the critical position of an insertion flowmeter in a 24-inch pipe that contains liquid flowing at a Reynolds number of 750,000.

21.4 Calculate the relationship between the centerline velocity and the average velocity when fluid is flowing with a Reynolds number of 6,000.

21.5 Calculate the error introduced by a perfectly round 16-inch pipe in which the wall thickness has been manufactured 1/8-inch larger than the nominal 1/2-inch thickness.

22

Bypass Flowmeters

Introduction Bypass flowmeter technology represents a viable flow measurement technique in large pipes, where a flowmeter that is the same size as the pipe becomes impractical or uneconomical. Increased turndown of a primary flow element can often be achieved using this approach to flow measurement. However, in most cases, performance is dependent upon manufacturer claims and calculations, as opposed to actual test data.

Principle of Operation The basic bypass flowmeter principle is to employ an element in the flowstream that generates an output signal that is measured with another flowmeter. Most such devices are differential pressure producers that utilize a secondary flowmeter to measure the flow developed by the differential pressure across the primary flowmeter. As the flow through the bypass of a differential producing device is linear with the flow through the total flowmeter system, the turndown of the system is limited by the turndown of the secondary flowmeter. Therefore, linear flowmeters are usually applied as secondary flowmeters to effect a turndown of approximately 10:1 from a differential producer that, when designed with a differential pressure transmitter, would achieve an approximate 3.5:1 turndown.

Both the primary and the secondary flowmeters must be correctly applied and installed to effect the flow measurement. The accuracies of both flowmeters should be considered to obtain the overall accuracy of the measurement. Calculations performed in designing these flowmeters are often not published or confirmed by independent sources; therefore, manufacturer accuracy claims and sizing techniques must be relied upon to predict flowmeter performance.

Types of Bypass Flowmeters

Orifice Plate/Turbine The turbine flowmeter secondary with an orifice plate flowmeter primary, often called a shunt flowmeter, is an in-line flowmeter in the 1 to 4-inch size that internally uses the bypass flow principle.

This flowmeter can achieve accuracies of ±2 percent of rate over a 10:1 range in some applications and is specifically designed to be applied for steam service. The flowmeter can also be applied to other gases.

For line sizes 4 inches and larger, these flowmeters can be used as the secondary flowmeter with a differential producing primary flowmeter, as shown in Figure 22-1.

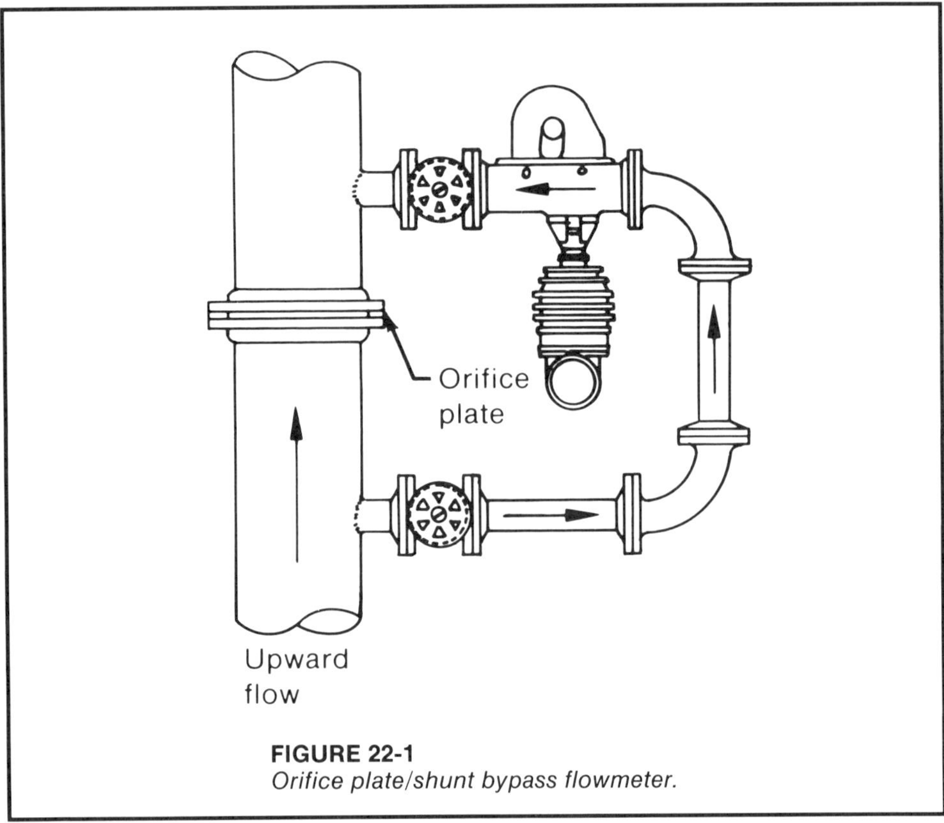

FIGURE 22-1
Orifice plate/shunt bypass flowmeter.

Orifice Plate/Rotameter This configuration is commonly applied to achieve economical local indication over a wider turndown than would be possible with a differential pressure indicator. The rotameter can also be specified with a transmitter.

Orifice Plate/Thermal Thermal flowmeters can be used as secondary flowmeters, bypassing an orifice plate in certain applications, as illustrated in Figure 22-3.

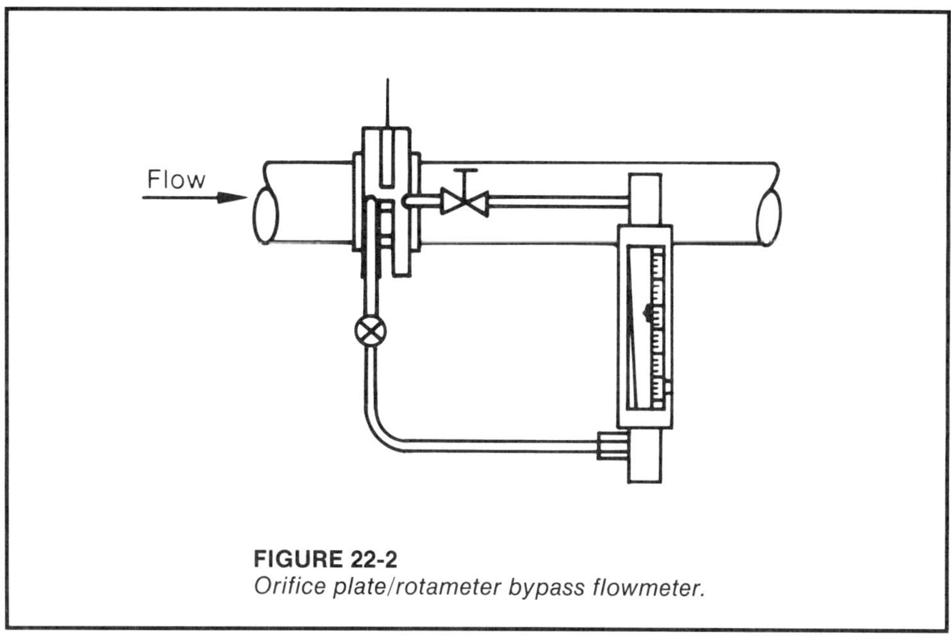

FIGURE 22-2
Orifice plate/rotameter bypass flowmeter.

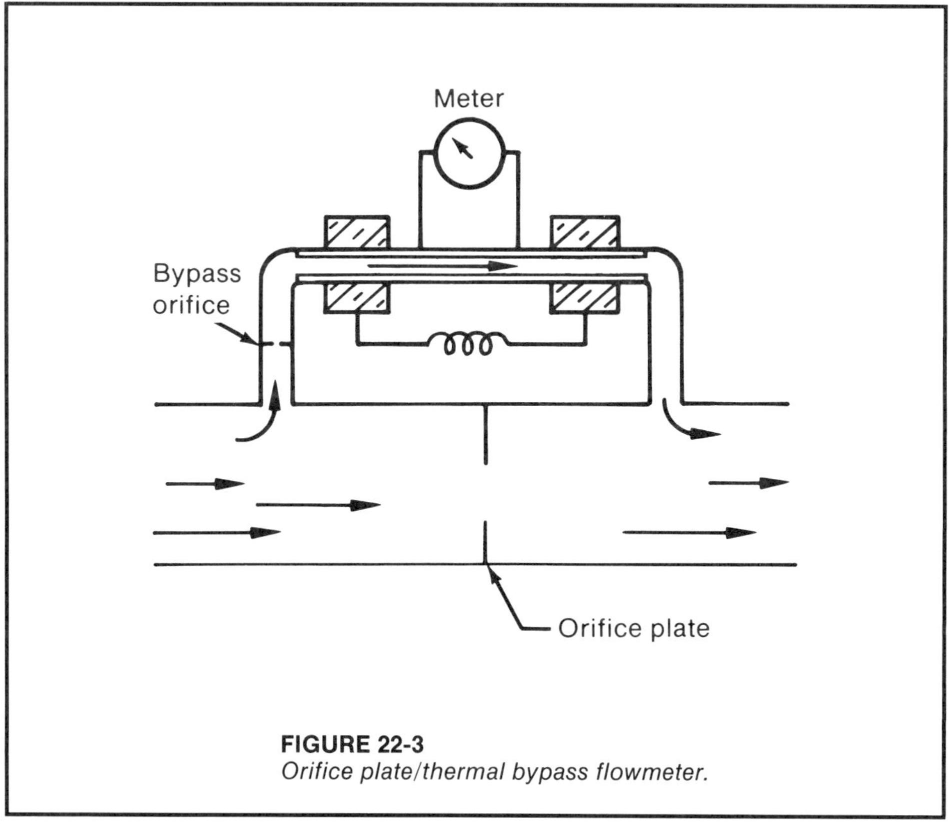

FIGURE 22-3
Orifice plate/thermal bypass flowmeter.

Laminar Flow Element/Thermal Flowmeter Thermal flowmeters can be used as secondary flowmeters, bypassing a laminar flow element in certain applications, as shown in Figure 22-4.

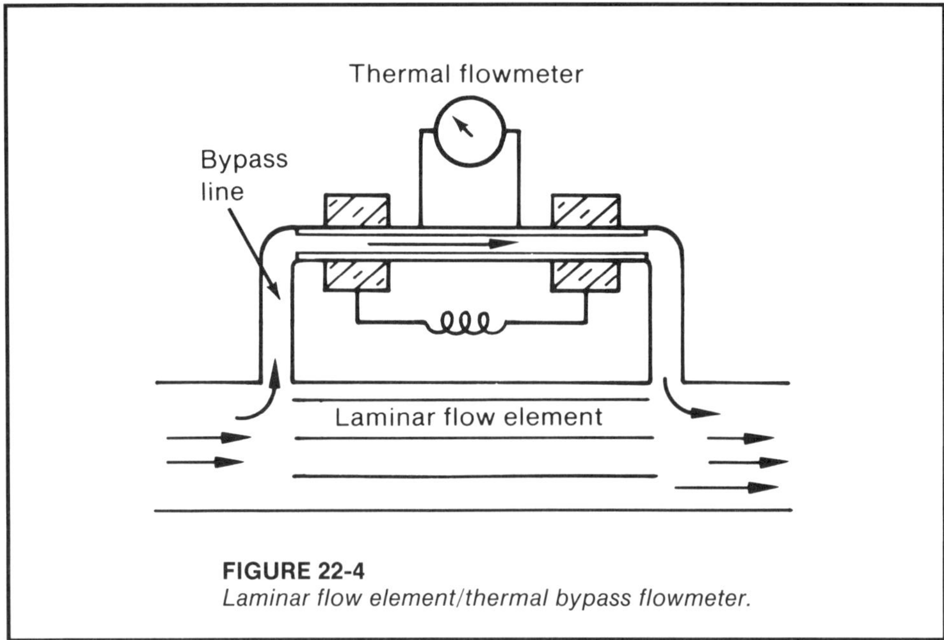

FIGURE 22-4
Laminar flow element/thermal bypass flowmeter.

Restriction/Oscillating Vane This flowmeter utilizes a restriction of fluid flow to develop a differential pressure that causes a portion of the flow to pass through the oscillating vane pathway in the restriction (see Figure 22-5). The frequency of oscillation is proportional to flow through the pathway, which is, in turn, proportional to the total flow.

Accuracy and repeatability are ±0.5% of meter capacity and ±0.2% rate over a temperature range of -25 to 95°C up to 300 psig. Reynolds number for accurate flow measurement must be greater than 3500 to 10,000, depending on size. Care should be taken to maintain the required downstream pressure for proper operation.

Other Many other bypass flowmeter combinations not presented herein nor published in manufacturer literature can be applied to liquid and gas applications. When bypass technology appears to be applicable, the user is well advised to consult with flowmeter vendors as to the feasibility of each application.

EXERCISES

22.1 Can a bypass flowmeter configuration result in a larger turndown than is possible from the primary flowmeter alone? Why or why not?

22.2 Can the shunt flowmeter be considered a true bypass flowmeter? Why or why not?

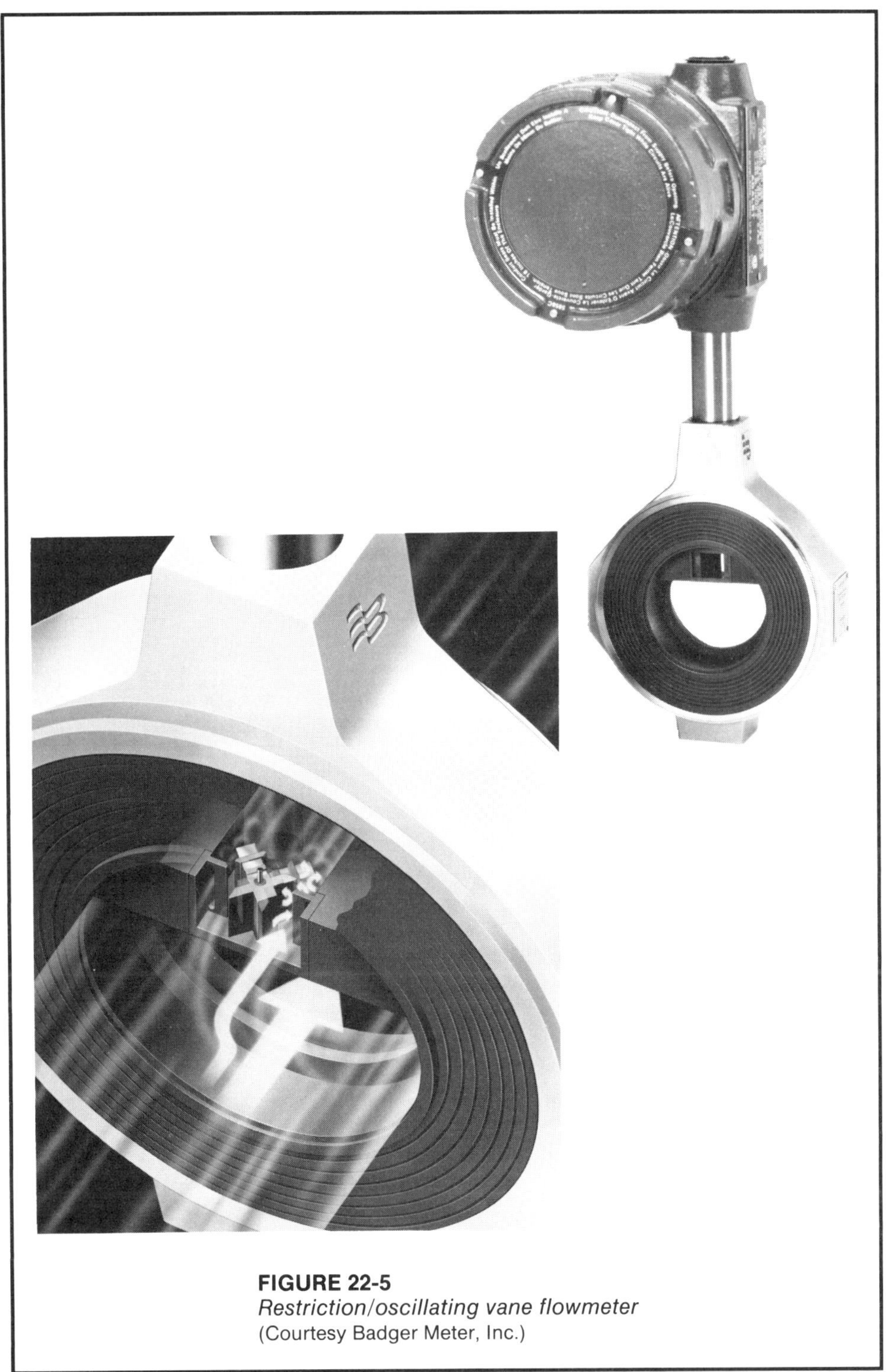

FIGURE 22-5
Restriction/oscillating vane flowmeter
(Courtesy Badger Meter, Inc.)

23

Factors in Flowmeter Selection

Introduction Flowmeter selection is a complex decision-making procedure, which, as a minimum, takes into account physical properties of the fluid to be measured, the process measurement needs, and the technical feasibility of the various flowmeter technologies. Many other factors, such as accuracy, cost, desired output, and the like, must also be included.

Flowmeter Categories Flowmeter applications can be categorized by the type of fluid to be measured. It should be noted that specific applications may be appropriate to more than one category of flowmeters.

Class I flowmeters with wetted moving parts are generally applied to clean fluids, while Class II flowmeters with no wetted parts can be applied to flows that may contain solids. Obstructionless Class III flowmeters are applicable to abrasive fluids, slurries, and applications where low pressure drop across the flowmeter is desired. Class IV flowmeters, which have non-wetted sensors are applicable in existing installations where pipe modifications are difficult or uneconomical, where exotic materials make other flowmeters uneconomical, in large pipes, and for temporary installations.

EXAMPLE 23-1

Problem: Select the flowmeter categories in the following applications.

1. Abrasive liquid with some solids
2. Clean liquid
3. Clean liquid, but very low flowmeter pressure drop

Solution:

1. Class III as well as Class IV flowmeters would generally be applicable due to the abrasiveness of the liquid and the presence of solids.

2. All flowmeter classifications would be applicable.
3. Class III or Class IV flowmeters would be applicable due to the differential pressure constraint.

Flowmeter Types Flowmeter applications may be further grouped by the type of measurement required. Volumetric and velocity flowmeters can be used to measure volumetric flow while, when applicable, mass flowmeters measure mass flow directly. Inferential flowmeters can be used to infer the volumetric or mass flow, as desired. Although the output of each type of flowmeter can be converted to mass or volumetric units as necessary, variations in fluid density may be sufficient to warrant a direct measurement of mass flow or on-line compensation of the volumetric or velocity measurement.

EXAMPLE 23-2

Problem: Flowmeter types are to be determined by the measurement requirements in the following simplified applications.

1. Fill a tank to a given level.
2. Add a reactant to complete a chemical reaction.
3. Measure different liquids with the same flowmeter.

Solution:

1. This requires that a given volume of liquid be put into a tank, so a volumetric or velocity flow measurement would be applicable if a level measurement were not possible. An inferential flowmeter could be used; however, changes in the operating conditions of the flowmeter will result in error in the volumetric measurement.
2. Chemical reactions require a molar balance of the various chemicals in order to react properly. If composition of the reactant is directly proportional to its density, then a mass flow element or compensated volumetric flowmeter might be used. Since changes in composition have a direct impact on the reaction, no flowmeter will necessarily yield a signal indicative of the molar addition rate.
3. Assuming that a mass measurement is desired and the specific gravities of the liquids are different, a mass flowmeter may be considered if compatible with the application and the fluids being measured. If a volumetric measurement is desired, a volumetric or a velocity flowmeter might be used. An inferential flowmeter may be applied, but it would require compensation for changes in specific gravity.

Performance Flowmeter performance can vary significantly with the technology employed to affect the flow measurement, as well as from one flowmeter to another that uses the same technology. Some industrial flowmeters are claimed to measure liquid and gas flow with accuracies as high at ± 0.25 percent of rate and ± 0.5 percent of rate, respectively, with turndowns that approach 100:1 in some applications.

Technology	Typical accuracy*	Turndown**	Range-ability	Straight run	2-Wire operation	Exotic materials of con-struction	Sensitive to density changes
Differential pressure							
Concentric orifice plate	±0.75% rate	3.5:1	2.5:1	Long	Yes	Yes	Yes
Other	Up to ±4% rate	3.5:1	2.5:1	Varies	Yes	Yes	Yes
Magnetic							
Conventional	±0.5 to 1% FS	Up to 10:1	10:1	Short	No	Yes	No
DC	±0.5 to 2% rate	Up to 10:1	10:1	Short	No	Yes	No
Mass							
Coriolis	±0.2–0.47% rate	Up to 10:1	10:1 or more	None	No	Yes	No
Hydraulic Wheatstone bridge	±0.5% rate	Up to 100:1	6:1	None	No	No	No
Oscillatory							
Fluidic	±1% rate	17 to 50:1	3.5:1	Short	Yes	No	No
Vortex	±0.75% rate	7:1	3:1	Short	Yes	Yes	No
Positive displacement	±0.2 to 2% rate	Over 3:1	Over 3:1	None	Yes	Some	No
Target	±1 to 6% rate	3.5 to 10:1	3:1	Long	Yes	Yes	Yes
Thermal	±2% rate	Over 40:1	None	Short	No	Yes	No
Turbine	±1% rate or better	Up to 10:1	10:1	Short	Yes	Yes	No
Ultrasonic	±0.5 to 10% FS	Up to 10:1	10:1	Long	No	Yes	No
Variable area	±1 to 2% FS	10:1	None	None	Yes	Yes	Yes

*Primary accuracy, when properly installed.
**From 7 fps (liquid service), assuming R_D constraints are satisfied.

FIGURE 23-1
Typical flowmeter specifications.

These specifications usually refer to the performance of the flowmeter in laboratory tests with the flowmeter in "perfect condition," using a well defined fluid such as water or air under carefully controlled conditions. The expected performance of the flowmeter in an industrial environment is generally much less. Sources of error include the effects of flow pulsation, hydraulic compromises, and variations in fluid properties such as composition density, viscosity, temperature, etc. Another sometimes overlooked source of error is the mismatch of flowmeter element or transmitter turndown with the actual process flow range. Even if a given flowmeter is capable of a 10:1 flow range and the process flow covers only the lower 20 percent of the range, the effective turndown of the meter is less than 2:1. Uncertainties in the physical properties of the fluid can contribute large errors to the flow measurement.

Installation errors, such as lack of concentricity of the flowmeter in the pipe, incorrect tap location or lack of the proper pipe interior finish in the area of the flowmeter, lack of sufficient straight run, and failure to follow all manufacturer mounting recommendations can be other sources of significant error. Therefore, the accuracy of the flow measurement is dependent not only on the flowmeter proper, but also on the entire flowmeter system.

End Use Flowmeter performance should be calculated on a system basis and not limited to considerations of the primary flow element. More than one of the various end uses for the flow signals may be required for a particular application; they are:

- Rate indication
- Control
- Totalization
- Alarm

Assuming that the accuracies of all flowmeter primaries are equal, which is not generally the case in practice, analog indication and/or control is best achieved by a linear analog flowmeter, which eliminates conversion errors due to linearization or conversion of a frequency to an analog signal. As few linear analog flowmeters exist, calculating the errors associated with each prospective flowmeter system over the desired range of operation can determine whether an analog or a digital flowmeter is superior for the application. Linear digital flowmeters are usually preferred: the errors associated with the digital-to-analog conversion are usually smaller than those associated with the transmitter and linearizer associated with a nonlinear analog flowmeter. However, the uncertainty of an analog meter movement may be significantly larger than that introduced by either linearization or conversion.

When digital indication and/or control is of primary importance and the digital system does not accept a frequency input signal, the analysis is identical to that of an analog indication except that the linearization and indication errors are virtually zero when performed digitally. If the digital system can accept frequency or analog inputs, use of a frequency input from a linear digital flowmeter will typically result in less error than that from a nonlinear analog flowmeter due to the error associated with the transmitter. Errors associated with calculations performed within the digital system to linearize or convert a frequency to a flow signal are typically insignificant and ignored.

Digital flowmeters lend themselves to totalization applications because conversion errors are not present in the flowmeter system. In a conventional analog instrumentation system the totalization of a nonlinear analog flowmeter introduces uncertainties involving transmitter, linearizer, and totalizer. When a linear digital flowmeter is applied, pulses can be counted to effect the totalization, thereby eliminating uncertainties introduced by other instruments. This results in totalization accuracy that is virtually equal to the accuracy of the flowmeter.

EXAMPLE 23-3

Problem: Assuming that each flowmeter primary measures with an accuracy of ±0.75 percent rate, determine whether a squared output analog flowmeter or a linear digital flowmeter is more accurate for:

1. Analog indication/control
2. Digital indication/control on a digital control system
3. Totalization with discrete instrumentation

Solution: In case 1, the indicator error is large compared to the other errors. The approximate flow error associated with the linear transmitter of 0.1 percent FS compared to that associated with the squared transmitter and square root extractor of 0.1 percent FS and 0.25 percent FS, respectively, is not significantly different.

There is no indicator error in case 2; however, the transmitter and the analog-to-digital converter associated with the squared analog flowmeter introduce flow errors of approximately 0.1 percent FS each. The linear digital flowmeter utilizes a frequency-to-analog converter, which would have an accuracy of 0.1 to 0.2 percent FS, in addition to the analog-to-digital converter in the digital control system. There is no significant difference between these systems.

In case 3 where totalization is required, the flow error associated with the squared output transmitter is typically 0.1 percent FS. That associated with the square root extractor and the integral totalizer is typically 0.5 percent rate. As the digital flowmeter produces pulses, each of which corresponds to a volume of flow, there is no flow error associated with the counting of these pulses to effect totalization of flow.

Power Requirements Flowmeters are available as 2-wire, 3-wire, or 4-wire designs; however, 2-wire designs are usually preferred if moderate transmission distances are involved.

Field installation costs are generally lower for 3-wire and 4-wire transmitters that operate on low voltage (for example, 24 V dc) than for 4-wire transmitters that require a separate conduit for power wiring. The choice of system is a decision that must consider the plantwide controls.

Safety In the case of hazardous fluids, flowmeters should be specified on the basis of fluid compatibility, electrical area classification, and any other codes or standards that may apply; for example, special cleaning in the case of oxygen service, special welding inspections in the case of flowmeters in high-pressure steam service, special enclosure ratings for meters located in areas where flammable dust, vapors, etc., are normally present.

Rangeability The ability to sufficiently adjust the range of a flowmeter in the field may mean the difference between starting up a facility and being delayed as a result of last minute design changes. Most manufacturers try to build in this ability as a convenience to the user (and to themselves) in order to minimize the number of problems that must be handled on a rush basis and the number of parts that must be manufactured and stocked as spares.

It should be noted that transmitters used for flowmeters with a squared output are adjustable over a flow range that is the square root of the adjustment range of the transmitter, which reduces the effective rangeability (see Figure 23-1).

Materials of Construction Flowmeter materials of construction must be compatible with the fluid being measured, or the flowmeter will fail prematurely or lose accuracy. If material selection is not established by previous experience or piping specifications, it is essential that materials recommendations be made by a qualified materials or corrosion specialist. Generally, the piping specifications for the process offer a good starting point for identifying the appropriate materials of construction and the type of inspections required, as long as it is understood that corrosion allowances for piping are much greater than can be tolerated by most flowmeter elements and transmitter components. Bearing, seal, and gasket materials must also be considered.

Maintainability The maintenance aspects of flowmeters should be examined in considering the purchase of a flowmeter, if possible. Typically, this is an extremely complex area of evaluation and is best judged on the basis of actual operating experience. If regular maintenance is expected, bypass piping may be appropriate.

Ease of Application Ease of application is a relatively intangible factor that affects the amount of time, effort, and technical expertise necessary to select and specify a flowmeter. Selection of a flowmeter based upon this factor alone is not optimal from technical or economic considerations.

While the experience gained from the operation of previous installations may be invaluable in selecting flowmeters for additional installations, new technologies may offer superior performance. Each application, even if it can be conveniently copied, should be investigated in detail to ensure that copying a flowmeter installation because it was used in the past is not the repetition of a mistake or the use of outmoded technology.

Ease of Installation Since installation requirements vary significantly with flowmeter technology, no particular flowmeter is clearly superior. There are, for instance, multiple trade-offs between the various piping and electrical requirements that should be considered in context with each application. It often appears that flowmeters with the fewest piping installation requirements have the most electrical or mechanical requirements.

An example of this is a Class IV ultrasonic flowmeter. It requires adequate straight run upstream of the flowmeter (which can be a problem in large pipe sizes) and no penetrations into the pipe; but it necessitates proper sensor attachment and special electrical items (such as coaxial cable) for the transducer electronics.

Installed Cost

Cost is one more input in the flowmeter selection process. Acceptable operation and technical correctness should be the primary factory in flowmeter selection, with cost as a secondary but nonetheless important factor. When the strategy of flowmeter selection based upon technical evaluation is followed, price/performance/maintenance comparisons can be made between the flowmeter with the best performance and others that are more economical. Installed cost is generally better than purchase price in evaluating the true cost of a flowmeter.

Figure 23-2 graphically shows flowmeter purchase cost as a function of liquid flow in gallons per minute for some flowmeter technologies. The installed cost of various flowmeters is shown as a function of flow in Figure 23-3, assuming the flowmeter will be operated at a full-scale velocity of 7 feet per second. The flowmeter size and relative installed cost can be estimated for the desired maximum flow by using the following procedure for liquid applications:

- Select size.
- Calculate full-scale velocity in the selected size,

$$v_{\text{full scale}} = \frac{7\ Q_{\text{full scale}}}{Q_{@\ 7\ \text{fps}}}$$

- Read estimate of relative installed cost for selected size when the full-scale velocity is within the velocity range of the flowmeter.
- If the full scale velocity is lower (higher) than velocity range of the flowmeter, select the next size smaller (larger) and repeat the procedure.

EXAMPLE 23-4

Problem: A magnetic flowmeter is to be installed to measure flows in the range of 0 to 100 gpm. Estimate the relative cost.

Solution: Consider a 3-inch magnetic flowmeter through which 160 gpm of liquid would flow at a velocity of 7 feet per second. The velocity at 100 gpm is $7 \times 100/160$, or 4.375 feet per second, and is within the 3 to 30 feet per second limitation of the flowmeter. The 3-inch flowmeter is applicable and its estimated installed cost is 1.07. Similarly, if a 2-inch magnetic flowmeter were selected, the velocity at 100 gpm would be 7 × 100/74, or 9.5 feet per second, which is acceptable and has an estimated cost of 1.04. A 1-inch magnetic flowmeter results in a velocity at 100 gpm of 7 × 100/19, or 36.8 feet per second, which is not acceptable.

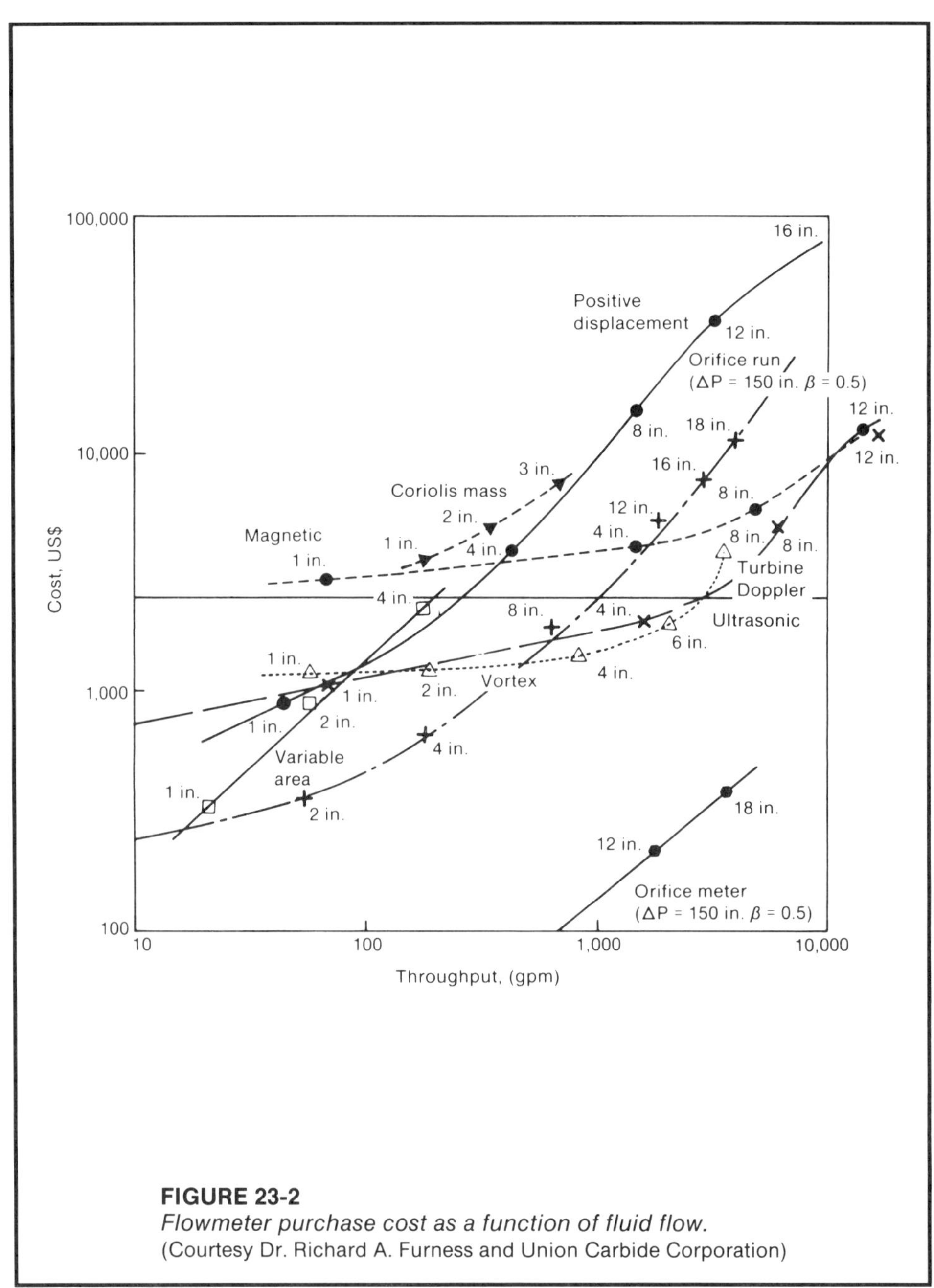

FIGURE 23-2
Flowmeter purchase cost as a function of fluid flow.
(Courtesy Dr. Richard A. Furness and Union Carbide Corporation)

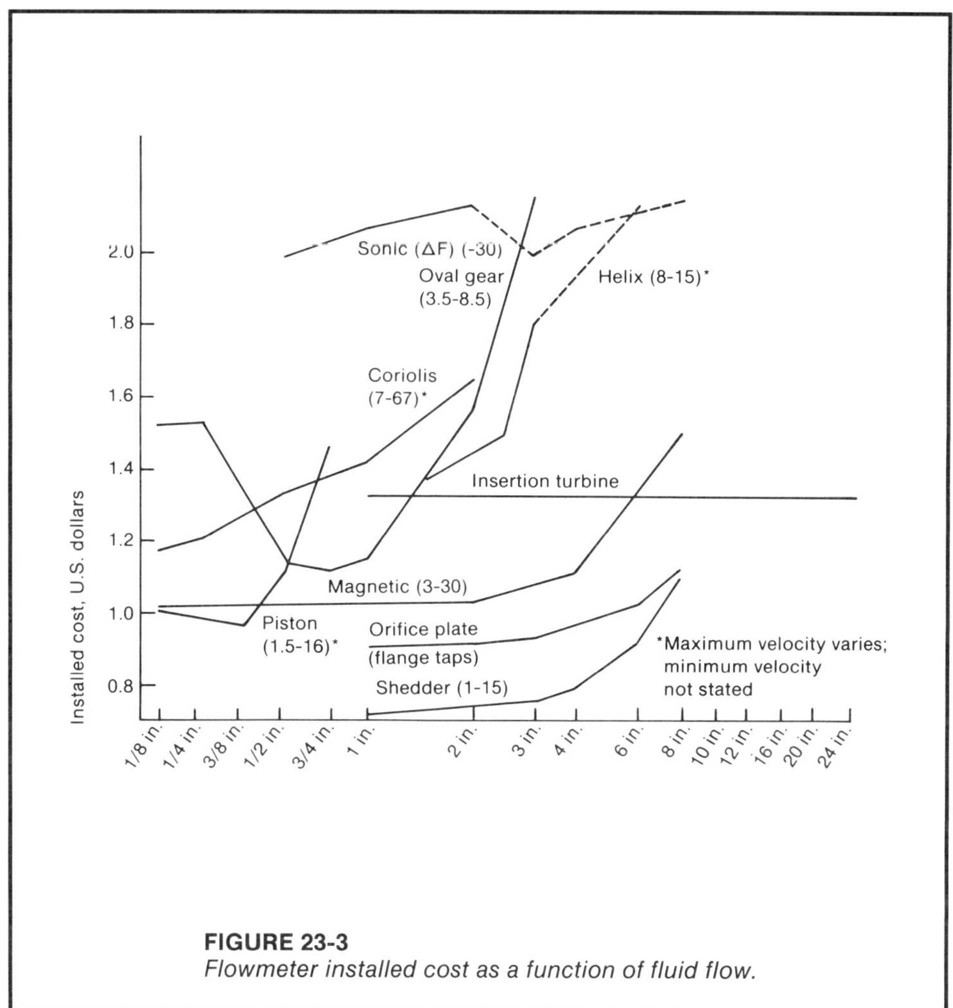

FIGURE 23-3
Flowmeter installed cost as a function of fluid flow.

Operating Cost Flowmeter operating costs are associated with the energy required to operate the flowmeter. This includes the electric energy required to operate the electronic components of the flowmeter, which is negligible in the case of 2-wire flowmeters, as well as the hydraulic energy required to operate the flowmeter. The hydraulic energy required is dependent upon the non-recoverable pressure loss the flowmeter causes, which can be significant in some applications.

Some applications allow energy savings by selection of a flowmeter with a lower non-recoverable pressure loss. Before operating costs are considered as a major factor in flowmeter selection, each application should be examined to determine if the energy savings can be realized.

Equations for the estimated non-recoverable pressure loss (h) of various flowmeters are shown in Figure 23-4.

Flowmeter	*Liquid*	*Gas (vapor)*		*Liquid/gas (vapor); mass flow*
Venturi:				
15° exit cone	→	$h_l = (0.436 - 0.86\beta + 0.59\beta^2)\,\Delta P$	(a)	←
7° exit cone	→	$h_l = (0.218 - 0.42\beta + 0.38\beta^2)\,\Delta P$	(b)	←
Universal Venturi Tube†	→	$h_l = (0.065 + 0.092\beta - 0.167\beta^2)\,\Delta P$	(c)	←
Lo-Loss tube‡	→	$h_l = (0.151 - 0.304\beta + 0.182\beta^2)\,\Delta P$	(d)	←
Nozzle	→	$h_l = (1 + 0.014\beta - 2.06\beta^2 + 1.18\beta^3)\,\Delta P$	(e)	←
Orifice	→	$h_l = (1 - 0.24\beta - 0.52\beta^2 - 0.16\beta^3)\,\Delta P$	(f)	←
Annubar: §		$h_l = \frac{k}{D}\,\Delta P$	(g)	
		Type 15/16 $k = 0.478$; 35/36 $k = 1.59$ 25/26 $k = 1.11$; 45/46 $k = 2.55$		
Pitot	→	$h_l = \frac{0.6}{D}\,\Delta P$	(h)	←
Target*	→	$h_l = 0.000467 \frac{\rho \bar{V}^2}{(1 - \beta_T)^{2.75}}$	(i)	←

FIGURE 23-4
Permanent pressure-loss equations (U.S. Units)

(continued)

Flowmeter	*Liquid*		*Gas (vapor)*		*Liquid/gas (vapor); mass flow*	
	$h_l = \frac{F_p G_F}{(1-\beta_T)^{2.75}} \left(\frac{q_{gpm}}{14.3D^2}\right)^2$	(j)	$h_l = \frac{Z\,GT}{p\,(1-\beta_T)^{2.75}} \left(\frac{q_{SCFH}}{19{,}520D^2}\right)^2$	(q)	$h_l = \frac{1}{p\,(1-\beta_T)^{2.75}} \left(\frac{q_{PPH}}{909D^2}\right)^2$	(v)
	$h_l = \frac{G_b^2}{F_p G_F (1-\beta_T)^{2.75}} \left(\frac{q_{GPM}}{14.3D^2}\right)^2$	(k)				
Turbine*	$h_l = 0.00577\,\rho \bar{V}^2$	(l)	$h_l = 0.0129\,\rho \bar{V}^2$	(r)	Liquid: $h_l = \frac{1}{\rho} \left(\frac{q_{PPH}}{259D^2}\right)^2$	(w)
	$h_l = F_p G_F \left(\frac{q_{gpm}}{4.08D^2}\right)^2$	(m)	$h_l = \frac{Z\,GT}{p} \left(\frac{q_{SCFH}}{3714D^2}\right)^2$	(s)	Gas (vapor): $h_l = \frac{1}{\rho} \left(\frac{q_{PPH}}{173D^2}\right)^2$	(x)
	$h_l = \frac{G_b^2}{F_p G_F} \left(\frac{q_{GPM}}{4.08D^2}\right)^2$	(n)				
Vortex*	⟶		$h_l = 0.00554\,\rho \bar{V}^2$	(t)	⟵	
	$h_l = F_p G_F \left(\frac{q_{gpm}}{4.17D^2}\right)^2$	(o)	$h_l = \frac{Z\,GT}{p} \left(\frac{q_{SCFH}}{5669D^2}\right)^2$	(u)	$h_l = \frac{1}{\rho} \left(\frac{q_{PPH}}{264D^2}\right)^2$	(y)
	$h_l = \frac{G_b^2}{F_p G_F} \left(\frac{q_{GPM}}{4.17D^2}\right)^2$	(p)				

† The manufacturer (BIF) should be consulted for exact information.
‡ The manufacturer (Badger Meter Inc.) should be consulted for exact information.
§ The manufacturer (Dieterich Standard Corp.) should be consulted for exact information.
* Foxboro flowmeter *gas*-turbine flowmeter equation based on Rockwell literature; the manufacturer should be consulted for exact information.

(From Miller, *Flow Measurement Engineering Handbook* ©1989, McGraw-Hill Book Company. Used with permission.)

FIGURE 23-4 (continued)

EXAMPLE 23-5

Problem: Consider a flowmeter that is used to control a flow of 0 to 1000 gpm of a liquid with a specific gravity of 1.13 and at a pressure of 50 psi through a control valve into a reactor that is vented to atmosphere. Determine whether any energy savings are realized by using a flowmeter that has a non-recoverable pressure drop of 3 psi at the nominal flow of 800 gpm rather than a flowmeter that has a non-recoverable pressure drop of 10 psi at the same operating conditions.

Solution: As both systems result in a total pressure drop of 50 psi, no energy savings are realized. The difference between the systems is that the pressure drop across the control valve is 47 psi in one case and 40 psi in the other.

When the flowmeter can be used to control the speed of the feed pump, energy savings can be realized. The speed of the pump (and hence the power input to the pump) is varied to generate only the amount of motive energy necessary to produce the desired flow. The energy loss of the flowmeter in horsepower is given by:

$$\text{hp} = (\Delta P_{\text{in. WC}} \times Q_{\text{lb/hr}}) \,/\, (3.8 \times 10^5 \times \eta \times \rho_{\text{lb/ft}^3})$$

and the energy cost as

$$\text{energy cost (\$/yr)} = 0.746 \times \text{hp} \times \text{operating hours/yr} \times \text{cost}_{\$/\text{kwh}}$$

Assuming the following at a nominal flow of 800 gpm:

Pump efficiency (η_p)	70%
Motor efficiency (η_m)	80%
Motor control efficiency (η_{mc})	95%
Hours of operation per year	7200 (300 days @ 24 hrs/day)
Electricity cost	$0.07/kwh

The energy loss and energy cost of the flowmeter with the 10 psi pressure drop at nominal flow is

$$\text{hp} = \frac{(10 \text{ psi} \times 27.71 \text{ in. WC/psi})\,(800 \text{ gpm} \times 60 \text{ min/hr} \times 1.13 \times 8.34 \text{ lb/gal})}{3.8 \times 10^5 \times (0.70 \times 0.80 \times 0.95) \times (1.13 \times 62.336 \text{ lb/ft}^3)}$$

$$= 8.8 \text{ hp}$$

$$\text{energy cost} = 0.746 \times 8.8 \text{ hp} \times 7200 \text{ hrs/yr} \times 0.07\text{/kwh} = \$3308.66\text{/yr}$$

The energy cost for the flowmeter with the 3-psi pressure drop at nominal flow is $3308.66 × 3/10, or $992.40. The energy saving realized in choosing the flowmeter with the lower pressure drop is $2316.26/year. Other savings may be realized if the pump were reduced in size because of the lower pressure requirements.

Maintenance Cost Flowmeter maintenance costs are those associated with keeping the flowmeter in service after it is in operation and includes not only the cost of parts but also the cost of labor. While some technologies by their nature have higher maintenance costs than others, it is not uncommon to find flowmeters utilizing the same technology but having significant differences in the amount of required maintenance. While the amount of maintenance required may be obvious in some flowmeter designs, the true test is the maintenance record of the flowmeter over its service life.

Foremen often keep records of maintenance performed on all instruments. One use of these data is to assess which instruments require excessive maintenance. Examination of these records can aid in replacement flowmeter selection. Other technologies and recent improvements that may also address the problems encountered should also be considered.

In assessing the maintenance requirements of a particular flowmeter, it should be realized that only very qualitative information is available. However, taken on the whole, it may point up a deeper problem of misapplication or even process considerations that have been previously overlooked.

EXERCISES

23.1 Estimate the relative installed cost of a vortex shedding flowmeter for a flow range of 0 to 100 gpm.

23.2 Estimate the relative installed cost of an orifice plate flowmeter for a flow range of 0 to 100 gpm.

23.3 Calculate the energy cost of an orifice plate flowmeter with a beta ratio of 0.5 and a full scale differential pressure of 100 inches of water column for a flow of 10,000 lbs/min of gas in a 24-inch diameter pipe operating at a density of 0.82 lbs/ft^3 for 24 hrs/day, 365 days/yr, when electrical energy costs are $0.07 per kwh. Compressor and motor efficiencies are assumed to be 80 percent.

24

Data Required for Flowmeter Selection

Introduction Obtaining accurate data for flowmeter selection is essential for effective selecting and sizing of flowmeters. Many applications involve substances on which little physical property information is available, and a best guess often is the only method available. Even when fluid data are available, the operating conditions and flows encountered during operation can be significantly different from those determined when the flowmeter was selected, as many processes are not well enough defined to operate close to the process conditions and flow ranges for which they were designed.

The above illustrates what is probably the most serious difficulty of flowmeter selection: a lack of accurate definition of sizing conditions and fluid properties. Specification of such data often involves judgment calls based on familiarity with the process fluids being used. Physical property data from handbooks is adequate for common substances, but process fluids are often mixtures or intermediates whose properties may not be adequately defined.

The person selecting or specifying the flowmeter should use caution in developing the process data without independent review by a qualified process engineer.

The flowmeter selection process is challenging in its own right, even if the fluid properties are well known. Attempts to combine the evaluation of process data and operating conditions result in such a divergent set of demands that either the flowmeter selection may not be given adequate attention or the process data may be inadequately substantiated. If the process conditions are not well defined, this fact should be clearly documented. For example, often measurement accuracy is not deemed important until after startup, so the design conditions should be documented as completely as possible.

Performance Flowmeter performance is often judged by the accuracy of the flow measurement that is achieved. Therefore, the most appropriate measure that should be considered is the overall accuracy of the flowmeter system, including any devices necessary to achieve the final end use. Differences between flowmeter accuracies may be examined; however, they may have an insignificant effect on the overall system accuracy due to errors introduced by other devices in the flowmeter system and uncertainties for which physical properties and operating conditions are known.

In some applications, repeatability, as opposed to overall system accuracy, is thought to be a sufficient measure of flowmeter performance. Some persons argue that in virtually all applications the numerical value of the flow is considered unimportant as long as the measurement is repeatable. This line of reasoning may be valid in some applications; however, it should be noted that flowmeter repeatability can be a function of Reynolds number or other factors. Variations in temperature, viscosity, pressure, density, and the like can alter where the flowmeter operates on its repeatable but perhaps nonlinear characteristic curve, thereby affecting the repeatability of the overall measurement. Therefore, repeatability should not be used as the absolute criterion in determining performance.

EXAMPLE 24-1

Problem: Consider a flowmeter that is ideally repeatable and linear above a Reynolds number of 10,000 and nonlinear but repeatable between Reynolds numbers of 3000 to 10,000. Determine whether the flowmeter is repeatable as a function of flow.

Solution: When Reynolds number is above 10,000, the flowmeter will be repeatable as a function of flow, as the flowmeter is both linear and repeatable. When the viscosity can vary, perhaps due to normal temperature fluctuations, uncertainty is introduced into the calculation of Reynolds number, as the viscosity is not well defined. When part of the range of possible Reynolds numbers is below 10,000, the flowmeter ceases to be repeatable as a function of flow; the flowmeter output for the same flow will change as Reynolds number changes due to differences in viscosity as a result of normal temperature variations.

Response of the flowmeter to flow, often measured by the time required to measure step changes in flow, is often an important factor, especially in applications where the flow is to be turned on for a relatively short period of time.

Fluid Properties

Fluid Name In the case of a commonly known fluid, knowledge of the name of the fluid can be beneficial in gathering sufficient physical property information from public sources such as handbooks, which can be used as a primary source of information or as a way to verify data obtained elsewhere. In many processes, however, the exact composition of the fluid stream is not known and specific property data are unavailable. This places a greater burden on the person specifying flowmeters. Physical property data and material compatibility cannot be specified with certainty.

Type of Fluid Types of fluids include liquids, gases, and vapors. These designations alone are not sufficient for accurate flowmetering. The fluid may be clean, dirty, or a liquid/solid slurry, or it may contain other combinations of states. A slurry should be defined in terms of the percentage, type, and particle size of solids and whether it is abrasive or fibrous or has unusual flow characteristics. The viscosity characteristics of liquids may be Newtonian or non-Newtonian in nature.

Compatibility of Materials The corrosiveness of the fluid at operating conditions will determine compatible materials of construction. Compatibility of metals can often be determined by the materials of construction used for piping, while compatibility of other materials, such as Teflon®, Viton®, and the like are determined by experience, calculations based upon chemical properties, or tests performed by manufacturers. Note that pipe material selection is based on corrosion allowances that are generally much larger than allowable for flowmeter services, so the piping material should be used as a starting point. The resources of a materials specialist and actual operating experience should be used. Materials decisions generally involve more than looking up corrosion rates in a handbook and should be made as early as possible.

Pressure and Temperature The operating pressure and temperature ranges at the flowmeter should be defined, especially in gas service where flowmeter selection may be affected by variations in pressure and temperature. Pressure and/or temperature compensation may be required for accurate flow measurement when variations are sufficiently large. Flowmeter flange ratings are determined by the maximum operating pressure and temperature of the fluid and can often be determined from the rating of other flanges in the same pipe, barring any special considerations that may be required for the flowmeter or process considerations.

Specific Gravity and Density The operating specific gravity of a liquid is required to size and calibrate many flowmeters. While the specific gravity of a liquid is virtually always assumed to be constant, temperature compensation may be required in some applications where sufficiently large temperature variations occur that can significantly affect accuracy. In gas and vapor applications, the density at standard and at operating conditions is generally required to apply most flowmeter technologies. This enables accurate conversion to standard units in the case of non-ideal gases and vapors. Compensation for density fluctuations is usually performed using pressure and/or temperature compensation or densitometer measurement.

Viscosity The viscosity of the fluid is needed (though not always accurately known) to estimate Reynolds number and determine the applicability of the various flowmeter technologies. Knowledge of the behavior of viscosity in liquid applications is valuable since large variations in viscosity may occur due to relatively small temperature changes. Large changes in Reynolds number may affect flowmeter selection. If these data are not available, it may be necessary to have tests run or to select a flowmeter that is not affected by Reynolds number.

Precise knowledge of gas viscosity is usually not as critical as it is with liquids. Small variations in viscosity between gases, the relative insensitivity of gas viscosity to operating conditions, and the relatively low viscosities involved effectively minimize the Reynolds number effect on gas measurement.

Operating Range The operating range of a flowmeter is the range of flows over which the flowmeter will perform accurately, from which the required turndown can be calculated. Because digital flowmeters may fail to operate below a minimum flow, the minimum flow should be carefully determined sufficiently low to handle all operating conditions. Analog flowmeters generally operate over a range from zero flow to full scale flow, although at reduced accuracies at the low end.

Other Physical Properties When certain technologies are being considered, knowledge of additional physical properties of the fluid may be necessary. For example, magnetic flowmeter applications require the electrical conductivity; thermal flowmeter applications require the thermal capacity and conductivity of the fluid. Other information such as solids content, tendency to deposit crystals, cleaning fluids, vapor pressure, etc., should also be noted.

Installation

Pipe Size The pipe size and schedule in which the flowmeter is to be installed are usually known, although the flowmeter size will be determined by the operating conditions and the desired flow range and may therefore differ from the nominal pipe size.

Differential Pressure Any constraints on the maximum allowable pressure drop that the flowmeter can develop should be defined before flowmeter selection. Often, process constraints limit the allowable unrecovered pressure drop across the flowmeter. In some liquid applications, excessive pressure drop can result in flashing, cavitation, and unnecessary energy loss.

Pipe Vibration Pipe vibration should not be overlooked. Special installation requirements entail coordination with piping design to ensure a proper installation.

Pulsating Flow Whether flow will be pulsating or steady in nature should be defined; some technologies are more immune to pulsation effects than others. It may be necessary to coordinate with process and mechanical resources to have pulsation dampers installed in some cases.

Straight Run Many flowmeters require that the velocity profile be properly developed upstream and downstream of the flowmeter. The common method is to use sufficient straight run upstream and downstream of the flowmeter (perhaps with a flow conditioner), to establish symmetrical flow profiles, to eliminate swirl, and to allow developed flow to occur.

The ability to incorporate sufficient straight run into the piping system without undue expense is important in flowmeter selection. Recognition that sufficient straight run cannot be designed into all piping systems can avoid a misapplication through selection of a technology that does not require excessive straight run.

Ambient Conditions Some knowledge of ambient conditions around the flowmeter can aid in selecting the proper one for the application, as some are not suitable for harsh environments. This can be illustrated by comparing flowmeters designed for a laboratory environment and those designed for a dirty industrial environment such as an outdoor installation.

Operation

Maintenance The amount of maintenance that must be performed on the flowmeter can be of paramount importance in selection. Sensors that must be cleaned on a daily basis or replaced weekly are a maintenance headache that cannot normally be tolerated in an industrial continuous flow environment. Some flowmeter technologies are more prone to failure than others, and reliability may be improved by using another flowmeter. Trade-offs may be required to achieve the increased reliability.

Availability of Parts and Service The availability of spare parts on site, at local service centers, and the general trend towards standardization are often valid considerations in flowmeter selection. However, selection greatly influenced by availability of parts and service centers can create problems. Selection should proceed based upon technical considerations, after which flowmeters for which spare parts and service are available should be considered. Particular attention should be paid to what trade-offs are made, if any, in selecting an alternate flowmeter.

Economic Considerations Installed flowmeter cost, which includes the costs of the flowmeter, miscellaneous parts, piping, and labor to effect a complete functional flowmeter system, should be considered in the selection process. The cost of auxiliary devices such as converters should be considered where applicable. The flowmeter selection process emphasizes technical considerations over economic considerations. It should be realized that a misapplied flowmeter, however inexpensive, represents an economic liability. For this reason, technical evaluation should be a prelude to an economic evaluation, from which trade-offs made in selecting a more economic flowmeter can be evaluated.

Operating Cost The unrecovered pressure loss across the flowmeter is the parameter from which an economic measure of energy consumption can be calculated. Annual energy consumption can be significant, and proper flowmeter selection can significantly reduce energy costs. It should be noted, however, that in many applications no economic advantage is gained from selecting a flowmeter technology that would conserve energy, as the energy that is saved cannot be recovered. Determination of the feasibility of energy recovery should be made; annual energy savings can be significantly higher than the cost differential to purchase a flowmeter with a lower pressure loss, especially in larger pipe sizes.

Future Considerations Consideration of the future uses of the flowmeter can influence the selection procedure. Possible considerations include plant expansion, discrepancies between design and actual operating conditions, anticipated process changes, alternate operating philosophies, and the like. Rangeability and the ability to measure accurately over a wider range of operating conditions should be considered for any foreseeable changes, so as to avoid replacement of the flowmeter at a later date.

Risk A certain amount of risk exists when a flowmeter is selected, and this can be accentuated when a new technology is applied for the first time. Due to inexperience, unfamiliar technologies are more prone to misapplication than the more established technologies, so there is more risk perceived in applying unfamiliar technologies. Unfamiliar flowmeter technologies can often be judged by evaluation of the principle of operation in conjunction with comments from users. Obtaining input from two or more users allows a broader perspective of the advantages and problems that were encountered.

Risks should only be taken after detailed investigation and analysis of an application so that it is identified and not haphazard. Decisions should not be based upon manufacturer marketing literature but rather on a technical basis. It may be appropriate to apply the flowmeter on a development basis with a follow-up evaluation.

Flowmeter Information Sheet The form shown in Figure 24-1 can be used to tabulate the information that is useful in flowmeter selection. For future reference, it is suggested that the form be completed, including documentation of the source of information and any assumptions that were made.

Fluid

Name ________________ Type ________________

Composition ________________________________

Compatible materials ________________________________

Operating conditions

Nominal line size ________________ Nominal pipe ID ________________

Density or SG, min/max ________________ Design ________________

Flow, min/max ________________ Design ________________

Pressure, min/max ________________ Design ________________

Temperature, min/max ________________ Design ________________

Viscosity, min/max ________________ Design ________________

Other ________________ Design ________________

Other ________________ Design ________________

Flowmeter characteristics

Nominal velocity, min/max ________________________________

Nominal Reynolds number, min/max ________________________________

Desired performance ________________ Desired range ________________

Selected flowmeter and size ________________________________

Actual velocity, min/max ________________________________

Actual Reynolds number, min/max ________________________________

Actual performance ________________________________

Actual range ________________________________

Reasons for flowmeter selection ________________________________

____

____

$$R_D = \frac{3160\, Q_{gpm}\, SG}{\mu_{cP}\, D_{in.}} = \frac{379\, Q_{acfm}\, \rho_{lb/ft^3}}{\mu_{cP}\, D_{in.}}$$

FIGURE 24-1
Flowmeter selection data sheet.

25

Flowmeter Selection Procedure

Introduction Increased emphasis on tighter control and closer material balance is continually increasing the number of flowmeter applications and putting increased emphasis on flowmeter performance. As a result, flowmeter selection should be performed and documented in a logical manner in order to achieve the best flowmeter installation, after considering technical and non-technical constraints.

Flowmeter Selection Procedure Flowmeter selection is generally a process of elimination based on technical criteria. In this way, all flowmeter technologies are considered possible solutions until a specific reason is found to eliminate one or more of them from consideration. Once this has been done, other less tangible constraints can be used to establish the final selection.

This procedure requires thorough familiarity with flowmetering in order to assess the various technical constraints.

Details of some technical and non-technical criteria and their applications to the flowmeter selection procedure are presented in this section. The graphs and data should be considered as typical and should be used as guidelines but not as absolute references, because these parameters will change as technologies develop and flowmeters improve.

Technical Criteria There are a considerable number of technical criteria such as pressure, temperature, specific gravity or density, viscosity, flow range, and the like. Beyond this, the flow characteristics of the meter are often Reynolds number-dependent, and this fact may be used to further identify those flowmeters not likely to perform well in a given application.

Figure 25-1 illustrates flowmeter constraints as a function of Reynolds number for various flowmeter technologies. Other constraints such as pressure drop, sensitivity to solids accumulation, etc., which are not covered in detail by such a graph, illustrate the nature of the multiple constraints in flowmeter selection.

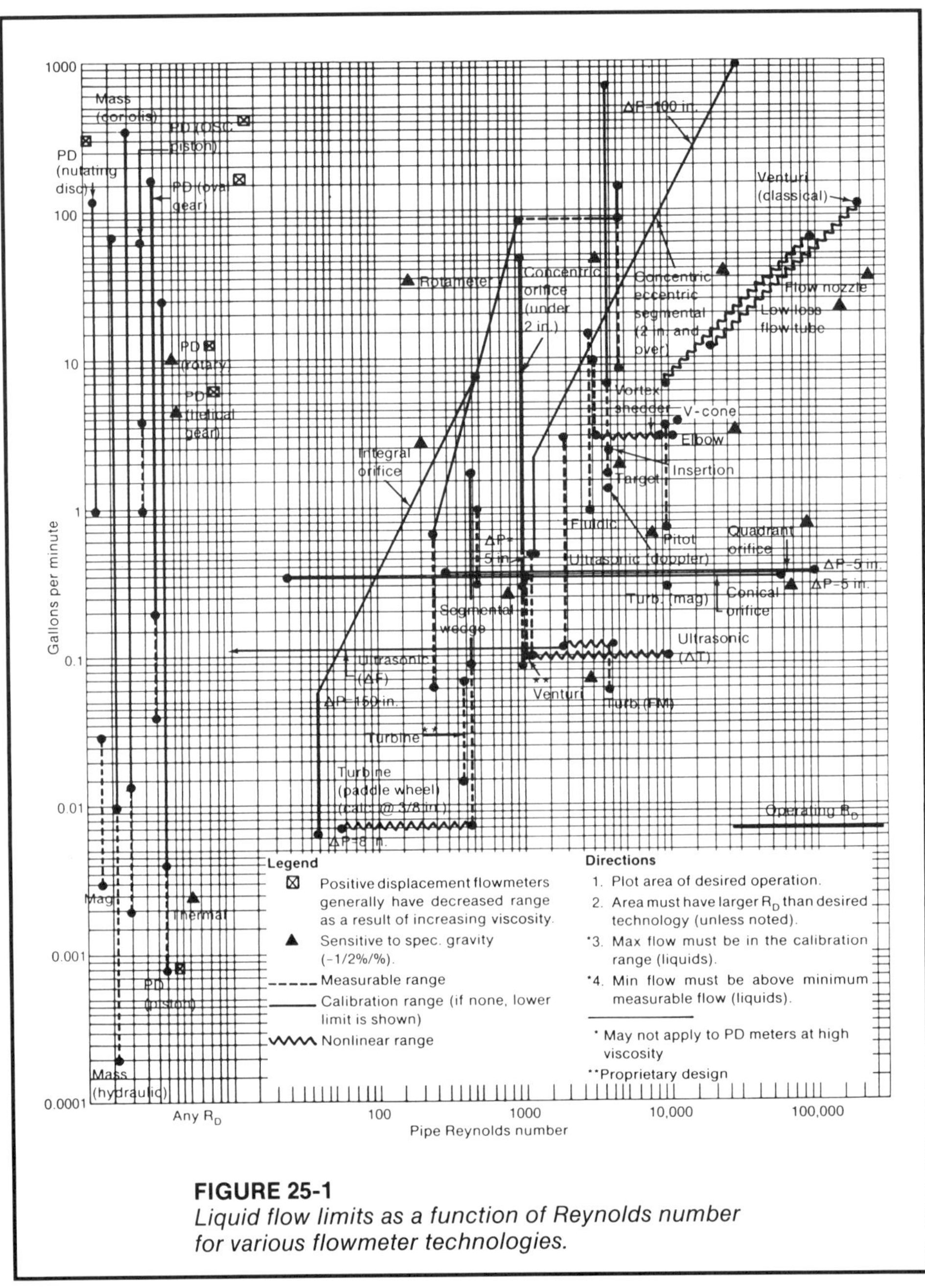

FIGURE 25-1
Liquid flow limits as a function of Reynolds number for various flowmeter technologies.

Solid lines in Figure 25-1 indicate ranges over which the flowmeter can be calibrated, while the dotted lines indicate ranges over which the flowmeter will operate accurately. The minimum required Reynolds number at various flows for accurate flowmeter operation is plotted and, unless otherwise indicated graphically by an arrow, applicable Reynolds numbers are assumed to continue to infinity and applicable flows are assumed to be in excess of 1000 gpm. Wavy lines indicate a nonlinear operating region, which in the case of digital flowmeters is indicative of nonlinearities present before Reynolds number decreases sufficiently that the flowmeter ceases to operate and turns off.

Part of the flowmeter selection procedure for liquid service is to identify the various Reynolds number constraints for the given operating conditions. Rangeability, accuracy, sensitivity to Reynolds number, etc., should all be considered in identifying those technologies that need no further consideration.

Operating Reynolds numbers can be plotted on the graph using worst case extremes of viscosity and other physical properties. As flowmeter sizing may require that the flowmeter be different from the pipe in which the fluid is flowing, Reynolds number may be plotted for the nominal pipe size, one size smaller than the nominal pipe size, and other pipe sizes determined by experience. Flowmeters that cannot be calibrated to the desired full scale flow and whose minimum flow is greater than the desired minimum flow should be eliminated, with the exception of positive displacement flowmeters operating at high viscosities (a condition that can decrease flowmeter range). These applications should be investigated on an individual basis. Flowmeters that do not operate accurately in the Reynolds number range graphed can also be eliminated.

As Reynolds number constraints are similar for liquid and gas applications, the procedure for gas service is identical to that for liquid service, with the exception that Reynolds number should be plotted as a horizontal line and should be used only as a criterion for elimination of flowmeter technologies, as flow is difficult to define due to the compressibility of gas.

Figure 25-2 shows the relationship between flowmeter rangeability and Reynolds number. The solid horizontal lines indicate the flow range over which each flowmeter technology can be applied. The dotted lines indicate the flow range over which the flowmeter can measure but can not be calibrated. Wavy lines indicate ranges of nonlinear flowmeter operation. Other technical criteria for consideration are included within columns on the graph. When the minimum and maximum Reynolds numbers are plotted on this graph, technologies can be eliminated based upon Reynolds number and other technical criteria.

Non-Technical Criteria There are a considerable number of non-technical criteria to be considered during the flowmeter selection process. These include but are not limited to cost (initial and installed), maintainability, spare parts availability, and vendor support after installation.

Flowmeters not eliminated by technical criteria should be further evaluated using manufacturer literature and available operating experience. Selection can be further refined by considering non-technical criteria such as cost, maintainability, delivery, etc. The final selection will generally embody a number of trade-offs but at the same time providing acceptable technical performance.

Applications

The flowmeter selection procedure presented above, while simple in concept, requires careful attention and evaluation of considerable detail. The applications presented below are not intended to show the best overall

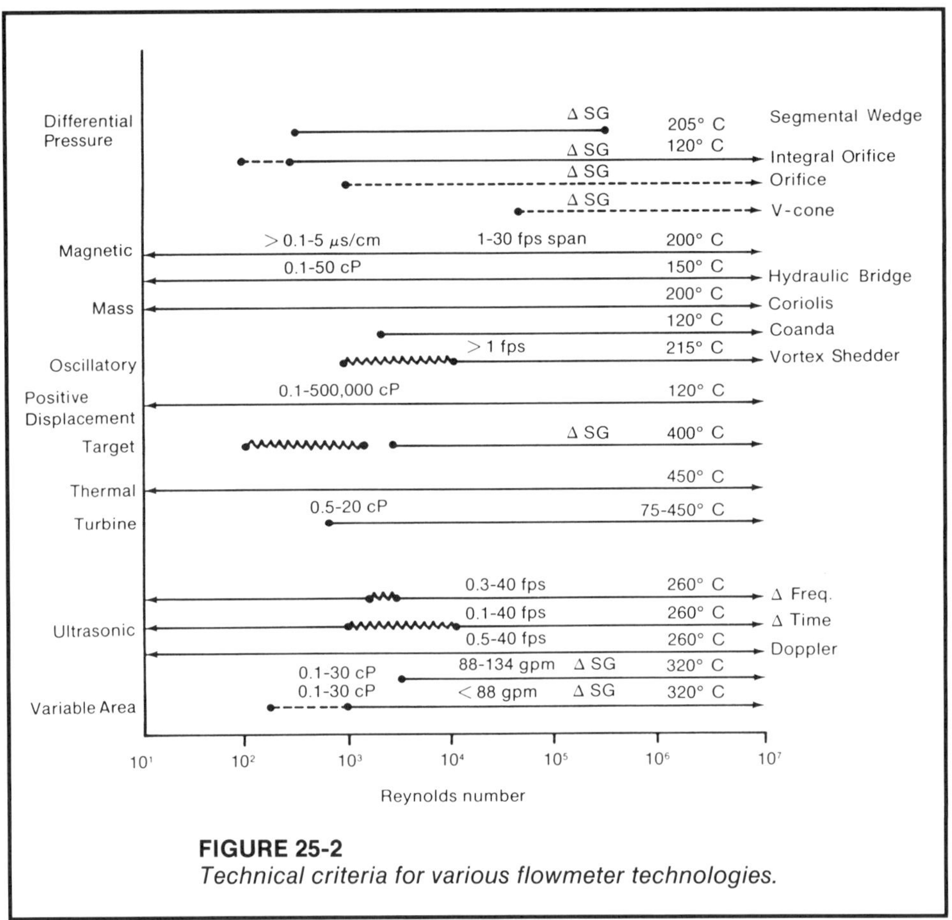

FIGURE 25-2
Technical criteria for various flowmeter technologies.

flowmeter for each service, but rather to illustrate the use of the flowmeter selection procedure on a given set of operating conditions.

Low Viscosity Liquids Low viscosity liquids such as water and some light hydrocarbons comprise a great many flowmeter applications.

EXAMPLE 25-1

Problem: Using the flowmeter selection procedure, eliminate technologies that are not applicable to a flow of 150 gallons per minute of water in a 4-inch schedule 40 pipe when the viscosity is assumed to be 1.0 cP and specific gravity is assumed to be 1.0. Accuracy requirements are ±1 percent of rate. The flowmeter must be mounted in a 60-inch straight section of pipe downstream of 2 elbows.

Solution: Reynolds number at maximum flow in the 4-in. pipe is given by

$$R_D = (3160 \times 150 \text{ gpm} \times 1.0)/(1.0 \text{ cP} \times 4.026 \text{ in.}) = 117{,}735$$

Using an arbitrary 10:1 turndown for calculating Reynolds number, Reynolds number at 10 percent flow is 11,774 in the 4-inch pipe. Similarly, Reynolds number can be calculated for a 3-inch flowmeter installation as 154,498 and 15,450 for full scale and 10 percent of full scale, respectively. The operating Reynolds numbers are graphed as shown in Figure 25-3.

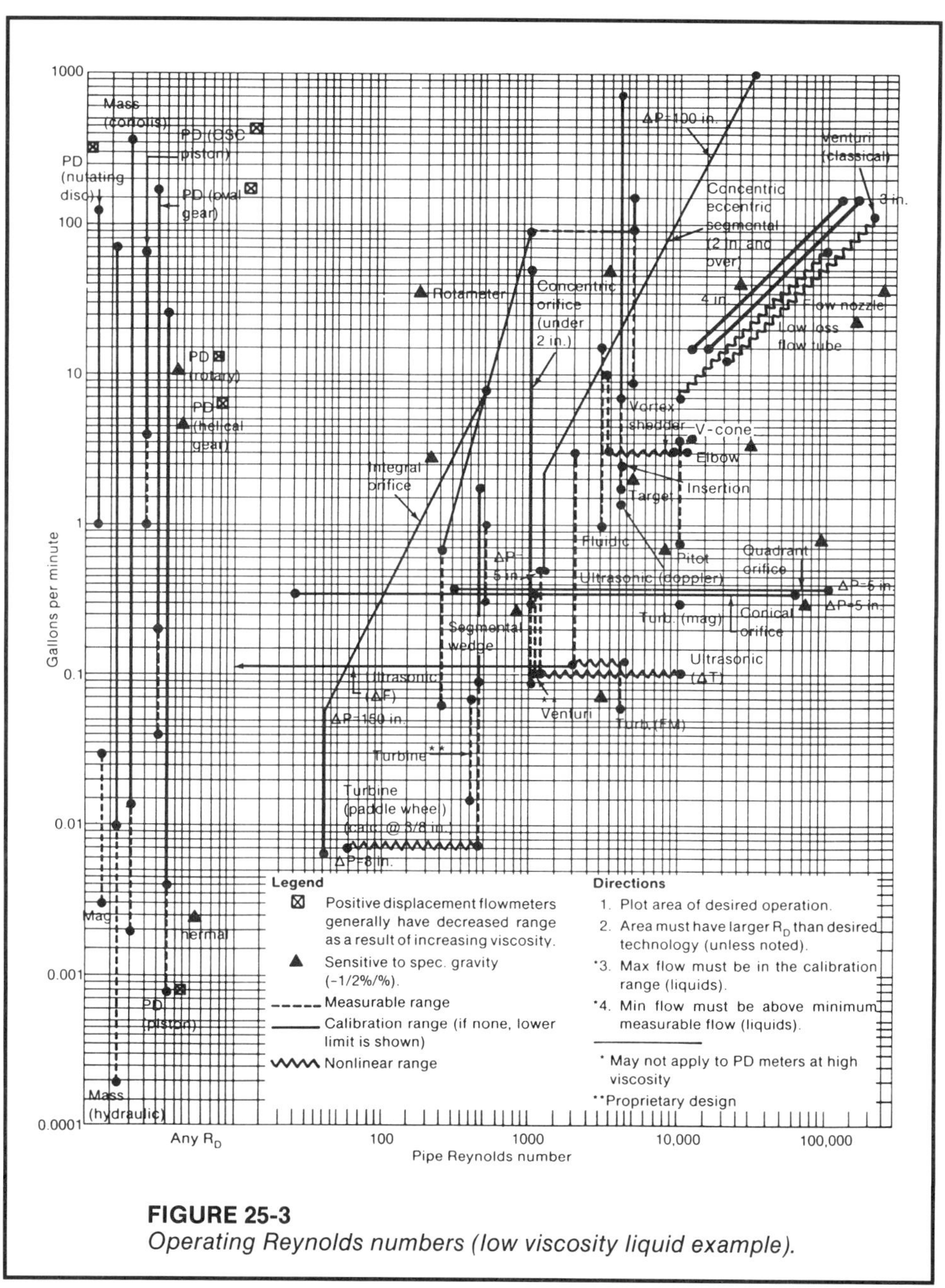

FIGURE 25-3
Operating Reynolds numbers (low viscosity liquid example).

The graph illustrates that Reynolds number is not sufficiently large to operate venturi, flow nozzle, and low-loss flow tube technologies. Using the flowmeter selection procedure, the various technologies are listed and technical criteria are first used to eliminate those that are not applicable. Nontechnical criteria can then be applied to eliminate still others. A summary of the above procedure is illustrated in Figure 25-4.

	Technical	Non-technical
Differential pressure		
Orifice		Insufficient straight run
Elbow		Insufficient accuracy
Flow nozzle	R_D too low	X
Flow tube	R_D too low	X
Laminar flow element	R_D too high	X
Segmental wedge		
Venturi	R_D too low	X
V-cone		
Magnetic		
Mass		
Coriolis		Expensive
Hydraulic	Out of range	X
Oscillatory		
Fluidic		
Vortex shedding		
Positive displacement		
Helical		Expensive, slippage
Nutating disc	Out of range	X
Oscillating piston	Out of range	X
Oval gear		Expensive, slippage
Piston	Out of range	X
Rotary		Expensive, slippage
Target		Insufficient accuracy
Thermal		Insufficient accuracy
Turbine		Moving parts not preferred
Ultrasonic		
Doppler		Insufficient accuracy
Time of flight		Expensive
Variable area	Out of range	X
Insertion	Hydraulic error too large	X
Bypass		Insufficient accuracy

FIGURE 25-4
Flowmeter selection summary (low viscosity liquid example).

The remaining technologies can be considered individually to determine the optimum flowmeter, given the technical criteria of the application on hand. In this example, if the installed cost were of prime concern, a vortex shedding flowmeter might be selected. If energy consumption of the flowmeter is important, when pressure drop is limited by process constraints, or when the stream contains solids that might settle out, a magnetic flowmeter might be considered because of its obstructionless design. As can be seen from the above discussion, flowmeter selection must be tailored to each application. The flowmeter selection procedure presents a goal and aids in organizing the data used to arrive at the final flowmeter selection.

Medium and High Viscosity Liquids Medium and high viscosity liquids, which have viscosities greater than a few centipoise, are considered together due to the overlap that exists in some flowmeter technologies because of sizing considerations. Extremely viscous materials are much more complex than the so-called Newtonian fluids. For example, asphalt, toothpaste, peanut butter, and other foodstuffs, or various types of organic and inorganic slurries often cannot be described by a single viscosity number and exhibit a variety of generally unfamiliar flow behaviors.

EXAMPLE 25-2

Problem: Using the flowmeter selection procedure, eliminate technologies that are not applicable to a flow of 30 gallons per minute of an organic liquid with a specific gravity of 1.17 in a 2-inch schedule 40 pipe when the viscosity can vary from 5 to 100 cP over the operating temperature range. Accuracy requirements are ±1 percent of rate.

Solution: The maximum value of Reynolds number at maximum flow and minimum viscosity in the 2-inch pipe is given by

$$R_D = (3160 \times 30 \text{ gpm} \times 1.17) / (5 \text{ cP} \times 2.067 \text{ in.}) = 10{,}732$$

However, at maximum flow it can be as low as

$$R_D = (3160 \times 30 \text{ gpm} \times 1.17) / (100 \text{ cP} \times 2.067 \text{ in.}) = 536.6$$

Values of Reynolds number at 10 percent of full scale flow are 1073 and 54, respectively. The operating Reynolds numbers are graphed as shown in Figure 25-5.

The graph illustrates that it is not possible to operate any of the flowmeters dependent upon Reynolds number over all operating conditions, even if the flowmeter size were reduced to 1 inch, which would effectively double Reynolds number. Elimination of flowmeter technologies is shown in Figure 25-6.

Remaining technologies include the mass flowmeter and various positive displacement flowmeters that can be considered individually to determine the optimum one, given the technical criteria of the application at hand. From this example, it can be seen how the flowmeter selection procedure eliminates many technologies when the proper criteria are considered.

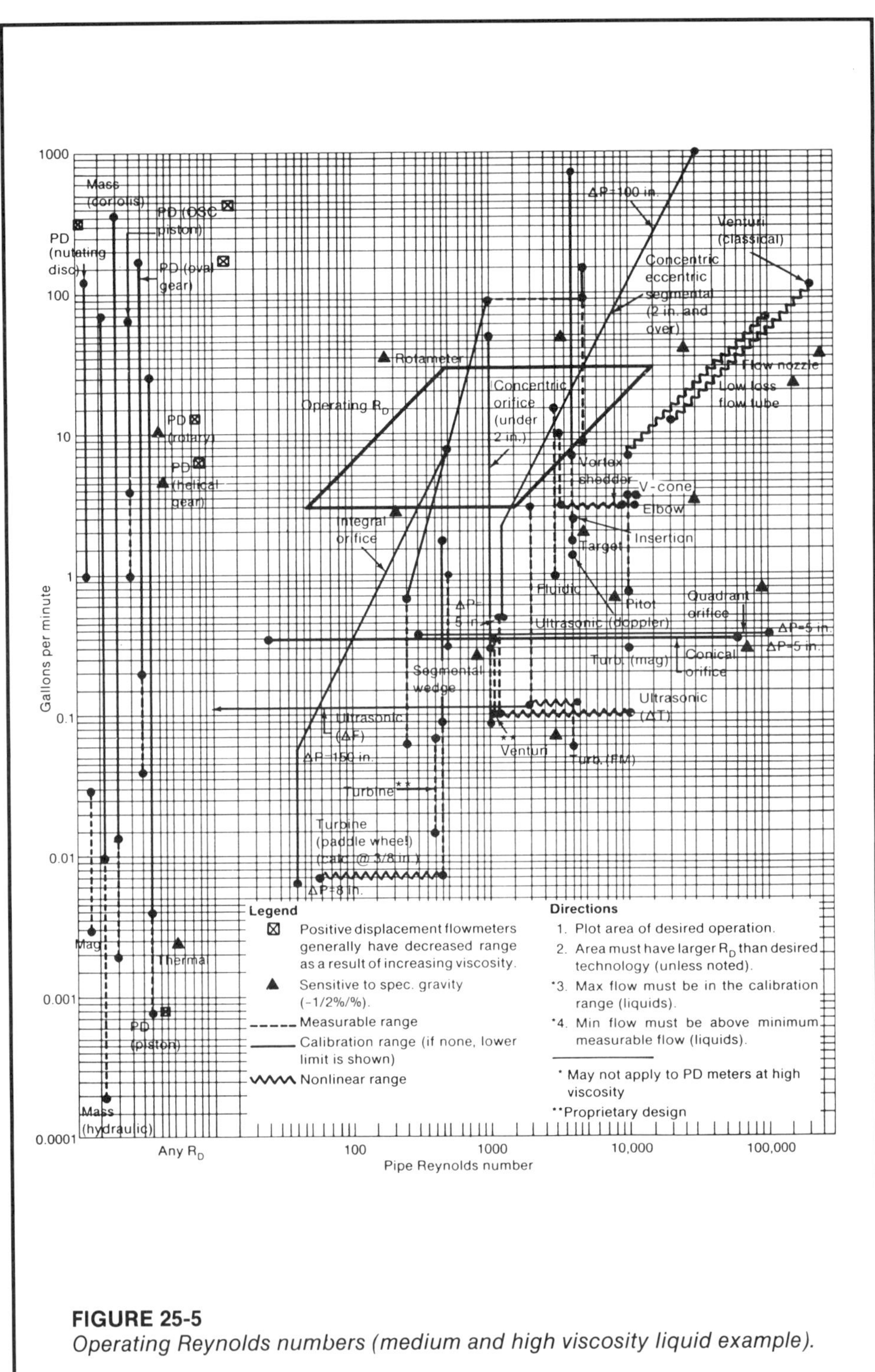

FIGURE 25-5
Operating Reynolds numbers (medium and high viscosity liquid example).

	Technical	Non-technical
Differential pressure		
Orifice	R_D too low	X
Elbow	R_D too low	X
Flow nozzle	R_D too low	X
Flow tube	R_D too low	X
Laminar flow element	R_D too high	X
Segmental wedge	R_D too low	X
Venturi	R_D too low	X
V-cone	R_D too low	X
Magnetic	Non-conductive	X
Mass		
Coriolis		
Hydraulic	Viscosity too high	X
Oscillatory		
Fluidic	R_D too low	X
Vortex shedding	R_D too low	X
Positive displacement		
Helical		
Nutating disc		
Oscillating piston		
Oval gear		
Piston	Out of range	X
Rotary		
Target	R_D too low	X
Thermal		Insufficient accuracy
Turbine	R_D too low	X
Ultrasonic		
Doppler	R_D too low	X
Time of flight	Operates in transition regime	X
Variable area	R_D too low	X
Insertion	R_D too low	X
Bypass		Not preferred in small pipe

FIGURE 25-6
Flowmeter selection summary (medium and high viscosity liquid example).

EXAMPLE 25-3

Problem: Using the flowmeter selection procedure, eliminate technologies that are not applicable to a flow of 0.2 to 0.6 gallons per minute of an organic liquid in a 1-inch schedule 40 pipe when the viscosity can vary from 4 to 15 cP over the operating temperature range and specific gravity is 1.0. Accuracy requirements are ±1 percent of rate.

Solution: The maximum value of Reynolds number at maximum flow and minimum viscosity in the 1-inch pipe is given by

$$R_D = (3160 \times 0.6 \text{ gpm} \times 1.0) / (4 \text{ cP} \times 1.049 \text{ in.}) = 452$$

However, at maximum flow it can be as low as

$$R_D = (3160 \times 0.6 \text{ gpm} \times 1.0) / (15 \text{ cP} \times 1.049 \text{ in.}) = 120$$

At the minimum flow of 0.2 gpm, Reynolds numbers at minimum and maximum viscosity are 151 and 40, respectively. As the liquid velocity in the 1-inch pipe is not excessive, Reynolds number calculations can be performed for a 1/2-inch pipe size in an attempt to increase Reynolds number such that more flowmeters may be applicable. The maximum value of Reynolds number at maximum flow and minimum viscosity in the 1/2-inch pipe is given by

$$R_D = (3160 \times 0.6 \text{ gpm} \times 1.0) / (4 \text{ cP} \times 0.622 \text{ in.}) = 762$$

However, at maximum flow it can be as low as

$$R_D = (3160 \times 0.6 \text{ gpm} \times 1.0) / (15 \text{ cP} \times 0.622 \text{ in.}) = 203$$

Values of Reynolds number at 0.2 gpm are 254 and 68, respectively. The operating Reynolds numbers calculated above are graphed in Figure 25-7.

The graph illustrates that for a 1-inch flowmeter, it is not possible to operate any of the flowmeters dependent upon Reynolds number over all operating conditions. When the flowmeter is reduced to 1/2 inch, effectively doubling Reynolds number, Reynolds number constraints are satisfied for an integral orifice plate, with the exception of extreme operating conditions of high viscosity at low flow conditions, which may only occur during short periods of time. This example illustrates how a change in flowmeter size can bring Reynolds number within the operating limits of a given flowmeter technology. Elimination of flowmeter technologies is shown in Figure 25-8.

Remaining technologies include various positive displacement flowmeters and the Coriolis mass flowmeter, which can be considered individually to determine the optimum one, given the non-technical criteria of the application at hand.

Gases Operating gas density can vary significantly. Applications can often be divided into low, medium, and high density categories. The operating density of the gas is dependent upon the combination of molecular weight of the gas and its operating pressure and temperature.

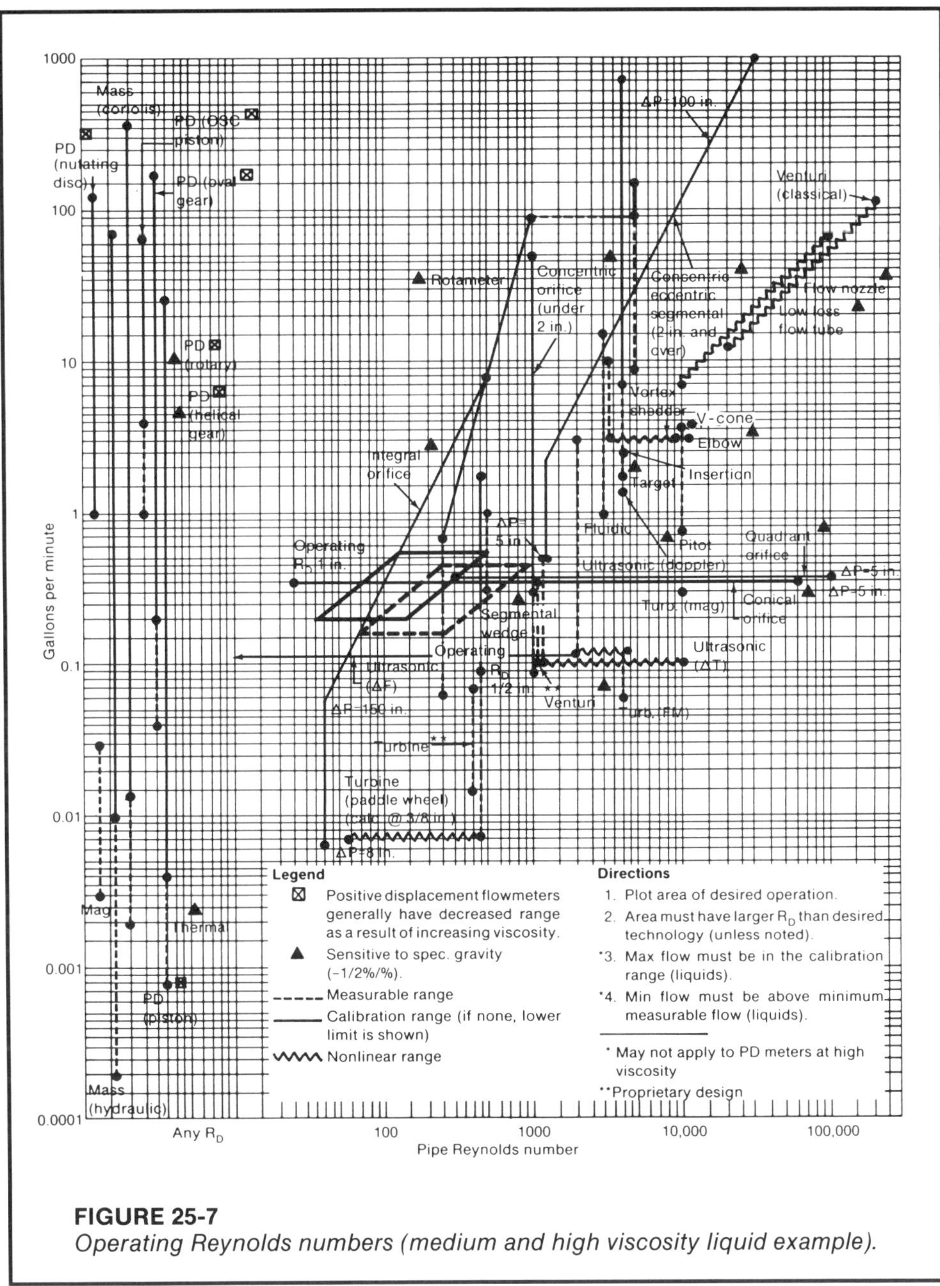

FIGURE 25-7
Operating Reynolds numbers (medium and high viscosity liquid example).

Low density applications present difficulties for flowmeters that utilize the momentum of the gas to operate the flowmeter, as the momentum may be insufficient. Low density applications are typified by most pure and process gases at low pressures or vacuum, as well as light gases such as hydrogen at low and medium pressure.

	Technical	Non-technical
Differential pressure		
Orifice		Insufficient accuracy
Elbow	R_D too low	X
Flow nozzle	R_D too low	X
Flow tube	R_D too low	X
Laminar flow element	R_D too low	X
Segmental wedge	R_D too low	X
Venturi	R_D too low	X
V-cone	R_D too low	X
Magnetic	Non-conductive	X
Mass		
Coriolis		
Hydraulic		Pump not preferred
Oscillatory		
Fluidic	R_D too low	X
Vortex shedding	R_D too low	X
Positive displacement		
Helical	Out of range @ operating viscosity	X
Nutating disc		
Oscillating piston		
Oval gear		
Piston		
Rotary	Out of range @ operating viscosity	X
Target	R_D too low	X
Thermal		Insufficient accuracy
Turbine	R_D too low	X
Ultrasonic		
Doppler	R_D too low	X
Time of flight		Expensive
Variable area	R_D too low	X
Insertion	R_D too low	X
Bypass		Not preferred

FIGURE 25-8
Flowmeter selection summary (medium and high viscosity liquid example).

EXAMPLE 25-4

Problem: Use the flowmeter selection procedure to eliminate technologies that are not applicable to a 50-acfm flow of hydrogen in a 2-inch schedule 40 pipe where the operating pressure is nominally 5 psi at a nominal operating temperature of 80°F.

Solution: From physical property tables, the density of hydrogen at 14.7 psi and 68°F is 0.00523 pound per cubic foot. The density at nominal operating conditions can be calculated as

$$\rho = 0.00523 \times [(460°\text{F} + 68°\text{F}) / (460°\text{F} + 80°\text{F})] \times [(14.7 \text{ psi} + 5 \text{ psi})/14.7 \text{ psi}]$$

$$= 0.00685 \text{ pound per cubic foot}$$

Similarly, the viscosity is 0.009 cP.

Reynolds number is calculated as

$$R_D = (379 \times 50 \text{ acfm} \times 0.00685 \text{ lb/ft}^3) / (0.009 \text{ cP} \times 2.067 \text{ in.}) = 6980$$

which corresponds to a full scale velocity of approximately 35.8 feet per second. The value of Reynolds number at 10 percent of full-scale flow is 698, which is relatively low for gas service and will play a significant role in flowmeter selection. The operating Reynolds numbers are graphed as a horizontal line as shown in Figure 25-9.

Technologies can be eliminated by using the technical and non-technical criteria, the results of which are summarized in Figure 25-10.

Remaining technologies include thermal profile and laminar flow elements, which can be evaluated individually to determine the optimum flowmeter for the application.

Most flowmeter applications are in the medium density category, which includes most gases at medium pressures and light gases such as hydrogen at high pressures. Commonly measured gases include air, nitrogen, process gases, and the like.

EXAMPLE 25-5

Problem: Using the flowmeter selection procedure, eliminate technologies that are not applicable to a 500-scfm flow of air in a 3-inch schedule 40 pipe where the operating pressure is nominally 50 psi at a nominal operating temperature of 100°F.

Solution: From physical property tables, the density of air at 50 psi and 100°F is 0.312 pound per cubic foot. Similarly, the viscosity is 0.017 cP. The flow in actual cubic feet per minute can be calculated as:

$$Q_{acfm} = 500 \times [(460°\text{F} + 100°\text{F})/(460°\text{F} + 60°\text{F})] \times [14.7 \text{ psi}/(14.7 \text{ psi} + 50 \text{ psi})]$$

$$= 122.4 \text{ acfm}$$

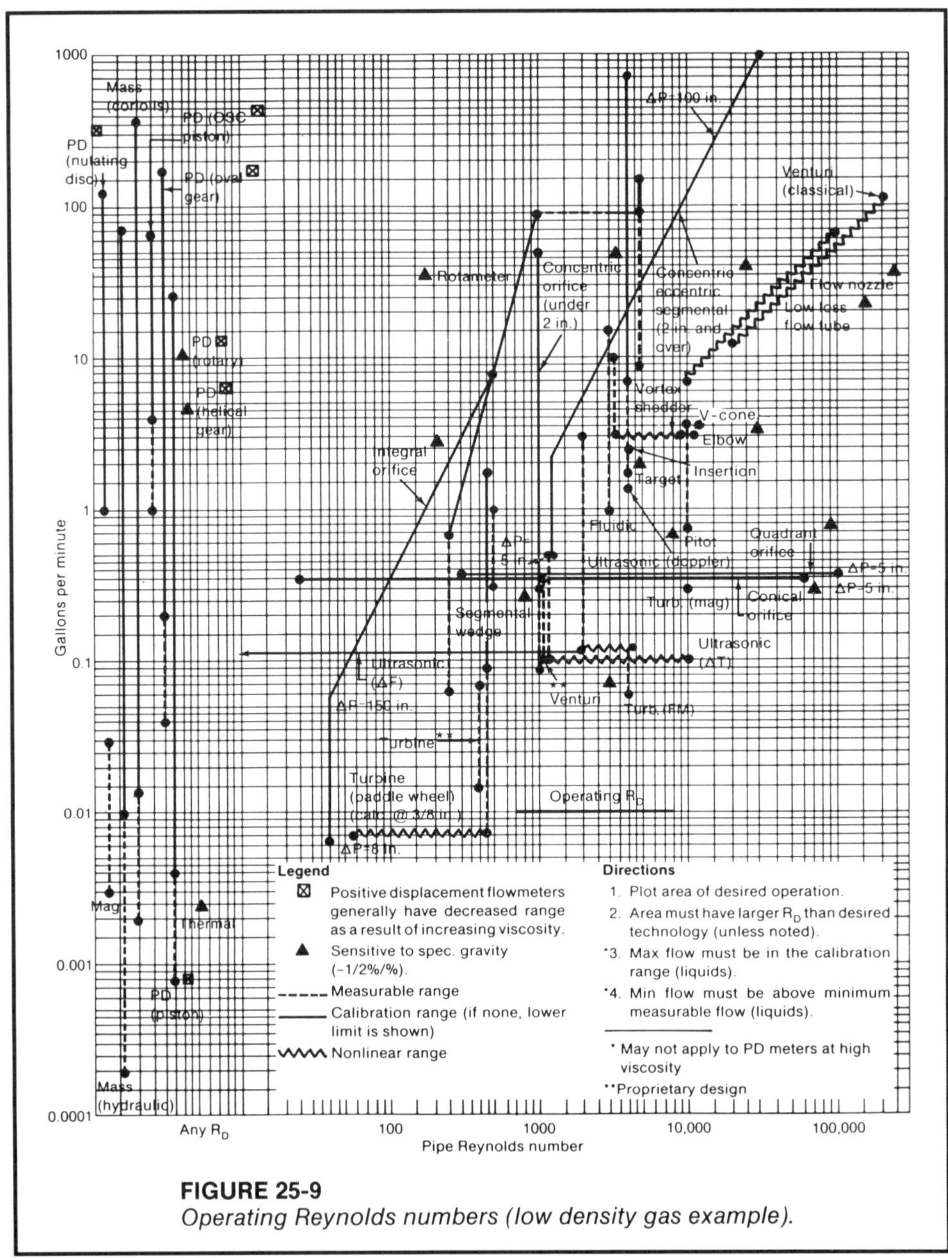

FIGURE 25-9
Operating Reynolds numbers (low density gas example).

Reynolds number can be calculated as

$$R_D = (379 \times 122.4 \text{ acfm} \times 0.312 \text{ lb/ft}^3)/(0.017 \text{ cP} \times 3.068 \text{ in.}) = 277{,}505$$

which corresponds to a full-scale velocity of approximately 39.8 feet per second. The value of Reynolds number at 10 percent of full scale flow is 27,751, which is sufficiently high that many flowmeter technologies are not

	Technical	Non-technical
Differential pressure		
Orifice		Requires pressure compensation
Elbow	R_D too low	X
Flow nozzle	R_D too low	X
Flow tube	R_D too low	X
Laminar flow element		
Segmental wedge		
Venturi	R_D too low	X
V-cone		Requires pressure compensation
Magnetic	Liquids only	X
Mass		
Coriolis		Insufficient mass flow
Hydraulic	Liquids only	X
Oscillatory		
Fluidic	Liquids only	X
Vortex shedding	R_D too low	X
Positive displacement		
Helical	Liquids only	X
Nutating disc	Liquids only	X
Oscillating piston	Liquids only	X
Oval gear	Liquids only	X
Piston	Liquids only	X
Rotary	Liquids only	X
Target	R_D too low	X
Thermal		
Turbine	R_D too low	X
Ultrasonic		
Doppler	Liquids only	X
Time of flight	Liquids only	X
Variable area		Requires pressure compensation
Insertion	R_D too low	X
Bypass		Not preferred

FIGURE 25-10
Flowmeter selection summary (low density gas example).

eliminated. The operating Reynolds numbers are graphed as a horizontal line as illustrated in Figure 25-11.

Technologies can be eliminated by using the technical and non-technical criteria, the results of which are summarized in Figure 25-12.

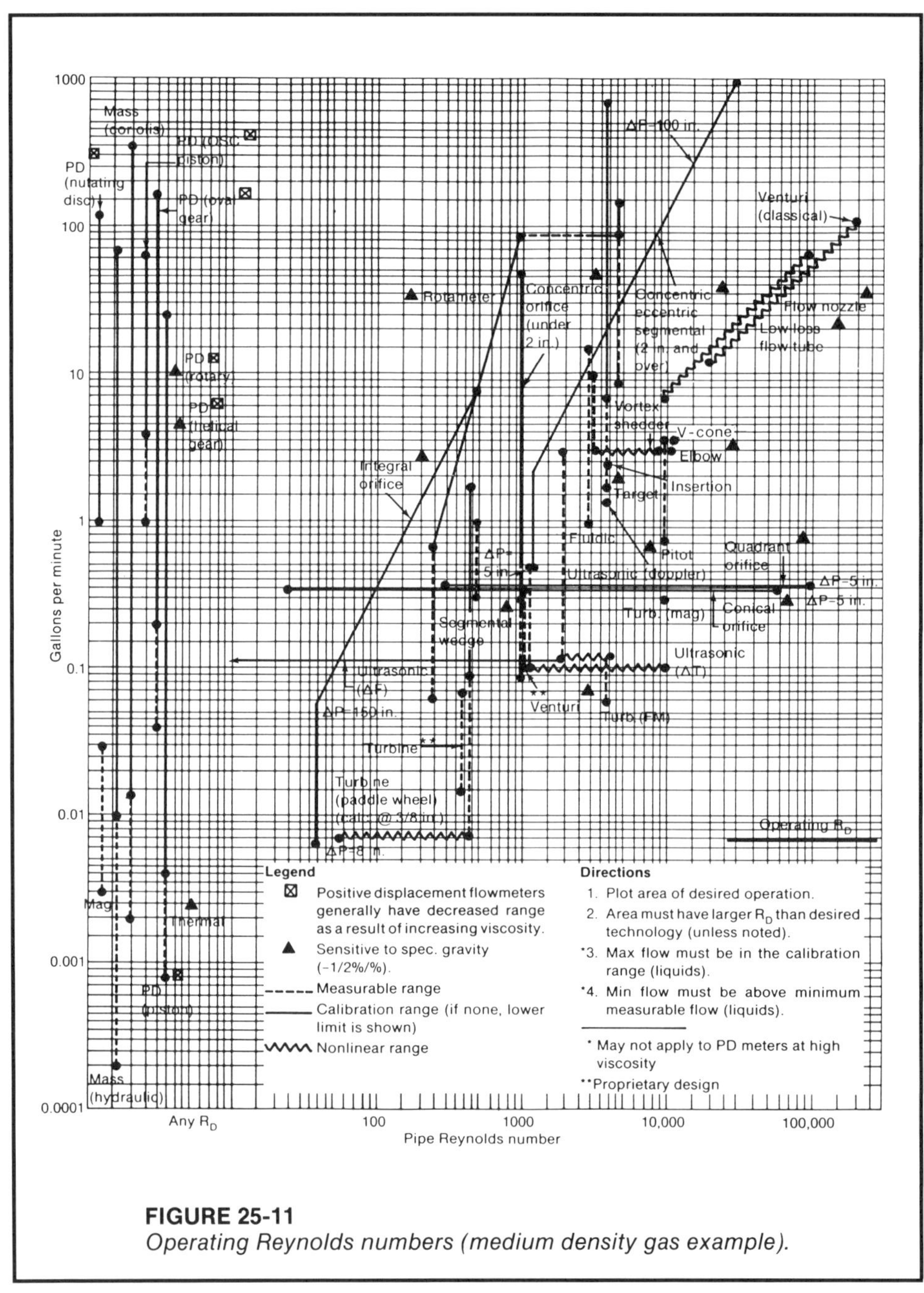

FIGURE 25-11

Operating Reynolds numbers (medium density gas example).

	Technical	Non-technical
Differential pressure		
Orifice		
Elbow		Insufficient accuracy
Flow nozzle	R_D too low	X
Flow tube	R_D too low	X
Laminar flow element	R_D too high	X
Segmental wedge		
Venturi	R_D too low	X
V-cone		
Magnetic	Liquids only	X
Mass		
Coriolis		Expensive
Hydraulic	Liquids only	X
Oscillatory		
Fluidic	Liquids only	X
Vortex shedding		
Positive displacement		
Helical	Liquids only	X
Nutating disc	Liquids only	X
Oscillating piston	Liquids only	X
Oval gear	Liquids only	X
Piston	Liquids only	X
Rotary	Liquids only	X
Target		Insufficient accuracy
Thermal		Insufficient accuracy
Turbine		Moving parts
Ultrasonic		
Doppler	Liquids only	X
Time of flight	Liquids only	X
Variable area		Moving parts
Insertion	Hydraulic errors too large	
Bypass		Not preferred

FIGURE 25-12
Flowmeter selection summary (medium density gas example).

Remaining technologies include orifice plate and vortex shedding technologies, which can be evaluated to determine the optimum one for the application. Note that this application is relatively straightforward and virtually all technologies applicable to gas flow are acceptable; therefore, many non-technical criteria can be used to determine the order in which the remaining technologies can be evaluated. It should be noted that due to minimum velocity constraints, flowmeters that use the momentum of the fluid to operate may have to be less than 3 inches in size in order to achieve a reasonable turndown.

High density gas applications involve the measurement of gases at high pressures. The commonly measured gases may be categorized with medium density flowmeter applications.

EXAMPLE 25-6

Problem: Use the flowmeter selection procedure to eliminate technologies that are not applicable to a 5000-scfm flow of nitrogen in a 2-inch schedule 80 pipe where the operating pressure is nominally 1000 psi at a nominal operating temperature of 70°F.

Solution: From physical property tables, the density of nitrogen at standard conditions of 60°F and 14.7 psi is 0.0727 pound per cubic foot. Similarly, the viscosity at operating conditions is 0.033 cP. The density at 1000 psi is calculated to be

$$\rho = 0.0727\ \text{lb/ft}^3 \times [(460°\text{F} + 60°\text{F})/(460°\text{F} + 70°\text{F})] \times [(14.7\ \text{psi} + 1000\ \text{psi})/14.7\ \text{psi psi}]$$

$$= 4.92\ \text{pounds per cubic foot}$$

The flow in actual cubic feet per minute can be calculated as

$$Q_{\text{acfm}} = 5000\ \text{scfm} \times [(460°\text{F} + 70°\text{F})/(460°\text{F} + 60°\text{F})] \times [14.7\ \text{psi}/(14.7\ \text{psi} + 1000\ \text{psi})]$$

$$= 73.8\ \text{acfm}$$

Reynolds number can be calculated as

$$R_D = (379 \times 73.8\ \text{acfm} \times 4.92\ \text{lb/ft}^3) / (0.033\ \text{cP} \times 1.939\ \text{in.}) = 2{,}150{,}646$$

which corresponds to a full-scale velocity of approximately 59.9 feet per second. Reynolds number can be graphed; however, it is sufficiently large that no technologies can be eliminated by Reynolds number constraints. They can, however, be eliminated by using the technical and non-technical criteria, the results of which are summarized in Figure 25-13.

Various technologies remain, including various differential pressure technologies, turbine flowmeters, and vortex shedding technology. Orifice plate technology is commonly applied in this type of application. However, other technologies such as vortex shedding and turbine flowmeters cannot be discounted in these applications as they offer equivalent performance in

	Technical	Non-technical
Differential pressure		
Orifice		
Elbow		Insufficient accuracy
Flow nozzle		
Flow tube		
Laminar flow element	R_D too high	X
Segmental wedge		
Venturi		
V-cone		
Magnetic	Liquids only	X
Mass		
Coriolis		
Hydraulic	Liquids only	X
Oscillatory		
Fluidic	Liquids only	X
Vortex shedding		
Positive displacement		
Helical	Liquids only	X
Nutating disc	Liquids only	X
Oscillating piston	Liquids only	X
Oval gear	Liquids only	X
Piston	Liquids only	X
Rotary	Liquids only	X
Target		Insufficient accuracy
Thermal		Insufficient accuracy
Turbine		
Ultrasonic		
Doppler	Liquids only	X
Time of flight	Liquids only	X
Variable area		Moving parts
Insertion		Not desired
Bypass		Not desired

FIGURE 25-13
Flowmeter selection summary (high density gas example).

many applications. It should be noted that even though the operating pressure is 1000 psi, a 600-pound flanged flowmeter will handle this service because of the low operating temperature.

Steam Steam is often classified as a gas flowmeter application. However, due to the quantity of flow measurements required in this service and the difficulties associated with condensation, it is considered separately. Steam flow measurement is performed under operating conditions that typically include both medium to high pressures and temperatures as well as various degrees of superheat. These relatively extreme operating conditions tend to eliminate many technologies.

EXAMPLE 25-7

Problem: Using the flowmeter selection procedure, eliminate technologies that are not applicable to a 0 to 60,000-pound per hour flow of 225-pound saturated steam in a 6-inch schedule 80 pipe.

Solution: From steam tables, the 225-psi saturated steam has an operating temperature of 397°F and an operating density of 0.521 pound per cubic foot. Similarly, the viscosity at operating conditions is 0.016 cP. The flow in actual cubic feet per minute can be calculated as

$$Q_{acfm} = (60{,}000 \text{ lbs/hr}) \times (\text{hr}/60 \text{ min}) \times (\text{ft}^3/0.521 \text{ lb/ft}^3) = 1919.4 \text{ acfm}$$

Reynolds number can be calculated as:

$$R_D = (379 \times 1919.4 \text{ acfm} \times 0.521 \text{ lb/ft}^3)/(0.016 \text{ cP} \times 5.761 \text{ in.}) = 4{,}111{,}730$$

which corresponds to a full-scale velocity of approximately $159.5 \times (6.065/5.761)^2$, or 176.8 feet per second. Reynolds number is sufficiently large that no technologies are eliminated by Reynolds number constraints. Technologies can be eliminated by using the technical and non-technical criteria, the results of which are summarized in Figure 25-14.

Various technologies remain, including various differential pressure, vortex shedding, and insertion flowmeter technologies. Orifice plate and vortex shedding technologies are the most economical choices, although certain insertion technologies may also be economical. However, errors inherent in centerline-positioned insertion flowmeters should be taken into account.

Large Pipe Flowmeters applied to large pipes merit special attention because of the complexities of piping layout and the difficulties of applying flowmeter technology to large piping.

EXAMPLE 25-8

Problem: Using the flowmeter selection procedure, eliminate technologies that are not applicable to a 0 to 10,000 gallons per minute flow of water flowing in a 24-inch, 1/2-inch wall pipe at 100°F.

	Technical	Non-technical
Differential pressure		
Orifice		
Elbow		Insufficient accuracy
Flow nozzle		
Flow tube		
Laminar flow element	R_D too high	X
Segmental wedge		
Venturi		Expensive compared to orifice
V-cone		Expensive compared to orifice
Magnetic	Liquids only	X
Mass		
Coriolis		
Hydraulic	Liquids only	X
Oscillatory		
Fluidic	Liquids only	X
Vortex shedding		
Positive displacement		
Helical	Liquids only	X
Nutating disc	Liquids only	X
Oscillating piston	Liquids only	X
Oval gear	Liquids only	X
Piston	Liquids only	X
Rotary	Liquids only	X
Target		Insufficient accuracy
Thermal		Insufficient accuracy
Turbine		Moving parts not desirable
Ultrasonic		
Doppler	Liquids only	X
Time of flight	Liquids only	X
Variable area		Moving parts not desirable; difficult to change range
Insertion		
Bypass		Not preferred

FIGURE 25-14
Flowmeter selection summary (steam example).

Solution: From property tables, the operating specific gravity of water is 0.994 and the operating viscosity is 0.67 cP. Reynolds number can be calculated as

$$R_D = (3160 \times 10{,}000 \text{ gpm} \times 0.994)/(0.67 \text{ cP} \times 23.0 \text{ in.}) = 2{,}038{,}313$$

Reynolds number is sufficiently high that no technologies will be eliminated by graphing Reynolds number. A summary of the technical and non-technical criteria used to eliminate technologies is presented in Figure 25-15.

Various technologies remain, including those of differential pressure (which generally exhibit significant energy losses at such high flows) and insertion flowmeters (which are typically very economical in large pipe diameter applications). Bypass flowmeters can also be applied.

	Technical	Non-technical
Differential pressure		
Orifice		Large energy loss
Elbow		Insufficient accuracy
Flow nozzle		
Flow tube		
Laminar flow element	R_D too high	X
Segmental wedge	Out of range	X
Venturi		
V-cone		Large energy loss
Magnetic		Expensive
Mass		
Coriolis	Out of range	X
Hydraulic	Out of range	X
Oscillatory		
Fluidic	Out of range	X
Vortex shedding	Out of range	X
Positive displacement		
Helical	Out of range	X
Nutating disc	Out of range	X
Oscillating piston	Out of range	X
Oval gear	Out of range	X
Piston	Out of range	X
Rotary	Out of range	X
Target	Out of range	X
Thermal	Out of range	X
Turbine		Expensive
Ultrasonic		
Doppler		Insufficient accuracy
Time of flight		Expensive
Variable area	Out of range	X
Insertion		
Bypass		

FIGURE 25-15
Flowmeter selection summary (large pipe example).

EXERCISES

25.1 Use the flowmeter selection procedure to eliminate technologies that are not applicable to a flow of 1000 gallons per minute of a liquid in an 8-inch schedule 40 pipe when the viscosity and specific gravity are assumed to be 1.4 cP and 0.89, respectively. Accuracy requirements are 1 percent of rate.

	Technical	Non-technical
Differential pressure		
Orifice		
Elbow		
Flow nozzle		
Flow tube		
Laminar flow element		
Segmental wedge		
Venturi		
V-cone		
Magnetic		
Mass		
Coriolis		
Hydraulic		
Oscillatory		
Fluidic		
Vortex shedding		
Positive displacement		
Helical		
Nutating disc		
Oscillating piston		
Oval gear		
Piston		
Rotary		
Target		
Thermal		
Turbine		
Ultrasonic		
Doppler		
Time of flight		
Variable area		
Insertion		
Bypass		

FIGURE Q1
Flowmeter selection summary.

25.2 Use the flowmeter selection procedure to eliminate technologies that are not applicable to a flow of 15 gallons per minute of an organic liquid in a 2-inch schedule 40 pipe when the viscosity can vary from 1000 to 1500 cP over the operating temperature range and specific gravity is 1.21. Accuracy requirements are ±1 percent of rate.

	Technical	Non-technical
Differential pressure		
Orifice		
Elbow		
Flow nozzle		
Flow tube		
Laminar flow element		
Segmental wedge		
Venturi		
V-cone		
Magnetic		
Mass		
Coriolis		
Hydraulic		
Oscillatory		
Fluidic		
Vortex shedding		
Positive displacement		
Helical		
Nutating disc		
Oscillating piston		
Oval gear		
Piston		
Rotary		
Target		
Thermal		
Turbine		
Ultrasonic		
Doppler		
Time of flight		
Variable area		
Insertion		
Bypass		

FIGURE Q2
Flowmeter selection summary.

25.3 Use the flowmeter selection procedure to eliminate technologies that are not applicable to a 300 scfm flow of an ideal gas in a 2-inch schedule 40 pipe where the operating pressure is nominally 100 psi at a nominal operating temperature of 40°F when the density at standard conditions is 0.110 pound per cubic foot and the operating viscosity is 0.015 cP.

	Technical	Non-technical
Differential pressure		
Orifice		
Elbow		
Flow nozzle		
Flow tube		
Laminar flow element		
Segmental wedge		
Venturi		
V-cone		
Magnetic		
Mass		
Coriolis		
Hydraulic		
Oscillatory		
Fluidic		
Vortex shedding		
Positive displacement		
Helical		
Nutating disc		
Oscillating piston		
Oval gear		
Piston		
Rotary		
Target		
Thermal		
Turbine		
Ultrasonic		
Doppler		
Time of flight		
Variable area		
Insertion		
Bypass		

FIGURE Q3
Flowmeter selection summary.

25.4 Use the flowmeter selection procedure to eliminate technologies that are not applicable to a 0 to 2000 pound per hour flow of 225-pound saturated steam in a 2-inch schedule 80 pipe.

	Technical	Non-technical
Differential pressure		
Orifice		
Elbow		
Flow nozzle		
Flow tube		
Laminar flow element		
Segmental wedge		
Venturi		
V-cone		
Magnetic		
Mass		
Coriolis		
Hydraulic		
Oscillatory		
Fluidic		
Vortex shedding		
Positive displacement		
Helical		
Nutating disc		
Oscillating piston		
Oval gear		
Piston		
Rotary		
Target		
Thermal		
Turbine		
Ultrasonic		
Doppler		
Time of flight		
Variable area		
Insertion		
Bypass		

FIGURE Q4
Flowmeter selection summary.

Appendix A
References

1. Crane Company, 1965. *Flow of Fluids through Valves, Fittings, and Pipe.* Technical Paper No. 410, Engineering Division.
2. Marks, 1967. *Standard Handbook for Mechanical Engineers*, 7th edition. McGraw-Hill Book Company.
3. C. E. Miller, 1978. On the application and performance of insertion turbine meters for steam flow measurement. Presented at the International District Heating Association Annual Conference.
4. Richard W. Miller, 1989. *Flow Measurement Engineering Handbook*, 2nd edition. McGraw-Hill Book Company.
5. John H. Perry, 1950. *Chemical Engineers' Handbook*, 3rd edition. McGraw-Hill Book Company.
6. H. Schlichting, 1955. *Boundary Layer Theory.* Pergamon Press.
7. Francis W. Sears and Mark W. Zemansky, 1955. *University Physics*, 2nd edition. Addison-Wesley Publishing Co., Inc.
8. L. K. Spink, 1967. *Principles and Practice of Flow Meter Engineering*, 9th edition. The Foxboro Company.

Appendix B Answers to Exercises

Chapter 2

1. 37.8°C; 25°C; 4.44°C
2. 11.696 psia; 12.762 psia; 31.696 psia; 32.885 psia; 232.253 psia
3. R_D = 249,442, 59.29 ft WC or 25.68 psi
4. 0.708 psi. There is no flow as the pressure is not sufficient to push the liquid 2 feet above the flowmeter.
5. SG = 1.231, - 9.56 ft WC. There is no flow.
6. 13.08 gpm
7. 7.65 ft/sec
8. 265.2 acfm; 751.3 scfm; 777,600.
9. 287.4 acfm; 1202.6 scfm

Chapter 3

1. No. There is always some limitation on the flowmeter.
2. 2 percent of rate accuracy minimizes the flow measurement error.
3. Nothing.
4. No, because the flowmeter will repeat well but may not be accurate and therefore not capable of accurately adding a precise amount of fluid.
5. Composite accuracy of 1 percent rate yields the lowest error among the choices given.
6. 3.33
7. 3.5:1
8. Answers will depend on further assumptions of individual students.

Chapter 4

1. 25 percent; 75 percent; 90 percent
2. The flow is greater than the measurement by approximately 21.7 percent and 10.3 percent for linear and squared flowmeters, respectively.
3. Yes.
4. No. Depends on flowmeter design.
5. No. Due to large uncertainties introduced by the lack of accurate process data.

Chapter 5

1. Digital flowmeters can turn off completely at low flow conditions.
2. A quantity of fluid has passed through the flowmeter.
3. Meter capacity and operating conditions.

Chapter 6

1. Usually yes.
2. Physical properties and operating conditions are better defined.
3. Up to a maximum of approximately 7×10^6.
4. Full scale error, rate error.

Chapter 7

1. Reference conditions.
2. Conversion error is typically 0.1 to 0.2 percent full scale.
3. Zero calibration error and zero drift can occur.
4. No zero calibration error or zero drift error occur.
5. Introduce additional error.
6. 0.098 percent full scale.
7. Linear flowmeter has superior accuracy due to direct pulse counting without converters.

Chapter 8

1. Yes. See typical corrosion chart.
2. All wetted parts.

3. The ability of a fluid to erode another material due to mechanical contact.
4. Yes. Flange pressure ratings are not fixed and are higher at low temperatures.
5. To minimize the effects of distorted flow profiles on the performance of the flowmeter.
6. Oxygen in contact with hydrocarbons can cause a fire.
7. Yes, since the expansion characteristics of the material from which the flowmeter is constructed are usually well defined.
8. 2-wire installations generally lower the cost of electrical installation.

Chapter 9

1. I — Flowmeters with moving parts typically have good long-term accuracy.
   II — Flowmeters with no moving parts are usually subject to fewer mechanical failures.
   III — Obstructionless flowmeters offer no restriction to flow.
   IV — Flowmeters with externally mounted sensors can usually be easily retro-fitted and have no wetted parts.

2. a. II; III; IV (solids can plug finely machined parts in Class I).
   b. III; IV (abrasive liquids are more apt to change tolerances in Classes I and II).
   c. All, but Class I and Class II may present material of construction problems.
   d. All.

3.

	Advantages	Disadvantages
Volumetric	Measures volume directly	Not density compensated
Velocity		Not density compensated
Inferential		Not density compensated
Mass	Density compensated	

4. a. Mass flowmeter is density compensated.
   b. Volumetric or velocity flowmeter due to the nature of the measurement.
   c. Volumetric or velocity flowmeter due to the nature of the measurement.
   d. All flowmeters.

Chapter 10

1.

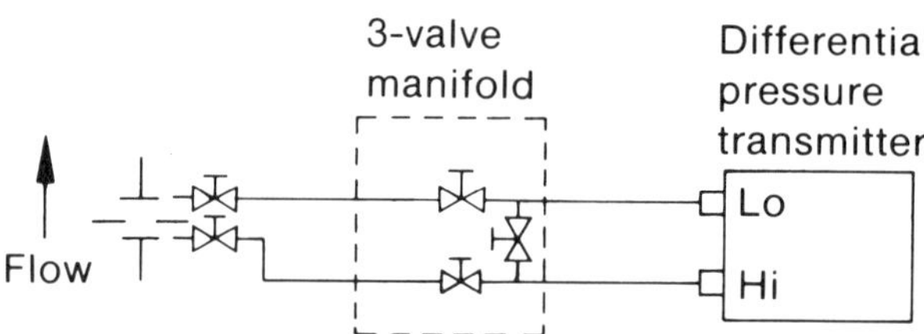

2. Small pipes — Flange or corner taps are preferred as dimensional stability is maintained by machined flanges.
Large pipes — Radius taps are preferred as changes in the bore do not affect the tap location. Also, flange-tapped flanges can be expensive.
3. 55.4 ft/sec (schedule 40 pipe). Beta is larger to produce the same differential.
4. Reads 3.8 percent low (for an ideal gas).
5. Easy to machine; experimental data available.
6. Taps must be moved for other than minor beta ratio changes.
7. 0 to 64 in. WC.
8. To obtain stable and predictable velocity profile upstream of the flowmeter.
9. Sloped downward towards transmitter.
10. Orifice wear, but process uncertainties can obscure the error.
11. Step 1. Flow range, 0 to 600 gpm @ 0 to 100 in. WC. Design flow, 480 gpm @ 64 in. WC.
Step 2. $R_D = 200{,}992$ — sufficiently high.
Step 3. $Q_{lb/hr} = 262{,}341$
$F_a = 1.001$
$T_r = 0.601$ and $Z_L = 0.0055$; $F_p = 1.00063$
$D^2 = 36.78$ in.^{2}
$\rho_{upstream} = 68.13$ lbs/ft^3
$S_M = 0.30120$
Step 4. $\beta_0 = 0.66187$
Step 5. $C_{inf} = 0.60415$; $b = 32.68$; $n = 0.75$
$C_0 = 0.60759$
Step 6. $Y = 1$
Step 7. $\beta_1 = 0.66645$
Step 8. $C_1 = 0.60755$
$\beta_2 = 0.66646$
Step 9. Bore = 4.042 inches

12. Step 1. Flow range, 0 to 2500 acfm @ 0 to 100 in. WC. Design flow, 2000 scfm @ 64 in. WC.
    Step 2. R_D = 750,071 — sufficiently high.
    Step 3. $Q_{lb/hr}$ = 12606 lb/hr
    F_a = 1.00004
    F_p = 1.0
    $\rho_{upstream}$ = 0.642
    S_M - 0.14895
    Step 4. β_0 = 0.48744
    Step 5. C_{inf} = 0.60246; b = 15.21; n = 0.75
    C_0 = 0.60306
    Step 6. X_1 = 0.02012
    $(Y_1)_0$ = 0.99364
    Step 7. β_1 = 0.49138
    Step 8. C_1 = 0.60316 C_2 = 0.60315
    $(Y_1)_1$ = 0.99363 $(Y_1)_2$ = 0.99363
    β_2 = 0.49112 β_3 = 0.49112
    Step 9. Bore = 2.979 inches

13. Step 1. Flow range, 0 to 10,000 lb/hr @ 0 to 100 in. WC.
    Step 2. R_D = 41,060,754 — sufficiently high.
    Step 3. F_a = 1.0063
    F_p = 1.0
    S_M = 0.26196
    Step 4. β_0 = 0.62570
    Step 5. C_{inf} = 0.60533; b = 28.40; n = 0.75
    C_0 = 0.60539
    Step 6. X_1 = 0.00962
    $(Y_1)_0$ = 0.99654
    Step 7. β_1 = 0.63110
    Step 8. C_1 = 0.60544
    $(Y_1)_1$ = 0.99653
    β_2 = 0.63108
    Step 9. Bore = 2.415 inches

Chapter 11

1. Automatic zero adjustment and improved accuracy.
2. Reduced size and reduced cost; reduced accuracy.
3. 3 to 30 ft/sec full scale velocity, 5 μs/cm min. con.
4. 0 to 30 gpm; 1 in. @ 11.1 fps or 1-1/2 in. @ 4.7 fps.
   0 to 50 gpm; 1-1/2 in. @ 7.9 fps or 2 in. @ 4.8 fps.
   0 to 1000 gpm; 6 in. @ 11.1 fps, 8 in. @ 6.4 fps, or 10 in. @ 4.1 fps.
5. DC

Chapter 12

1. Yes (approximately 3 psi pressure drop).
2. No (↓15 ↓↓↓).
3. Orifices can plug.
4. Constant volume pump characteristics vary over 50 cP.

Chapter 13

1. Rectangular weir — 2 ft @ 3.75 inches of head
   Cipoletti weir — 2 ft @ 3.7 inches of head
   Triangular weir — 45° @ 12 inches of head
2. 6 in. @ 8.4 in. of head
   9 in. @ 6 in. of head
   1 ft @ 5.2 in. of head
   1.5 ft @ 4 in. of head
   2 ft @ 3.4 in. of head

Chapter 14

1. 316 SS; 316 SS
2. Fluidic flowmeters operate accurately at lower Reynolds numbers.
3. There is no zero adjustment of the primary element.
4. High viscosities reduce Reynolds numbers to where the flowmeter is non-linear or ceases to operate.
5. Condensate will not strike the shedder.
6. 6-in. vortex shedder; turndown is approximately 7.8:1.
7. 4-in. vortex shedder; turndown is approximately 10:1.
   6-in. vortex shedder; turndown is approximately 3.8:1.
8. 2-in. vortex shedder; turndown is approximately 32.7:1.
   3-in. vortex shedder; turndown is approximately 13.8:1.
   4-in. vortex shedder; turndown is approximately 6.4:1.
   6-in. vortex shedder; turndown is approximately 2.4:1.

Chapter 15

1. 6-in. helical gear flowmeter is close, such that a 10-in. flowmeter may be necessary.
2. 1-1/2 in. nutating disc flowmeter.
3. 1-in. oscillating piston, which has a pressure drop of approximately 15 psid at full scale flow.

4. 1-in. small capacity oval gear flowmeter, which has a pressure drop of approximately 7.5 psid @ max flow and a turndown of approximately 14.3:1.
5. 1/2-in. piston flowmeter has a pressure drop that may be excessive so that the 3/4-in. flowmeter is applicable.
6. 2-in. rotary flowmeter.

Chapter 16

1. Reynolds number must be greater than 4000.
2. Yes.
3. 2-in. target flowmeter with 0.658 target-to-bore ratio.
4. 1-in. target flowmeter.

Chapter 17

1. Usually not.
2. 500 sccm flowmeter will measure 0 to 160 sccm of propane.

Chapter 18

1. 10-in. turbine flowmeter.
2. The maximum flow corresponds to the maximum flow allowed through a 4-in. turbine flowmeter.
3. 6-in. turbine flowmeter.
4. Typically accuracy and repeatability are ±1 percent rate and ±0.05 percent rate, respectively, for liquid service.

Chapter 19

1. 8-in. ultrasonic flowmeter @ 8.98 ft/sec full scale
   10-in. ultrasonic flowmeter @ 5.70 ft/sec full scale
   12-in. ultrasonic flowmeter @ 4.01 ft/sec full scale
   A Doppler flowmeter could be used if bubbles or particles are present, but the minimum velocity constraint is 0.5 ft/sec, which reduces turndown.
2. 6-in. differential frequency ultrasonic flowmeter @ 6.66 ft/sec full scale. Reynolds number is 2381 at full scale, so accurate measurement will occur above approximately 504 gpm. Reynolds number is not sufficiently high to apply a Doppler ultrasonic flowmeter.
3. Particles and bubbles reflect ultrasonic energy.

Chapter 20

1. When it ruptures.
2. 169.3 gpm water
3. 49.0 scfm air
4. 1541.5 scfm air

Chapter 21

1. 405.4 gpm
2. $n = 9.59$; $V_{ave} = 83.1$ ft/sec; $V_o = 96.55$ ft/sec; $V_y = 86.29$ ft/sec
3. $n = 7.46$; $(Y/R)_{crit} = 0.7589$; critical position is 9.49 inches from the pipe wall.
4. $n = 6.13$; $V_{ave} = 0.795\ V_o$
5. 3.4 percent.

Chapter 22

1. Yes, when the turndown of the primary flow element without the transmitter has a larger turndown than with the transmitter, and the secondary flowmeter with transmitter has a larger turndown than the primary element with the transmitter.
2. Yes, because it is designed with primary and secondary flow elements.

Chapter 23

1. 2-inch vortex shedding flowmeter relative cost is 0.74.
2. 2-inch orifice plate flowmeter relative cost is 0.92.
3. Permanent pressure loss is 72.5 in. WC or 2.62 psi. The energy loss is 174.5 hp at a cost of $80,252.55/yr.

Chapter 25

1. R_D @ max flow in an 8-inch pipe is 251,705.

	Technical	Non-technical
Differential pressure		
Orifice		
Elbow		Insufficient accuracy
Flow nozzle	R_D too low	X
Flow tube	R_D too low	X
Laminar flow element		Insufficient accuracy
Segmental wedge	R_D too low	X
Venturi		
V-cone		
Magnetic		
Mass		
Coriolis	Out of range	X
Hydraulic	Out of range	X
Oscillatory		
Fluidic		
Vortex shedding		
Positive displacement		
Helical		Excessive slippage
Nutating disc	Out of range	X
Oscillating piston	Out of range	X
Oval gear	Out of range	X
Piston	Out of range	X
Rotary		Excessive slippage
Target	Out of range	X
Thermal		Insufficient accuracy
Turbine		Moving parts
Ultrasonic		
Doppler		Insufficient accuracy
Time of flight		
Variable area	Out of range	X
Insertion		Insufficient accuracy
Bypass		Insufficient accuracy

FIGURE Q-1
Flowmeter selection summary.

2. R_D @ max flow in a 2-inch pipe is 18.5 at max viscosity. All R_D dependent flowmeters except the time of flight ultrasonic type are eliminated by graphing Reynolds number.

	Technical	Non-technical
Differential pressure		
Orifice	R_D too low	X
Elbow	R_D too low	X
Flow nozzle	R_D too low	X
Flow tube	R_D too low	X
Laminar flow element		Insufficient accuracy
Segmental wedge	R_D too low	X
Venturi	R_D too low	X
V-cone	R_D too low	X
Magnetic	Non-conductive	X
Mass		
Coriolis		
Hydraulic	Viscosity too high	X
Oscillatory		
Fluidic	R_D too low	X
Vortex shedding	R_D too low	X
Positive displacement		
Helical		
Nutating disc		
Oscillating piston		
Oval gear		
Piston		
Rotary		
Target	R_D too low	X
Thermal		Insufficient accuracy
Turbine	R_D too low	X
Ultrasonic		
Doppler	R_D too low	X
Time of flight		Expensive
Variable area	R_D too low	X
Insertion	R_D too low	X
Bypass		Not preferred

FIGURE Q-2
Flowmeter selection summary.

3. The flow is 36.97 acfm at an operating density of 0.893 lb/ft^3. R_D @ max flow in a 2-inch pipe is 403,560.

	Technical	Non-technical
Differential pressure		
Orifice		
Elbow		Insufficient accuracy
Flow nozzle	R_D too low	X
Flow tube	R_D too low	X
Laminar flow element	R_D too high	X
Segmental wedge		
Venturi	R_D too low	X
V-cone		
Magnetic	Liquids only	X
Mass		
Coriolis		Expensive
Hydraulic	Liquids only	X
Oscillatory		
Fluidic	Liquids only	X
Vortex shedding		
Positive displacement		
Helical	Liquids only	X
Nutating disc	Liquids only	X
Oscillating piston	Liquids only	X
Oval gear	Liquids only	X
Piston	Liquids only	X
Rotary		X
Target		Insufficient accuracy
Thermal		Insufficient accuracy
Turbine		Moving parts
Ultrasonic		
Doppler	Liquids only	X
Time of flight	Liquids only	X
Variable area		Moving parts
Insertion	Hydraulic error too high	X
Bypass		Not preferred

FIGURE Q-3
Flowmeter selection summary.

4. The flow is 63.98 acfm. R_D @ max flow in a 2-inch pipe is 407,215.

	Technical	Non-technical
Differential pressure		
Orifice		
Elbow		Insufficient accuracy
Flow nozzle	R_D too low	X
Flow tube	R_D too low	X
Laminar flow element	R_D too high	X
Segmental wedge		
Venturi	R_D too low	X
V-cone		
Magnetic	Liquids only	X
Mass		
Coriolis		
Hydraulic	Liquids only	X
Oscillatory		
Fluidic	Liquids only	X
Vortex shedding		
Positive displacement		
Helical	Liquids only	X
Nutating disc	Liquids only	X
Oscillating piston	Liquids only	X
Oval gear	Liquids only	X
Piston	Liquids only	X
Rotary	Liquids only	X
Target		
Thermal		Insufficient accuracy
Turbine		Moving parts
Ultrasonic		
Doppler	Liquids only	X
Time of flight	Typically, liquids only	X
Variable area		Moving parts
Insertion	Hydraulic error too high	X
Bypass		Not preferred

FIGURE Q-4
Flowmeter selection summary.

Index